Indian Statistical Institute Series

The **Indian Statistical Institute Series,** a Scopus-indexed series, publishes high-quality content in the domain of mathematical sciences, bio-mathematics, financial mathematics, pure and applied mathematics, operations research, applied statistics and computer science and applications with primary focus on mathematics and statistics.

Launched at the 125th birth Anniversary of P.C. Mahalanobis, the series will publish high-quality content in the form of textbooks, monographs, lecture notes, and contributed volumes. Literature in this series are peer-reviewed by global experts in their respective fields, and will appeal to a wide audience of students, researchers, educators, and professionals across mathematics, statistics and computer science disciplines.

Raghunath Arnab

Indirect Methods of Data Collection and Analysis from Surveys

STATISTICAL INDIAN INSTITUTE
UNITY IN DIVERSITY

Springer

Raghunath Arnab
Department of Statistics
University of KwaZulu-Natal
Durban, South Africa

ISSN 2523-3114 ISSN 2523-3122 (electronic)
Indian Statistical Institute Series
ISBN 978-981-97-6004-6 ISBN 978-981-97-6005-3 (eBook)
https://doi.org/10.1007/978-981-97-6005-3

This Springer imprint is published by the registered company Springer Nature Singapore Pte Ltd.
The registered company address is: 152 Beach Road, #21-01/04 Gateway East, Singapore 189721, Singapore

Dedicated to the memory of my parents late Gopendra Krishna Arnab and late Amia Lekha Arnab

Preface

Surveys of the human population often involve questionnaires relating to highly personal and sensitive issues such as domestic violence, the history of induced abortion, involvement in fraud, plagiarism, affiliation with political parties, opinion of government policies, and religious practices, to name a few. In these human surveys, the direct method of interviewing participants regarding sensitive or personal topics does not provide good-quality data; respondents often provide false, socially approved responses or even refuse to answer due to the social stigma or fear that their personal information might be revealed to a third party. Consequently, estimates based on such data yield unreliable and misleading estimates because of the large number of nonsampling errors and nonresponses.

Warner (1965) introduced an ingenious indirect method of data collection known as the randomized response (RR) technique. In this method, respondents do not need to answer the sensitive question directly to the interviewer. Instead, the respondent should answer one of two questions by using the outcome of a randomization device performed by the respondent in the absence of the interviewer. The interviewer will never be able to identify which one of the questions was answered. The privacy of the respondents was thereby protected. As a result, RR techniques provide better quality data, boost respondents' cooperation by protecting their privacy, and prevent an unacceptable rate of nonresponses.

Warner's method has been generalized by several researchers in various directions to collect better-quality data that provides efficient estimates of the parameters of interest. The validity of the proposed RR techniques and their application in real surveys were rigorously studied by various authors. However, some weaknesses were also identified, and a few alternative indirect methods of data collection were proposed. The alternative methods include the item count (ICT) or item sum (IST) technique, the non-randomized response technique (NRR), the nominative technique, and the negative questioning method, amongst others.

The books of Fox and Tracy (1986), and Chaudhuri and Mukerjee (1988) provide insight into the early development of the subject, while the more recent books of Chaudhuri (2011), and Chaudhuri and Chirstofides (2013) provide recent developments. Most proposed RR methodologies are limited to the simple random sampling with replacement method (SRSWR) only. But in reality, most large surveys are based on complex survey designs involving stratification and unequal probability sampling procedures. Generally, these surveys are multicharacter surveys where data are collected on several items, a few of which are sensitive. For example, in a household survey, one may collect data relating to non-sensitive topics such as age, gender, family size, and level of education, along with sensitive variables that might include family income and incidents of HIV infection or domestic violence. Hence, analysis of RR survey data based on complex survey designs plays an important role.

This book aims to provide a chronological, comprehensive, and up-to-date review of indirect methods of data collection, along with methods of drawing inferences from such data. The methods provided here are applicable to any complex survey design and a wide class of estimators. It also provides avenues for further research in this field. This research monograph is based on the materials published in various monographs and journals, a few unpublished original works, and up-to-date research on the subject. The book is expected to benefit, amongst others, survey statisticians, researchers, and social scientists who collect and analyze data relating to sensitive matters in the areas of criminology, marketing, health, and political science.

I wish to thank my wife, Mrs. Rita Arnab, for her moral support and encouragement. My gratitude also goes to Mr. W. M. Thupeng, University of Botswana, and Prof. John Olaomi, University of South Africa, for editing the manuscript. I would like to thank the Dean, School of Mathematics, Computer Science, and Statistics, University of the KwaZulu-Natal, South Africa for his support.

August 2023

Raghunath Arnab
University of KwaZulu-Natal
Durban, South Africa

References

Chaudhuri, A. (2011). *Randomized response and indirect questioning techniques in surveys*. Chapman & Hall, CRC Press, Taylor & Francis Group.

Chaudhuri, A., & Christofides, T. (2013). *Indirect questioning in sample surveys*. Springer.

Chaudhuri, A., & Mukerjee, R. (1988). *Randomized response: Theory and techniques*. Marcel Dekker.

Fox, J. A., & Tracy, P. E. (1986). *Randomized response: A method for sensitive surveys*. Sage.

Warner, S. L. (1965). Randomize response: A survey technique for eliminating evasive answer bias. *Journal of the American Statistical Association, 60*, 63–69.

Contents

About the Author

Raghunath Arnab is an Honorary Professor at the Department of Statistics, University of KwaZulu-Natal, South Africa. Earlier, he served as Professor at various universities including the University of Botswana; the University of KwaZulu-Natal, South Africa; the Indian Statistical Institute, Kolkata, India; the R. K. Mission Residential College, Narendrapur, India; Sultan Qaboos University, Oman; the University of Nirobi, Kenya; the National University of Lesotho, Lesotho; the University of Ibadan, Nigeria; and the University of Ife, Nigeria. He also visited the Department of Mathematics at Texas A&M University, USA; the University of Granada, Spain; Middle East Technical University, Ankara, Turkey; and the University of Alberta, Canada, among others.

Born in India, Prof. Arnab obtained his Ph.D. in Statistics from the Indian Statistical Institute, Calcutta, India, in 1981; M.Sc. in Statistics; and B Sc. Honors in Statistics from the University of Calcutta, respectively, in 1972 and 1970. With more than 47 years of teaching and research experience, he has taught several courses in statistics, mathematics, and business statistics and supervised several Ph.D. and master's students. He has published more than 100 papers in peer-reviewed journals and is serving as associate editor of a few international journals. He served as External Examiner and Assessor for several universities. He also served as External Examiner for several Ph.D. and master's theses.

Chapter 1
Randomized Response Techniques: Early Developments

Abstract Social researchers often face enormous problems collecting truthful data relating to socially sensitive issues in human surveys. Warner (Journal of the American Statistical Association 60:63–69, 1965) introduced an ingenuous indirect questioning method of data collection known as the randomized response (RR) technique for such human surveys. The proposed RR technique can be used to collect better quality data relating to sensitive subjects as well as boost cooperation from the respondents. This chapter provides a comprehensive chronological review of the early developments of RR techniques or models. Details of the methods for the estimation of parameters from the RR survey data based on the simple random sampling with replacement (SRSWR) method are discussed.

Keywords Item sum technique · Maximum likelihood estimator (MLE) · Nominative techniques · Nonrandomized response technique · Randomized response · Simple random sampling with replacement (SRSWR) · Three-card method

1.1 Introduction

In various areas of applied research, such as sociological, ecological, economic, and medical studies, researchers frequently gather information relating to highly personal, sensitive, and stigmatized issues such as induced abortion, drug addiction, sexual behavior, the incidence of domestic violence, political affiliation, and non-compliance with laws and regulations, among others. Respondents in such surveys who use direct questioning methods in an interview either give false or socially desirable answers or refuse to respond at all due to social stigma and/or fear of their personal information being disclosed to third parties. In such cases, indirect questioning methods such as randomized response (RR), non-randomized response (NRR), item sum techniques (IST), the three-card method, nominative techniques, and surveys with negative questions may be used to collect better quality data, protect respondents' privacy, and reduce an unacceptably high rate of nonresponse. The RR

R. Arnab, *Indirect Methods of Data Collection and Analysis from Surveys*, Indian Statistical Institute Series, https://doi.org/10.1007/978-981-97-6005-3_1

technique was introduced by Warner (1965). Warner's technique was extended by several researchers, including Arcos et al. (2015), Arnab (1990, 1994, 1996), Arnab and Rueda (2016), Franklin (1989), Greenberg et al. (1969), Horvitz et al. (1967), Kuk (1990), Mangat and Singh (1990) and Singh and Singh (1992, 1993), among others.

The validity of RR techniques over traditional methods of data collection, such as direct questioning (DQ) methods involving face-to-face interviews, filling out questionnaires, and the mail inquiry method, was studied by van der Heijden et al. (2000), Lara et al. (2004), and Lensvelt-Mulders et al. (2005), among others. According to studies, the RR method produces more reliable data on sensitive items than the DQ method (Rueda et al., 2015).

Applications of RR techniques in surveys of sensitive characteristics were reported by various researchers: Abernathy et al. (1970) concerning the incidence of induced abortions; Folsom et al. (1973) regarding drinking and driving; Goodstadt and Gruson (1975); and Kerkvliet (1994) concerning drug use. Tamhane (1981) used RR techniques in estimating the prevalence of the use of hard drugs and seeking psychiatric help; van der Heijden et al. (2000) concerning Social Security fraud; Arnab and Singh (2010) concerning HIV/AIDS infection; Krumpal (2012) concerning opinions of xenophobia and anti-Semitism in Germany; Arnab and Mothupi (2015) concerning the sexual behavior of university students; and Arnab and Rodríguez (2019) concerning cheating in exams and bullying.

Comprehensive reviews of RR techniques are given by Arnab (2017), Chaudhuri and Mukherjee (1988), Chaudhuri (2011), Chaudhuri and Christofides (2013), and Fox and Tracy (1986). In the sequel, we will discuss RR techniques introduced by various researchers.

1.2 Warner's Model

Warner (1965) introduced the randomized response (RR) model for estimating π, the proportion of individuals in a population belonging to a group, say A, that possesses a sensitive characteristic such as being HIV positive. The members of the group $\overline{A}$ that are complementary to the group A, may or may not possess the sensitive attribute. For example, being HIV positive is sensitive, but being HIV negative is not. While both affiliation and non-affiliation with a political party may or may not be sensitive.

A sample s of size n was selected from a population U by the simple random sampling with replacement (SRSWR) method. In Warner's method, the selected respondents were not asked directly to disclose their membership in the group A. Instead, they were asked to perform a randomized response (RR) trial and to disclose truthfully the outcome of the trial performed. In Warner's RR trial, each of the selected respondents is directed to draw a card at random from a pack of well-shuffled cards of two types. One type bears the statement.

Q_A: I am a member of the group A with proportions, $P(\neq 1/2)$, while the other type bears the statement

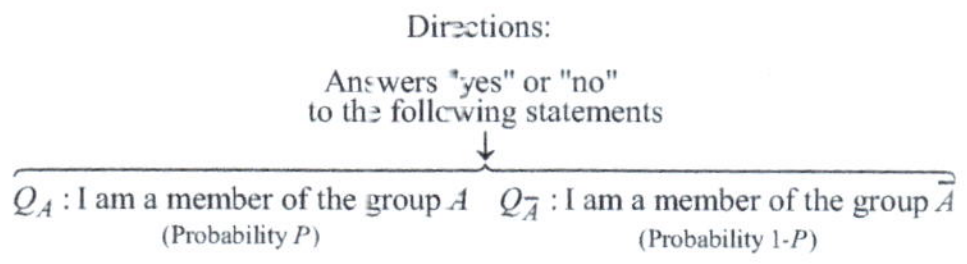

Fig. 1.1 Warner's model

$Q_{\overline{A}}$: I am a member of the group $\overline{A}$ with proportion $1 - P$, where $\overline{A}$ is the complement of the group A.

The respondents were asked to report "yes" if the statement written on the card drawn corresponded to their own status; if the statement did not correspond to their own status, they should report "no". In order to protect the secrecy of the response, the whole experiment was performed in the absence of the interviewers. The privacy of the respondents was maintained, as the interviewer would not know which type of card the respondent selected (Fig. 1.1).

1.2.1 Estimation of the Proportion π

For the Warner's model, the probability of obtaining a "yes" response from a respondent is

$$\theta_W = (2P - 1)\pi + (1 - P) \tag{1.2.1}$$

Let n_W be the total number of "yes" answers in the sample s and $\hat{\theta}_W = n_W/n.$, Warner (1965) derived the following results:

Theorem 1.2.1

$$\text{(i)}\hat{\pi}_W = \frac{\hat{\theta}_W - (1 - P)}{2P - 1}, P \neq 1/2 \tag{1.2.2}$$

is an unbiased estimator of.

(ii) The variance of $\hat{\pi}_W$ is

$$\begin{aligned}\text{Var}(\hat{\pi}_W) &= \frac{\theta_W(1 - \theta_W)}{n(2P - 1)^2}\\ &= \frac{\pi(1 - \pi)}{n} + \frac{P(1 - P)}{n(2P - 1)^2}\end{aligned} \tag{1.2.3}$$

(iii) An unbiased estimator of $\text{Var}(\hat{\pi}_W)$ is

$$\hat{\text{Var}}(\hat{\pi}_W) = \frac{\hat{\theta}_W(1 - \hat{\theta}_W)}{(n - 1)(2P - 1)^2} \tag{1.2.4}$$

1.2.2 True MLE of π

The likelihood function of obtaining n_W "yes" answers and $n - n_W$ "no" answers from a sample of n respondents is

$$L = \binom{n}{n_W} \theta_W^{n_W} (1 - \theta_W)^{n-n_W} \tag{1.2.5}$$

However, Singh (1976, 1978) observed that $\hat{\pi}_W$ given by Eq. (1.2.2) is not the maximum likelihood estimator of π as $\hat{\pi}_W$ can take values outside of [0, 1], the parameter space of π. For example, $P = 0.4$ and $\hat{\theta}_w = 0.3$ yields $\hat{\pi}_W = \frac{0.3-0.6}{-0.2} = 1.5$ as an inadmissible value of π. Singh (1976, 1978) obtained the MLE of π as follows:

$$\hat{\pi}_{MLE} = \begin{cases} \hat{\pi}_W, \text{ if } 0 \leq \hat{\pi}_W \leq 1 \\ 1, \text{ if } \hat{\pi}_W > 1 \\ 0, \text{ if } \hat{\pi}_W < 0 \end{cases} \tag{1.2.6}$$

For further commentary on this issue, readers are referred to Devore (1977) and Flinger et al. (1977) among others.

1.2.3 Efficiency of Warner's Estimator

1.2.3.1 Warner Estimator Versu Direct Estimator

Consider a direct response (DR) survey where a sample of n respondents is selected by the SRSWR method, and each respondent provides the true direct response about their membership of the sensitive characteristic A. Let n_1 be the number of respondents who answer "yes". Then, the estimated proportion of π is $\hat{\pi} = n_1/n$ and the variance of $\hat{\pi}$ is

$$\text{Var}(\hat{\pi}) = \frac{\pi(1-\pi)}{n} \tag{1.2.7}$$

The efficiency of the Warner estimator $\hat{\pi}_W$ compared to the direct estimator $\hat{\pi}$ is

$$\begin{aligned} E_W &= \frac{\text{Var}(\hat{\pi})}{\text{Var}(\hat{\pi}_W)} \\ &= \left[1 + \frac{P(1-P)}{\pi(1-\pi)(2P-1)^2}\right]^{-1} \\ &\leq 1 \end{aligned} \tag{1.2.8}$$

Clearly, the Warner estimator $\hat{\pi}_W$ is less efficient than the direct estimator $\hat{\pi}$. The amount of loss of efficiency $1 - E_W$ is compromised for the protection of the respondents' privacy. The efficiency E_W increases as the deviation of P from $1/2$ increases. So, the optimum value of P should be close to 0 or equal to 1. The efficiency E_W attains a maximum when $P = 1$ or $P = 0$. In this case, Warner's technique is reduced to the DR technique, and the respondents' privacy is jeopardized.

1.2.3.2 Warner Estimator Versus MLE Estimator

The MLE estimator $\hat{\pi}_{MLE}$ is a biased estimator for π. The expression of bias and mean-square error (MSE) of $\hat{\theta}_W$ were obtained by Singh (1976).

The bias of $\hat{\pi}_{MLE}$ for $P > 1/2$ is

$$\begin{aligned} B_{MLE} &= E(\hat{\pi}_{MLE} - \pi) \\ &= \sum_{n_W \geq nP} (1 - \hat{\pi}_W) b(n_W, n, \pi) - \sum_{n_W \leq n(1-P)} \hat{\pi}_W b(n_W, n, \pi) \end{aligned} \tag{1.2.9}$$

The magnitude and the sign of the bias B_{MLE} in Eq. (1.2.9) depend on the values of the parameters n,P and π. The amount of bias was found to be very small (Singh, 1976).

The MLE estimator $\hat{\pi}_{MLE}$ is more efficient than $\hat{\pi}_W$ as the mean-square error (MSE) of $\hat{\pi}_{MLE}$ is smaller than that of $\hat{\pi}_W$. This is because $|\hat{\pi}_{MLE} - \pi| \leq |\hat{\pi}_W - \pi|$ in all situations (Chaudhuri & Mukherjee, 1988).

1.3 Unrelated Question Model (U-Model)

In Warner's RR model, respondents are asked to answer the question of whether they are members of a stigmatizing group A or its complementary group $\bar{A}$. Since the questions are perfectly negatively related, a respondent may not feel safe answering them. Further, sometimes both questions are sensitive to the respondents. To overcome such difficulties, Greenberg et al. (1969) and Horvitz et al. (1967) introduced the U-model, which is also known as Simmon's model.

In this U-model, the respondents select a card at random from a pack of cards, as in Warner's method. The pack consists of two types of cards, with proportions P_1 and $1 - P_1$. The card type-1 bears the statement Q_A :"I am a member of the group A" and the card type-2 bears the statement Q_B : "I am a member of the group B". The group A is sensitive, such as drug users, while the group B is completely innocuous and unrelated to the characteristic A, such as vegetarians. The respondent answers truthfully, "yes" if the statement written on the card drawn matches his or her status, and "no" if the statement does not match his or her status (Fig. 1.2)

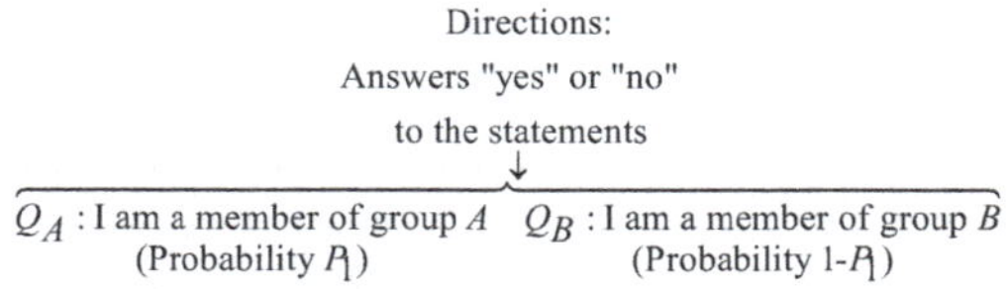

Fig. 1.2 Unrelated question model

The whole experiment is performed in the absence of the interviewer. The privacy of a respondent is guaranteed, as the interviewer will not know which of the two statements the respondent has answered.

1.3.1 *Estimation of* π

Let π_B be the proportion of persons belong to the innocuous group B. Then, the probability of getting a "yes" answer from a respondent selected at random is

$$\theta_{\mathrm{G}} = \pi P_1 + \pi_B (1 - P_1) \tag{1.3.1}$$

Let n_1 be the total number of "yes" answers obtained from a sample of n respondents selected by the SRSWR method. Consider two scenarios as follows.

1.3.1.1 π_B is Known

Let $\hat{\theta}_{\mathrm{G}} = \frac{n_1}{n}$ be the proportion of "yes" answers in the sample. The mean, variance, and an unbiased estimator of the variance of $\hat{\theta}_{\mathrm{G}}$ are given as follows:

Theorem 1.3.1

(i) $\hat{\pi}_{\mathrm{G}} = \frac{\hat{\theta}_{\mathrm{G}} - (1-P_1)\pi_B}{P_1}$ is an unbiased estimator of π

(ii) The variance of $\hat{\pi}_{\mathrm{G}}$ is

$$\mathrm{Var}(\hat{\pi}_{\mathrm{G}}) = \frac{\theta_{\mathrm{G}}(1 - \theta_{\mathrm{G}})}{nP_1^2}$$

(iii) An unbiased estimator of $\mathrm{Var}(\hat{\pi}_{\mathrm{G}})$ is

$$\widehat{\mathrm{Var}}(\hat{\pi}_{\mathrm{G}}) = \frac{\hat{\theta}_{\mathrm{G}}(1 - \hat{\theta}_{\mathrm{G}})}{(n-1)P_1^2}$$

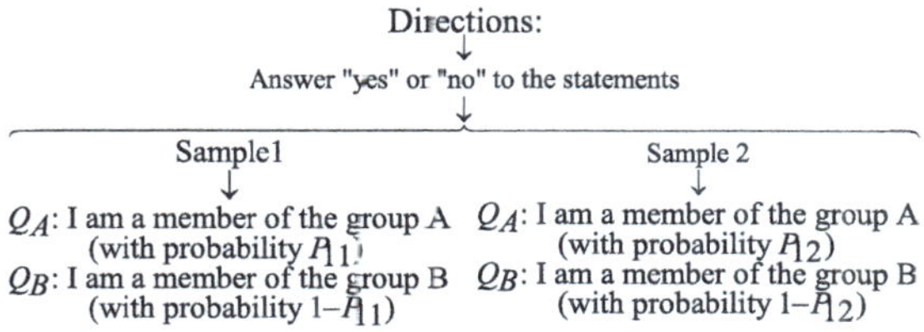

Fig. 1.3 U-model with π_B unknown

1.3.1.2 π_B is Unknown

In case π_{B} is unknown, two independent samples s_1 and s_2 of sizes n_1 and $n_2(= n-n_1)$ are selected from the population by the SRSWR method. Respondents selected in the sample $s_i(i = 1, 2)$ are asked to perform the RR technique described in Sect. 1.3 with parameter P_{1i} (Fig. 1.3).

The probability of obtaining a "yes" answer from a respondent belonging to the sample s_i is

$$\theta_{\mathrm{G}i} = \pi P_{1i} + \pi_B(1 - P_{1i}) \tag{1.3.2}$$

Let $\hat{\theta}_{\mathrm{G}i} = \frac{n_i}{n}$ be the proportion of "yes" answers obtained from the respondents in the sample s_i. Now, noting that $E(\hat{\theta}_{\mathrm{G}i}) = \theta_{\mathrm{G}i}$, $\mathrm{Var}(\hat{\theta}_{\mathrm{G}i}) = \frac{\theta_{\mathrm{G}i}(1-\theta_{\mathrm{G}i})}{n_i}$ and $\hat{\mathrm{Var}}(\hat{\theta}_{\mathrm{G}i}) = \frac{\hat{\theta}_{\mathrm{G}i}(1-\hat{\theta}_{\mathrm{G}i})}{n_i-1}$, the following results are obtained.

Theorem 1.3.2

(i) $\hat{\pi}_{\mathrm{G}}^* = \frac{(1-P_{12})\hat{\theta}_{\mathrm{G}1}-(1-P_{11})\hat{\theta}_{\mathrm{G}2}}{P_{11}-P_{12}}$; $P_{11} \neq P_{12}$, is an unbiased estimator of π

(ii) The variance of $\hat{\pi}_{\mathrm{G}}^*$ is

$$\mathrm{Var}(\hat{\pi}_{\mathrm{G}}^*) = \frac{1}{(P_{11} - P_{12})^2}\left[\frac{(1 - P_{12})^2\theta_{\mathrm{G}1}(1 - \theta_{\mathrm{G}1})}{n_1} + \frac{(1 - P_{11})^2\theta_{\mathrm{G}2}(1 - \theta_{\mathrm{G}2})}{n_2}\right]$$

(iii) An unbiased estimator of $\mathrm{Var}(\hat{\pi}_{\mathrm{G}}^*)$ is

$$\hat{\mathrm{Var}}(\hat{\pi}_{\mathrm{G}}^*) = \frac{1}{(P_{11} - P_{12})^2}\left[\frac{(1 - P_{12})^2\hat{\theta}_{\mathrm{G}1}(1 - \hat{\theta}_{\mathrm{G}1})}{n_1 - 1} + \frac{(1 - P_{11})^2\hat{\theta}_{\mathrm{G}2}(1 - \hat{\theta}_{\mathrm{G}2})}{n_2 - 1}\right]$$

1.3.1.3 Optimum Choices of Parameters

The optimum value of n_1 for a fixed value of n is obtained by minimizing $\mathrm{Var}(\hat{\pi}_{\mathrm{G}}^*)$ subject to the constraint $n_1 + n_2 = n$. The optimum value of n_1 and the corresponding value of $\mathrm{Var}(\hat{\pi}_{\mathrm{G}}^*)$ come out as

$$\text{opt}\ (n_1) = n\frac{(1-P_{12})\sqrt{\theta_{G1}(1-\theta_{G1})}}{(1-P_{12})\sqrt{\theta_{G1}(1-\theta_{G1})}+(1-P_{11})\sqrt{\hat{\theta}_{G2}(1-\theta_{G2})}} \tag{1.3.3}$$

and

$$V_{\text{opt}}(\hat{\pi}_G^*) = \frac{\left[(1-P_{12})\sqrt{\theta_{G1}(1-\theta_{G1})}+(1-P_{11})\sqrt{\hat{\theta}_{G2}(1-\theta_{G2})}\right]^2}{n(P_{11}-P_{12})^2} \tag{1.3.4}$$

In practice, the optimum value of n_1 cannot be used since it involves unknown parameters π and π_B. However, one can estimate the parameters from a pilot survey or past experiences. For the choice of π_B, no clear solution is available. A moderate value of π_B between 0.25 and 0.75 was proposed by Chaudhuri and Mukherjee (1988). Regarding the choice of P_{11} and P_{12}, Moors (1971) suggested one should choose P_{11} as large as possible and P_{12} as small as possible. The choice of $P_{12} = 0$ is acceptable as the characteristic B is not sensitive. However, in this case, the respondents selected in the sample s_2 must disclose the true status of the innocuous characteristics.

1.3.2 Moors's Estimator

For $P_{11} = P$ and $P_{12} = 0$, we have that

$$\theta_{G1} = \theta_{GM} = \pi P + \pi_B(1-P) \ \text{ and } \ \theta_{G2} = \pi_B \tag{1.3.5}$$

So, in this situation, $\hat{\pi}_G^*$ reduces to Moors's estimator

$$\hat{\pi}_M^* = \frac{\hat{\theta}_{G1}-(1-P)\hat{\theta}_{G2}}{P} \tag{1.3.6}$$

The Moors's estimator $\hat{\pi}_M^*$ is unbiased for π and has the variance

$$\text{Var}(\hat{\pi}_M^*) = \frac{\left[\sqrt{\theta_{GM}(1-\theta_{GM})}+\sqrt{(1-P)^2\pi_B(1-\pi_B)}\right]^2}{nP^2} \tag{1.3.7}$$

Moors has shown that $\hat{\pi}_M^*$ is more efficient than $\hat{\pi}_G^*$ in the sense $\text{Var}(\hat{\pi}_M^*) \leq V_{\text{opt}}(\hat{\pi}_G^*)$.

1.3.2.1 Problems of Moors's Estimator

Mangat et al. (1997) pointed out that Moors's model fails to protect the privacy of those respondents who appear in both samples s_1 and s_2,and report either "yes" in the first sample s_1 and "no" in the second sample s_2 or "no" in s_1 and "yes" in s_2. Mangat et al. (1997) and Munir et al. (1998) provided alternative methods that are free from such drawbacks but remain less efficient than the Moors' estimator $\hat{\pi}_M^*$. Further improvements were provided by Mangat, Singh and Singh (1993), and Singh et al. (2000). Arnab (2007) extended Singh et al.'s (2000) results to general sampling designs.

1.3.3 Comparison of the U-Model with the Warner Model

Dowling and Shachtman (1975) compared the performances of Warner's estimator $\hat{\pi}_W$ with the U-model estimators $\hat{\pi}_G$,$\hat{\pi}_G^*$, and Moors's estimator $\hat{\pi}_M^*$ theoretically. However, such a comparison yielded no elegant results. Details are given by Chaudhuri and Mukherjee (1988).

1.4 U-Model with Two Innocuous Questions

Folsom et al. (1973) proposed a RR technique involving the sensitive characteristic A of interest along with two non-sensitive characteristics B(such as religion: Christian) and C(a food habit: vegetarian). The population proportions π_B, π_C of the characteristics B and C are assumed to be unknown. Here, two independent samples s_1 and s_2 of sizes n_1 and n_2 are selected from the population by the SRSWR method. Respondents in the sample $s_i(i = 1, 2)$ are asked to answer two questions. One is an indirect question using a randomized devise R_i, and the other is a direct question $Q_i(i = 1, 2)$ relating to an innocuous characteristic B or C.

More specifically, respondents in the sample s_1 are to answer the following questions:

(i) Based on R_1: "Do you possess the sensitive characteristic A?", with probability P, or the innocuous question "Are you a Christian?", with probability $1 - P$,

and

(ii) Direct question Q_1: "Are you a vegetarian?"

On the other hand, the respondents in the sample s_2 are to answer the following two questions:

(i) Based on R_2: "Do you possess the sensitive characteristic A?", with probability P, or the innocuous question "Are you a vegetarian?", with probability $1 - P$.
(ii) Direct question Q_2: Are you a Christian?

Then, the probability of obtaining "yes" answers from a respondent from R_1 and Q_1 are

$$\theta_{\mathrm{F1}} = P\pi + (1-P)\pi_B \text{ and } \pi_C.$$

Similarly, the probability of getting a "yes" answer from a respondent to the questions R_2 and Q_2 are (Fig. 1.4)

$$\theta_{\mathrm{F2}} = P\pi + (1-P)\pi_C \text{ and } \pi_B.$$

Let $\hat{\theta}_{\mathrm{F1}}$, $\hat{\theta}_{\mathrm{F2}}$, $\hat{\pi}_C$, and $\hat{\pi}_B$ be the proportion of "yes" answers for the questions R_1, R_2, Q_1, and Q_2, respectively. The unbiased estimators of π based on the samples s_1 and s_2 are as follows:

$$\hat{\pi}_{\mathrm{F1}} = \frac{\hat{\theta}_{F1} - (1-P)\hat{\pi}_B}{P} \text{ and } \hat{\pi}_{\mathrm{F2}} = \frac{\hat{\theta}_{\mathrm{F2}} - (1-P)\hat{\pi}_C}{P} \tag{1.4.1}$$

The variances and covariance of $\hat{\pi}_{\mathrm{F1}}$ and $\hat{\pi}_{\mathrm{F2}}$ are

$$\mathrm{Var}(\hat{\pi}_{\mathrm{F1}}) = \frac{1}{P^2}\left[\frac{\theta_{\mathrm{F1}}(1-\theta_{\mathrm{F1}})}{n_1} + (1-P)^2\frac{\pi_B(1-\pi_B)}{n_2}\right],$$

$$\mathrm{Var}(\hat{\pi}_{\mathrm{F2}}) = \frac{1}{P^2}\left[\frac{\theta_{\mathrm{F2}}(1-\theta_{\mathrm{F2}})}{n_2} + (1-P)^2\frac{\pi_C(1-\pi_C)}{n_1}\right] \text{ and}$$

$$\mathrm{Cov}(\hat{\pi}_{\mathrm{F1}}, \hat{\pi}_{\mathrm{F2}}) = -\frac{(1-P)}{P^2}\left[\begin{array}{c}\dfrac{P(\pi_{AC} - \pi\pi_C) + (1-P)(\pi_{BC} - \pi_B\pi_C)}{n_1} \\ +\dfrac{P(\pi_{AB} - \pi\pi_B) + (1-P)(\pi_{BC} - \pi_B\pi_C)}{n_2}\end{array}\right] \tag{1.4.2}$$

where π_{AB}, π_{AC} and π_{BC} denote the proportion of individuals in the population possessing both the characteristics A and B, A and C, and B and C, respectively.

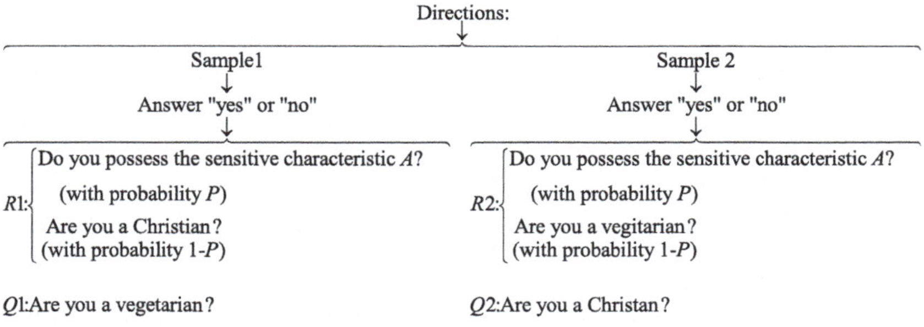

Fig. 1.4 U-model with two questions

1.5 Forced Randomized Response Model (F-Model)

The Forced randomized response (F-model) was proposed by Boruch (1972) and Horvitz et al. (1976). In the F-model, the respondent answers "yes" or "no" truthfully to the sensitive question "Do you possess the sensitive attribute A?" with probability P, is forced to answer "yes" with probability α and is forced to answer "no" with probability $1 - P - \alpha$. The RR technique may involve selecting a marble at random from a box containing three colors of marbles; green, yellow, and red, say, with proportions P, α and $\beta = 1 - P - \alpha(0 < P, \alpha, 1 - P - \alpha < 1)$, respectively. In the event the respondent chooses a green marble, he/she answers "yes" if he/she possesses the sensitive attribute A; otherwise, he/she answers "no". If the respondent chooses a yellow marble, he/she is to answer "yes". In case the respondent selects a red marble, then the respondent should answer "no". The whole experiment is performed in the absence of the interviewer. The confidentiality of the response is protected, as the interviewer will not know which marble the respondent has chosen (Fig. 1.5).

In this method, the probability of getting an answer "yes" from a respondent is

$$\theta_B = \pi P + \alpha \tag{1.5.1}$$

where π is the proportion of people possessing the sensitive attribute A in the population.

Let $\hat{\theta}_B$ be the proportion of "yes" answers obtained from a sample of size n selected by the SRSWR method. Then, we derive the following results easily: The proof is omitted because it is easy.

Theorem 1.5.1

(i) $\hat{\pi}_{\mathrm{B}} = \frac{\hat{\theta}_{\mathrm{B}} - \alpha}{P}$ is an unbiased estimator of π
(ii) The variance of $\hat{\pi}_{\mathrm{B}}$ is

$$\mathrm{Var}(\hat{\pi}_{\mathrm{B}}) = \frac{\theta_B(1 - \theta_B)}{nP^2}$$

(iii) An unbiased estimator of $\mathrm{Var}(\hat{\pi}_{\mathrm{B}})$ is

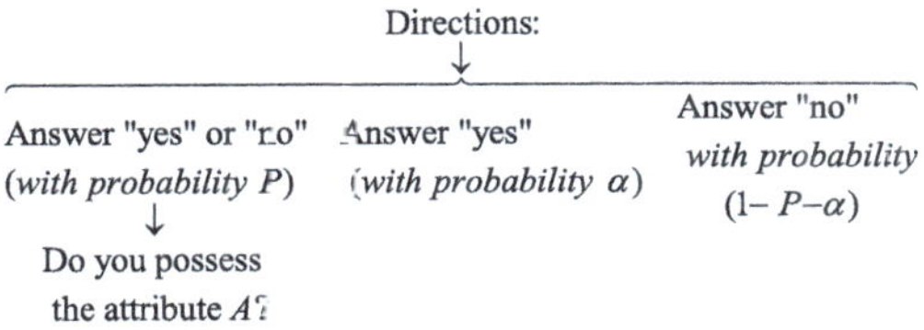

Fig. 1.5 Forced randomized response

$$\hat{\text{Var}}(\hat{\pi}_B) = \frac{\hat{\theta}_B(1 - \hat{\theta}_B)}{(n - 1)P^2}$$

Singh (1993) and Singh and Mathur (2003) extended the Forced randomized response (F-model) as follows:

1.5.1 Singh's (1993) Model

In Singh's (1993) model, respondents provide true "yes" or "no" answers to the question Q: "Do you belong to the sensitive group A?" with probability P and answer "yes" with probability $\alpha = 1 - P$ irrespective of his or her membership to A. Thus, Singh's (1993) model is a particular case of the F-model with $\beta = 0$ (Fig. 1.6).

1.5.2 Singh and Mathur's Model

In Singh and Mathur's (2003) model, respondents provide direct responses of "yes" or "no" to the question Q: "Do you belong to the sensitive group A?" with probability T and provide RR using Singh (1993), mentioned above, with probability $1 - T$. In this case, the probability of obtaining a "yes" answer from a respondent is (Fig. 1.7)

$$\lambda_{SM} = T\pi + (1 - T)\lambda_S$$

where $\lambda_S = \pi P + (1 - P)$.

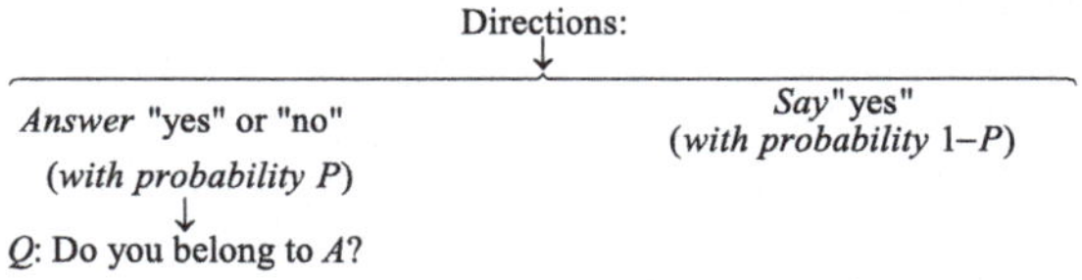

Fig. 1.6 Singh's model (1993)

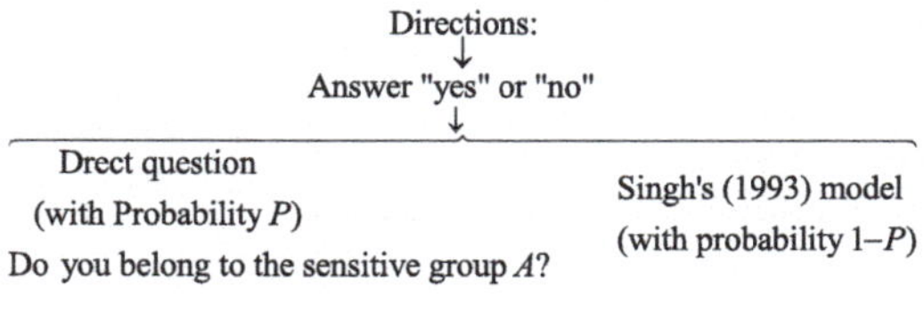

Fig. 1.7 Singh and Mathur's model (2003)

Let n_1 be the number of "yes" answers in a sample of size n selected by the SRSWR method. Singh and Mathur (2003) proposed a biased estimator of π as

$$\hat{\pi}_{\text{SM}} = \frac{n_1}{n}$$

They derived the expressions of the Bias and MSE of $\hat{\pi}_{\text{SM}}$ as follows:

$$B(\hat{\pi}_{\text{SM}}) = (1-T)(1-P)(1-\pi)$$

$$\text{MSE}(\hat{\pi}_{\text{SM}}) = \frac{1}{n} \left[\pi(1-\pi)\{P + T(1-P)\}^2 + (1-T)(1-P)(1-\pi)\{1-(1-P)(1-T)\}\right] + [(1-\pi)(1-T)(1-P)]^2$$

Singh and Mathur (2003) compared the performance of their estimator with Singh and Singh (1993)'s estimator numerically but provided no theoretical result.

1.6 Mangat and Singh's Model

Mangat and Singh (1990) proposed a two-stage RR method where, in the first stage, each of the selected respondents is asked to perform a randomized device R consisting of two statements, namely, (i) "I belong to the sensitive group A" and (ii) "Go to the random device $R*$" with probabilities T and $1-T$, respectively. If the respondent selects statement (i), then he or she should answer "yes" if he or she belongs to A and answer "no" otherwise. On the other hand, if the respondent selects statement (ii), then he or she should perform the random device $R*$. The random device $R*$ is Warner's (1965) technique involving two statements, (i) "I belong to the sensitive group A" and "I do not belong to the sensitive group A" with probabilities P and $1-P$, respectively. The whole procedure is completed by the respondents in the absence of the interviewer to their privacy.

Let a sample s of size n be selected from a population by the SRSWR method. The probability of obtaining a "yes" answer is (Fig. 1.8)

$$\begin{aligned}\theta_{\text{MS}} &= T\pi + (1-T)\{P\pi + (1-P)(1-\pi)\} \\ &= \pi\{T + (1-T)(2P-1)\} + (1-T)(1-P) \qquad (1.6.1)\end{aligned}$$

Let $\hat{\theta}_{\text{MS}}$ be the proportion of "yes" answers obtained in sample s. Noting that $\hat{\theta}_{\text{MS}}$ follows a binomial distribution $B(n, \theta_{\text{MS}})$ with parameters n and θ_{MS}, we have that

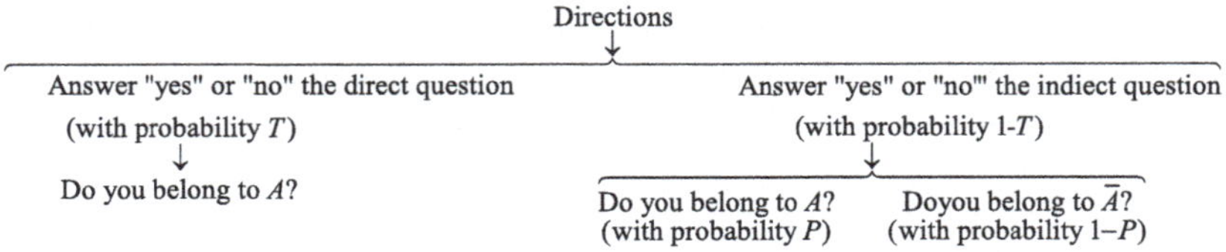

Fig. 1.8 Mangat Singh's model (1990)

$$E(\hat{\theta}_{\mathrm{MS}}) = \theta_{\mathrm{MS}};\ \mathrm{Var}(\hat{\theta}_{\mathrm{MS}}) = \frac{\theta_{\mathrm{MS}}(1-\theta_{\mathrm{MS}})}{n} \text{ and } \hat{\mathrm{Var}}(\hat{\theta}_{\mathrm{MS}}) = \frac{\hat{\theta}_{\mathrm{MS}}(1-\hat{\theta}_{\mathrm{MS}})}{n-1} \tag{1.6.2}$$

as an unbiased estimator of the $\mathrm{Var}(\hat{\theta}_{\mathrm{MS}})$.

Equation (1.6.2) yields Theorem 1.6.1.

Theorem 1.6.1

(i) $\hat{\pi}_{\mathrm{MS}} = \frac{\hat{\theta}_{\mathrm{MS}}-(1-T)(1-P)}{D}$ is an unbiased estimator of π

where $D = T + (1-T)(2P-1) \neq 0$.

(ii) The variance of $\hat{\pi}_{\mathrm{MS}}$ is

$$\mathrm{Var}(\hat{\pi}_{\mathrm{MS}}) = \frac{\theta_{\mathrm{MS}}(1-\theta_{\mathrm{MS}})}{nD^2}$$

(iii) An unbiased estimator of is

$$\hat{\mathrm{Var}}(\hat{\pi}_{\mathrm{MS}}) = \frac{\hat{\theta}_{\mathrm{MS}}(1-\hat{\theta}_{\mathrm{MS}})}{(n-1)D^2}$$

1.6.1 Comparison with Warner's Model

The estimator $\hat{\pi}_{\mathrm{MS}}$ is more efficient than Warner's estimator $\hat{\pi}_{\mathrm{W}}$ in (1.2.2) if $\mathrm{Var}(\hat{\pi}_{\mathrm{W}}) < \mathrm{Var}(\hat{\pi}_{\mathrm{MS}})$

$$\text{i.e., if } T > \frac{1-2P}{1-P} \tag{1.6.3}$$

The above inequality shows that the Mangat-Singh RR model can be made more efficient than the Warner model by choosing a suitable value of T for a given value P. Alternatively, Warner's model can be made more efficient than the Magat-Singh model by choosing P appropriately for a given T.

1.6.2 *Untruthful Reporting*

Consider the situation where the respondents do not report completely truthfully. Let T_1 and T_2 be the probabilities that a respondent in the sensitive group A answers truthfully at the first and second stages, respectively. Then the probability of obtaining a "yes" answer from a respondent selected from the sample is

$$\theta'_{MS} = T\pi T_1 + (1 - T)\{P\pi T_2 + \pi(1 - P)(1 - T_2) + (1 - P)(1 - \pi)\} \quad (1.6.4)$$

In the case of untruthful reporting, $\hat{\pi}_{MS}$ is not an unbiased estimator of π. The expressions of bias and mean-square error of $\hat{\pi}_{MS}$ were provided by Mangat-Singh (1990) as follows:

$$B(\hat{\pi}_{MS}) = \pi\left\{\frac{T(T_1 - T_2)}{2P - 1 + 2T(1 - P)} + T_2 - 1\right\} \quad (1.6.5)$$

and

$$\text{MSE}(\hat{\pi}_{MS}) = \frac{\pi T_2(1 - \pi T_2)}{n} + \frac{(1 - T)(1 - P)\{1 - (1 - T)(1 - P)\}}{n\{2P - 1 + 2T(1 - P)\}^2} + \pi^2(1 - T_2)^2$$
$$+\pi T(T_1 - T_2)$$
$$\frac{[1 + \pi(n - 1)\{T(T_1 - T_2) + 4TT_2(1 - P) + 2T_2(2P - 1)\} - 2(1 - T)(1 - P) - 2\pi n\{2T(1 - P) + (2P - 1)\}]}{n\{2P - 1 + 2T(1 - P)\}^2} \quad (1.6.6)$$

1.7 Mangat Model A

In Mangat's (1994) Model A, the respondent is asked to report the answer "yes" if he or she possesses the sensitive attribute A. If the respondent does not have an attribute A, the respondent is to perform Warner's RR technique with a parameter P as described in Sect. 1.2. and provide either a "yes" or "no" answer. The whole procedure is completed in the absence of the interviewer. It should be noted that a "yes" answer may come from a respondent if he or she belongs to the sensitive group A as well as from those who do not belong to the group A. The confidentiality of the respondent reporting "yes" will not be violated. However, the respondent reporting "no" will be identified since he or she always comes from the group $\bar{A}$, the complement of group A. Mangat argued that cooperation from the respondents will not be diminished as respondents belong to $\bar{A}$ do not carry any perceived stigma (Fig. 1.9).

Suppose a sample of size n is selected from a population by the SRSWR method. Then, the probability of obtaining a "yes" answer from a respondent selected from the sample is

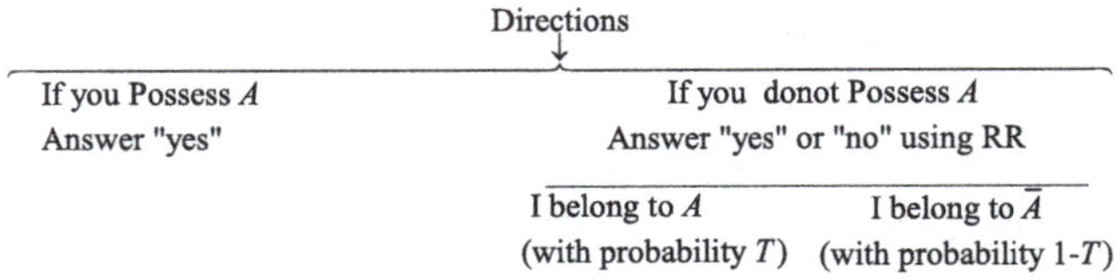

Fig. 1.9 Mangat model A

$$\theta_M = \pi + (1 - \pi)(1 - P) \tag{1.7.1}$$

Let $\hat{\theta}_M$ be the proportion of "yes" answers in the sample. Then, $\hat{\theta}_M$ follows a $B(n, \theta_M)$ and we will get the following results:

Theorem 1.7.1

(i) $\hat{\pi}_M = \frac{\hat{\theta}_M - (1-P)}{P}$ is an unbiased estimator of π

(ii) The variance of $\hat{\pi}_M$ is

$$\text{Var}(\hat{\pi}_M) = \frac{\theta_M(1 - \theta_M)}{nP^2}$$

(iii) An unbiased estimator of $\text{Var}(\hat{\pi}_M)$ is

$$\hat{\text{Var}}(\hat{\pi}_M) = \frac{\hat{\theta}_M(1 - \hat{\theta}_M)}{(n - 1)P^2}$$

1.7.1 Comparison with Mangat-Singh's and Warner's Model

The estimator $\hat{\pi}_M$ will be more efficient than the Mangat-Singh (1990) estimator if

$$\text{Var}(\hat{\pi}_M) < \text{Var}(\hat{\pi}_{MS})$$

$$\text{i.e., if } \pi > 1 - \frac{P(1 - T)\{1 - (1 - T)(1 - P)\}}{\{2P - 1 + 2T(1 - P)\}^2},\ P \neq 0.5 \tag{1.7.2}$$

The estimator $\hat{\pi}_M$ will be more efficient than $\hat{\pi}_W$ if

$$\pi > 1 - \left(\frac{P}{2P - 1}\right)^2 \tag{1.7.3}$$

The conditions (1.7.2) and (1.7.3) show that none of Warner's (1965), Mangat and Singh's (1990), or Mangat's (1994) models is the best in all situations. One model can be made better than the other by choosing the parameters suitably, and vice versa.

1.7.2 Untruthful Reporting

Let T_3 be the probability that the respondent belonging to the sensitive group reports a truthful answer. Since the respondents in the non-sensitive group have no reason to tell a lie, the probability of obtaining a "yes" answer from a selected respondent is

$$\theta'_{\mathrm{M}} = \pi T_3 + (1 - \pi)(1 - P) \tag{1.7.4}$$

Let $\hat{\theta}'_{\mathrm{M}}$ = proportion of "yes" answers in the sample. Then $\hat{\pi}'_{\mathrm{M}} = \frac{\hat{\theta}'_{\mathrm{M}} - (1-P)}{P}$ is a biased estimator of π. The magnitudes of bias and mean-square error of $\hat{\pi}'_{\mathrm{M}}$ were obtained by Mangat and Singh (1990) are as follows:

$$B(\hat{\pi}'_{\mathrm{M}}) = \frac{\pi(T_3 - 1)}{P}$$

and

$$\mathrm{MSE}(\hat{\pi}'_{\mathrm{M}}) = \frac{1}{P^2}\left[\{\pi(1 - T_3)\}^2 + \frac{\pi T_3(1 - \pi T_3) + (1 - \pi)(1 - P)\{1 - (1 - \pi)(1 - P) - 2\pi T_3\}}{n}\right] \tag{1.7.5}$$

1.8 Mangat's Model B (1992) and Chang and Liang's Model (1996)

Mangat Model B (1992) is the same as the Chang and Liang Model (1996). In these models, the selected respondent is requested to provide either the direct response "yes" or "no" relating to his or her membership of the sensitive group A with probability T as in Mangat and Singh's (1990) model, or an RR "yes" or "no" using Greenberg's U-model with probability $1-T$. In the U-model, the respondent reports either his or her membership in the sensitive group A with probability P or membership in an innocuous group B with probability $1 - P$. The proportion of people belonging to the innocuous group π_B is assumed to be known.

For these models, the probability of obtaining a "yes" answer from a respondent selected at random is (Fig. 1.10)

$$\lambda^*_{\mathrm{MS}} = T\pi + (1 - T)\{P\pi + (1 - P)\pi_{\mathrm{B}}\} \tag{1.8.1}$$

Let $\hat{\lambda}^*_{\mathrm{MS}}$ be the proportion of "yes" answers obtained from a sample of size n selected by the SRSWR method. Then one gets the following results:

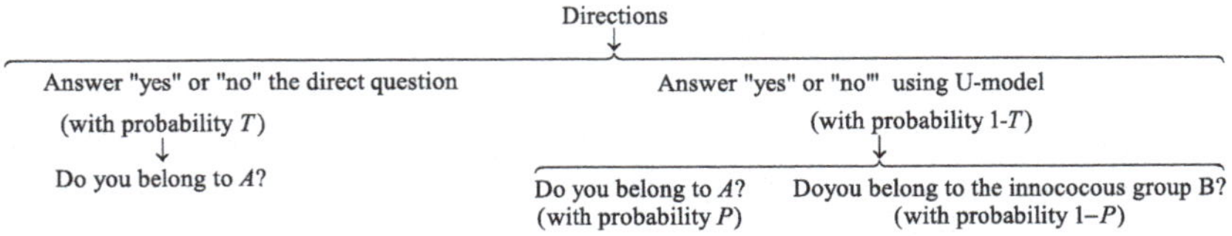

Fig. 1.10 Mangat model B (1992) and Chang and Liang model (1996)

Theorem 1.8.1 Under SRSWR sampling,

(i) $\hat{\pi}^*_{\text{MS}} = \frac{\hat{\lambda}^*_{\text{MS}} - (1-T)(1-P)\pi_{\text{B}}}{T+(1-T)P}$ is an unbiased estimator of π.

(ii) The variance of $\hat{\pi}^*_{\text{MS}}$ is $\text{Var}(\hat{\pi}^*_{\text{MS}}) = \frac{\lambda^*_{\text{MS}}(1-\lambda^*_{\text{MS}})}{n\{T+(1-T)P\}^2}$

(iii) $\hat{\text{Var}}(\hat{\pi}^*_{\text{MS}}) = \frac{\hat{\lambda}^*_{\text{MS}}(1-\hat{\lambda}^*_{\text{MS}})}{(n-1)\{T+(1-T)P\}^2}$

1.9 Mangat, Singh and Singh Model

In Mangat et al. (1992) model, the respondent reports "yes" if he or she belongs to the sensitive group A. If the respondent does not belong to the sensitive group A, he or she is instructed to use a randomization device $R1$. In the randomization device $R1$, the respondent selects a card at random from a box containing two types of cards bearing letters A and B with proportions P and $1 - P$. The respondent should report truthfully, "yes", only if he or she selects the card B and possesses the innocuous characteristic B. Otherwise, the respondent should answer "no". The proportion of persons belonging to the group B in the population π_{B} is known. The respondent should not disclose the type of card drawn to the interviewer to protect his or her privacy.

For this model, the probability of obtaining "yes" answer from a respondent selected at random is

$$\lambda_{\text{MSS}} = \pi + (1-\pi)(1-P)\pi_{\text{B}} \tag{1.9.1}$$

Let $\hat{\lambda}_{\text{MSS}}$ be the proportion of "yes" answers in a sample of size n selected by the SRSWR method (Fig. 1.11).

Theorem 1.9.1

(i) $\hat{\pi}_{\text{MSS}} = \frac{\hat{\lambda}_{\text{MSS}} - (1-P)\pi_{\text{B}}}{1-(1-P)\pi_{\text{B}}}$ is an unbiased estimator of π

(ii) The variance of $\hat{\pi}_{\text{MSS}}$ is $\text{Var}(\hat{\pi}_{\text{MSS}}) = \frac{\lambda_{\text{MSS}}(1-\lambda_{\text{MSS}})}{n[1-(1-P)\pi_{\text{B}}]^2}$

(iii) An unbiased estimator of $\text{Var}(\hat{\pi}_{\text{MSS}})$ is

$$\hat{\text{Var}}(\hat{\pi}_{\text{MSS}}) = \frac{\hat{\lambda}_{\text{MSS}}(1-\hat{\lambda}_{\text{MSS}})}{(n-1)[1-(1-P)\pi_{\text{B}}]^2}$$

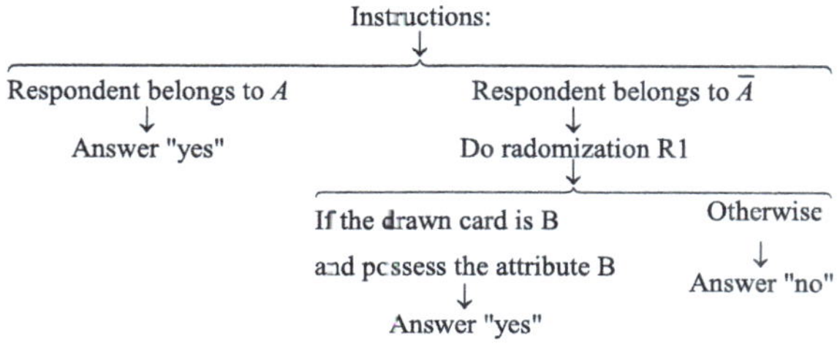

Fig. 1.11 Mangat, Singh and Singh's model

1.10 Singh, Horn, Singh and Mangat Model

Singh et al. (2003) proposed a forced RR model where three types of cards A, B, and C with proportions P_1, P_2, and $P_3(= 1 - P_1 - P_2)$ are taken in a box. Card types A and B bear the statements "I belong to the sensitive group A" and "I belong to the innocuous group B", respectively, while card type-C is blank. A respondent is asked to draw a card at random and is asked to report "yes" if the statement on the card matches his or her status; otherwise, they should report "no" (Fig. 1.12).

For this model, the probability of obtaining a "yes" answer is

$$\lambda_{S1} = P_1\pi + P_2\pi_B$$

where π_B is the proportion of persons in the innocuous group which is known.

Singh et al. (2003) derived the following results:

(i) $\hat{\pi}_{S1} = \frac{\hat{\lambda}_{S1} - (1-P_1)\pi_B}{P_1}$ is an unbiased estimator of π

where $\hat{\lambda}_{S1}$ is the proportion of "yes" answers in the sample.

(ii) $\text{Var}(\hat{\pi}_{S1}) = \frac{\pi(1-\pi)}{n} + \frac{\pi(1-P_1-2P_2\pi_B)}{nP_1} + \frac{P_2\pi_B(1-P_2\pi_B)}{nP_1^2}$

(iii) $\hat{\text{V}}\text{ar}(\hat{\pi}_{S1}) = \frac{\hat{\lambda}_{S1}(1-\hat{\lambda}_{S1})}{(n-1)P_1^2}$ is an unbiased estimator of $\text{Var}(\hat{\pi}_{S1})$.

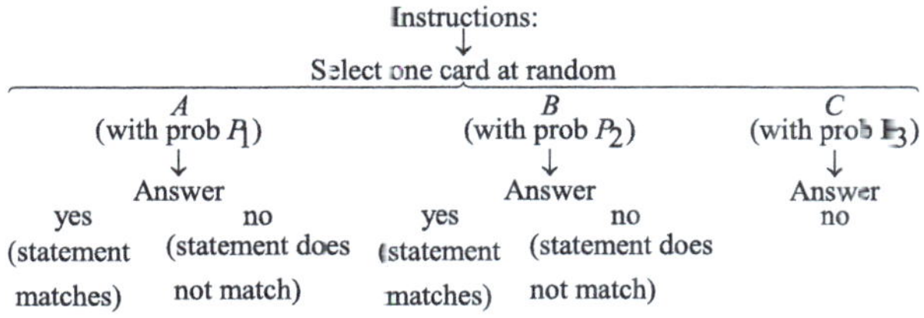

Fig. 1.12 Singh, Horn, Singh and Mangat model

1.11 Chang, Wang and Huang's Model

In Chang, Wang and Huang's (2004) model, four types of cards A, B, C, and D with proportions P_1, P_2,P_3, and $P_4 (= 1 - P_1 - P_2 - P_3)$ are placed in a box. Card types A and B bear the statements "I belong to the sensitive group A" and "I do not belong to the sensitive group A", while card types C and D bear the statements "yes" and "no", respectively. Each of the selected respondents is asked to choose one card at random from the box. If the card types A and B are chosen, then respondents should report "yes" or "no" if the statement written on the card matches or does not match their actual statuses. In the case that cards C and D are chosen, respondents should report the statement written on the chosen card.

In this model, the probability of getting a "yes" answer is $\lambda_c = \pi P_1 + (1-\pi)P_2 + P_3$ and an unbiased estimator of population proportion π is

$$\hat{\pi}_c = \frac{\hat{\lambda}_c - (P_2 + P_3)}{P_1 - P_2}; P_1 \neq P_2 \tag{1.11.1}$$

where $\hat{\lambda}_c$ is the proportion of "yes" answers in an SRSWR of size n.

The variance and unbiased estimator of the variance of $\hat{\pi}_c$ are given by (Fig. 1.13)

$$\text{Var}(\hat{\pi}_c) = \frac{\lambda_c(1 - \lambda_c)}{n(P_1 - P_2)^2} \text{and} \hat{\text{Var}}(\hat{\pi}_c) = \frac{\lambda_c(1 - \lambda_c)}{(n-1)(P_1 - P_2)^2} \tag{1.11.2}$$

1.12 Singh and Joarder's (1977) Model

In Singh and Joarder's (1997a, 1997b) model, a sample s of n respondents is selected by the SRSWR method. Each of the selected respondents belonging to the sensitive group A is asked to perform the Warner technique with a parameter P. Respondents belonging to the non-sensitive group $\bar{A}$ are asked to provide an answer of "yes" or "no" according to the Warner trial. Respondents belonging to the sensitive group A, are asked to report "yes" if the drawn card bears the statement "I belong to the sensitive group A". If the drawn card bears the statement "I belong to the non-sensitive group $\bar{A}$", the respondents are asked to repeat Warner's technique again and report the usual response of "yes" or "no" (Fig. 1.14).

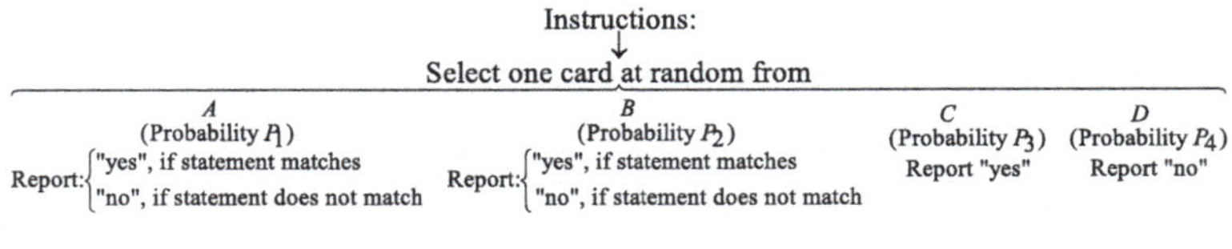

Fig. 1.13 Chang et al. model

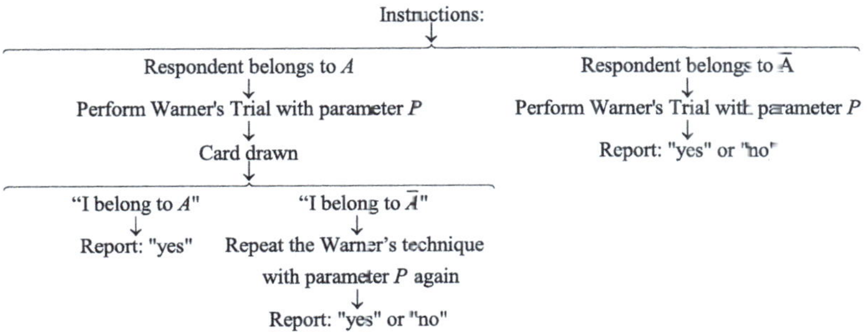

Fig. 1.14 Singh and Joarder's model

For this RR technique, the probability of obtaining a "yes" answer from a respondent is

$$\begin{aligned}\lambda_{SJ} &= \pi\{P + (1 - P)P\} + (1 - \pi)(1 - P) \\ &= \pi\{2P - 1 + P(1 - P)\} + (1 - P)\end{aligned}$$

Let $\hat{\lambda}_{SJ}$ be the proportion of "yes" answers in the sample s. Singh and Joarder derived the following results.

Theorem 1.12.1

(i) $\hat{\pi}_{JS} = \frac{\hat{\lambda}_{SJ} - (1-P)}{2P-1+P(1-P)}$ is an unbiased estimator of π
(ii) The variance of $\hat{\pi}_{JS}$ is

$$\mathrm{Var}(\hat{\pi}_{JS}) = \frac{\pi(1-\pi)}{n} + \frac{P(1-P)}{n[(2P-1) + P(1-P)]}\left[\frac{1}{(2P-1) + P(1-P)} - \pi\right]$$

(iii) An unbiased estimator of $\hat{\pi}_{JS}$ is

$$\hat{\mathrm{Var}}(\hat{\pi}_{JS}) = \frac{\hat{\lambda}_{JS}(1 - \hat{\lambda}_{JS})}{(n-1)[(2P-1) + P(1-P)]^2}$$

Singh and Joarder (1997a, 1997b) have shown that for $P > 1/2$, $\mathrm{Var}(\hat{\pi}_W) \geq \mathrm{Var}(\hat{\pi}_{JS})$, i.e., Singh and Joarder's (1997a, 1997b) RR model is more efficient than Warner's model (Chaudhuri, 2011, pages 154–155). However, the comparison is not fair because Warner's model is based on only one trial, whereas respondents in Singh and Joarder's RR technique may often need to perform more than one trial, making it more expensive and complex than Warner's model.

1.13 Implicit Randomized Response Technique

The implicit RR technique was introduced by Takahasi and Sakasegawe (1977). The proposed RR technique does not require any randomizing device. This technique is applicable not only to face-to-face interview surveys but also to telephone interviews and mail surveys. Three independent samples s_1, s_2, and s_3 of sizes n_1,n_2, and n_3 are selected from a population by the SRSWR method. Each of the selected units (respondents) in the sample is assigned to one of the three mutually exclusive and exhaustive categories, say B_1,B_2, and B_3. Categories might be months of birth, the last two digits of the mobile number, days of birth in a week, etc. For example, respondents are classified into categories B_1, B_2, and B_3 if they were born between January and April, May and August, and September and December, respectively. After choosing his or her group, the respondent is asked to report 0 or 1 according to the scheme given in the following Table 1.1.

For example, if a respondent in the sample s_1 belongs to B_1, then the respondent will answer "0" if he or she possesses the sensitive characteristic A and answer "1" if he or she possesses the characteristic $\overline{A}$.

Estimation of π_A

Let π_A be the proportion of individuals in the population who possess the attribute A and $\pi_{Ai}(\pi_{\overline{A}i})$ be the proportion of the persons in the population that possess the attribute $A(\overline{A})$ and belong to the category B_i, $i = 1, 2, 3$. Now, noting that

$$\pi_A = \pi_{A1} + \pi_{A2} + \pi_{A3} \text{and} \, \pi_{\overline{A}} = \pi_{\overline{A}1} + \pi_{\overline{A}2} + \pi_{\overline{A}3} \qquad (1.13.1)$$

we find the probability that a respondent answers "1" in the samples s_1, s_2, and s_3, respectively, are as follows:

$$\begin{aligned} \theta_1 &= \pi_{\overline{A}1} + \pi_{A2} + \pi_{A3} = \pi_A + \pi_{\overline{A}1} - \pi_{A1} \\ \theta_2 &= \pi_{A1} + \pi_{\overline{A}2} + \pi_{A3} = \pi_A + \pi_{\overline{A}2} - \pi_{A2} \\ \theta_3 &= \pi_{A1} + \pi_{A2} + \pi_{\overline{A}3} = \pi_A + \pi_{\overline{A}3} - \pi_{A3} \end{aligned} \qquad (1.13.2)$$

Table 1.1 Implicit RR Technique by Takahasi and Sakasegawe

Categories	Sample s_1		Sample s_2		Sample $s_3 s_3$	
	Attribute		Attribute		Attribute	
	A	$\overline{A}$	A	$\overline{A}$	A	$\overline{A}$
B_1	0	1	1	0	1	0
B_2	1	0	0	1	1	0
B_3	1	0	1	0	0	1

Equations (1.13.1) and (1.13.2) yield

$$\theta_1 + \theta_2 + \theta_3 = \pi_A + 1$$

$$\text{i.e., } \pi_A = \theta_1 + \theta_2 + \theta_3 - 1 \tag{1.13.3}$$

Let $\hat{\theta}_i$ be the proportion of "1" answer in sample s_i,$i = 1, 2, 3$. Then, we obtain the following results.

Theorem 1.13.1

(i) $\hat{\pi}_{\text{ATS}} = \hat{\theta}_1 + \hat{\theta}_2 + \hat{\theta}_3 - 1$ is an unbiased estimator of π_A
(ii) The variance of $\hat{\pi}_{\text{ATS}}$ is

$$\text{Var}(\hat{\pi}_{\text{ATS}}) = \frac{\theta_1(1-\theta_1)}{n_1} + \frac{\theta_2(1-\theta_2)}{n_2} + \frac{\theta_3(1-\theta_3)}{n_3}$$

(iii) An unbiased estimator of $\text{Var}(\hat{\pi}_{\text{ATS}})$ is

$$\hat{\text{Var}}(\hat{\pi}_{\text{ATS}}) = \frac{\hat{\theta}_1(1-\hat{\theta}_1)}{n_1 - 1} + \frac{\hat{\theta}_2(1-\hat{\theta}_2)}{n_2 - 1} + \frac{\hat{\theta}_3(1-\hat{\theta}_3)}{n_3 - 1}$$

Proof The proof is straightforward and, hence, is omitted.

Remark 1.13.1 It is advisable to use the groups B_i's such that the probability of classification of individuals into the groups is approximately equal, and they should not be correlated with the sensitive characteristic A of interest.

Remark 1.13.2 Takahasi and Sakasegawe (1977) claimed that their proposed estimator $\hat{\pi}_{\text{ATS}}$ is the maximum likelihood estimator (MLE) of π_{A}. However, Singh (1978) pointed out that the estimator $\hat{\theta}_i$ is not the MLE of θ_i. If P_{Bi} the probability that an individual would choose B_i is known, then the MLE of θ_i is

$$\hat{\theta}_{i\text{MLE}} = \begin{cases} \hat{\theta}_i, & \textit{if } P_{Bi} < \theta_i < 1 - P_{Bi} \\ P_{Bi}, & \textit{if } \quad \theta_i \leq P_{Bi} \\ 1 - P_{Bi}, & \textit{if } \quad \theta_i \geq 1 - P_{Bi} \end{cases} \tag{1.13.4}$$

However, if P_{Bi} is not known, it is not possible to find the MLE of π_{A}.

1.14 Cheater Detection Model (CDM)

The basic assumption of the RR model is that respondents answer truthfully "yes" or "no" following instructions given by the investigators. Although neither a "yes" nor a "no" can reveal the true status of the respondent, some of the researchers, cf.

Böckenholt et al. (2009), Becker (2010), and Wolter (2012), indicated that some of the respondents do not trust the anonymity of the method and deliberately provide wrong answers. Clark and Desharnais (1998), and Dawes and Moore (1979) proposed the Cheater Detection Model (CDM) for the Forced Randomization Model (F-model), while Reiber et al. (2020) extended it to the Unrelated Question Model (U-model). In the CDM, the prevalence of the sensitive attribute π cannot be unbiasedly estimated. However, it enables us to estimate the lower and upper bounds of π.

1.14.1 Clark and Desharnais (1998) CDM

Consider a population classified into two groups: sensitive (A) and non-sensitive ($\overline{A}$) with proportions π and $1-\pi$, respectively. Clark and Desharnais considered an RR model where respondents are asked either to answer the sensitive question S: "Do you belong to the group A?" with a probability p, or to provide a forced response (F) "yes" with a probability $1-p$. Clark and Desharnais classified the respondents into four groups. Honest yes (HY), who answer truthfully "yes" to the question S if they belong to the group A, Honest no (HN), who answer truthfully "no" to the question S if they belong to the group $\overline{A}$, Cheaters (C), who ignore instructions and, to be safe, answer "no" regardless of the outcome of the RR trials. The fourth group comprises those who answer "yes", ignoring instructions regardless of the outcome of the trial. Elbe and Pitsch (2018) reported that a clean athlete may provide a "yes" response to implementing stricter anti-doping policies. Here, we assume that such a group is so rare that it can be ignored. Let the probability of the respondents belonging to the HY, HN, and C groups be $\tilde{\pi}$, β, and $\gamma(\tilde{\pi}+\beta+\gamma=1)$, respectively. Here $\tilde{\pi}$ and $\tilde{\pi}+\gamma$ are the lower and upper bounds of π. The probability of obtaining a "yes" answer from a respondent is

$$\begin{aligned}\lambda &= p\tilde{\pi}+(1-p)(\tilde{\pi}+\beta)\\ &= \tilde{\pi}+(1-p)\beta\end{aligned} \tag{1.14.1}$$

In order to estimate $\tilde{\pi}$, Clark and Desharnais selected two independent SRSWR samples of sizes n_1,n_2 and requested all the selected respondents to perform the RR trials independently with $p=p_1$ and $p=p_2$. Let m_1 and m_2 be the number of "yes" answers. The proposed unbiased estimators of $\tilde{\pi}$, β and γ are

$$\hat{\tilde{\pi}}=\frac{(1-p_2)\hat{\lambda}_1-(1-p_1)\hat{\lambda}_2}{p_1-p_2};\ \hat{\beta}=\frac{\hat{\lambda}_2-\hat{\lambda}_1}{p_1-p_2}\text{and } \hat{\gamma}=1-(\hat{\pi}+\hat{\beta}) \tag{1.14.2}$$

where $\hat{\lambda}_1=m_1/n_1$ and $\hat{\lambda}_2=m_2/n_2$.

Since $\hat{\tilde{\pi}}$,$\hat{\beta}$ and $\hat{\gamma}$ are moment estimates, they can take inadmissible values outside of the parameter space (0, 1).

The variances of $\hat{\tilde{\pi}}$ and $\hat{\beta}$ are obtained as follows:

$$\operatorname{Var}(\hat{\tilde{\pi}}) = \frac{1}{(p_1 - p_2)^2}\left[(1-p_2)^2\frac{\lambda_1(1-\lambda_1)}{n_1} + (1-p_1)^2\frac{\lambda_2(1-\lambda_2)}{n_2}\right]$$

$$\operatorname{Var}(\hat{\beta}) = \frac{1}{(p_1 - p_2)^2}\left[\frac{\lambda_1(1-\lambda_1)}{n_1} + \frac{\lambda_2(1-\lambda_2)}{n_2}\right]$$

1.14.2 Reiber et al. (2020) CDM

In this model, respondents are directed to answer "yes" or "no" by using the U-model. In this model, respondents answer either the sensitive question **S** with a probability p or an innocuous question **N** relating to the non-sensitive attribute **B** with a probability $1 - p$. Here also, it is assumed that a certain proportion γ of the respondents (C), called cheaters, always answer "no" to both the **S** and **N** questions. The remaining are honest respondents (H) with proportion $1 - \gamma$ who always answer truthfully both questions. Let $\tilde{\pi} = \pi(1-\gamma)$ be the population proportion of respondents belonging to the sensitive group who answer "yes" to the sensitive question **S**. For this model, the probability of obtaining a "yes" response from a respondent selected at random is

$$\lambda = p\tilde{\pi} + (1-p)(1-\gamma)\pi_B \tag{1.14.3}$$

where π_B be the proportion of individuals who carry the neutral attribute B.

In order to estimate $\tilde{\pi}$, two independent samples of sizes n_1 and n_2 with parameters $p = p_1$ and $p = p_2$ are selected by the SRSWR method, and each of the respondents is asked to answer "yes" or "no" by using the U-model. Let $\hat{\lambda}_1$ and $\hat{\lambda}_2$ be the proportion of "yes" answers. In this case, unbiased estimators of $\tilde{\pi}$ and γ, and their variances are obtained as follows:

$$\hat{\tilde{\pi}} = \frac{\hat{\lambda}_2(1-p_1) - \hat{\lambda}_1(1-p_2)}{p_1 - p_2}, \hat{\gamma} = 1 - \frac{\hat{\lambda}_2 p_1 - \hat{\lambda}_1 p_2}{(p_1 - p_2)\pi_B} \tag{1.14.4}$$

$$\operatorname{Var}(\hat{\tilde{\pi}}) = \frac{1}{(p_1 - p_2)^2}\left[(1-p_1)^2\frac{\lambda_2(1-\lambda_2)}{n_2} + (1-p_2)^2\frac{\lambda_1(1-\lambda_1)}{n_1}\right];$$

$$\operatorname{Var}(\hat{\gamma}) = \frac{1}{(p_1 - p_2)^2\pi_B^2}\left[p_1^2\frac{\lambda_2(1-\lambda_2)}{n_2} + p_2^2\frac{\lambda_1(1-\lambda_1)}{n_1}\right]$$

1.15 Discussions

The collection of truthful data relating to sensitive items in human surveys by the direct method and mail inquiry surveys is not easy. Respondents often refuse to respond or provide wrong and/or socially acceptable responses because of the social stigma and/or fear that their personal information may be disclosed to third parties. As a result, the quality of the collected data is susceptible to high non-sampling and nonresponse errors. To improve the quality of data, Warner (1965) provided an ingenious indirect method of data collection, known as the "randomized response" (RR) technique (or model), where respondents need not provide any direct response but need to answer based on the outcome of an experiment, known as the "randomization device", performed by the respondent in the absence of the interviewer. The privacy of the respondents is maintained because the outcome of the experiment and the basis for answering the questions remain unknown to the interviewer.

In Warner's (1965) RR model, respondents are required to choose one card at random from a pack containing two types of cards bearing the statements Q1: "I belong to the sensitive group A" and Q2: "I do not belong to the sensitive group A" with proportions P and $1 - P$, respectively. The respondent should report truthfully "yes" if the statement written on the card drawn matches his or her status. Otherwise, the respondent should report "no" as his or her response. The confidentiality of the respondent is maintained, as the interviewer will not be able to know which of the questions Q1 and Q2 the respondent has answered.

The main drawback of Warner's (1965) RR method is that statements Q1 and Q2 are complementary to each other. Apart from that, both Q1 and Q2 may be sensitive. To overcome such difficulties, several alternative RR models have been proposed. In Greenberg et al.'s (1969) Unrelated Question Model (U-model) or Simmon's model, the statement Q2 was replaced by: "I belong to group B", where group B is totally innocuous or unrelated to group A. It is expected that the respondents may feel more comfortable with the U-model. In Boruch (1972) Forced RR model (F-model), respondents are required to answer "yes" or "no" to the direct question Q1: "Are you a member of the sensitive attribute?" with a probability P, forced to answer "yes" with a probability α, and forced to answer "no" with a probability $1 - P - \alpha$. The majority of the RR models are combinations of Warner's, Symmon's, and Brouch's models. The implicit RR model proposed by Takahasi and Sakasegawe (1977) is referred to as the non-randomized response (NRR) model in Chap. 10. In this method, no randomization device is required. The method can be used for face-to-face interviews, telephone interviews, and mail surveys.

Böckenholt et al. (2009), Becker (2010), and Wolter (2012) indicated that some of the respondents do not trust the anonymity of the RR methods and deliberately provide wrong answers. To cope with this problem, the Cheater Detection Model (CDM) has been proposed. The proposed CDM enables us to estimate the lower and upper bounds of prevalence π.

For each of the RR models discussed in this chapter, expressions of the unbiased estimators of the population proportion, variances, and unbiased estimators of the variances of the estimators are obtained assuming that the sample is selected by the SRSWR method. All the proposed unbiased estimators of the proportion are moment estimators and can take inadmissible values outside of the parameter space [0,1]. The maximum likelihood estimators are very complex, and no explicit expression of their variance can be obtained.

For comparison of the efficiencies of the proposed RR models in the chapter, the relative variances of the estimators are compared, ignoring the level of privacy protection. The variance of an estimator depends on the model parameters. For a given sample size, the conventional estimator based on a direct survey, where information is obtained directly from respondents, possesses a lower variance than the RR surveys. The amount of loss is compromised for the protection of the privacy of the respondents. However, the quality of the data obtained by the RR surveys is much better than that of the DR surveys. As a result, non-sampling errors in RR surveys are expected to be much lower than in DR surveys. None of the proposed RR models is found to be the best in all situations. Some of the proposed models are very difficult to administer in practice because of their complexity, and respondents are prone to providing wrong answers because of misunderstandings of the randomization method.

References

Abernathy, J. R., Greenberg, B. G., & Horvitz, D. G. (1970). Estimates of induced abortion in urban North Carolina. *Demography, 7*, 19–29.

Arcos, A., Rueda, M., & Singh, S. A. (2015) Generalized approach to randomized response for quantitative variables. *Quality & Quantity, 49*, 1239–1256.

Arnab, R. (1990). On commutativity of design and model expectations in randomized response surveys. *Communications in Statistics -Theory and Methods, 19*, 3751–3757.

Arnab, R. (1994). Non-negative variance estimation in randomized response surveys. *Communications in Statistics - Theory and Methods, 23*, 1743–1752.

Arnab, R. (1996). Randomized response trials: A unified approach for qualitative data. *Communications in Statistics - Theory and Methods, 25*(6), 1173–1183.

Arnab, R. (2007). Randomized response technique for complex survey designs. *Statistical Papers, 48*, 131–141.

Arnab, R. (2017). *Survey sampling theory and applications*. Academic Press.

Arnab, R., & Rodríguez, B. C. (2019). Variance Jackknife estimation for randomized response surveys: A simulation study and an application to explore cheating in exams and bullying. To appear in *Computational and Mathematical Methods, 2*(5). https://doi.org/10.1002/cmm4.1073

Arnab, R., & Rueda, M. (2016). Optional randomized response: A critical review. In A. Chaudhuri, T. C. Christofides, & C. R. Rao (Eds.), *Handbook of statistics* (Vol. 34, pp. 253–271). Elsevier, U.K.

Arnab, R., & Singh, S. (2010). Randomized response techniques: An application to the Botswana AIDS impact survey. *Journal of Statistical Planning and Inference, 140*, 941–953.

Arnab, R., & Mothupi, T. (2015). Randomized response techniques: A case study of the risky behaviors' of students of a certain University. *Model Assisted Statistics and Applications, 10*, 421–430.

Becker, C. (2010). Exploring sensitive topics: Sensitivity, jeopardy and cheating. In H. Locarek-Junge & C. Weihs (Eds.), *Classification as a Tool for Research: Proceedings of the 11th IFCS*

Biennial Conference and 33rd Annual Conference of the Gesellschaft für Klassifikation (p. 299-305).

Böckenholt, U., Barlas, S., & van der Heijden, P. (2009). Do randomized response designs eliminate response biases? An empirical study of noncompliance behavior. *Journal of Applied Economics, 24*, 377–392. 10.1002/ jae.1052.

Boruch, R. F. (1972). Relations among statistical methods for assuring confidentiality of social research data. *Social Science Research, 1*, 403–413.

Chang, H.-J., & Liang, D.-H. (1996). A two-stage unrelated randomized response procedure. *Australian Journal of Statistics, 38*, 43–58.

Chang, H.-J., Wang, C. L., & Huang, K.-C. (2004). On estimating the proportion of a quantitative sensitive character using randomized response sampling. *Quality and Quantity, 38*, 675–680.

Chaudhuri, A. (2011). *Randomized response and indirect questioning techniques in surveys*. Chapman & Hall, CRC Press, Taylor & Francis Group.

Chaudhuri, A., & Mukerjee, R. (1988). *Randomized response: Theory and techniques*. Marcel Dekker.

Chaudhuri, A., & Christofides, T. (2013). *Indirect questioning in sample surveys*. Springer.

Clark, S., & Desharnais, R. (1998). Honest answers to embarrassing questions: Detecting cheating in the randomized response model. *Psychological Methods, 3*(2), 160–168.

Dawes, R. M., & Moore, M. (1979). Die Guttman skalierung orthodoxer und randomisierter reaktionen [Traditional Guttman-scaling and randomized response]. In F. Peterman (Ed.), *Attitude measurement* (pp. 117–132). Verlag fur Psychologie

Devore, J. L. (1977). A note on RR techniques. *Communications in Statistics -Theory and Methods, 6*, 1525–1529.

Dowling, T. A., & Shachtman, R. (1975). On the relative efficiency of RR models. *Journal of the American Statistical Association, 68*, 525–530.

Elbe, A., & Pitsch, W. (2018). Doping prevalence among Danish elite athletes. *Performance Enhancement and Health, 6*, 28–32. https://doi.org/10.1016/j.peh

Flinger, M. A., Policello, G. E., & Singh, J. (1977). A comparison of two RR survey methods with consideration for the level of respondent protection. *Communications in Statistics -Theory and Methods, 6*, 1511–1524.

Folsom, R. E., Greenberg, B. G., Horvitz, D. G., & Abernathy, J. R. (1973). The two alternative questions randomized response model for human surveys. *Journal of the American Statistical Association, 68*, 525–530.

Fox, J. A., & Tracy, P. E. (1986). *Randomized response: A method for sensitive surveys*. Sage.

Franklin, L. A. (1989). A comparison of estimators for randomized response sampling with continuous distribution from dichotomous populations. *Communications in Statistics -Theory and Methods, 18*, 489–505.

Goodstadt, M. S., & Gruson, V. (1975). The randomized response technique; a test on drug use. *Journal of the American Statistical Association, 70*, 814–818.

Greenberg, B. D., Abul-Ela, A. L. A., Simmons, W. R., & Horvitz, D. G. (1969). The U-question randomized response model: Theoretical framework. *Journal of the American Statistical Association, 64*, 520–539.

Horvitz, D. G., Greenberg, B. G., & AberAbernathy, J. R. (1976). RR: A data gathering device for sensitive questions. *International Statistical Review, 44*, 181–196.

Horvitz, D. G., Shah, B. V., & Simmons, W. R. (1967). The U-question randomized response model. In *Proceedings of the Survey Research Methods Section of the American Statistical Association* (pp. 65–72).

Kerkvliet, J. M. (1994). Estimating a logit model with randomized data; the case of cocaine use. *Australian Journal of Statistics, 36*, 9–20.

Krumpal, I. (2012). Estimating the prevalence of xenophobia and anti-Semitism in Germany: Comparison of randomized response and direct questioning. *Social Science Research, 41*, 1387–1403.

Kuk, A. Y. C. (1990). Asking sensitive question indirectly. *Biometrika, 77*, 436–438.

Lara, D., Strickler, J., Olavarrieta, C. D., & Ellertson, C. (2004). Measuring induced abortion in Mexico: A comparison of four methodologies. *Sociological Methods & Research, 32*, 529–558.

Lensvelt-Mulders, G., van der Hox, J., & Heijden, P. (2005). How to improve the efficiency of randomized response designs. *Quality and Quantity, 39*, 253–263.

Mangat, N. S. (1992). Two stage randomized response sampling procedure using U-question. *Journal of the Indian Society of Agricultural Statistics, 44*, 82–88.

Mangat, N. S. (1994). An improved Randomized response strategy. *Journal of the Royal Statistical Society, Series B, 56*, 93–95.

Mangat, N. S., & Singh, R. (1990). An alternative randomized response procedure. *Biometrika, 77*, 349–442.

Mangat, N. S., Singh, R., & Singh, B. (1993). On Moor's randomized response model. *Biometrical Journal, 5*, 727–732.

Mangat, N. S., Singh, R., & Singh, S. (1992). An improved U-question randomized response strategies. *Calcutta Statistical Association Bulletin, 42*, 277–281.

Mangat, N. S., Singh, R., & Singh, S. (1997). Violation of respondent's privacy in Moors model-its rectification through a random group strategy. *Communications in Statistics -Theory and Methods, 26*, 743–754.

Moors, J. J. A. (1971). Optimization of the U- question RR model. *Journal of the American Statistical Association, 66*, 627–629.

Munir, M., Singh, S., & Horn, S. (1998). On the confidentiality guaranteed under randomized response sampling: A comparison with several new techniques. *Biometrical Journal, 40*, 237–242.

Reiber, F., Pope, H., & Ulrich, R. (2020). Cheater detection using the unrelated question model. *Sociological Methods & Research, 20*, 1–23. https://doi.org/10.1177/0049124120914919

Rueda, M., Cobo, B., & Arcos, A. (2015). Package 'RRTCS': Randomized response techniques for complex surveys. Retrieved from http://cran.r-project.org/web/packages/RRTCS

Singh, J. (1976). A note on RR techniques. *In Proceedings of the Survey Research Methods Section of the American Statistical Association* (p. 772).

Singh, J. (1978). A note on maximum likelihood estimation from randomized response models. In *Proceedings of the Survey Research Methods Section of the American Statistical Association* (pp. 282–283).

Singh, S. (1993). An alternative to Warner's randomized response technique. *Statistica, anno* LIII, 67–71.

Singh, S., & Joarder, A. H. (1997a). Unknown repeated trials in randomized response sampling. *Journal of the Indian Society of Agricultural Statistics, 50*, 70–74.

Singh, S., & Singh, R. (1992). Improved Franklin's model for randomized response sampling. *Journal of the Indian Society of Agricultural Statistics, 30*, 109–122.

Singh, S., & Singh, R. (1993). Generalized Franklin's model for randomized response sampling. *Communications in Statistics -Theory and Methods, 22*, 741–755.

Singh, S., & Joarder, A. (1997b). Unknown repeated trials in randomized response sampling. *Journal of the Indian Society of Agricultural Statistics, 50*, 103–105.

Singh, S., Singh, R., & Mangat, N. S. (2000). Some alternative strategies to Moor's model in randomized response sampling. *Journal of the Statistical Planning and Inference. 83*, 243–255.

Singh, H. P., & Mathur, N. (2003). Modified optional randomized response sampling techniques. *Journal of the Indian Society of Agricultural Statistics, 56*, 199–206.

Singh, S., Horn, H., Singh, R., & Mangat, N. S. (2003). On the use of modified randomization device for estimating the prevalence of the sensitive attribute. *Statistics in Transition, 6*, 515–522.

Takahasi, K., & Sakasegawa, H. (1977). An RR technique without use of any randomizing device. *Annals of the Institute of Statistical Mathematics, 29*, 1–8.

Tamhane, A. C. (1981). Randomized response techniques for multiple sensitive attributes. *Journal of the American Statistical Association, 76*, 916–923.

van der Heijden, P. G. M., Van Gils, G., Bouts, J., & Hox, J. J. (2000). A comparison of randomized response, computer assisted self interview, and face to face direct questioning, eliciting sensitive information in the context of welfare and unemployment benefit. *Sociological Methods and Research, 28*, 505–537.

Warner, S. L. (1965). Randomize response: A survey technique for eliminating evasive answer bias. *Journal of the American Statistical Association, 60*, 63–69.

Wolter, F. (2012). Heikle Fragen in Interviews. In *Heikle Fragen in Interviews*. VS Verlag für Sozialwissenschaften. https://doi.org/10.1007/978-3-531-19371-7-2s

Chapter 2
Polychotomous Randomized Response Model

Abstract Various RR techniques for estimating the proportion of individuals belonging to a sensitive category were presented in the previous chapter. In this chapter, such procedures are extended to the situation where the population is classified into several disjoint categories, at least one of which is sensitive and at most one is non-sensitive. Most of the polychotomous techniques are based on the applications of Warner, Unrelated Question, Forced RR, and their combinations. Details of a few important randomization techniques and methods of estimating population proportions from such techniques are discussed here.

Keywords Multiple sensitive attributes · Randomized response · SFSWR method

2.1 Introduction

In Chap. 1, various RR techniques are proposed for estimating π, the proportion of individuals in a population that belong to a certain sensitive group A. The entire population was divided into two mutually exclusive and exhaustive groups A and its complement $\overline{A}$. Using a certain RR technique, respondents provided indirect information relating to their membership in either group A or $\overline{A}$ on a probability basis without revealing their actual membership directly to the interviewer. Thus, the privacy of the respondents was protected, and the response rate was much improved than if respondents had been questioned directly on such sensitive issues.

In this chapter, RR techniques for more than one sensitive characteristic are reviewed. Here we consider a population classified into $t(\geq 2)$ mutually exclusive and exhaustive categories such that at least one and at most $t-1$ of them are identified as stigmatizing. For example, patients in a hospital can be classified into three classes: A_1: HIV positive and TB negative; A_2: HIV positive and TB positive; and A_3: HIV negative. Here, classes A_1 and A_2 are stigmatizing, while A_3 is non-stigmatizing.

R. Arnab, *Indirect Methods of Data Collection and Analysis from Surveys*, Indian Statistical Institute Series, https://doi.org/10.1007/978-981-97-6005-3_2

2.2 Abul-Ela et al.'s Model

Let a population be classified into $t(\geq 2)$ mutually exclusive and exhaustive groups $A_1, \ldots, A_t$ and the proportion of individuals belonging to the j th group A_j be π_j, $0 < \pi_j < 1$, and $\sum_{j=1}^{t} \pi_j = 1$. The objective is to estimate the unknown proportion π_j, $j = 1, \ldots, t$. In Abul-Ela et al.'s (1967) model, $t - 1$ independent samples $s_1, \ldots, s_k, \ldots., s_{t-1}$ of sizes $n_1, \ldots, n_k, \ldots., n_{t-1}$ are selected from the population by the SRSWR method. The randomization device consists of $t - 1$ decks of cards. Each deck contains t different types of cards. Card type-$j(= 1, \ldots, t)$ says "I belong to group A_j". The $r(= 1, \ldots, t - 1)$ th deck consists of j th type of cards with proportion $P_{rj}(P_{rj} > 0, \sum_{j=1}^{t} P_{rj} = 1)$. Each of the respondents in the sample s_k $(k = 1, \ldots, t - 1)$ is asked to draw one card at random from the k th deck and read the statement on the card unnoticed by the interviewer. The respondent should answer "yes" if the statement matches his or her status and answer "no" otherwise (Fig. 2.1).

2.2.1 *Estimation of Proportions*

The probability of obtaining a "yes" response from a respondent belonging to the k th sample s_k is

$$\theta_k = \sum_{j=1}^{t} P_{kj}\pi_j, k = 1, \ldots, t - 1 \tag{2.2.1}$$

Noting that $\sum_{j=1}^{t-1} \pi_j = 1 - \pi_t$, , we express (2.2.1) as

Directions:

Samples

s_1 ↓ *Draw one card at random from* ↓	.....	s_k ↓ *Draw one card at random from* ↓	.....	s_{t-1} ↓ *Draw one card at random from* ↓
Card type-1:I belong to A_1 (Probability P_{11})	.....	Card type-1: I belong to A_1 (Probability P_{k1})	.....	Card type-1: I belong to A_1 (Probability $P_{t-1,1}$)
Card type- j : I belong to A_j (Probability P_{1j})	.....	Card type- j : I belong to A_j (Probability P_{kj})	.....	Card type- j : I belong to A_j (Probability $P_{t-1,j}$)
	.....	.....	.....	.....
Card type- t : I belong to A_t (Probability P_{1t})	.....	Card type- t : I belong to A_t (Probability P_{kt})	.....	Card type- t : I belong to A_t (Probability $P_{t-1,t}$)

Fig. 2.1 Abul-Ela et al.'s model

$$\theta_k = \sum_{j=1}^{t-1} P_{kj}\pi_j + P_{kt}\left(1 - \sum_{j=1}^{t-1}\pi_j\right)$$
$$= \sum_{j=1}^{t-1}(P_{kj} - P_{kt})\pi_j + P_{kt}$$

i.e., $\theta_k - P_{kt} = (P_{k1} - P_{kt})\pi_1 + \cdots + (P_{kj} - P_{kt})\pi_j + \cdots + (P_{kt-1} - P_{kt})\pi_{t-1}$ (2.2.2)

In matrix notation, Eq. (2.2.2) becomes

$$\boldsymbol{P\pi} = \boldsymbol{\delta} \tag{2.2.3}$$

where

$$\boldsymbol{P} = \begin{bmatrix} P_{11} - P_{1t} & \cdots & P_{1j} - P_{1t} & \cdots P_{1t-1} - P_{1t} \\ P_{k1} - P_{kt} & \cdots & P_{kj} - P_{kt} & \cdots P_{kt-1} - P_{kt} \\ P_{t-11} - P_{t-1t} & \cdots & P_{t-1j} - P_{t-1t} & \cdots \; P_{t-1t-1} - P_{t-1t} \end{bmatrix},$$

$$\boldsymbol{\pi} = \begin{pmatrix} \pi_1 \\ \cdot \\ \pi_j \\ \cdot \\ \pi_{t-1} \end{pmatrix} \text{and } \boldsymbol{\delta} = \begin{pmatrix} \theta_1 - P_{1t} \\ \cdot \\ \theta_j - P_{jt} \\ \cdot \\ \theta_{t-1} - P_{t-1t} \end{pmatrix}$$

Assuming the chosen matrix **P** is non-singular, Eq. (2.2.3) reduces to

$$\boldsymbol{\pi} = \mathbf{P}^{-1}\boldsymbol{\delta} \tag{2.2.4}$$

Suppose that in the k th sample, n_{k1} persons report "yes" and $n_k - n_{k1}$ persons report "no". Noting that $\hat{\theta}_k = \frac{n_{k1}}{n_k}$ is a $\mathrm{B}(n_k, \theta_k)$ variate, we have that

$$E(\hat{\theta}_k) = \theta_k, \; \mathrm{Var}(\hat{\theta}_k) = \frac{\theta_k(1-\theta_k)}{n_k} \text{ and } \mathrm{Cov}(\hat{\theta}_k, \hat{\theta}_l) = 0 \text{ for } k \neq l \tag{2.2.5}$$

From Eqs. (2.2.4) and (2.2.5), it is easy to verify the results in Theorem 2.2.1.

Theorem 2.2.1

(i) $\hat{\pi} = \mathbf{P}^{-1}\hat{\delta} = (\hat{\pi}_1 \cdots \hat{\pi}_{t-1})'$ is an unbiased estimator of $\boldsymbol{\pi}$ and $\hat{\pi}_t = 1 - \sum_{j=1}^{t-1}\hat{\pi}_j$ is an unbiased estimator of π_t, where $\hat{\delta} = \left(\hat{\theta}_1 - P_{1t}, \cdots, \hat{\theta}_k - P_{kt}, \cdots, \hat{\theta}_{t-1} - P_{t-1t}\right)'$ and **P** is non-singular.

(ii) The variance of $\hat{\pi}$ is

$$\mathrm{Var}(\hat{\pi}) = \mathbf{P}^{-1}\boldsymbol{D}_{\theta}^{-1}(\mathbf{P}^{-1})\prime$$

where $\boldsymbol{D}_{\theta}$ is a $k-1 \times k-1$ diagonal matrix with jth diagonal element is $\frac{\theta_j(1-\theta_j)}{n_j}$, $j = 1, \ldots, k-1$

(iii) An unbiased estimator of $\mathrm{Var}(\hat{\pi})$ is

$$\mathrm{V\hat{a}r}(\hat{\pi}) = \mathbf{P}^{-1}\hat{\mathrm{D}}_{\theta}^{-1}(\mathbf{P}^{-1})\prime$$

where $\hat{\mathrm{D}}_{\theta}$ is a $k-1 \times k-1$ diagonal matrix with jth diagonal element is $\frac{\hat{\theta}_j(1-\hat{\theta}_j)}{n_j-1}$, $j = 1, \ldots, k-1$.

2.2.2 Choice of Design Parameters

Abul-Ela et al. (1967) pointed out that the variance $\mathrm{Var}(\hat{\pi}_j)$ decreases rapidly as P_{kj}'s deviates from equality $1/t$. In fact, $\mathrm{Var}(\hat{\pi}_j)$ can be minimized by choosing $P_{kj} = 1$ and $P_{kl} = 0$ for $k \neq l$. In this case, the RR procedure is reduced to a direct response survey. On the other hand, if one chooses a very small value for P_{kj}, then the degree of cooperation from respondents will be reduced, as the cards represented by that proportion might not be easily observed, and, hence, the purpose of the RR survey will be defeated. At this stage, P_{kj} should be determined empirically to increase the cooperation from the respondents and at the same time, minimize the variances of the estimators.

2.2.3 Trichotomous Randomized Response Model

Substituting $t = 3$ in Eq. (2.2.1) and (2.2.2), one finds

$$\begin{aligned} &\pi_1 + \pi_2 + \pi_3 = 1 \\ &P_{k1} + P_{k2} + P_{k3} = 1 \\ &\theta_k = (P_{k1} - P_{k3})\pi_1 + (P_{k2} - P_{k3})\pi_2 + P_{k3} \\ &k = 1, 2, 3 \end{aligned} \tag{2.2.6}$$

Equation (2.2.6) leads to

$$\begin{pmatrix} P_{11} - P_{13} & P_{12} - P_{13} \\ P_{21} - P_{23} & P_{22} - P_{23} \end{pmatrix} \begin{pmatrix} \pi_1 \\ \pi_2 \end{pmatrix} = \begin{pmatrix} \frac{n_{11}}{n_1} - P_{13} \\ \frac{n_{21}}{n_2} - P_{13} \end{pmatrix} \tag{2.2.7}$$

Assuming

$$\begin{aligned}\Delta =&\begin{vmatrix} P_{11}-P_{13} & P_{12}-P_{13} \\ P_{21}-P_{23} & P_{22}-P_{23}\end{vmatrix}\\ =&(P_{11}-P_{13})(P_{22}-P_{23})-(P_{12}-P_{13})(P_{21}-P_{23}) \neq 0,\end{aligned} \tag{2.2.8}$$

we obtain

Theorem 2.2.2

(i) $\hat{\pi}_1 = \frac{(\hat{\theta}_1-P_{13})(P_{22}-P_{23})-(\hat{\theta}_2-P_{23})(P_{12}-P_{13})}{\Delta}$,

$$\hat{\pi}_2 = -\frac{(\hat{\theta}_1-P_{13})(P_{21}-P_{23})-(\hat{\theta}_2-P_{23})(P_{11}-P_{13})}{\Delta}$$

and

$$\hat{\pi}_3 = 1-\hat{\pi}_1-\hat{\pi}_2$$

are unbiased estimators of π_1, π_2, and π_3.

(ii) The variances of $\hat{\pi}_1$, $\hat{\pi}_2$, and $\hat{\pi}_3$ are

$$\text{Var}(\hat{\pi}_1) = \frac{1}{\Delta^2}\left[\frac{(P_{22}-P_{23})^2\theta_1(1-\theta_1)}{n_1}+\frac{(P_{12}-P_{13})^2\theta_2(1-\theta_2)}{n_2}\right]$$
$$\text{Var}(\hat{\pi}_2) = \frac{1}{\Delta^2}\left[\frac{(P_{21}-P_{23})^2\theta_1(1-\theta_1)}{n_1}+\frac{(P_{11}-P_{13})^2\theta_2(1-\theta_2)}{n_2}\right]$$
$$\text{Var}(\hat{\pi}_3) = \frac{1}{\Delta^2}\left[\frac{(P_{22}-P_{21})^2\theta_1(1-\theta_1)}{n_1}+\frac{(P_{12}-P_{11})^2\theta_2(1-\theta_2)}{n_2}\right]$$

(iii) Unbiased estimators of the variances of $\text{Var}(\hat{\pi}_1)$, $\text{Var}(\hat{\pi}_2)$, and $\text{Var}(\hat{\pi}_3)$ are

$$\widehat{\text{Var}}(\hat{\pi}_1) = \frac{1}{\Delta^2}\left[\frac{(P_{22}-P_{23})^2\hat{\theta}_1(1-\hat{\theta}_1)}{n_1-1}+\frac{(P_{12}-P_{13})^2\hat{\theta}_2(1-\hat{\theta}_2)}{n_2-1}\right]$$
$$\widehat{\text{Var}}(\hat{\pi}_2) = \frac{1}{\Delta^2}\left[\frac{(P_{21}-P_{23})^2\hat{\theta}_1(1-\hat{\theta}_1)}{n_1-1}+\frac{(P_{11}-P_{13})^2\hat{\theta}_2(1-\hat{\theta}_2)}{n_2-1}\right]$$
$$\widehat{\text{Var}}(\hat{\pi}_3) = \frac{1}{\Delta^2}\left[\frac{(P_{22}-P_{21})^2\hat{\theta}_1(1-\hat{\theta}_1)}{n_1-1}+\frac{(P_{12}-P_{11})^2\hat{\theta}_2(1-\hat{\theta}_2)}{n_2-1}\right]$$

Also, Abul-Ela et al. (1967) considered the effect of incompletely truthful reporting on the regular trinomial and the trichotomous randomized estimators under a response probability model. The authors obtained expressions of the biases and mean square errors of the estimators. The proposed trichotomous RR model was applied in the "Population Change Survey" conducted by the Research Triangle

Institute and the Department of Biostatistics of the University of North Carolina for the National Center for Health Statistics. The aim was to estimate the proportion of unwed mothers who gave birth between October 1964 and October 1965 in North Carolina. The unwed mothers were classified into three groups: Group A: married at the time of conception; Group B: married during pregnancy; and Group C: unmarried at the time of delivery. The survey revealed that the RR survey in general provides increased cooperation from the respondents, and at the same time, the estimates show a substantial gain in efficiency compared to the direct survey estimates.

2.3 Eriksson's Models

In this section, we will consider two RR models, Model A and Model B, proposed by Eriksson (1973a, 1973b).

2.3.1 *Model A*

Eriksson's (1973a) Model A is an extension of Abul-Ela et al.'s (1967) model. In this method $t-1$ independent samples, $s_1, \ldots, s_i, \ldots., s_{t-1}$ of sizes $n_1, \ldots, n_i, \ldots, n_{t-1}$ are selected by the SRSWR method. Each of the respondents belonging to the ith sample s_i draws a card at random from a pack containing two types of cards bearing the statements "I belong to the sensitive group A_i" and "I do not belong to the sensitive group A_i" with proportions P_i and $1-P_i$; $i = 1, \ldots, t-1$, respectively. The respondents answer "yes" if the statement matches their status; otherwise, they answer "no" (Fig. 2.2).

The probability of obtaining a "yes" answer from a respondent in the i th sample s_i is

$$\theta_i = \pi_i P_i + (1-\pi_i)(1-P_i) \tag{2.3.1}$$

Directions:

Samples

s_1 *Draw one card at random* ↓	⋮	s_i *Draw one card at random* ↓	⋮	s_{t-1} *Draw one card at random* ↓
Card type-1: I belong to A_1 (Probability P_1)	⋮	Card type-1: I belong to A_i (Probability P_i)	⋮	Card type-1: I belong to A_{t-1} (Probability P_{t-1})
Card type-2: I do not belong to A_1 (Probability $1-P_1$)	⋮	Card type-2: I do not belong to A_i (Probability $1-P_i$)	⋮	Card type-2: I do not belong to A_{t-1} (Probability $1-P_{t-1}$)

Fig. 2.2 Eriksson's model A

Let $\hat{\theta}_i$ be the proportion of "yes" answers in the ith sample. Then, we get the following results:

Theorem 2.3.1

(i) $\hat{\pi}_i = \frac{\hat{\theta}_i - (1-P_i)}{2P_i - 1}$; $P_i \neq 1/2$ is an unbiased estimator of π_i for $i = 1, \ldots, t-1$ and $\hat{\pi}_t = 1 - \sum_{i=1}^{t-1} \hat{\pi}_i$ is an unbiased estimator of π_t.

(ii) The variances of $\hat{\pi}_i$ and $\hat{\pi}_t$ are

$$\mathrm{Var}(\hat{\pi}_i) = \frac{\theta_i(1-\theta_i)}{n_i(2P_i - 1)^2} \text{ and } \mathrm{Var}(\hat{\pi}_t) = \sum_{i=1}^{t-1} \frac{\theta_i(1-\theta_i)}{n_i(2P_i - 1)^2}$$

(iii) Unbiased estimators of $\mathrm{Var}(\hat{\pi}_i)$ and $\mathrm{Var}(\hat{\pi}_t)$ are

$$\hat{\mathrm{Var}}(\hat{\pi}_i) = \frac{\hat{\theta}_i(1-\hat{\theta}_i)}{(n_i - 1)(2P_i - 1)^2} \text{ and } \hat{\mathrm{Var}}(\hat{\pi}_t) = \sum_{i=1}^{t-1} \frac{\hat{\theta}_i(1-\hat{\theta}_i)}{(n_i - 1)(2P_i - 1)^2}$$

2.3.2 *Model B*

In Eriksson's (1973a, 1973b) Model B, the population is classified into t mutually exclusive and exhaustive sensitive groups $A_1, \ldots, A_t$. The randomization device for this model is as follows: Each of the selected respondents in the sample is asked to draw one card at random from a deck containing $t+1$ different types of cards. Type-i card bears the statement "Say, you belong to group A_i", $i = 2, \ldots, t$, while type-$t+1$ bears statement "Say your true group". The proportions of cards of type-$t+1$ and type-i in the deck are $P(0 < P < 1)$ and $P_i(1-P)\left(P_i > 0, \sum_{i=1} P_i = 1\right)$, respectively. The respondent must report truthfully about his or her group if he or she draws the card "type-$t+1$". Otherwise, he or she should report the group as directed on the selected card (Fig. 2.3).

Suppose that a sample of size n is selected from the population by the SRSWR method, and let π_i be the proportion of persons belonging to the i th group. Then,

λ_i = Probability of obtaining answer "A_i" from a respondent selected at random.
= Probability that the person belongs to the group A_i and draws a card type-$t+1$.
+ Probability that the person draws a card type-i written A_i

$$= P\pi_i + P_i(1-P) \tag{2.3.2}$$

Let $\hat{\lambda}_i$ = proportion of respondents selected in the sample who reported the group A_i. Since $\hat{\lambda}_i$ is a binomial variable with parameters n and λ_i, we have the following results:

Directions:

Draw one card at random from:

Card type 1:		Card type j:		Card type t		Card type $t+1$
Say: "You belong to group A_1" (Probability $P_1(1-P)$)	...	Say: "You belong to group A_j" (Probability $P_j(1-P)$)	...	Say: "You belong to group A_t" (Probability $P_t(1-P)$)	...	Disclose truthfully your group (Probability P)

Fig. 2.3 Eriksson's model B

Theorem 2.3.2

(i) $\hat{\pi}_i = \frac{\hat{\lambda}_i - P_i(1-P)}{P}$ is an unbiased estimator of π_i

(ii) The variance of $\hat{\pi}_i$ is

$$\text{Var}(\hat{\pi}_i) = \frac{\lambda_i(1-\lambda_i)}{nP^2}$$
$$= \frac{1}{n}\left[\pi_i(1-\pi_i) + \frac{(1-P)}{P}\left\{\frac{P_i(1-P_i)}{P} + \pi_i(1-\pi_i) + (P_i - \pi_i)^2\right\}\right]$$

(iii) An unbiased estimator of $\text{Var}(\hat{\pi}_i)$ is

$$\hat{\text{Var}}(\hat{\pi}_i) = \frac{\hat{\lambda}_i(1-\hat{\lambda}_i)}{(n-1)P^2}$$

Remark 2.3.1 In the Abul-Ela et al. (1967) and Greenberg et al. (1969) models, one needs to draw t independent samples for estimating the proportion of persons belonging to the t different classes, while, in Eriksson's (1973a, 1973b) model, one needs to draw only one sample.

Remark 2.3.2 The estimator $\hat{\pi}_j$ becomes negative if $\hat{\lambda}_j < P_j$.

2.4 Unrelated–Question Model

Greenberg et al. (1969) extended the unrelated question model, where a population was classified into three ($t = 3$) mutually exclusive and exhaustive categories A_1,A_2, and A_3, each of which could be stigmatizing. Then, three independent samples of sizes n_1, n_2, and n_3 were selected by the SRSWR method. Respondents in the ith sample s_i were asked to draw a card at random from a deck of three different types of cards, two of which bore sensitive statements such as "I belong to the group A_1"

Directions:

Sample S_1 Draw one card at random and answer "yes" or "no" ↓	Sample S_2 Draw one card at random and answer "yes" or "no" ↓	Sample S_3 Draw one card at random and answer "yes" or "no" ↓
Type1: I belong to the group A_1 (probability P_{11}) Type2: I belong to the group A_2 (probability P_{12}) Type3: I was born in April (probability P_{13})	Type1: I belong to the group A_1 (probability P_{21}) Type2: I belong to the group A_2 (probability P_{22}) Type3: I was born in April (probability P_{23})	Type1: I belong to the group A_1 (probability P_{31}) Type2: I belong to the group A_2 (probability P_{32}) Type3: I was born in April (probability P_{33})

Fig. 2.4 Greenberg et al.'s unrelated question model

and "I belong to the group A_2", and the third bearing an innocuous statement "I was born in April" with proportions P_{i1}, P_{i2}, and $P_{i3}(= 1 - P_{i1} - P_{i2})$, respectively. A respondent should answer "yes" if the statement matches his or her status and answer "no" otherwise (Fig. 2.4).

Let π_B, the proportion of individuals in the population born in April be unknown. Then, the probability of getting a "yes" answer from a respondent selected from the i th sample is

$$\theta_i = P_{i1}\pi_1 + P_{i2}\pi_2 + P_{i3}\pi_B;\ i = 1, 2, 3 \tag{2.4.1}$$

Now, if $\hat{\theta}_i$.is the proportion of "yes" answers in the sample, then the unbiased estimators of π_1, π_2, π_3, and π_B are obtained as

$$\begin{aligned}\begin{pmatrix}\hat{\pi}_1\\ \hat{\pi}_2\\ \hat{\pi}_B\end{pmatrix} &= \begin{pmatrix}P_{11} & P_{12} & P_{13}\\ P_{21} & P_{22} & P_{23}\\ P_{31} & P_{32} & P_{33}\end{pmatrix}^{-1}\begin{pmatrix}\hat{\theta}_1\\ \hat{\theta}_2\\ \hat{\theta}_3\end{pmatrix}\\ &= \begin{pmatrix}P_{11} & P_{12} & 1 - P_{11} - P_{12}\\ P_{21} & P_{22} & 1 - P_{21} - P_{22}\\ P_{31} & P_{32} & 1 - P_{31} - P_{32}\end{pmatrix}^{-1}\begin{pmatrix}\hat{\theta}_1\\ \hat{\theta}_2\\ \hat{\theta}_3\end{pmatrix}\end{aligned} \tag{2.4.2}$$

and $\hat{\pi}_3 = 1 - \hat{\pi}_1 - \hat{\pi}_2$

For a unique solution, the design parameters should satisfy

$$\begin{vmatrix}P_{11} & P_{12} & 1 - P_{11} - P_{12}\\ P_{21} & P_{22} & 1 - P_{21} - P_{22}\\ P_{31} & P_{32} & 1 - P_{31} - P_{32}\end{vmatrix} \neq 0$$

So, the expressions of the variance of $\hat{\pi}* = (\hat{\pi}_1, \hat{\pi}_2, \hat{\pi}_B)'$ and its unbiased estimator are as follows:

$$\text{Var}(\hat{\pi}*) = \mathbf{P}^{*-1} diag\left(\frac{\theta_1(1-\theta_1)}{n_1}, \frac{\theta_2(1-\theta_2)}{n_2}, \frac{\theta_3(1-\theta_3)}{n_3}\right)(\mathbf{P}^{*-1})'$$

$$\hat{\text{Var}}(\hat{\pi}*) = \mathbf{P}^{*-1} diag\left(\frac{\hat{\theta}_1(1-\hat{\theta}_1)}{n_1-1}, \frac{\hat{\theta}_2(1-\hat{\theta}_2)}{n_2-1}, \frac{\hat{\theta}_3(1-\hat{\theta}_3)}{n_3-1}\right)(\mathbf{P}^{*-1})'$$

where $\mathbf{P}* = \begin{pmatrix} P_{11} & P_{12} & 1-P_{11}-P_{12} \\ P_{21} & P_{22} & 1-P_{21}-P_{22} \\ P_{31} & P_{32} & 1-P_{31}-P_{32} \end{pmatrix}$.

2.5 Liu, Chow, and Mosley's Model

Liu et al. (1975) classified a population into t mutually exclusive and exhaustive groups, $A_1, ..., A_t$ at least one of which is sensitive. The objective was to estimate the proportion of individuals belonging to each of the sensitive groups. In Liu et al.'s (1975) randomized device, a flask with a long, narrow neck and t different colored balls are taken. Each color represents a group of people, e.g., color-1 represents the group A_1, color-2 represents group $A_2, \ldots,$ and color-t represents the group A_t.The number of balls of color-k is m_k and $m_1 \neq m_2 \neq, \cdots, \neq m_t$. The total number of balls is $m = \sum_{k=1}^{t} m_k$. The neck of the flask is such that only one ball can pass through it, and the neck is long enough to accommodate all the balls. The respondent is asked to hold the flask upside down, shake it thoroughly, and allow all the balls to fall into the neck. The neck has $\text{M} = \underset{k}{\text{Max}}(m - m_k + 1)$ locations, and it is marked from the bottom $1, 2, \cdots, M$. The respondent will report to which location the first ball from the bottom of the neck that identifies his or her attribute has fallen. For example, if the respondent belongs to group A_1, he should tell in which location the first ball of color-1 has fallen from the bottom. As usual, the whole experiment is done in the absence of the interviewer to maintain the respondents' confidentiality.

Estimation of Proportions

Let P_{kj} be the probability that the first ball of color-k appears in the j th location, $j = 1, ..., M$. Then,

$P_{k1} = \frac{m_k}{m}, P_{k2} = \left(\frac{m-m_k}{m}\right)\left(\frac{m_k}{m-1}\right), P_{k3} = \frac{m-m_k}{m}\left(\frac{m-m_k-1}{m-1}\right)\left(\frac{m_k}{m-2}\right)$, and in general

$$P_{kj} = \begin{cases} \frac{(m-j)!(m-m_k)!m_k}{m!(m-m_k+1-j)!} \text{ for } j = 1, 2, \ldots, m - m_i + 1 \\ 0 \text{ otherwise} \end{cases} \tag{2.5.1}$$

Let θ_k be the proportion of the population that belongs to the sensitive group A_k. The probability of receiving an answer "j" from a respondent selected at random using the SRSWR method of size n is given by

$$\lambda_j = \sum_{k=1}^{t} P_{kj}\theta_k, j = 1, 2, \ldots, M \tag{2.5.2}$$

Let n_j and $\hat{\lambda}_j = \frac{n_j}{n}$ be the number and proportion of respondents answering "j" in the selected sample. Then, we obtain

$$\begin{aligned}
E(\hat{\lambda}_j) &= \sum_{k=1}^{t} P_{kj}\theta_k \\
&= \sum_{k=1}^{t-1} P_{kj}\theta_k - P_{tj}\theta_t \\
&= \sum_{k=1}^{t-1} P_{kj}\theta_k + P_{tj}\left(1 - \sum_{k=1}^{t-1} \pi_k\right) \\
&= \sum_{k=1}^{t-1} (P_{kj} - P_{tj})\theta_k + P_{tj}
\end{aligned}$$

$$\text{i.e., } E(\hat{\lambda}_j - P_{tj}) = \sum_{k=1}^{t-1} (P_{kj} - P_{tj})\theta_k \text{ for } j = 1, 2, \ldots, M \tag{2.5.3}$$

The variance of $\hat{\lambda}_j$, co-variance of $\hat{\lambda}_j$ and $\hat{\lambda}_k$ $(j \neq k)$, and their unbiased estimators are given as follows:

$$\begin{aligned}
\text{Var}(\hat{\lambda}_j) &= \frac{\lambda_j(1 - \lambda_j)}{n}; \ \text{Cov}(\hat{\lambda}_j, \hat{\lambda}_k) = -\frac{\lambda_i\lambda_k}{n}; \\
\hat{\text{Var}}(\hat{\lambda}_j) &= \frac{\hat{\lambda}_i(1 - \hat{\lambda}_i)}{n - 1} \text{ and } \ \hat{\text{Cov}}(\hat{\lambda}_j, \hat{\lambda}_k) = -\frac{\hat{\lambda}_j\hat{\lambda}_k}{n - 1}
\end{aligned} \tag{2.5.4}$$

Equations (2.5.3) and (2.5.4) yield.

Theorem 2.5.1

(i) $\hat{\theta} = (\boldsymbol{X}'\boldsymbol{V}^{-1}\boldsymbol{X})^{-1}\boldsymbol{X}'\boldsymbol{V}^{-1}\boldsymbol{d}$ and $\hat{\theta}_t = 1 - \sum_{i=1}^{t-1} \hat{\theta}_i$ are unbiased estimators of $\boldsymbol{\theta}$ and θ_t,

(ii) The variance of $\hat{\theta}$ is $\text{Var}(\hat{\theta}) = (\boldsymbol{X}'\boldsymbol{V}^{-1}\boldsymbol{X})^{-1}$,

(iii) An unbiased estimator of $\text{Var}(\hat{\theta})$ is $\hat{\text{Var}}(\hat{\theta}) = (\boldsymbol{X}'\hat{\text{V}}^{-1}\boldsymbol{X})^{-1}$,

where

$$\theta = \begin{pmatrix} \theta_1 \\ \theta_2 \\ \cdot \\ \theta_{t-1} \end{pmatrix}, \boldsymbol{d} = \begin{pmatrix} \hat{\lambda}_1 - P_{t1} \\ \hat{\lambda}_2 - P_{t2} \\ \cdot \\ \hat{\lambda}_{t-1} - P_{tt-1} \end{pmatrix},$$

$$\boldsymbol{X} = \begin{pmatrix} P_{11} - P_{t1} & P_{21} - P_{t1} & \cdot & P_{t-11} - P_{t1} \\ P_{12} - P_{t2} & P_{22} - P_{t2} & \cdot & P_{t-12} - P_{t2} \\ \cdot & \cdot & \cdot & \cdot \\ P_{1t-1} - P_{tt-1} & P_{2t-1} - P_{tt-1} & \cdot & P_{t-1t-1} - P_{tt-1} \end{pmatrix}$$

$$\boldsymbol{V} = \frac{1}{n} \begin{pmatrix} \lambda_1(1-\lambda_1) & -\lambda_1\lambda_2 & \cdot & -\lambda_1\lambda_{t-1} \\ -\lambda_1\lambda_2 & \lambda_2(1-\lambda_2) & \cdot & -\lambda_2\lambda_{t-1} \\ \cdot & \cdot & \cdot & \cdot \\ -\lambda_1\lambda_m & -\lambda_2\lambda_m & \cdot & \lambda_m(1-\lambda_{t-1}) \end{pmatrix} \text{and}$$

$$\hat{\mathrm{V}} = \frac{1}{n-1} \begin{pmatrix} \hat{\lambda}_1(1-\hat{\lambda}_1) & -\hat{\lambda}_1\hat{\lambda}_2 & \cdot & -\hat{\lambda}_1\hat{\lambda}_{t-1} \\ -\hat{\lambda}_1\hat{\lambda}_2 & \hat{\lambda}_2(1-\hat{\lambda}_2) & \cdot & -\hat{\lambda}_2\hat{\lambda}_{t-1} \\ \cdot & \cdot & \cdot & \cdot \\ -\hat{\lambda}_1\lambda_m & -\hat{\lambda}_2\hat{\lambda}_m & \cdot & \hat{\lambda}_m(1-\hat{\lambda}_{t-1}) \end{pmatrix}$$

Liu et al.'s (1975) RR technique is obviously very complex. A simpler RR technique is proposed by Liu and Chow (1976), and it is given as follows.

2.6 Liu and Chow Model

Liu and Chow (1976) considered Eriksson B's (1973) model using a different randomization device. In their device, two different colored (red and black) balls of the same size are put inside a flask with a transparent, narrow neck, through which only one ball can pass. The red balls are marked $1, 2, \ldots, k$. The proportion of black balls in the jar is $P(0 < P < 1)$ while the proportion of red balls, marked "j" is $P_j(P_j > 0, \sum_{j=1}^{k} P_j = 1 - P)$. A respondent is asked to shake the flask thoroughly and then turn it upside down. Then, he or she was asked to report his or her category truthfully if the bottom-most ball is black. If the bottommost ball is red, then the respondent reports the number written on the ball. For this technique, the probability of obtaining a response "j" from a respondent selected at random is

$$\lambda_j = \pi_j P + P_j, j = 1, \ldots, k \tag{2.6.1}$$

Let a sample of size n be selected from the population by the SRSWR method, and n_j be the number of times the response 'j' was reported ($\sum n_j = n$). Since $n_j = n\hat{\lambda}_j$ follows a multinomial distribution, one obtains.

Theorem 2.6.1

(i) $\hat{\pi}_j = \frac{\hat{\lambda}_j - P_j}{P}$ is an unbiased estimator of π_j
(ii) Variance of $\hat{\pi}_j$ is $\text{Var}(\hat{\pi}_j) = \frac{\lambda_j(1-\lambda_j)}{nP^2}$
(iii) An unbiased estimator of $\text{Var}(\hat{\pi}_j)$ is

$$\hat{\text{Var}}(\hat{\pi}_j) = \frac{\hat{\lambda}_j(1-\hat{\lambda}_j)}{(n-1)P^2}$$

2.7 Augmentation Response Model

In the Kim and Flueck (1978) augmentation response model, observations are coded to protect the privacy of the actual values. Let x be a random variable that takes the value j, i.e., $x = j$, if a respondent belongs to j th sensitive group $A_j, j = 1, 2, \ldots, t$. The respondent is asked to select one number y from $1, 2, \ldots, t$ with probability $\text{Prob}(y = i) = P_i$, ($P_i > 0, \sum_{i=1}^{t} P_i = 1$) and to provide his or her response to the interviewer as

$$z = \begin{cases} x - y, & \text{if } \; x + y \le t \\ x - y - t, & \text{if } \; x + y > t \end{cases} \tag{2.7.1}$$

Let λ_j be the probability of obtaining a response $j(= 1, 2, \ldots, t)$ from a respondent selected at random from a population by the SRSWR method, and $\hat{\lambda}_j$ be the sample proportion of responses "j". Then,

$$\begin{aligned} \lambda_1 =& \text{Prob}(x = 1, y = 3) + \text{Prob}(x = 2, y = 2) + \text{Prob}(x = 3, y = 1) \\ =& \pi_1 P_3 + \pi_2 P_2 + \pi_3 P_1 \\ \lambda_2 =& \text{Prob}(x = 1, y = 1) + \text{Prob}(x = 2, y = 3) + \text{Prob}(x = 3, y = 2) \\ =& \pi_1 P_1 + \pi_2 P_3 + \pi_3 P_2 \\ \lambda_3 =& \text{Prob}(x = 1, y = 2) + \text{Prob}(x = 2, y = 1) + \text{Prob}(x = 3, y = 3) \\ =& \pi_1 P_2 + \pi_2 P_1 + \pi_3 P_3 \end{aligned} \tag{2.7.2}$$

Noting that $\pi_1 + \pi_2 + \pi_3 = 1$, $P_1 + P_2 + P_3 = 1$ and that $\hat{\lambda}_j$ is an unbiased estimator of λ_j, one finds unbiased estimators of π_1, π_2, and π_3 as

$$\begin{aligned} \hat{\pi}_1 &= \frac{(P_3 - P_2)(\hat{\lambda}_1 - P_1) + (P_1 - P_2)(\hat{\lambda}_2 - P_2)}{D}, \\ \hat{\pi}_2 &= \frac{(P_2 - P_1)(\hat{\lambda}_1 - P_1) + (P_3 - P_1)(\hat{\lambda}_2 - P_2)}{D}, \\ \hat{\pi}_3 &= 1 - \hat{\pi}_1 - \hat{\pi}_2 \text{ where } D = (P_3 - P_1)(P_3 - P_2) - (P_2 - P_1)^2 \neq 0 \end{aligned} \tag{2.7.3}$$

Taking into account that the variance of $\hat{\lambda}_j = \text{Var}(\hat{\lambda}_j) = \frac{\lambda_j(1-\lambda_j)}{n}$ and co-variance of $\hat{\lambda}_1$ and $\hat{\lambda}_2 = \text{Cov}(\hat{\lambda}_1, \hat{\lambda}_2) = -\frac{\lambda_1\lambda_2}{n}$, one finds

$$
\begin{aligned}
&\text{Var}(\hat{\pi}_1) \\
&=\frac{(P_3-P_2)^2\lambda_1(1-\lambda_1)-2(P_3-P_2)(P_1-P_2)\lambda_1\lambda_2+(P_1-P_2)^2\lambda_2(1-\lambda_2)}{nD^2};\\
&\text{Var}(\hat{\pi}_2) \\
&=\frac{(P_2-P_1)^2\lambda_1(1-\lambda_1)-2(P_2-P_1)(P_3-P_1)\lambda_1\lambda_2+(P_3-P_1)^2\lambda_2(1-\lambda_2)}{nD^2};\\
&\text{Var}(\hat{\pi}_3) \\
&=\frac{(P_3-P_1)^2\lambda_1(1-\lambda_1)-2(P_3-P_1)(P_3-P_2)\lambda_1\lambda_2+(P_3-P_2)^2\lambda_2(1-\lambda_2)}{nD^2}
\end{aligned} \tag{2.7.4}
$$

Because the unbiased estimators of $\text{Var}(\lambda_j)$ and $\text{Cov}(\lambda_i, \lambda_j)$ are $\frac{\hat{\lambda}_j(1-\hat{\lambda}_j)}{n-1}$ and $-\frac{\hat{\lambda}_1\hat{\lambda}_2}{n-1}$, respectively, an unbiased estimator of $\text{Var}(\hat{\pi}_j)$ is obtained by substituting λ_j and n by $\hat{\lambda}_j$ and $n-1$, respectively, in the expression of $\text{Var}(\hat{\pi}_j)$.

The General Case

For the general case, we obtain

$$
\begin{aligned}
\lambda_1 - P_1 &= (P_t - P_1)\pi_1 + \cdots + (P_2 - P_1)\pi_{t-1}\\
\lambda_2 - P_2 &= (P_1 - P_2)\pi_1 + \cdots + (P_3 - P_2)\pi_{t-1}\\
\cdots &= \cdots\cdots\cdots\\
\lambda_{t-1} - P_{t-1} &= (P_{t-2} - P_{t-1})\pi_1 + \cdots + (P_t - P_{t-1})\pi_{t-1}
\end{aligned} \tag{2.7.5}
$$

Now, by writing, $\boldsymbol{\pi} = (\pi_1, \pi_2, \cdots, \pi_{t-1})'$, $\hat{\pi} = (\hat{\pi}_1, \hat{\pi}_2, \cdots, \hat{\pi}_{t-1})'$,$\hat{\lambda}* = (\hat{\lambda}_1 - P_1, \hat{\lambda}_2 - P_2, \cdots, \hat{\lambda}_{t-1} - P_{t-1})'$,

$$
\boldsymbol{P} = \begin{pmatrix} P_t - P_1 & P_{t-1} - P_1 & \cdots & P_2 - P_1 \\ P_1 - P_2 & P_t - P_2 & \cdots & P_3 - P_2 \\ \cdots & \cdots & \cdots & \cdots \\ P_{t-2} - P_{t-1} & P_{t-3} - P_{t-1} & \cdots & P_t - P_{t-1} \end{pmatrix},
$$

$$
\boldsymbol{V}(\hat{\lambda}) = \frac{1}{n}\begin{pmatrix} \lambda_1(1-\lambda_1) & -\lambda_1\lambda_2 & \cdots & -\lambda_1\lambda_{t-1} \\ -\lambda_2\lambda_1 & \lambda_2(1-\lambda_2) & \cdots & -\lambda_2\lambda_{t-1} \\ \cdots & \cdots & \cdots & \cdots \\ -\lambda_{t-1}\lambda_1 & -\lambda_{t-1}\lambda_2 & \cdots & \lambda_{t-1}(1-\lambda_{t-1}) \end{pmatrix}
$$

$$\text{and } \hat{V}(\hat{\lambda}) = \frac{1}{n-1}\begin{pmatrix} \hat{\lambda}_1(1-\hat{\lambda}_1) & -\hat{\lambda}_1\hat{\lambda}_2 & \cdots & -\hat{\lambda}_1\hat{\lambda}_{t-1} \\ -\hat{\lambda}_2\hat{\lambda}_1 & \hat{\lambda}_2(1-\hat{\lambda}_2) & \cdots & -\hat{\lambda}_2\hat{\lambda}_{t-1} \\ \cdots & \cdots & \cdots & \cdots \\ -\hat{\lambda}_{t-1}\hat{\lambda}_1 & -\hat{\lambda}_{t-1}\hat{\lambda}_2 & \cdots & \hat{\lambda}_{t-1}(1-\hat{\lambda}_{t-1}) \end{pmatrix}, \quad (2.7.6)$$

the following theorem is obtained.

Theorem 2.7.1

(i) $\hat{\pi} = \mathbf{P}^{-1}\hat{\lambda}*$ and $\hat{\pi}_t = 1-\sum_{i=1}^{t-1}\hat{\pi}_i$ are unbiased estimators of $\boldsymbol{\pi}$ and π_t provided $\mathbf{P}$ is non-singular.
(ii) The variance of $\hat{\pi}$ is $V\hat{a}r(\hat{\pi}) = \mathbf{P}^{-1}\boldsymbol{V}(\mathbf{P}^{-1})'$
(iii) An unbiased estimator of $\mathrm{Var}(\hat{\pi})$ is $V\hat{a}r(\hat{\pi}) = \mathbf{P}^{-1}\hat{V}(\mathbf{P}^{-1})'$

Remark 2.7.1 The Augmentation model reduces to the Warner (1965) model for $t = 2$.

2.8 Multi-stages Randomization

Burke and Dalenius (1976) observed that two or more stages of randomization may provide respondents with a greater sense of confidentiality than a single stage of randomization. Chaudhuri and Mukerjee (1988) described multi-stage randomization as follows: Suppose a population is classified into t categories with respect to a certain sensitive characteristic A. The population is also classified into t categories according to a certain unrelated, non-sensitive characteristic B. The proportions of the non-sensitive categories of the characteristic B in the population $\pi_1^*, \pi_2^*, \cdots, \pi_t^*$ are known. At the first stage of randomization, a person selects a ball from a box containing two colored balls, black and white, with proportions $\beta(0 < \beta < 1)$ and $1-\beta$, respectively. In the second stage, the respondent should draw a card at random from a deck containing two types of cards, e.g., type-1 and type-2 with proportions $\alpha(0 < \alpha < 1)$ and $1-\alpha$, respectively. Card type-1 says, "Disclose your sensitive category A, such as $j(= 1, \cdots, t)$ if you pick a black ball in stage-1 or disclose your non-sensitive category B, namely $l(= 1, \cdots, t)$ if you pick a white ball in stage-1". Card type-2 says, "Disclose your sensitive category A if you pick a white ball in stage-1 or disclose your non-sensitive category if you pick a black ball in stage-1". Fig. 2.5

The respondent will report the true category, such as "j" to which he or she belongs. The interviewer will not know whether the respondent is a member of the sensitive or non-sensitive category "j" as the interviewer does not know which card and which ball the respondent was drawn. Let a sample of size n be selected from the population by the SRSWR method and $\hat{\lambda}_j$ be the portion of the response "j" is obtained. For the proposed RR method, the probability of obtaining a response "j" is

Stage – 1 ↓	Stage – 2 ↓
Select a ball from a box containing black balls (proportion β) and white balls (proportion $1-\beta$)	Select a card from a pack: Type-1 (proportion α) and Type-2 (proportion $1-\alpha$)

Outcomes of trials Stage-1 Stage-2	Randomized response: Disclose your
Black ball type-1 card	Sensitive Category
White ball type-1 card	Non-sensitive Category
Black ball type-2 card	Non-sensitive Category
White ball type-2 card	Sensitive Category

Fig. 2.5 Multi-stage randomization model

$$\begin{aligned}\lambda_j &= \beta\{\alpha\pi_j + (1-\alpha)\pi_j^*\} + (1-\beta)\{(1-\alpha)\pi_j + \alpha\pi_j^*\}\\ &= \{\alpha\beta + (1-\alpha)(1-\beta)\}\pi_j + \{(1-\alpha)\beta + \alpha(1-\beta)\}\pi_j^*\\ &= \delta\pi_j + (1-\delta)\pi_j^* \qquad (2.8.1)\end{aligned}$$

where $\delta = \alpha\beta + (1-\alpha)(1-\beta)$.

Noting that $\hat{\lambda} = (\hat{\lambda}_1, \cdots, \hat{\lambda}_{t-1})$ follows a multinomial distribution with $E(\hat{\lambda}_j) = \lambda_j$, $\text{Var}(\hat{\lambda}_j) = \lambda_j(1-\lambda_j)/n$, and $\text{Cov}(\hat{\lambda}_j, \hat{\lambda}_k) = -\lambda_j\lambda_k/n$ for $j \neq k = 1, \cdots, t-1$, the following theorem is obtained.

Theorem 2.8.1 For $j = 1, 2, \ldots, t$,

(i) $\hat{\pi}_j = \frac{\hat{\lambda}_i - (1-\delta)\hat{\pi}_j^*}{\delta}$ is an unbiased estimator of π_j,

(ii) The variance of $\hat{\pi}_j$ is $\text{Var}(\hat{\pi}_j) = \frac{\lambda_j(1-\lambda_j)}{n\delta^2}$,

(iii) An unbiased estimator of $\text{Var}(\hat{\pi}_j)$ is $\hat{\text{Var}}(\hat{\pi}_j) = \frac{\hat{\lambda}_j(1-\hat{\lambda}_j)}{(n-1)\delta^2}$.

2.9 Multiple Sensitive Attributes

Tamhane (1981) proposed a RR technique involving $t(\geq 2)$ attributes $A_1, A_2, \ldots, A_t$; all attributes need not be sensitive. Let $\theta_{i_1 i_2,\ldots,i_u}$ denotes the unknown proportion of individuals in the population who possess the attribute $A_{i_1}, A_{i_2}, \ldots, A_{i_u} (1 \leq i_1 < i_2 < \cdots < i_u \leq t)$. Our interest is to make inferences concerning marginal proportions $\theta_j, j = 1, \cdots, t$, and bivariate proportions θ_{jk}, $1 \leq j < k \leq t$. In the Tamhane (1981) RR technique, the statements are phrased such that the first $r(< t)$ statements "I am a member of A_i"$(1 \leq i \leq r)$, a 'no" response to each one is non-stigmatizing. The remaining $t - r$ statements, phrased "I am not a member of A_i" $(r + 1 \leq i \leq t)$ a "yes" answer to each one is non-stigmatizing. An appropriate choice r would be approximately $t/2$. Let us define modified attributes B_i s are such that $B_i = A_i$ for $i = 1, 2, ..., r$ and $B_i = \overline{A}_i$ (complement of A_i) for $i = r + 1, r + 2, ..., t$. Thus, a "yes" response to each of B_i's is stigmatizing, and a "no" answer is non-stigmatizing. Further, let's $\pi_{i_1 i_2 \cdots i_u}$ denote the proportion of individuals in the population who possess the attributes $B_{i_1}, B_{i_2}, \cdots, B_{i_u}$. Obviously, $\theta_{i_1 i_2 \cdots i_u}$ can be obtained from $\pi_{i_1 i_2 \cdots i_u}$ and vice versa. Let samples $s_1, s_2, \ldots, s_b$ of sizes $n_1, n_2, \ldots, n_b, \sum_{i=1}^{b} n_i = n$ be selected independently from the population by the SRSWR method. Each of the respondents selected in a sample is asked to pick one of the t statements (unnoticed by the interviewer) at random using a randomization device and then record the response. The respondents are asked to repeat the procedure independently. Let $P_{hi}^{(l)} \left(\sum_{i=1}^{t} P_{hi}^{(l)} = 1 \right)$ be the probability that a respondent selected in the sample $h(= 1, 2, \ldots, b)$ picks the $i(= 1, 2, \ldots, t)$ th statement at the $l(= 1, 2)$ th trial.

2.9.1 Estimation of Proportions

Here, responses are coded so that a score 2^{l-1} is assigned to a "yes" answer on the $l(= 1, 2)$ th trial, i.e., sores "1" and "2" are attached to the "yes" answer of trial 1 and trial 2, respectively. The score "0" is assigned to the "no" answer in each of the trials. Let ν be the total scores. Then,

$$\text{scores Responses}$$
$$\nu = \begin{cases} 0 \text{ no(trial - 1) no(trial - 2)} \\ 1 \text{ yes(trial - 1) no(trial - 2)} \\ 2 \text{ no(trial - 1) yes(trial - 2)} \\ 3 \text{ yes(trial - 1) yes(trial - 2)} \end{cases} \tag{2.9.1}$$

Let $\lambda_{h\nu}$ be the probability of obtaining a score ν by a respondent selected in the sample s_h,

$$\boldsymbol{\lambda} = (\lambda_{11}, \lambda_{12}, \lambda_{13}, \cdots, \lambda_{b1}, \lambda_{b2}, \lambda_{b3})' \text{and } \boldsymbol{\pi} = (\pi_1, \pi_2, \cdots, \pi_t, \pi_{12}, \pi_{13}, \cdots, \pi_{t-1,t})'$$

Tamhane (1981) obtained

$$\boldsymbol{\lambda} = \boldsymbol{R}\boldsymbol{\pi} \tag{2.9.2}$$

where elements of the matrix $\boldsymbol{R} = ((R_{hi}))$ are

$$\left.\begin{aligned} R_{3h-2,i} &= P_{hi}^{(1)}(1 - P_{hi}^{(2)}) \\ R_{3h-1,i} &= (1 - P_{hi}^{(1)})P_{hi}^{(2)} \\ R_{3h,i} &= P_{hi}^{(1)}P_{hi}^{(2)} \end{aligned}\right\} \text{for } h = 1, ..., b;\ i = 1, ..., t \tag{2.9.3}$$

and

$$\begin{aligned} R_{3h-2,k} &= R_{3h-1,k} = -R_{3h,k} \\ &= -\left(P_{hi}^{(1)}P_{hj}^{(2)} + P_{hj}^{(1)}P_{hi}^{(2)}\right) \text{for } 1 \le i < j \le t;\ k = it - i(i+1)/2 + j \end{aligned} \tag{2.9.4}$$

For estimation of t marginal proportions π_i, and $t(t-1)/2$ bivariate proportions π_{ij}, one requires $b \ge t(t-1)/2$. Let n_{hv} be the number of respondents in the hth sample who reported a score of $v(= 0, 1, 2, 3)$ and $\hat{\lambda}_{hv} = n_{hv}/n_h$. Now, choosing $P_{hi}^{(l)}$ suitably to make $\boldsymbol{R}$ a full column rank, one may obtain an unbiased estimator of $\boldsymbol{\pi}$ by solving the equation

$$\hat{\lambda} = \boldsymbol{R}\hat{\boldsymbol{\pi}} \tag{2.9.5}$$

where $\hat{\lambda} = (\hat{\lambda}_{11}, \hat{\lambda}_{12}, \hat{\lambda}_{13}, \ldots.., \hat{\lambda}_{b1}, \hat{\lambda}_{b2}, \hat{\lambda}_{b3})'$and
$\hat{\pi} = (\hat{\pi}_1, , \ldots.., \hat{\pi}_t, \hat{\pi}_{12}, \hat{\pi}_{13}, \ldots, \hat{\pi}_{t-1,t})'$.

However, for $t > 2$, Eq. (2.9.5) may not yield a unique solution of $\hat{\pi}$. Further, the resulting solution may be inadmissible, as was observed by Devore (1977) and Singh (1976, 1978) for Warner's (1965) RR technique. To obtain an admissible estimator of $\boldsymbol{\pi}$, Tamhane (1981) recommends the restricted maximum likelihood method, which maximizes the likelihood function

$$L \propto \prod_{h=1}^{b}\prod_{v=0}^{3}(\lambda_{hv})^{n_{hv}},\ \lambda_{h0} = 1 - \lambda_{h1} - \lambda_{h2} - \lambda_{h3} \tag{2.9.6}$$

subject to constraints (i) $0 \le \pi_i \le 1$, $i = 1, \ldots, t$ and (ii) $\max(0, \pi_i + \pi_j - 1) \le \pi_{ij} \le \max(\pi_i, \pi_j)$, $1 \le i < j \le t$.

The main demerit of this method is that one cannot get analytical expressions of $\hat{\pi}$ and their variances.

2.9.2 Testing of Hypothesis

For testing the hypothesis $H_F : \pi_{ij} = \pi_i\pi_j$ for $i < j \in F = \{1, \ldots, t\}$, Tamhane (1981) proposed a generalized likelihood ratio test as follows. Let L be the value of the likelihood function obtained in (2.9.6) and L_F^* be the value maximum likelihood function (2.9.6) with additional constraints $\pi_{ij} = \pi_i\pi_j$,$i < j \in F$. Then, under H_F,

$-2\log\left(L_F^*/L\right)$ asymptotically follows chi-square distribution with f (= number of pairs in F) degrees of freedom.

Tamhane (1981) studied the asymptotic properties of the proposed estimator $\hat{\pi}$, the optimal choice of design parameters, and measures of the Jeopardy function of the respondents. He also used the proposed RR technique to estimate the prevalence of hard drug use and seeking psychiatric help among North Western University Industrial Psychology students.

2.10 Mukerjee's Model

Chaudhuri and Mukerjee (1988) presented a general theory of the multi-attribute RR technique, which was originally developed by Mukerjee (1981). In this technique, m stigmatizing attributes $A_1, A_2, \ldots, A_m$ were considered. The j th attribute A_j possesses t_j mutually exclusive and exhaustive categories $A_1^j, \ldots, A_{t_j}^j, j = 1, \ldots, m$. The proportion of individuals belonging to the i_1 category of A_1, i_2 category of A_2,... i_m category of A_m in the population is denoted by $\pi(i_1, i_2, \ldots, i_m)$. To estimate $\pi(i_1, i_2, \ldots, i_m)$, a random sample of size n is selected from the population by the SRSWR method. For the proposed RR technique, each of the respondents selected in the sample was asked to perform m independent RR trials, one for each attribute. For the j th attribute A_j, the respondent reports $k(k = 1, \ldots, t_j)$ if he or she would really belong to A_k^j, the k th category of the attribute A_j; otherwise, the respondent is asked to report a number $r(= 1, 2, \ldots, t_j)$ with probability $P_r^j(\sum_{r=1}^{t_j} P_r^j = 1, j = 1, \ldots, m)$ as his or her response. Thus, from each respondent, we get a vector of response $\boldsymbol{x} = (x_1, \ldots, x_j, \ldots, x_m)$ where $x_j(= 1, \ldots, t_j)$ denotes the response for the category of the attribute A_j $(j = 1, \ldots, m)$. So, from a sample of n respondents, we get data comprising n vectors of responses. From the data thus obtained, one can find an unbiased estimator of $\pi(i_1, i_2, \ldots, i_m)$. The derivations of the results are complex and, hence, are not presented here. Readers are referred to Chaudhuri and Mukerjee (1988) for further details.

2.11 Discussions

Randomized response techniques for estimating proportions of individuals belonging to the several sensitive attributes were proposed by Abul-Ela et al. (1967), Eriksson (1973, 1973a), Greenberg et al. (1969), Liu et al. (1975), Liu and Chow (1976), Kim and Flueck (1978), Burke and Dalenius (1976), Tamhane (1981), and Mukerjee (1988), among others, and are discussed in this chapter. The proposed methods cover major contributions in the area of polychotomous RR models.

In Aubul-Ela et al.'s (1967) model, $t - 1$ independent samples are selected from a population classified into t categories, of which at least one and at most $t - 1$

categories are sensitive. Respondents selected in the sample are asked to draw a card at random from a deck consisting of t types of cards. The Eriksson (1973a, 1973b) model A is almost similar to Aubul-Ela et al.'s model. The only difference is that for Eriksson's (1973a, 1973b) Model A, respondents selected one card from a deck consisting of two types of cards. As a result, the privacy of the respondents appears to be greater in Aubul-Ela et al.'s model than in Eriksson's, but the complexity of the estimation procedure in Aubul-Ela et al.'s model is greater than in Eriksson's model. Abul-Ela considered the effect of incompletely truthful reporting in a real survey, the 'Population Change Survey".

Eriksson's (1973a) model B requires only one sample to be drawn, and respondents are required to draw only one card from $t + 1$ types of cards, with one of the cards asking respondents to disclose truthfully their membership in their category. The estimation procedure is relatively simple for the Eriksson model B.

Greenberg's proposed unrelated question model involving three categories is expected to elicit more cooperation from the respondents, and the computation procedure is not tedious. Liu et al. (1975) method is very easy to administer and maintains the confidentiality of the respondents, but it involves difficult computation procedures.

Liu and Chow's (1976) procedure is the same as Erisson's (1973) procedure and much easier to administer. Kim and Fluek's (1978) procedure is easier to administer, but respondents may face difficulty decoding their responses. Burke and Dalenius's (1976) procedure provides more anonymity to the respondents, but it is not easy to carry out the instructions given to them. Mukerjee's (1981) and Tamhane's (1981) RR techniques provide better scope but are difficult to administer and computationally more difficult. Most of the proposed estimators are moment estimators and hence can take inadmissible values.

References

Abul-Ela, Abdel-Latif, A., Greenberg, B.G., & Horvitz, D.G. (1967). A multi proportions RR model. *Journal of the American Statistical Association, 62*, 990–1008.

Burke, P. D., & Dalenius, T. (1976). Some new ideas in the realm of randomized inquiries. *International Statistical Review, 44*, 219–221.

Chaudhuri, A., & Mukerjee, R. (1988). *Randomized response: Theory and techniques*. Marcel.

Devore, J. L. (1977). A note on the RR techniques. *Communications in Statistics—Theory and Methods, 6*, 1525–1529.

Eriksson, S. (1973a). *Randomized interviews for sensitive questions*. Ph.D. thesis, University Gothenburg.

Eriksson, S. A. (1973b). A new model for randomized response. *International Statistical Review, 41*, 101–113.

Greenberg, B. G., Abul-Ela, Abdel-Latif, A., Simmons, W. R., Horvitz, B. G. (1969). The unrelated question RR model: Theoretical frame work. *Journal of the American Statistical Association, 64*, 520–539

Kim, J., & Flueck, J. A. (1978). An additive randomized response model. In *Proceedings of the American Statistical Association, Social Survey Research Methods* (pp. 351–355).

Liu, P. T., Chow, L. P., & Mosley, W. H. (1975). Use of randomized response technique with a new randomizing device. *Journal of the American Statistical Association, 70*, 329–332.

Liu, T. P., & Chow, L. P. (1976). A new discrete quantitative RR model. *Journal of the American Statistical Association, 71*, 72–73.

Mukerjee, R. (1981). Inference on confidential characters from survey data. *Calcutta Statistical Association Bulletin, 30*, 77–88.

Singh, J. (1976). A note on RR techniques. *In Proceedings of The Survey Research Methods Section of the American Statistical Association* (p. 772).

Singh, J. (1978). A note on maximum likelihood estimation from randomized response models. In s*Proceedings of the survey research methods section of the American Statistical Association* (pp. 282–283).

Tamhane, A. C. (1981). RR techniques for multi sensitive attributes. *Journal of the American Statistical Association, 76*, 916–923.

Warner, S. L. (1965). Randomize response: A survey technique for eliminating evas ve answer bias. *Journal of the American Statistical Association, 60*, 63–69.

Chapter 3
RR Techniques for Quantitative Characteristics

Abstract This chapter provides a chronological development of randomized response techniques for quantitative sensitive variables. The methods of estimation of the population mean, variance, and unbiased estimator of the variance are proposed under the SRSWR sampling scheme. Details of Greenberg et al., (Journal of the American Statistical Association. 66:243–250, 1971), Eriksson, (International Statistical Review. 48:29–32, 1973), Godambe, (International Statistical Review. 48:29–32, 1980), the additive model of Poole, (Journal of the American Statistical Association 69:1002–1005, 1974), the product model of Pollock and Bek, (Journal of the American Statistical Association 71:884–886, 1976), Eichhorn and Hayre's (1989) model, the distributive model, among others, are discussed. The general RR model proposed by Arnab, (Communications in Statistics e Theory and Methods 19:3751–3757, 1990), along with the method of estimation of population characteristics, is presented. It is also demonstrated that almost all the qualitative and quantitative RR models can be obtained as special cases of Arnab, (Communications in Statistics e Theory and Methods 19:3751–3757, 1990) model.

Keywords Additive model · Distributive model · Product model · Quantitative variable

3.1 Introduction

In the preceding chapters, we have discussed some of the important RR techniques that have been developed for qualitative sensitive characteristics. In this chapter, we will present a few important RR techniques that are applicable to quantitative sensitive characteristics such as duration of suffering from AIDS, expenditure on illegal drugs, amount of illegal income, and the number of sexual partners, among others. Applications of RR techniques for quantitative characteristics in real surveys were conducted by various researchers, such as Greenberg et al. (1971): concerning the number of abortions over a lifetime and the income of heads of households; Bouza (2009): concerning the number of persons with whom they had sexual contact

R. Arnab, *Indirect Methods of Data Collection and Analysis from Surveys*, Indian Statistical Institute Series, https://doi.org/10.1007/978-981-97-6005-3_3

previously and a posteriori were detected by an AIDS-infected person; Holbrook and Krosnick (2010): concerning voting turnout; Conteh et al. (2015): concerning the number of abortions over a lifetime and the income of heads of households; Cobo et al. (2016): related to the use of cannabis by students.

The RR technique for sensitive quantitative characteristics was first introduced by Greenberg et al. (1971). The chronological development of RR techniques for quantitative sensitive characteristics is provided in this chapter. Methods of estimation of the population mean, variance, and unbiased estimators of the variance are proposed for Eriksson's (1973), Eichhorn and Hayre's (1989), Greenberg et al.'s (1971), and Godambe's (1980) models. The most general RR model proposed by Arnab (1990), along with the estimation of population characteristics, is also presented. It is shown that most of the RR models that are used for both qualitative and quantitative sensitive characteristics may be considered special cases of this model. In this chapter, we will limit ourselves to the SRSWR sampling only.

3.2 Greenberg et al. Quantitative Gjestavang and Singh RR Model

In the Greenberg et al. (1971) RR model, a sample of respondents is selected by the SRSWR method. Each of the selected respondents is asked to answer one of the following two questions:

Q1: How much did the household's head earn last year?

Q2: How much do you think the average household head earns in a year?

The first question relating to an individual's household earnings is considered a sensitive characteristic, and it will be denoted by "y". The second question concerning the average earnings of a group of people is a non-sensitive characteristic, and it will be denoted by "x". Question Q2 is unrelated or innocuous. The respondents are asked to draw one ball at random from a box consisting of two types of balls of the same shape but different in proportions, e.g., blue with proportion $P(0 < P < 1)$ and red with proportion $1 - P$. If a respondent picks a blue ball, he or she is required to answer the sensitive Q1. Otherwise, if the respondent chooses a red ball, he or she answers the unrelated question Q2. Let the respondent provide an answer z. The whole experiment is performed in the respondent's absence. The interviewer will not be able to identify whether the response z represents the value of x or a value of y.

Let the density functions of x and y be $g(x)$ and $f(y)$; and let the means and variances of x and y be μ_x, μ_y and σ_x^2, σ_y^2, respectively.

Estimation of μ_y

Here we have two scenarios: (i) μ_x is known, and (ii) μ_x is unknown.

μ_x *is Known*

Theorem 3.2.1 Let $z_1, z_2, \ldots, z_n$ be the RRs obtained from the selected sampled respondents. Then,

(i) $\hat{\mu}_{yG} = \frac{\bar{z}-(1-P)\mu_x}{P}$ is an unbiased estimator of μ_y
(ii) The variance of $\hat{\mu}_{yG}$ is

$$\text{Var}(\hat{\mu}_{yG}) = \frac{\sigma_z^2}{nP^2}$$

(iii) An unbiased estimator of $\text{Var}(\hat{\mu}_{yG})$ is

$$\hat{\text{Var}}(\hat{\mu}_{yG}) = \frac{s_z^2}{nP^2}$$

where $\bar{z} = \frac{1}{n}\sum_{i=1}^{n} z_i$ is the sample mean, $\sigma_z^2 = P\sigma_y^2+(1-P)\sigma_x^2+P(1-P)(\mu_y-\mu_x)^2$ and $s_z^2 = \frac{1}{n-1}\sum_{i=1}^{n}(z_i - \bar{z})^2$ is the sample variance.

Proof

(i) Noting the density function of z is

$$\psi(z) = Pf(z) + (1-P)g(z) \tag{3.2.1}$$

one finds

$$E(z) = P\mu_y + (1-P)\mu_x \tag{3.2.2}$$

Eq. (3.2.2) yields

$$E(\hat{\mu}_{yG}) = \frac{E(\bar{z}) - (1-P)\mu_x}{P} = \mu_y$$

(ii) Further,

$$E(z^2) = PE(y^2) + (1-P)E(x^2)$$

$$= P(\sigma_y^2 + \mu_y^2) + (1-P)(\sigma_x^2 + \mu_x^2)$$

gives

$$\sigma_z^2 = \text{Var}(z) = P\sigma_y^2 + (1-P)\sigma_x^2 + P(1-P)(\mu_y - \mu_x)^2 \tag{3.2.3}$$

Eq. (3.2.3) yields

$$\begin{aligned}\text{Var}(\hat{\mu}_{yG}) &= \frac{\text{Var}(\overline{z})}{P^2}\\ &= \frac{\text{Var}(z)}{nP^2}\\ &= \frac{\sigma_z^2}{nP^2}\end{aligned}$$

(iii) Since $z_1, z_2, \ldots, z_n$ are independently and identically distributed random variables, we obtain $E\left[\hat{\text{Var}}(\hat{\mu}_{yG})\right] = \text{Var}(\hat{\mu}_{yG})$.

μ_x *is Unknown*

In case μ_x is unknown, two independent samples s_1 and s_2 of sizes n_1 and n_2 are selected from the population by the SRSWR method. Each of the selected respondents in the sample $s_i (i = 1, 2)$ is asked to answer one of the questions Q1 and Q2 with probabilities P_i and $1 - P_i$, respectively. Let $z_{i1}, z_{i2}, \ldots, z_{in_i}$ be the RRs obtained from the respondents of the sample s_i, $i = 1, 2$ and the sample means and variances of the randomized responses be denoted by

$$\overline{z}_i = \frac{1}{n_i}\sum_{j=1}^{n_i} z_{ij} \text{ and } s_{zi}^2 = \frac{1}{n_i - 1}\sum_{j=1}^{n_i}(z_{ij} - \overline{z}_i)^2 \tag{3.2.4}$$

Now noting that

$$E(z_{ij}) = P_i\mu_y + (1 - P_i)\mu_x \text{ and } \text{Var}(z_{ij}) = P_i\sigma_y^2 + (1 - P_i)\sigma_x^2 + P_i(1 - P_i)(\mu_y - \mu_x)^2$$

for $i = 1, 2; j = 1, \ldots, n_i$, one finds,

$$\begin{aligned}&E(\overline{z}_i) = P_i\mu_y + (1 - P_i)\mu_x \text{ and}\\ &\sigma_{zi}^2 = \text{Var}(\overline{z}_i) = \frac{P_i\sigma_y^2 + (1 - P_i)\sigma_x^2 + P_i(1 - P_i)(\mu_y - \mu_x)^2}{n_i}\end{aligned} \tag{3.2.5}$$

Further, noting s_{zi}^2 is an unbiased estimator $\text{Var}(z_{ij})$, the following theorem is obtained.

Theorem 3.2.2

(i) $\hat{\mu}_{yG}^* = \frac{(1-P_2)\overline{z}_1 - (1-P_1)\overline{z}_2}{P_1 - P_2}$ is unbiased estimator of μ_y provided $P_1 \neq P_2$
(ii) The variance of $\hat{\mu}_{yG}^*$ is

$$\text{Var}(\hat{\mu}_{yG}^*) = \frac{(1 - P_2)^2\sigma_{z1}^2/n_1 + (1 - P_1)^2\sigma_{z2}^2/n_2}{(P_1 - P_2)^2}$$

where $\sigma_{zi}^2 = P_i\sigma_y^2 + (1 - P_i)\sigma_x^2 + P_i(1 - P_i)(\mu_x - \mu_y)^2;\ i = 1, 2$

(iii) An unbiased estimator of $\mathrm{Var}(\hat{\mu}_{yG}^*)$ is

$$\hat{\mathrm{Var}}(\hat{\mu}_{yG}^*) = \frac{(1 - P_2)^2 s_{z1}^2/n_1 + (1 - P_1)^2 s_{z2}^2/n_2}{(P_1 - P_2)^2}$$

Optimum Choices of Design Parameters

The variance of $\hat{\mu}_{yG}^*$ decreases with the increase of the difference $P_1 - P_2$. The suggested working rule of Greenberg et al. (1971) is to choose $P_1 + P_2 = 1$ after choosing P_1 as far from 0.5 as practicable without giving the respondent the impression that the randomized device is favouring a particular question. Past surveys show that choosing P_1 between 0.7 and 0.8 or their complements provides satisfactory results. For a fixed set of values of (n_1, n_2), the variance of $\hat{\mu}_{yG}^*$ decreases with the decrease of σ_x^2, and $|\mu_y - \mu_x|$. Hence, one should choose μ_x close to μ_y as much as possible while keeping σ_x^2 very small. However, a choice of σ_x^2 much less than σ_y^2 may cause a loss of cooperation from respondents. As a result, it is recommended that σ_x^2 should be chosen as large as σ_y^2. In cases where no good knowledge of the mean and variance of the sensitive variable is available, it may be difficult to choose a suitable unrelated variable. The optimum value of n_1 that minimizes the variance of $\hat{\mu}_{yG}^*$ subject to a fixed total sample size $n = n_1 + n_2$ is

$$n_1 = n\frac{(1 - P_2)\sigma_{z1}}{(1 - P_2)\sigma_{z1} + (1 - P_1)\sigma_{z2}} \tag{3.2.6}$$

Greenberg et al. (1971) provided the Rao-Cramér (CR) lower bound of the variance an unbiased estimator $\hat{\theta}$ of $\vartheta(= \mu_x \text{ or } \mu_y)$ as

$$\mathrm{Var}(\hat{\theta}) \geq \left[n_1 E\left(\frac{\partial \ln\psi_1(z)}{\partial\theta}\right)^2 + n_2 E\left(\frac{\partial \ln\psi_2(z)}{\partial\theta}\right)^2\right]^{-1} \tag{3.2.7}$$

where $\psi_i(z) = P_i f(z) + (1 - P_i)g(z);\ i = 1, 2$

Greenberg et al. (1971) also derived the expression of the CR lower bounds for Poisson, Exponential, and Normal distributions. However, for the attainment of the lower bound, Greenberg et al. (1971) pointed out that there cannot exist an unbiased estimator of μ_A whose variance meets the CR bound, since $\frac{\partial \mathrm{Log}\psi_i(z)}{\partial\mu_A}$ cannot be put in a form to satisfy the necessary and sufficient condition for the existence of such an estimator, namely, $\frac{\partial \mathrm{Log}\psi_i(z)}{\partial\mu_A} = c(\hat{\mu}_A - \mu_A)$, where c is independent of $\hat{\mu}_A$ but may depend on μ_A.

3.3 Eriksson's Model

Let y_i be the value of the stigmatized characteristic y of the ith $(i = 1, \ldots, N)$ unit of a population U. In Eriksson's (1973) RR model, each of the respondents selected in the sample is asked to disclose their true value of y with a known probability $c(0 < c < 1)$ or a value $Q_j(j = 1, \ldots, M)$ with a probability $q_j(0 < q_j < 1, \sum_{j=1}^{M} q_j = 1 - c)$. The vector $\boldsymbol{Q} = (Q_1, \ldots, Q_M)$ is anticipated to cover the range of y values and the value M depends on the range of y. For example, let y be the expenditure on illegal drugs, which ranges between 0 and 2000. In this case, we may use the following RR device described by Arnab (2017). Take a jar consisting of several capsules that are identical in appearance. Each capsule contains only one ticket. Let q proportion of capsules contain tickets written with the statement "Disclose your true expenditure on drugs", and the remaining capsules contain tickets written numbers $0\,(= Q_1), 5(= Q_2), \ldots, 2000(= Q_{401})$ with proportions $q_1, \ldots, q_M; M = 401$. If a respondent selects a ticket that says, "Disclose your true expenditure on drugs," then he or she must report his or her true expenditure on drugs as a response. Otherwise, he or she will report the amount written on the ticket as his or her response. The whole experiment is performed in the absence of the interviewer. The interviewer will receive only a number as a response and will not be able to identify whether the number represents the true expenditure on the drug or the number written on the ticket. Hence, the confidentiality of the respondent is maintained.

Estimation of the Mean

Let z_i be the randomized response obtained from the ith $(i = 1, \ldots, N)$ respondent; E_R, V_R and C_R, respectively, denote expectation, variance, and co-variance with respect to the randomized device R.

For Eriksson's RR technique,

$$z_i = \begin{cases} y_i, \text{ with probability } c \\ Q_i, \text{ with probability } q_i \end{cases} \tag{3.3.1}$$

Denoting $q_j^* = q_j/(1 - c)$ and $\mu_Q = \sum_{j=1}^{M} q_j^* Q_j$, one finds

$$E_R(z_i) = cy_i + (1 - c)\mu_Q \text{ and } E_R(z_i^2) = cy_i^2 + (1 - c)\sum_{j=1}^{M} q_j^* Q_j^2 \tag{3.3.2}$$

Eq. (3.3.2) yields the variance of z_i as

$$V_R(z_i) = cy_i^2 + (1-c)\sum_{j=1}^{M} q_j^* Q_j^2 - \{cy_i + (1-c)\mu_Q\}^2$$
$$= (1-c)\left[c(y_i - \mu_Q)^2 + \sigma_Q^2\right] \quad (3.3.3)$$

where $\sigma_Q^2 = (\sum_{j=1}^{M} q_j^* Q_j^2 - \mu_Q^2)$.

Denoting $r_i = \frac{z_i-(1-c)\mu_Q}{c}$, we have.

Theorem 3.3.1

(i) $E_R(r_i) = \frac{E_R(z_i)-(1-c)\mu_Q}{c} = y_i$
(ii) The variance of r_i is

$$V_R(r_i) = \frac{V_R(z_i)}{c^2} = \frac{(1-c)}{c}\left[(y_i - \mu_Q)^2 + \frac{\sigma_Q^2}{c}\right] = \phi_i.$$

(iii) An unbiased estimator of $V_R(r_i)$ is

$$\hat{\phi}_i = (1-c)\left[(r_i - \mu_Q)^2 + \frac{\sigma_Q^2}{c}\right]$$

Let a sample s of n units be selected from the population U by the SRSWR method, and let z_i be the randomized response obtained from the ith unit using Eriksson's (1973) RR model. Then, using Theorem 3.3.1, one obtains.

Theorem 3.3.2

(i) $\overline{r} = \frac{1}{n}\sum_s r_i$ is an unbiased estimator of the population mean $\overline{Y} = \frac{1}{N}\sum_{i\in U} y_i$

where $\sum_s$ and $\sum_{i\in U}$ denote, respectively, the sum of the units in s including repetition and distinct units in U.

(ii) The variance of $\overline{r}$ is

$$\text{Var}(\overline{r}) = \frac{1}{n}(\sigma_y^2 + \overline{\phi})$$

where $\sigma_y^2 = \frac{1}{N}\sum_{i\in U}(y_i - \overline{Y})^2$, $\overline{Y} = \sum_{i\in U} y_i/N$ and $\overline{\phi} = \frac{1}{N}\sum_{i\in U}\phi_i$

(iii) An unbiased estimator of $\text{Var}(\overline{r})$ is

$$\hat{\text{Var}}(\overline{r}) = \frac{1}{n(n-1)}\sum_s (r_i - \overline{r})^2$$

Proof Let $E_p(V_p)$ denotes expectation (variance) with respect to the sampling design $p($ SRSWR in this case). Then,

(i)
$$E(\bar{r}) = E_p\left[E_R\left(\frac{1}{n}\sum_s r_i\right)\right] = E_p\left(\frac{1}{n}\sum_s y_i\right) = \overline{Y}$$

(ii)
$$\mathrm{Var}(\bar{r}) = V_p[E_R(\bar{r})] + E_p[V_R(\bar{r})] = V_p\left(\frac{1}{n}\sum_s y_i\right) + E_p\left(\frac{1}{n^2}\sum_s \phi_i\right) = \frac{1}{n}(\sigma_y^2 + \bar{\phi})$$

(iii)
$$E[\hat{\mathrm{Var}}(\bar{r})] = \frac{1}{n(n-1)}E[\sum_s r_i^2 - n\bar{r}^2]$$
$$= \frac{1}{n(n-1)}\left[E_p\{\sum_s E_R(r_i^2)\} - nE(\bar{r}^2)\right]$$
$$= \frac{1}{n(n-1)}\left[E_p\left\{\sum_s (y_i^2 + \phi_i)\right\} - n(V(\bar{r}) + \overline{Y}^2)\right]$$
$$= \frac{1}{(n-1)}\left[\frac{1}{N}\sum_{i\in U} (y_i^2 + \phi_i) - \left\{V(\bar{r}) + \overline{Y}^2\right\}\right]$$
$$= \mathrm{Var}(\bar{r})$$

3.4 Multiplicative Model

Eichhorn and Hayre (1983), and Pollock and Bek (1976) proposed the multiplicative model. In this model, the ith respondent selected in the sample reports a randomized response

$$z_i = y_i\vartheta_i \tag{3.4.1}$$

where y_i is the true response of the sensitive variable y and ϑ_i is a scrambled response with a known mean μ_ϑ and known variance σ_ϑ^2. The scrambled response ϑ_i may be generated by drawing a random sample from a known distribution such as chi-square, normal, or binomial. The mean of the scrambled variable μ_ϑ should be equal to 1, otherwise, the range of z should be different from the range of the sensitive variable y. The main demerit of this method is that the respondents need to multiply two numbers y_i and ϑ_i to report their responses, which might be difficult and cause erroneous responses. Let μ_y and σ_y^2 be the mean and variance of y. Then

$$E_R(z_i) = y_iE_R(\vartheta_i) = y_i\mu_\vartheta \tag{3.4.2}$$

and $V_R(z_i) = y_i^2V_R(\vartheta_i)$

$$= y_i^2\sigma_\vartheta^2 = \sigma_z^2 \tag{3.4.3}$$

Let $z_1, \ldots, z_n$ be the RR obtained from a sample s of n respondents selected by the SRSWR method, and $\bar{z} = \frac{\sum_s z_i}{n}$ be the sample men. Then,

$$E\left(\frac{\bar{z}}{\mu_\vartheta}\right) = E_p\left(\frac{\sum_s E_R(z_i)}{n\mu_\vartheta}\right) = E_p\left(\frac{\sum_s y_i}{n}\right) = \mu_y \tag{3.4.4}$$

and

$$\begin{aligned}\text{Var}\left(\frac{\bar{z}}{\mu_\vartheta}\right) &= V_p\left\{E_R\left(\frac{\bar{z}}{\mu_\vartheta}\right)\right\} + E_P\left\{V_R\left(\frac{\bar{z}}{\mu_\vartheta}\right)\right\}\\ &= V_p\left(\frac{1}{n}\sum_s y_i\right) + \frac{\sigma_\vartheta^2}{\mu_\vartheta^2}E_P\left(\frac{1}{n}\sum_s y_i^2\right)\\ &= \frac{1}{n}\left[\sigma_y^2 + \frac{c_\vartheta^2}{\mu_\vartheta^2}(\sigma_y^2 + \mu_y^2)\right]\end{aligned} \tag{3.4.5}$$

where $\sigma_y^2 = \frac{1}{N}\sum_{i=1}^N (y_i - \mu_y)^2$.

Equations (3.4.4) and (3.4.5) yield.

Theorem 3.4.1

(i) $\hat{\mu}_y = \frac{\bar{z}}{\mu_\vartheta}$ is an unbiased estimator of μ_y

(ii) The variance of $\hat{\mu}_y$ is

$$\text{Var}(\hat{\mu}_y) = \frac{1}{n}(\sigma_z^2) = \frac{1}{n}\left[\sigma_y^2 + \frac{\sigma_\vartheta^2}{\mu_\vartheta^2}(\sigma_y^2 + \mu_y^2)\right]$$

(iii) An unbiased estimator of $\text{Var}(\hat{\mu}_y)$ is

$$\hat{\text{Var}}(\hat{\mu}_y) = \frac{s_z^2}{n}$$

where $s_z^2 = \frac{1}{n-1}\sum_s (z_i - \bar{z})^2$.

3.5 Additive Model

Duffy and Waterton (1984), Poole (1974), and Warner (1971) proposed the additive model. In this model, the ith respondent selected in the sample provides the RR.

$$z_i = y_i + x_i \tag{3.5.1}$$

where y_i is the true value of the sensitive characteristic y and x_i is a random sample from a population with a known mean μ_x and known variance σ_x^2.

For this model the revised RR $r_i = z_i - \mu_x$ satisfies

$$E_R(r_i) = y_i,\ V_R(r_i) = \sigma_x^2 \text{ and } C_R(r_i, r_j) = 0 \text{ for } i \neq j \tag{3.5.2}$$

3.6 Distributive Model

Combining the additive and multiplicative models stated above, we can propose the distributive model. In the proposed model, each of the selected respondents in the sample s selects two random samples x_{1i} and x_{2i} independently from two populations x_1 and x_2 with known means μ_{x1}, μ_{x2} and known variances $\sigma_{x1}^2, \sigma_{x2}^2$, respectively. The respondents are asked to provide their RR as

$$z_i = x_{1i}y_i + x_{2i} \tag{3.6.1}$$

For the proposed model,

$$E_R(z_i) = \mu_{x1}y_i + \mu_{x2},\ V_R(z_i) = \sigma_{x1}^2 y_i^2 + \sigma_{x2}^2 \text{ and } C_R(z_i, z_j) = 0 \text{ for } i \neq j \in U \tag{3.6.2}$$

In this case, the revised response $r_i = \frac{z_i - \mu_{x2}}{\mu_{x1}}$, $\mu_{x1} \neq 0$ satisfies the model

$$E_R(r_i) = y_i,\ V_R(r_i) = \frac{\sigma_{x1}^2}{\mu_{x1}^2} y_i^2 + \frac{\sigma_{x2}^2}{\mu_{x1}^2} = \phi_i \text{ and } C_R(r_i, r_j) = 0 \text{ for } i \neq j \tag{3.6.3}$$

An unbiased estimator of ϕ_i is

$$\hat{\phi}_i = \frac{\sigma_{x1}^2 r_i^2 + \sigma_{x2}^2}{\mu_{x1}^2 + \sigma_{x1}^2} \tag{3.6.4}$$

Particular Cases

The model (3.6.1) reduces to the additive model when $x_{1i} = 1$ and the multiplicative model when $x_{2i} = 0$ for every $i \in U$. The model (3.6.1) reduces to Chaudhuri's (1987) model when x_1 takes the values $a_1, \ldots, a_A$ each with probability $1/A$ and x_2 takes the values $b_1, \ldots, b_B$ each with probability $1/B$. In this case $\mu_{x1} = \frac{\sum a_i}{A} = \bar{a}$, $\mu_{x2} = \frac{\sum b_i}{B} = \bar{b}$, $\sigma_{x1}^2 = \frac{\sum (a_i - \bar{a})^2}{A}$ and $\sigma_{x2}^2 = \frac{\sum (b_i - \bar{b})^2}{B}$.

3.7 Bar-Lev et al.'s Model

Bar-lev et al. (2004) extended the Eichhorn and Hayre (1983) model slightly. In this model, the respondent chosen from the sample provides the true response y_i with probability p and a scrambled response $y_i x_i$ with probability $1 - p$, where x_i is a random sample from a distribution with known mean μ_x and variance σ_x^2, such as normal or chi-square.

Let z_i be RR obtained from the ith respondent, then

$$z_i = \begin{cases} y_i, \text{ with probability } p \\ y_i x_i, \text{ with probability } 1 - p \end{cases} \tag{3.7.1}$$

For this model,

$$E_R(z_i) = y_i p + y_i \mu_x (1 - p) \text{ and } E_R(z_i^2) = y_i^2 p + y_i^2 (\sigma_x^2 + \mu_x^2)(1 - p) \tag{3.7.2}$$

Equation (3.7.2) yields the revise RR response

$$r_i = \frac{z_i}{p + (1 - p)\mu_x} \tag{3.7.3}$$

For the Bar-lev et al.'s model

$$E_R(r_i) = y_i,\ V_R(r_i) = \frac{y_i^2 (1 - p)\left[p(1 - \mu_x)^2 + \sigma_x^2\right]}{\{p + (1 - p)\mu_x\}^2} = \phi_i \text{ and}$$
$$C_R(r_i, r_j) = 0 \text{ for } i \neq j \tag{3.7.4}$$

Bar-lev et al. (2004) didn't provide any unbiased estimator of ϕ_i. However, an unbiased estimator of ϕ_i is given as

$$\hat{\phi}_i = \frac{(1 - p)\left[p(1 - \mu_x)^2 + \sigma_x^2\right]}{(1 - p)\left[p(1 - \mu_x)^2 + \sigma_x^2\right] + \{p + (1 - p)\mu_x\}^2} r_i^2 \tag{3.7.5}$$

3.8 Gjestavang and Singh's Model

Gjestavang and Singh (2009) modified Bar-lev et al.'s (2004) model further. In their proposed model, using a suitable RR device, respondents were asked to report the true value of the sensitive characteristic y with probability p_1, scrambled response yx with probability p_2 as in Bar-lev et al.'s (2004) model, and a fixed value, say, k with probability $p_3 (= 1 - p_1 - p_2)$. Let z_i be the RR obtained from the ith response. Then,

$$z_i = \begin{cases} y_i, \text{ with probability } p_1 \\ y_i x_i, \text{ with probability } p_2 \\ k, \text{ with probability } p_3 \end{cases} \tag{3.8.1}$$

In this model,

$$E_R(z_i) = p_1 y_i + y_i \mu_x p_2 + k p_3$$

and $V_R(z_i) = y_i^2\left[p_1 + (\sigma_x^2 + \mu_x^2)p_2 - (p_1 + \mu_x p_2)^2\right] - 2kp_3(p_1 + \mu_x p_2)y_i + k^2 p_3(1 - p_3)$

The revised RR

$$r_i = \frac{z_i - kp_3}{p_1 + p_2\mu_x}$$

satisfies

$$E_R(r_i) = y_i,\ V_R(r_i) = \alpha y_i^2 + \beta y_i + \gamma = \phi_i \text{ and } C_R(r_i, r_j) = 0 \text{ for } i \neq j$$

where

$$\alpha = \frac{p_1 + (\sigma_x^2 + \mu_x^2)p_2 - (p_1 + p_2\mu_x)^2}{(p_1 + p_2\mu_x)^2},\ \beta = -\frac{2kp_3}{(p_1 + p_2\mu_x)} \text{ and } \gamma = \frac{k^2 p_3(1 - p_3)}{(p_1 + p_2\mu_x)^2} \tag{3.8.2}$$

An unbiased estimator of ϕ_i is

$$\hat{\phi}_i = \frac{\alpha r_i^2 + \beta r_i + \gamma}{1 + \alpha} \tag{3.8.3}$$

Remark 3.8.1 The main demerit of the Gjestavang and Singh's model is that the respondents' privacy is not protected, as the interviewer can easily identify that the response "k" is a forced response but not a genuine one.

3.9 Arnab's Model (A General Model)

Arnab (1990) proposed a very general RR model that covers almost all the RR models, irrespective of whether they are qualitative or quantitative, that are available in the literature. In this model, the randomized response z_i obtained from the ith respondent is transformed to $r_i = g(z_i)$, a suitable function of z_i so that the transformed RR r_i satisfies the model

$$E_R(r_i) = y_i,\ V_R(r_i) = \phi_i \text{ and } C_R(r_i, r_j) = 0 \text{ for } i \neq j \tag{3.9.1}$$

where $\phi_i = \phi_i(y_i)$ is a function of y_i only and the responses from the ith and jth respondents are independent.

Unbiased estimation of ϕ_i.

Theorem 3.9.1 In the case $\phi_i = \alpha y_i^2 + \beta y_i + \gamma$, an unbiased estimator ϕ_i is

$$\hat{\phi}_i = \frac{1}{\alpha+1}\{\alpha r_i^2 + \beta r_i + \gamma\}$$

Proof

$$\begin{aligned}E_R(\hat{\phi}_i) &= \frac{1}{\alpha+1}\{\alpha(y_i^2 + \phi_i) + \beta y_i + \gamma\}\\ &= \frac{1}{\alpha+1}\{\alpha y_i^2 + \beta y_i + \gamma + \alpha\phi_i\}\\ &= \frac{\phi_i + \alpha\phi_i}{\alpha+1}\\ &= \phi_i\end{aligned}$$

3.9.1 Estimation of the Population Mean

For a SRSWR sample of size n, an unbiased estimator of the mean $\overline{Y} = \frac{1}{N}\sum_{i\in U} y_i$ is

$$\overline{r} = \frac{1}{n}\sum_{i\in s} r_i \tag{3.9.2}$$

where $\sum_{i\in s}$ denotes sum of the units in s with repetition.

The model (3.9.1) yields

$$\begin{aligned}E(\overline{r}) &= E_p\left[E_R\left(\frac{1}{n}\sum_{i\in s} r_i\right)\right]\\ &= E_p\left(\frac{1}{n}\sum_{i\in s} y_i\right)\\ &= \overline{Y}\end{aligned} \tag{3.9.3}$$

The variance of $\overline{r}$ is

$$\begin{aligned}\text{Var}(\overline{r}) &= V_p\left[E_R\left(\frac{1}{n}\sum_{i\in s} r_i\right)\right] + E_p\left[V_R\left(\frac{1}{n}\sum_{i\in s} r_i\right)\right]\\ &= V_p\left(\frac{1}{n}\sum_{i\in s} y_i\right) + E_p\left(\frac{1}{n^2}\sum_{i\in s} \phi_i\right)\end{aligned}$$

$$= \frac{1}{n}(\sigma_y^2 + \overline{\phi}) \tag{3.9.4}$$

where $\overline{\phi} = \frac{1}{N}\sum_{i\in U}\phi_i$.

Further, noting r_i s independently distributed random variables, an unbiased estimator of $\text{Var}(\overline{r})$ is

$$\hat{\text{Var}}(\overline{r}) = \frac{1}{n(n-1)}\sum_{i\in s}(r_i - \overline{r})^2 \tag{3.9.5}$$

The above findings are summarised as follows:

Theorem 3.9.2

(i) $\overline{r} = \frac{1}{n}\sum_{i\in s} r_i$ is an unbiased estimator of $\overline{Y}$
(ii) $\text{Var}(\overline{r}) = \frac{1}{n}(\sigma_y^2 + \overline{\phi})$
(iii) An unbiased estimator of $\text{Var}(\overline{r})$ is

$$\hat{\text{Var}}(\overline{r}) = \frac{1}{n(n-1)}\sum_{i\in s}(r_i - \overline{r})^2$$

Now we will describe a few examples of RR models that can be treated as special cases of the model (3.9.1).

3.9.2 Qualitative Variables

For the qualitative characteristic y, we define $y_i = 1$, if the ith individual in the population possesses the attribute A and otherwise $y_i = 0$. We will denote $z_i = 1(0)$ if ith individual answers Yes (No). For the qualitative characteristics y, one finds

$$\begin{aligned} E_R\{r_i(r_i-1)\} &= V_R(r_i) + \{E_R(r_i)\}^2 - E_R(r_i) \\ &= \phi_i + y_i^2 - y_i \\ &= \phi_i (\text{as } y_i^2 = y_i) \end{aligned}$$

Hence, if ϕ_i is unknown, one may use the following simpler alternative formula for the unbiased estimator of ϕ_i as

$$\hat{\phi}_i = r_i(r_i - 1) \tag{3.9.6}$$

3.9.3 Special Cases

3.9.3.1 Warner's (1965) Model

For this model $P(z_i = 1) = \begin{cases} P, \; if \; i \in A \\ 1 - P, \; if \; i \notin A \end{cases}$

i.e. $P(z_i = 1) = Py_i + (1 - P)(1 - y_i)$.

Hence,

$$E_R(z_i) = Py_i + (1 - P)(1 - y_i) = (2P - 1)y_i + (1 - P) = E_R(z_i^2) \quad (3.9.7)$$

From Eq. (3.9.7), one finds $r_i = \frac{z_i-(1-P)}{2P-1}$ that satisfies

$$E_R(r_i) = y_i, \; V_R(r_i) = \frac{P(1-P)}{(2P-1)^2} = \phi_i$$

Here ϕ_i is known, so, one may use either $\hat{\phi}_i = \frac{P(1-P)}{(2P-1)^2}$ or $r_i(1 - r_i)$ as an unbiased estimator of ϕ_i.

Greenberg et al. Unrelated Question Model

For the unrelated question model (Described in 1.3.) with known π_x

$$y_i = \begin{cases} 1, \; if \; i \in A \\ 0, \; if \; i \notin A \end{cases}, \; x_i = \begin{cases} 1, \; if \; i \in B \\ 0, \; if \; i \notin B \end{cases}$$

and

$$P(z_i = 1) = P_1 y_i + (1 - P_1)x_i = E_R(z_i) = E_R(z_i^2) \quad (3.9.8)$$

In this case $r_i = \frac{z_i-(1-P_1)x_i}{P_1}$ yields $E_R(r_i) = y_i$ and $V_R(r_i) = \frac{(1-P_1)(y_i-x_i)^2}{P_1} = \phi_i$. Here ϕ_i is unknown as y_i and x_i are unknown. So, one may choose $\hat{\phi}_i = r_i(r_i - 1)$.

3.9.3.2 Mangat and Singh's Model

For the Mangat and Singh's (1990) model described in Sect. 1.6,

$$P(z_i = 1) = Ty_i + (1 - T)\{Py_i + (1 - P)(1 - y_i)\}$$

$$\text{i.e. } E(z_i) = E(z_i^2) = P* + (1 - P*)(1 - y_i) \text{ where } P* = T + (1 - T)P \quad (3.9.9)$$

Equation (3.9.9), yields $r_i = \frac{z_i-(1-P*)}{2P*-1}$ that satisfies

$$E_R(r_i) = y_i,\ V_R(r_i) = \frac{P*(1-P*)}{(2P*-1)^2} = \phi_i(\text{known}) = \hat{\phi}_i$$

3.9.3.3 Mangat's a Model

For the Mangat's model A (1994) described in Sect. 1.7

$$E(z_i) = E(z_i^2) = y_i + (1-P)(1-y_i)$$

In this case

$$r_i = \frac{z_i-(1-P)}{P},\ E_R(r_i) = y_i \text{ and } V_R(r_i) = \frac{(1-P)(1-y_i)}{P} = \phi_i \tag{3.9.10}$$

Here ϕ_i is unknown as it involves y_i. So, we may take $\hat{\phi}_i = r_i(r_i-1)$.

3.9.3.4 Brouch's Model

For the Boruch (1972) model cited in Sect. 1.5, we have $E(z_i) = E(z_i^2) = y_i P + \alpha$. In this case $r_i = \frac{z_i-\alpha}{P}$ yields

$$E_R(r_i) = y_i \text{ and } V_R(r_i) = \frac{\{P(1-P)-2P\alpha\}y_i + \alpha(1-\alpha)}{P^2} = \phi_i \tag{3.9.11}$$

Here also ϕ_i is unknown, so we take $\hat{\phi}_i = r_i(r_i-1)$.

3.9.4 *Quantitative Variables*

3.9.4.1 Greenberg et al.'s (1971) Quantitative Model

Let y_i and x_i be the values of the sensitive and non-sensitive characteristics y and x of the ith respondent. Denoting the RR obtained from the ith respondent by z_i, one finds

$$E_R(z_i) = Py_i + (1-P)x_i \text{ and } E_R(z_i^2) = Py_i^2 + (1-P)x_i^2 \tag{3.9.12}$$

In this case $r_i = \frac{z_i-(1-P)x_i}{P}$ yields

$$E_R(r_i) = y_i,\ V_R(r_i) = \frac{(1-P)(y_i - x_i)^2}{P} = \phi_i,\ \text{and}\ \hat{\phi}_i = (1-P)(r_i - x_i)^2 \tag{3.9.13}$$

3.9.5 A Tabular Presentation

A tabular representation of RR models that may be regarded as a particular case of Arnab's (1990) model is given below Table 3.1.

3.10 Godambe's Model

In Godambe's (1980) RR model, the ith respondent reports the true response y_i with probability, c and W_i with probability $1-c$ where W_i is the value of the non-sensitive auxiliary variable w of the ith respondent. The individual values of the auxiliary variable w are unknown to the investigator, but the total $\sum_{i\in U} W_i = W$ is known. In this case, the RR from the ith respondent is

$$z_i = \begin{cases} y_i, \text{ with probability } c \\ W_i, \text{ with probability } 1-c \end{cases} \tag{3.10.1}$$

Equation (3.10.1) yields

$$E_R(z_i) = cy_i + (1-c)W_i$$

and $V_R(z_i) = cy_i^2 + (1-c)W_i^2 - \{cy_i + (1-c)W_i\}^2 = c(1-c)(y_i - W_i)^2 = \sigma_{zi}^2$

Since the values of W_i s are unknown, Godambe's model (3.10.1) does not yield a revised RR r_i that satisfies Arnab's model (3.9.1). However, an unbiased estimator of the population mean $\overline{Y}$ based on an SRSWR sample of size n can be obtained as

$$\hat{\overline{Y}}_G = \frac{\overline{z} - (1-c)\overline{W}}{c} \tag{3.10.2}$$

where $\overline{z} = \frac{1}{n}\sum_{i\in s} z_i$ and $\overline{W} = W/N$.

The variance of $\hat{\overline{Y}}_G$ is given by

$$\begin{aligned}\text{Var}(\hat{\overline{Y}}_G) &= V_p\left[E_R(\hat{\overline{Y}}_G)\right] + E_p\left[V_R(\hat{\overline{Y}}_G)\right] \\ &= V_p\left(\frac{1}{n}\sum_{i\in s} y_i\right) + c(1-c)E_p\left(\frac{1}{n^2}\sum_{i\in s}\sigma_{zi}^2\right)\end{aligned}$$

Table 3.1 RR models that are particular case of Arnab's model

RR models	Randomized response	r_i	ϕ_i	$\hat{\phi}_i$
Qualitative variable: $y_i = 1(0)$ if $i \in A(i \notin A)$				
Warner (1965)	$z_i = \begin{cases} y_i & \text{with probability } P \\ 1-y_i & \text{wih probability } 1-P \end{cases}$	$\frac{z_i-(1-P)}{2P-1}$	$\frac{P(1-P)}{(2P-1)^2}$	$\frac{P(1-P)}{(2P-1)^2}$
Greenberg et al. (1969): U- model	$z_i = \begin{cases} y_i & \text{with probability } P_1 \\ x_i & \text{with probability } 1-P_1 \end{cases}$	$\frac{z_i-(1-P_1)x_i}{P_1}$	$\frac{(1-P_1)(y_i-x_i)^2}{P_1}$	$r_i(r_i-1)$
Mangat and Singh (1990)	$z_i = \begin{cases} y_i & \text{with probability } T \\ z_i^* & \text{with probability } 1-T \end{cases}$ where $z_i^* =$ $\begin{cases} y_i & \text{with probability } P \\ 1-y_i & \text{with probability } 1-P \end{cases}$	$\frac{z_i-(1-P*)}{2P*-1}$; where $P* = T+(1-T)P$	$\frac{P*(1-P*)}{(2P*-1)^2}$	$\frac{P*(1-P*)}{(2P*-1)^2}$
Mangat (1994)	$z_i = y_i + (1-y_i)z_i^*$ where $z_i^* =$ $\begin{cases} y_i & \text{with probability } P \\ 1-y_i & \text{with probability } 1-P \end{cases}$	$\frac{z_i-(1-P)}{P}$	$\frac{(1-P)(1-y_i)}{P}$	$\frac{(1-P)(1-r_i)}{P}$

(continued)

Table 3.1 (continued)

RR models	Randomized response	r_i	ϕ_i	$\hat{\phi}_i$
Boruch (1972)	$z_i = \begin{cases} y_i \text{ with probability } P \\ 1 \text{ with probability } \alpha \\ 0 \text{ with probability } 1-P-\alpha \end{cases}$	$\frac{z_i-\alpha}{P}$	$[\{P(1-P)-2\alpha P\}y_i$ $+\alpha(1-\alpha)]/P^2$	$[\{P(1-P)-2\alpha P\}y_i$ $+\alpha(1-\alpha)]/P^2$
Quantitative variable				
Greenberg et al. (1971)	$z_i = \begin{cases} y_i \text{ with probability } P \\ x_i \text{ with probability } (1-P) \end{cases}$	$\frac{z_i-(1-P)x_i}{P}$	$\frac{(1-P)(y_i-x_i)^2}{P}$	$(1-P)(r_i-x_i)^2$
Eriksson (1973)	$z_i = \begin{cases} y_i \text{ with probability } c \\ Q_j \text{ with probability } q_j \end{cases}$	$\frac{z_i-(1-c)\mu_Q}{c}$	$\frac{(1-c)}{c}\left[(y_i-\mu_Q)^2+\frac{\sigma_Q^2}{c}\right]$	$(1-c)\left[(r_i-\mu_Q)^2+\frac{\sigma_Q^2}{c}\right]$
Additive model	$z_i = y_i + x_i$	$z_i - x_i$	σ_x^2	σ_x^2
Multiplicative model)	$z_i = y_i\vartheta$	z_i/μ_ϑ	$y_i^2\sigma_\vartheta^2/\mu_\vartheta^2$	$r_i^2\sigma_\vartheta^2/(\sigma_\vartheta^2+\mu_\vartheta^2)$
Distributive model	$z_i = x_{1i}y_i + x_{2i}$	$\frac{z_i-\mu_{x2}}{\mu_{x1}}$	$\frac{\sigma_{x1}^2 y_i^2+\sigma_{x2}^2}{\mu_{x1}^2}$	$\frac{\sigma_{x1}^2 r_i^2+\sigma_{x2}^2}{\mu_{x1}^2+\sigma_{x1}^2}$
Bar-lev et al. (2004)	$z_i = \begin{cases} y_i \text{ with probability } p \\ y_i x_i \text{ with probability 1 -} p \end{cases}$	$\frac{z_i}{p+(1-p)\mu_x}$	$\frac{y_i^2(1-p)\left\{\begin{matrix} p(1-\mu_x)^2 \\ +\sigma_x^2 \end{matrix}\right\}}{\{p+(1-p)\mu_x\}^2}$	$\frac{(1-p)[p(1-\mu_x)^2+\sigma_x^2]r_i^2}{(1-p)[p(1-\mu_x)^2+\sigma_x^2]+\{p+(1-p)\mu_x\}^2}$
Gjestavang and Singh (2009)	$z_i = \begin{cases} y_i \text{ with probability } p_1 \\ y_i x_i \text{ with probability } p_2 \\ k \text{ with probability } p_3 \end{cases}$	$\frac{z_i-kp_3}{p_1+p_2\mu_x}$	$\alpha y_i^2+\beta y_i+\gamma$ where α, β and γ are given in (3.8.2)	$\frac{\alpha r_i^2+\beta r_i+\gamma}{1+\alpha}$

$$= \frac{\sigma_y^2}{n} + \frac{c(1-c)}{n}\overline{\sigma}_z^2$$

where $\sigma_y^2 = \frac{1}{N}\sum_{i\in U}(y_i - \overline{Y})^2$ and $\overline{\sigma}_z^2 = \frac{1}{N}\sum_{i\in U}\sigma_{zi}^2$.

An unbiased estimator of $\mathrm{Var}(\hat{\overline{Y}}_G)$ is

$$\hat{\mathrm{Var}}(\hat{\overline{Y}}_G) = \frac{1}{c^2}s_z^2$$

where $s_z^2 = \frac{1}{n-1}\sum_{i\in s}(z_i - \overline{z})^2$.

3.11 Discussions

The randomized response technique for the quantitative characteristic was proposed by Greenberg et al. (1971). It was extended by Eriksson (1973). Duffy and Waterton (1984), Poole (1974), and Warner (1971) proposed the additive model, while Eichhorn and Hayre (1983), and Pollock and Beck (1976) proposed the multiplicative model. Combining the additive model and the multiplicative model, the distributive model is proposed in this section. Chaudhuri's (1987) model is a particular case of the distributive model. Gjestavang and Singh's (2009) model does not protect the privacy of the respondents, as the interviewer can easily identify that the response "k" from a respondent is not a true response but a forced response. Most of the qualitative and quantitative RR models may be regarded as special cases of the general model proposed by Arnab (1990). However, Godambe's (1980) model does not belong to Arnab's model. From an efficiency point of view, none of the proposed models is the best for all situations in the sense of having the lowest variance. Hence, researchers compared efficiencies numerically. The appropriate choice of the model depends on the situations, costs, and administration of techniques in a real survey. For example, in the additive model, respondents need to do addition; in the multiplication model, respondents need to do multiplication; and in the distributive model, respondents need to do both multiplication and addition. As a result, the respondents may be prone to providing the wrong responses, making the model inefficient.

References

Arnab, R. (1990). On commutativity of design and model expectations in randomized response surveys. *Communications in Statistics e Theory and Methods, 19*, 3751–3757.

Arnab, R. (2017). *Survey sampling theory and applications*. Academic press.

Bar-Lev, S. K., Bobovitch, E., & Boukai, B. (2004). A note on randomized response model for quantitative data. *Metrika, 60*, 255–260.

Boruch, R. F. (1972). Relations among statistical methods for assuring confidentiality of social research. *Communications in Statistics e Theory and Methods, 19*, 3751–3757.

Bouza, C. N. (2009). Ranked set sampling and randomized response procedures for estimating the mean of a sensitive quantitative character. *Metrika, 60*, 255–260.

Chaudhuri, A. (1987). Randomized response surveys of finite populations: A unified approach with quantitative data. *Journal of the Statistical Planning and Inference, 113*, 25–34.

Cobo, B., Rueda, M., & Lopez-Torrecillas (2016). Application of randomized response techniques for investigating cannabis use by Spanish university students. *International Journal of Methods in Psychiatric Research, 26*, e1517.

Conteh, A., Gavin, M. C., & Solomon, J. (2015). Quantifying illegal hunting: A novel application of the quantitative randomised response technique. *Journal of Statistical Planning and Inference, 189*, 16–23.

Duffy, C. J., & Waterton, J. J. (1984). RR models for estimating the distribution function of a quantitative character. *International Statistical Review, 52*, 165–171.

Eichhorn, B. H., & Hayre, L. S. (1983). Scrambled randomized response methods for obtaining sensitive quantitative data. *Journal of Statistical Planning and Inference, 7*, 306–316.

Eriksson, S. A. (1973). A new model for randomized response. *International Statistical Review, 48*, 29–32.

Gjestavang, C. R., & Singh, S. (2009). An improved randomized response model: Estimation of mean. *Journal of Applied Statistics, 36*, 1361–1367.

Greenberg, B. D., Abul-Ela, A. L. A., Simmons, W. R., & Horvitz, D. G. (1969). The U- question randomized response model: Theoretical framework. *Journal of the American Statistical Association, 64*, 520–539.

Godambe, V. P. (1980). Estimation in RR trials. *International Statistical Review, 48*, 29–32.

Greenberg, B. G., Kuebler, R. R., Abernathy, J. R., & Horvitz, D. G. (1971). Application of the randomized 21 response technique in obtaining quantitative data. *Journal of the American Statistical Association, 66*, 243–250.

Holbrook, A. L., & Krosnick, J. A. (2010). Social desirability bias in voter turnout reports: Tests using the item count technique. *Public Opinion Quarterly, 74*(1), 37–67.

Mangat, N.S. (1994). An improved Randomized response strategy. *Journal of the Royal Statistical Society, Series B, 56*, 93–95.

Mangat, N. S., & Singh, R. (1990). An alternative randomized response procedure. *Biometrika, 77*, 349–442.

Pollock, K. H., & Bek, Y. (1976). A comparison of three RR models for quantitative data. *Journal of the American Statistical Association, 71*, 884–886.

Poole, W. K. (1974). Estimation of the distribution function for a continuous type random variable through RR. *Journal of the American Statistical Association, 69*, 1002–1005.

Warner, S. L. (1965). Randomized response: A survey technique for eliminating evasive answer bias. *Journal of the American Statistical Association, 60*, 63–69.

Warner, S. L. (1971). The linear randomized response model. *Journal of the American Statistical Association, 366*, 884–888.

Chapter 4
Unified Sampling Theory

Abstract In this chapter, the basic terminologies, definitions, concepts, and theories of sampling from a finite population are presented. Various sampling strategies based on equal and unequal probability sampling designs are also presented. The inferential aspects of sampling, such as design-based, model-based, and model-design-unbiased estimators, are discussed. Optimal sampling strategies under various superpopulation models are identified.

Keywords Design-based inference · Inference from finite population · Model-based inference · Model-design-based inference · Optimal sampling strategies · Sampling designs · Sampling strategies · Superpopulation model · Unbiased estimators

4.1 Terminologies and Definitions

4.1.1 Population, Unit, and Sampling Frame

A population is the totality or aggregate of objects or individuals in a given region at a particular point of time. Each element of the population is called a unit. Thus, a population is a collection of units. For example, a population of households in a certain city refers to the collection of all the households in that city, and each household is called a unit. The population size is the total number of units in a population, and it will be denoted by N. In our present discussion, we will consider a finite population, where N is a finite number and each unit of the population is distinct and identifiable. We further assume that the sampling frame, i.e., a complete list of all the units of the population with proper identification, is available without any error, i.e., no omission or duplication of a unit in the list. We will define a finite population as

$$U = (1, \ldots, i, \ldots, N) \tag{4.1.1}$$

R. Arnab, *Indirect Methods of Data Collection and Analysis from Surveys*, Indian Statistical Institute Series, https://doi.org/10.1007/978-981-97-6005-3_4

4.1.2 Parameter and Parameter Space

From the population, we collect information on various items. For example, from a population of households, one may be interested in collecting information relating to household size, gender, household income, household expenditure, etc. Each item of information is called a variable or characteristic of the population. We will denote such a variable by y and y_i will denote the value of the variable under study for the ith unit. For example, expenditure on education of the i th household of a population will be denoted by y_i. The population parameter for the characteristic y will be denoted by the vector:

$$\boldsymbol{y} = (y_1, \ldots, y_i, \ldots, y_N) \tag{4.1.2}$$

The set of all possible values of the vector $\boldsymbol{y}$ is known as the parametric space of the characteristic y and it will be denoted by

$$\Omega_y = \{-\infty < y_i < \infty \, \forall \, i \in U\} \tag{4.1.3}$$

In practice, we are not interested in knowing all the coordinates of the parameter vector $\boldsymbol{y}$ but we are often interested in various parametric functions of $\boldsymbol{y}$ such as $Y = \sum_{i \in U} y_i =$ the population total, $\overline{Y} = \sum_{i \in U} y_i / N =$ the population mean, the population variance $= S_y^2 = \sum_{i \in U} (y_i - \overline{Y})^2 / (N - 1)$, the population coefficient of variation (CV) $= S_y / \overline{Y}$, etc. We will denote a parametric function of $\boldsymbol{y}$ as $\theta = \theta(\boldsymbol{y})$.

4.1.3 Sample, Sample Size, and Effective Sample Size

Sample is an ordered sequence of elements from the population U. It will be denoted as

$$s = (i_1, i_2, \ldots, i_{n_s}) \tag{4.1.4}$$

where $i_k \in U$. The elements $i_1, i_2, \ldots, i_{n_s}$ may or may not be distinct. The total number of units n_s in s is called the sample size, while the total number of distinct units in s is called the effective sample size, which will be denoted by ν_s.

4.1.4 Sampling Design

In order to estimate the parametric function θ, we select a sample s from the population U with certain probability $p(s)$. Let S be the collection of all the possible samples s that can be selected from U. A sampling design p is a function defined on

S such that

$$0 \leq p(\mathrm{s}) \leq 1 \text{ and } \sum_{\mathrm{s}\in \mathrm{S}} p(\mathrm{s}) = 1 \quad (4.1.5)$$

A sampling design p is called a fixed sample size design (FS(n)) if each sample contains a fixed number of n units, i.e., $P(n_s = n) = 1$. A sampling design is called a fixed effective size ν (FES(ν)) design if ν_s, the number of distinct units in any sample s is fixed as ν, i.e., $P(\nu_s = \nu) = 1$. The expectation, variance, and covariance with respect to the sampling design p will be denoted by E_p, V_p, and C_p, respectively.

Example 4.1.1 Let $U = \{1, 2, 3, 4, 5\}$ be a finite population from which the samples $s_1 = (1, 2, 3)$, $s_2 = (2, 5)$, $s_3 = (1, 2, 4)$, and $s_4 = (5)$ are selected with respective probabilities $p(s_1) = 0.25$, $p(s_2) = 0.30$, $p(s_3) = 0.20$, and $p(s_4) = 0.25$. Here p is a sampling design since $0 \leq p(s) \leq 1$, $\sum_{s\in \mathrm{S}} p(s) = 1$, and $\mathrm{S} = \{s_1, s_2, s_3, s_4\}$.

4.1.5 Sampling Scheme

It is a method of selecting a sample from a population where units are selected one by one (repetition is allowed) from a population so that the probability of selecting a unit in a particular draw is predetermined. We will denote $p_i(k)$ as the probability of selection of ith unit at the kth draw. The probabilities $p_i(k)$ satisfy

$$\text{(i) } p_i(k) \geq 0,\ i = 1, \ldots, N;\ k = 1, \ldots, n_0, \text{ and (ii) } \sum_{i\in U} p_i(k) = \quad (4.1.6)$$

where n_0 is the number of draws.

For a without replacement (WOR) sampling scheme (design), all units are distinct, while for a with replacement (WR) sampling scheme, all units may not be distinct, i.e., some may be repeated more than once.

Hanurav (1966) established a correspondence between a sampling scheme and a sampling design. He showed that any sampling scheme results in a sampling design, and for a given sampling design, there exists at least one sampling scheme that implements the sampling design. Hence, we will treat the terms "sampling scheme" and "sampling design" as synonymous. For further details, readers are referred to Arnab (2017).

4.1.6 Inclusion Probabilities

Let $I_s(i) = \begin{cases} 1, \; if \; i \in s \\ 0, \; if \; i \notin s \end{cases}$ be an indicator variable. The inclusion probability of the ith unit is the probability that the unit i will be included in a sample and it is denoted by $\pi_i = \sum_{s \in S} I_{si} p(s) = \sum_{s \supset i} p(s)$. Similarly, the inclusion probability for the i and $j(\neq i)$th unit will be denoted by $\pi_{ij} = \sum_{s \in S} I_{si} I_{sj} p(s) = \sum_{s \supset i,j} p(s)$. The higher order inclusion probabilities π_{ijk} s are similarly defined.

In Example 4.1.1, $\pi_1 = p(s_1) + p(s_3) = 0.25 + 0.20 = 0.45$, $\pi_2 = p(s_1) + p(s_2) + p(s_3) = 0.25 + 0.30 + 0.20 = 0.75$, and $\pi_{12} = p(s_1) + p(s_3) = 0.25 + 0.20 = 0.45$.

4.1.7 Consistency Conditions of Inclusion Probabilities

$$\text{(i)} \sum_{i \subset U} \pi_i = E(\nu_s) = \nu, \;\; \text{and (ii)} \sum_{i \neq} \sum_{j \in U} \pi_{ij} = \nu_s(\nu_s - 1) + V(\nu_s) \tag{4.1.7}$$

For a FES(ν) design, the consistency conditions (4.1.7) reduce to

$$\text{(i)} \sum_{i \subset U} \pi_i = \nu, \;\; \text{(ii)} \sum_{j(\neq i) \in U} \pi_{ij} = (\nu - 1)\pi_i \; \text{and} \;\; \text{(iii)} \sum_{i \neq} \sum_{j \in U} \pi_{ij} = \nu(\nu - 1) \tag{4.1.8}$$

For detailed derivations of the consistency conditions, readers are referred to Cassel et al. (1977) and Arnab (2017).

4.2 Estimators and Their Properties

4.2.1 Data

For making any inference relating to the parameter of interest θ of a population, we draw a sample s from the population and collect information $(s, \mathbf{y}) = (y_i, i \in s)$ from all the units selected in the sample s. Such information is called data and it is denoted by $d = (i, y_i, i \in s)$.

4.2.2 Statistic

A statistic $e(s, \mathbf{y})$ is a function of data d such that for a given $(s, \mathbf{y})$ the value of $e(s, \mathbf{y})$ depends on only those y_i s that belong to s but it does not involve $y_j, j \notin s$.

4.2.3 Estimator, Unbiased Estimator, and Mean-Square Error

An estimator is a statistic that is used to estimate the parameter θ of interest. We will denote an estimator of θ from a sample s by $\hat{\theta} = \hat{\theta}(s, y)$. The estimator is a random variable, as its value depends on the selected sample s. The estimator $\hat{\theta}(s, y)$ is said to be unbiased of θ if

$$E\left[\hat{\theta}(s, y)\right] = E_p\left[\hat{\theta}(s, y)\right] = \sum_s \hat{\theta}(s, y)p(s) = \theta \; \forall \mathbf{y} \in \Omega \tag{4.2.1}$$

The difference $E\left[\hat{\theta}(s, y)\right] - \theta = B(\theta)$ is called the bias of the estimator $\hat{\theta}(s, y)$. The mean-square error of $\hat{\theta}(s, y)$ is denoted by

$$\begin{aligned} \text{MSE}\left[\hat{\theta}(s, y)\right] &= E\left[\hat{\theta}(s, y) - \theta\right]^2 \\ &= \sum_s \left[\hat{\theta}(s, y) - \theta\right]^2 p(s) \\ &= \text{Var}\{\hat{\theta}(s, y)\} + \left[B\{\hat{\theta}(s, y)\}\right]^2 \end{aligned} \tag{4.2.2}$$

4.2.4 Uniformly Minimum Variance Unbiased Estimator

Let $\hat{\theta}_1$ and $\hat{\theta}_2$ be two different unbiased estimators of θ. Then $\hat{\theta}_1$ is said to be better than $\hat{\theta}_2$ if

$$\begin{aligned} &\text{Var}(\hat{\theta}_1) \le \text{Var}(\hat{\theta}_2) \quad \forall \mathbf{y} \in \Omega \\ &\text{and} \\ &\text{Var}(\hat{\theta}_1) < \text{Var}(\hat{\theta}_2) \text{ at least one } \mathbf{y} \in \Omega \end{aligned} \tag{4.2.3}$$

The estimator $\hat{\theta}_0$ is said to be the uniformly minimum variance unbiased estimator (UMVUE) of θ in C_u, the class of unbiased estimators of θ if $\hat{\theta}_0$ is better than any other estimator in the class C_u, i.e., for any estimator $\hat{\theta} * (\neq \hat{\theta}_0) \in C_u$ the following hold:

$$\begin{aligned}&\text{Var}(\hat{\theta}_0) \le \text{Var}(\hat{\theta}*)\forall \boldsymbol{y} \in \Omega\\&\text{and}\\&\text{Var}(\hat{\theta}_0) < \text{Var}(\hat{\theta}*) \text{ at least one } \boldsymbol{y} \in \Omega\end{aligned} \quad (4.2.4)$$

4.2.5 Admissible Estimator

An estimator $\hat{\theta}_a$ is said to be admissible in the class of estimators C if there exists no other estimator better than $\hat{\theta}_a$ in the class C. An estimator that is not admissible is called an inadmissible estimator.

4.2.6 Linear Unbiased Estimators

In order to estimate a linear parametric function of $\boldsymbol{y}$ such as population total Y or population mean $\overline{Y}$, we often use a linear unbiased estimator of the form:

$$\tilde{t}(s, y) = w_s + \sum_{i \in s} w_{si} y_i \quad (4.2.5)$$

where w_s and w_{si} are known constants (weights) free from y-values, w_s may depend on the sample s while w_{si} may depend on both the s and units i that belong to s. $\sum_{i \in s}$ denotes the sum over the distinct units in s. The class of linear unbiased estimators will be denoted by L_u.

The estimator $\tilde{t}(s, y)$ will be unbiased for Y if and only if

$$E[\tilde{t}(s, y)] = \sum_s \left\{ w_s + \sum_{i \in s} w_{si} y_i \right\} p(s) = Y \quad \forall \boldsymbol{y} \in \Omega$$

i.e.,

$$\sum_s w_s p(s) = 0 \text{ and } \sum_{s \supset i} w_{si} p(s) = 1 \quad \forall i \in U \quad (4.2.6)$$

4.2.7 Linear Homogeneous Unbiased Estimators

A linear homogeneous estimator of the total Y is of the form:

$$t(s, y) = \sum_{i \in s} w_{si} y_i \tag{4.2.7}$$

The class of linear homogeneous unbiased estimators will be denoted by L_{hu}. The estimator $t(s, y) (\in L_{hu})$ will be unbiased for the total Y if and only if

$$\begin{aligned} E[t(s, y)] &= \sum_s t(s, y) p(s) \\ &= \sum_s \left(\sum_{i \in s} w_{si} y_i \right) p(s) = Y \; \forall \mathbf{y} \in \Omega \end{aligned}$$

i.e.,

$$\sum_{s \supset i} w_{si} p(s) = 1 \quad \forall i \in U \tag{4.2.8}$$

The variance of $t(s, y)$ is given by

$$\begin{aligned} \text{Var}[t(s, y)] =& E_p[t(s, y)]^2 - Y^2 \\ =& \sum_s \left\{ \sum_{i \in s} w_{si}^2 y_i^2 + \sum_{i \neq j} \sum_{\in s} w_{si} w_{sj} y_i y_j \right\} p(s) - Y^2 \\ =& \sum_{i \in U} y_i^2 \left(\sum_{s \supset i} w_{si}^2 p(s) \right) + \sum_{i \neq} \sum_{j \in U} y_i y_j \left(\sum_{s \supset i,j} w_{si} w_{sj} p(s) \right) - Y^2 \\ =& \sum_{i \in U} y_i^2 (\alpha_i - 1) + \sum_{i \neq} \sum_{j \in U} (\alpha_{ij} - 1) y_i y_j \end{aligned} \tag{4.2.9}$$

where

$$\alpha_i = \sum_{s \supset i} w_{si}^2 p(s) \text{ and } \alpha_{ij} = \sum_{s \supset i,j} w_{si} w_{sj} p(s) \tag{4.2.10}$$

To find an unbiased estimator of $\text{Var}[t(s, y)]$, consider the following homogeneous quadratic estimator of the form:

$$\hat{\text{Var}}[t(s, y)] = \sum_{i \in s} c_{si} y_i^2 + \sum_{i \neq j} \sum_{\in s} c_{sij} y_i y_j \tag{4.2.11}$$

where c_{si} and c_{sij} are known constants that depend on the data $d = (i, y_i, i \in s)$.

Now noting

$$
\begin{aligned}
E\{\hat{\text{Var}}[t(s, y)]\} &= \sum_{s} \left\{ \sum_{i \in s} c_{si} y_i^2 + \sum_{i \neq j} \sum_{\in s} c_{sij} y_i y_j \right\} p(s) \\
&= \sum_{i \in U} y_i^2 \left(\sum_{s \supset i} c_{si} p(s) \right) + \sum_{i \neq j} \sum_{\in U} y_i y_j \left(\sum_{s \supset ij} c_{sij} p(s) \right)
\end{aligned} \tag{4.2.12}
$$

The estimator $\hat{\text{Var}}[t(s, y)]$ is unbiased for $\text{Var}[t(s, y)]$ if and only if

$$
E\{\hat{\text{Var}}[t(s, y)]\} = \text{Var}[t(s, y)] \quad \forall \mathbf{y} \in \Omega
$$

i.e.,

$$
\text{(i)} \sum_{s \supset i} c_{si} p(s) = \alpha_i - 1 \text{ and (ii)} \sum_{s \supset ij} c_{sij} p(s) = \alpha_{ij} - 1 \tag{4.2.13}
$$

We summarize the above discussions in the following theorem:

Theorem 4.2.1

(i) The homogeneous linear estimator $t(s, y) = \sum_{i \in s} w_{si} y_i$ will be unbiased for Y if and only $\sum_{s \supset i} w_{si} p(s) = 1 \quad \forall i \in U$.

(ii) The variance of $t(s, y)$ is

$$
\text{Var}[t(s, y)] = \sum_{i \in U} y_i^2 (\alpha_i - 1) + \sum_{i \neq} \sum_{j \in U} (\alpha_{ij} - 1) y_i y_j
$$

where $\alpha_i = \sum_{s \supset i} w_{si}^2 p(s)$ and $\alpha_{ij} = \sum_{s \supset i, j} w_{si} w_{sj} p(s)$

(iii) A homogeneous quadratic unbiased estimator of $\text{Var}[t(s, y)]$ is

$$
\hat{\text{Var}}[t(s, y)] = \sum_{i \in s} c_{si} y_i^2 + \sum_{i \neq j} \sum_{\in s} c_{sij} y_i y_j
$$

where c_{si} and c_{sij} satisfy the conditions $\sum_{s \supset i} c_{si} p(s) = \alpha_i - 1$ and $\sum_{s \supset ij} c_{sij} p(s) = \alpha_{ij} - 1$.

One can choose c_{si} and c_{sij} satisfying conditions (4.2.13) in various ways. Two obvious choices are

(i) $c_{si} = (\alpha_i - 1)/\pi_i$, $c_{sij} = (\alpha_{ij} - 1)/\pi_{ij}$ and (ii) $c_{si} = (\alpha_i - 1)/\{M_i p(s)\}$ and $c_{sij} = (\alpha_{ij} - 1)/\{M_{ij} p(s)\}$ where $M_i = \sum_{s \supset i}$ and $M_{ij} = \sum_{s \supset i, j}$ are, respectively, the total number of samples that contain the unit i, and units i and $j (\neq i)$ with positive probabilities.

4.3 Nonexistence Theorems

It is easy to note that, for a given sampling design p, one can construct various unbiased estimators for Y in the classes L_h and L_{hu}. Hence, it is a natural question how to pick up the best estimator (UMVUE). Unlike sampling from an infinite population, sampling from a finite population, the UMVUE does not exist in general. We will state the following important results without providing proof relating to the non-existence of the best estimator. For detailed proof, readers are referred to Cassel et al. (1977) and Arnab (2017).

Theorem 4.3.1 Basu (1971): For a noncensus design p, there does not exist the UMVUE of $\theta = \theta(\mathbf{y})$ in the class of all unbiased estimators C_p.

Arnab (2017) showed the following result as a corollary of this theorem.

Theorem 4.3.2 For a noncensus design, p, the UMVUE of the population total, Y, does not exist in L_u, the class of linear unbiased estimators.

Theorem 4.3.3 Gadambe (1955) and Hanurav (1966): The UMVUE of the total Y for a noncensus and non-uni-cluster sampling design, p, with $\pi_i > 0$ for $\forall i = 1, .., N$ does not exist in the class of linear homogeneous unbiased estimators L_{hu}. However, the UMVUE does exist if p is a uni-cluster sampling design.

A sampling design p is called a uni-cluster design p_u, if samples s, $s*$ with $p(s) > 0, p(s*) > 0$ imply either (i) $s \cap s* = \phi$ (null set) or (ii) s and $s*$ are equivalent $(s \sim s*)$ in the sense that s and $s*$ contain the same set of distinct units.

4.4 Sampling Strategies

A sampling strategy is a combination of a sampling design p and an estimator t. A sampling strategy will be denoted by the pair $h = (p, t)$. A strategy h is called an unbiased strategy for estimating Y, if the associated estimator t is unbiased for Y. Now we will describe a few important strategies that are used in practice.

4.4.1 Arbitrary Sampling Design

The estimator $t(s, y)$ defined in (4.2.7) reduces to the Horvitz-Thompson estimator (HTE, 1952)

$$\hat{Y}_{\text{HT}} = \sum_{i \in s} \frac{y_i}{\pi_i} \tag{4.4.1}$$

if $w_{si} = 1/\pi_i$ and $\pi_i > 0\ \forall i \in U$.

Some of the important properties of $\hat{Y}_{\text{HT}}$ are stated as follows.

Theorem 4.4.1

(i) $\hat{Y}_{\text{HT}}$ is unbiased for Y, i.e.,

$$E(\hat{Y}_{\text{HT}}) = Y \tag{4.4.2}$$

(ii) The variance of $\hat{Y}_{\text{HT}}$ is

$$V_{\text{HT}} = \text{Var}(\hat{Y}_{\text{HT}}) = \sum_{i\in U} \frac{1-\pi_i}{\pi_i} y_i^2 + \sum_{i\neq j}\sum_{\in U} \frac{\pi_{ij}-\pi_i\pi_j}{\pi_i\pi_j} y_i y_j \tag{4.4.3}$$

(iii) An unbiased estimator of V_{HT} is

$$\hat{V}_{\text{HT}} = \hat{\text{Var}}(\hat{Y}_{\text{HT}}) = \sum_{i\in s} \frac{1-\pi_i}{\pi_i^2} y_i^2 + \sum_{i\neq j}\sum_{\in s} \frac{\pi_{ij}-\pi_i\pi_j}{\pi_i\pi_j\pi_{ij}} y_i y_j \tag{4.4.4}$$

For a fixed effective size sampling design (FES), V_{HT} can be expressed as

$$V_{\text{HT}} = V_{YG} = \frac{1}{2}\sum_{i\neq j}\sum_{\in U} (\pi_i\pi_j - \pi_{ij})\left(\frac{y_i}{\pi_i} - \frac{y_j}{\pi_j}\right)^2 \tag{4.4.5}$$

The expression (4.4.5) yields an unbiased estimator of V_{HT} as

$$\hat{V}_{\text{YG}} = \frac{1}{2}\sum_{i\neq j}\sum_{\in U} \frac{\pi_i\pi_j - \pi_{ij}}{\pi_{ij}}\left(\frac{y_i}{\pi_i} - \frac{y_j}{\pi_j}\right)^2 ; \pi_{ij} > 0 \tag{4.4.6}$$

The estimator $\hat{V}_{\text{YG}}$ is known as Yates & Grundy's (1953) variance estimator. $\hat{V}_{\text{YG}}$ is unbiased estimator of V_{HT} for a fixed effective size sampling design only, while $\hat{V}_{\text{HT}}$ is unbiased of V_{HT} for any sampling design. Both the estimators can take negative values in some situations. The sufficient condition for nonnegativity of $\hat{V}_{\text{YG}}$ is known as $\pi_i\pi_j - \pi_{ij} \geq 0$ while no sufficient condition for nonnegativity of $\hat{V}_{\text{HT}}$ is known.

4.4.2 Simple Random Sampling Without Replacement (SRSWOR) Design

The method of selection of a sample s of n units from a population U of N units by the SRSWOR is as follows: On the first draw, a unit is selected from the entire population U of N units at random, i.e., with probability $1/N$. On the second draw, another unit is selected from those $N-1$ units that were not selected in the first draw with an equal probability $1/(N-1)$. In general, at the $k(=1,\ldots,n)$th draw one unit is selected from those $N - k + 1$ units that were not selected in the earlier $k - 1$ draws with probability $1/(N - k + 1)$. So, for the SRSWOR sampling, all the selected units in the sample s are distinct. The inclusion probabilities for the first two orders are

$$\pi_i = n/N \text{ and } \pi_{ij} = n(n-1)/\{N(N-1)\} \tag{4.4.7}$$

Substituting (4.4.7) in Theorem 4.4.1, the following results are obtained.

Theorem 4.4.2

(i) $\overline{y}_s = \sum_{i \in s} y_i/n$ is an unbiased estimator of $\overline{Y} = \frac{1}{N}\sum_{i \in U} y_i$.
(ii) The variance of $\overline{y}_s$ is

$$\text{Var}(\overline{y}_s) = \left(\frac{1}{n} - \frac{1}{N}\right) S_y^2$$

where $S_y^2 = \frac{1}{N-1}\sum_{i \in U}(y_i - \overline{Y})^2$.

(iii) An unbiased estimator of $\text{Var}(\overline{y}_s)$ is

$$\hat{\text{Var}}(\overline{y}_s) = \left(\frac{1}{n} - \frac{1}{N}\right) s_y^2$$

where $s_y^2 = \frac{1}{n-1}\sum_{i \in s}(y_i - \overline{y}_s)^2$

4.4.3 *Simple Random Sampling with Replacement (SRSWR) Design*

In the SRSWR method, n units are selected independently from the population U so that the probability of selection of a unit in each of the n draws is equal to $1/N$, i.e., $p_i(k) = 1/N$ (defined in (4.1.6)). Thus, for the SRSWR method of sampling, a unit may be repeated more than once in a sample. Let $y_{(r)}$ be the y value of the unit that was selected at the rth draw, i.e., $y_{(r)} = y_j$ if the rth draw selects jth unit with probability $1/N$; $r = 1, \ldots, n$; $j = 1, \ldots, N$.

Theorem 4.4.3.

(i) $\overline{y}_{wr} = \frac{1}{n}\sum_{r=1}^{n} y_{(r)}$ is an unbiased estimator of $\overline{Y}$,
(ii) $\text{Var}(\overline{y}_{wr}) = \sigma_y^2/n$,

where $\sigma_y^2 = \frac{1}{N}\sum_{i \in U}(y_i - \overline{Y})^2 = \frac{N-1}{N}S_y^2$.

(iii) An unbiased estimator of $\text{Var}(\overline{y}_{wr})$ is

$$\hat{\text{Var}}(\overline{y}_{wr}) = \overline{s}_y^2/n$$

where $\overline{s}_y^2 = \frac{1}{n-1}\sum_{r=1}^{n}(y_{(r)} - \overline{y}_{wr})^2$.

Proof

(i) $E(\overline{y}_{wr}) = \frac{1}{n}\sum_{r=1}^{n} E(y_{(r)})$

$= \overline{Y}$ (since $E(y_{(r)}) = \overline{Y}$)

(ii) $\text{Var}(\overline{y}_{wr}) = \frac{1}{n^2}\sum_{r=1}^{n}\text{Var}(y_{(r)})$

$= \sigma_y^2/n$ (since $\text{Var}\{y_{(r)}\} = \sigma_y^2$ and the draws are independent).

(iii)
$$E\{\hat{\text{Var}}(\overline{y}_{wr})\} = \frac{1}{n(n-1)}E\left\{\sum_{r=1}^{n} y_{(r)}^2 - n\overline{y}_{wr}^2\right\}$$
$$= \frac{1}{n(n-1)}\left[\sum_{r=1}^{n} E(y_{(r)}^2) - nE(\overline{y}_{wr}^2)\right]$$
$$= \frac{1}{n(n-1)}\left\{\sum_{r=1}^{n}(\sigma_y^2 + \overline{Y}^2) - n\left(\overline{Y}^2 + \frac{\sigma_y^2}{n}\right)\right\}$$
$$= \sigma_y^2/n$$

4.4.3.1 Sample Mean Based on Distinct Units

In a SRSWR sampling scheme, all the units selected in the sample s may not be distinct. Some of the units may be repeated more than once. Basu (1958) pointed out that the unordered sample $s*$ based on the distinct units of s is a sufficient statistic. Let ν_{s*} be the number of distinct units in s and $\overline{y}_{s*} = \frac{1}{\nu_{s*}}\sum_{i\in s*} y_i$ be the sample mean of the distinct units. The following results derived by Basu (1958) are stated below without proof. The proof is available in Arnab (2017).

Theorem 4.4.4.

(i) $\overline{y}_{s*}$ is an unbiased estimator of $\overline{Y}$.
(ii) The variance of $\overline{y}_{s*}$ is

$$\text{Var}(\overline{y}_{s*}) = \left[E_p\left(\frac{1}{\nu_{s*}}\right) - \frac{1}{N}\right]S_y^2 < \text{Var}(\overline{y}_{wr}) = \sigma_y^2/n$$

(iii) An unbiased estimator of $\text{Var}(\overline{y}_{s*})$ is

$$\hat{\text{Var}}(\overline{y}_{s*}) = \left(\frac{1}{\nu_{s*}} - \frac{1}{N}\right)s^2_{\nu*}$$

where $s^2_{\nu*} = \frac{1}{\nu^*_s - 1}\sum_{i\in s*}(y_i - \overline{y}_{s*})^2$ is the sample variance of the distinct units in s.

The probability distribution of ν_{s*} was obtained by Feller (1957) as

$\text{Prob}(\nu_{s*} = j) = \frac{1}{N^n}\binom{N}{j}\sum_{r=0}^{j}(-1)^r\binom{j}{r}(j-r)^n; j = 1, .., min(n, N)$.

Pathak (1961) provided the exact expression of $E(1/\nu_{s*})$ and obtained the approximate expression of $\text{Var}(\overline{y}_{s*})$ as follows:

$$E\left(\frac{1}{\nu_{s*}}\right) = \frac{1^{n-1} + 2^{n-1} + \cdots + N^{n-1}}{N^n}$$

$$\begin{aligned}\text{Var}(\overline{y}_{s*}) &= \frac{1^{n-1} + 2^{n-1} + \cdots + (N-1)^{n-1}}{N^n}S^2_y \\ &\cong \left(\frac{1}{n} - \frac{1}{2N} + \frac{n-1}{12N^2}\right)S^2_y\end{aligned}$$

Remark 4.4.1 Theorem 4.4.4 shows that $\overline{y}_{s*}$ has a smaller variance than that of $\overline{y}_{wr}$ $\forall \mathbf{y} \in \Omega_y$. Hence, for the SRSWR sampling scheme, the sample mean $\overline{y}_{wr}$ based on all the units is inadmissible.

4.4.4 Probability Proportional to Size with Replacement (PPSWR) Sampling Design

Here we assume that each unit of the population is associated with an auxiliary (or measure of size) variable x. The variable x is assumed to be nonnegative and well related to the study variable y. Let x_i be the value of the size variable x for the $i(\in U)$th unit and $X = \sum_{i\in U} x_i$ be its total. The normed size measure for the ith unit is denoted by $p_i = x_i/X$, $p_i \geq 0$ and $\sum_{i\in U} p_i = 1$. In a PPSWR sampling of size n design, the units are selected independently one by one from the population U so that the probability of selection of the ith unit at the kth draw is $p_i(k) = p_i$; $i = 1, .., N$; $k = 1, .., n$. Let $y_{(r)}$ and $p_{(r)}$ be the value of the study variable y and the normed size measure of the unit that was selected at the $r(= 1, .., n)$th draw. Thus, $y_{(r)} = y_j$ and $p_{(r)} = p_j$ if the rth draw produces jth unit with probability p_j, $j(\in U)$. The Hansen & Hurwitz (1943) estimator for the population total Y is given by

$$\hat{Y}_{\text{HH}} = \frac{1}{n}\sum_{r=1}^{n}\frac{y_{(r)}}{p_{(r)}} \tag{4.4.8}$$

The estimator $\hat{Y}_{\text{HH}}$ has the following properties:

Theorem 4.4.5

(i) $\hat{Y}_{\text{HH}} = \frac{1}{n}\sum_{r=1}^{n}\frac{y_{(r)}}{p_{(r)}}$ with $p_{(r)} > 0$ is an unbiased estimator of the total Y.

(ii) The variance of $\hat{Y}_{\text{HH}}$ is

$$\text{Var}(\hat{Y}_{\text{HH}}) = \frac{1}{n}\sum_{i\in U}\left(\frac{y_i}{p_i} - Y\right)^2$$

(iii) An unbiased estimator of $\text{Var}(\hat{Y}_{\text{HH}})$ is

$$\hat{\text{Var}}(\hat{Y}_{\text{HH}}) = \frac{1}{n(n-1)}\sum_{r=1}^{n}\left(\frac{y_{(r)}}{p_{(r)}} - \hat{Y}_{\text{HH}}\right)^2$$

Proof Noting,

$$\begin{aligned}
E_p\left(\frac{y_{(r)}}{p_{(r)}}\right) &= \sum_{j\in U}\frac{y_j}{p_j}p_j = Y, \\
V_p\left(\frac{y_{(r)}}{p_{(r)}}\right) &= \sum_{j\in U}p_j\left(\frac{y_j}{p_j} - Y\right)^2 \text{ and} \\
C_p\left(\frac{y_{(r)}}{p_{(r)}}, \frac{y_{(k)}}{p_{(k)}}\right) &= 0 \text{ for } r \neq k
\end{aligned} \tag{4.4.9}$$

one can verify the theorem easily.

4.4.5 Probability Proportional to Size Without Replacement (PPSWOR) Sampling Design

For the PPSWOR sampling scheme, on the first draw, a unit $i_1 (\in U)$ is selected from the entire population U with probability proportional to its measure of size x_{i_1}, i.e., with probability $p_{i_1}(1) = x_{i_1}/X = p_{i_1}$. On the second draw, another unit $i_2 (\neq i_1)$ is selected with probability proportional to x_{i_2} from $U - i_1$, the set of units which was not selected in the first draw, i.e., the probability of selection $i_2 (\neq i_1)$ in the second draw given that the unit i_1 was selected in the first draw is $p_{i_2}(2) = x_{i_2}/(X - x_{i_1}) = p_{i_2}/(1 - p_{i_1})$. In general, at the kth draw a unit $i_k (\neq i_1, .., i_{k-1})$ is selected from the set of units $U - i_1 - .. - i_{k-1}$ that were not selected in the earlier $k - 1$ draws with probability proportional to x_{i_k}, i.e., with probability

$$\begin{aligned}
p_{i_k}(k) &= x_{i_k}/(X - x_{i_1} - .. - x_{i_{k-1}}) \\
&= p_{i_k}/(1 - p_{i_1} - \ldots - p_{i_{k-1}});
\end{aligned}$$

$$i_k = 1, 2, .., N, k = 1, .., n.$$

Hence, for the PPSWOR sampling, all the units in a sample are distinct, and the probability of selecting an ordered sample $\vec{s} = (i_1, \to i_2, \to, .., \to i_n)$ where the unit i_k is selected at the kth draw is

$$\begin{aligned} p(i_1, \to i_2, \to, .., \to i_n) &= p_{i_1}(1) \times p_{i_2}(2) \times ... \times p_{i_n}(n) \\ &= p_{i_1} \times \frac{p_{i_2}}{1 - p_{i_1}} \times ..... \times \frac{p_{i_n}}{1 - p_{i_1} - ... - p_{i_{n-1}}} \end{aligned} \quad (4.4.10)$$

4.4.5.1 Raj's (1956) Estimator

From the ordered sample $(i_1, \to i_2, \to, .., \to i_n)$, one finds unbiased estimators for the total Y based on the units selected in the draws $1, 2, .., r, .., n$ as follows:

$$\begin{aligned} t(1) &= \frac{y_{i_1}}{p_{i_1}(1)} = \frac{y_{i_1}}{p_{i_1}} \\ t(2) &= y_{i_1} + \frac{y_{i_2}}{p_{i_2}(2)} = y_{i_1} + \frac{y_{i_2}}{p_{i_2}} \times (1 - p_{i_1}) \\ &\vdots \\ t(r) &= y_{i_1} + ... + y_{i_{r-1}} + \frac{y_{i_r}}{p_{i_r}(r)} = y_{i_1} + ... + y_{i_{r-1}} + \frac{y_{i_r}}{p_{i_r}} \times (1 - p_{i_1} - ... - p_{i_{r-1}}) \\ &\vdots \\ t(n) &= y_{i_1} + ... + y_{i_{n-1}} + \frac{y_{i_n}}{p_{i_n}(n)} = y_{i_1} + ... + y_{i_{n-1}} + \frac{y_{i_r}}{p_{i_n}} \times (1 - p_{i_1} - ... - p_{i_{n-1}}) \end{aligned} \quad (4.4.11)$$

The properties of the ordered estimators (4.4.11) are given without derivation as follows:

Theorem 4.4.6

(i) $E\{t(r)\} = Y$.
(ii) $\text{Var}\{t(r)\} \leq \text{Var}\{t(r-1)\}$; $r = 2, .., n$.
(iii) $\text{Cov}\{t(i), t(j)\} = 0$ for $i \neq j \in U$.

From the ordered estimators (4.4.11), Raj (1956) proposed an unbiased estimator of Y as

$$\hat{Y}_{\text{RAJ}} = \frac{1}{n}\sum_{r=1}^{n} t(r) \quad (4.4.12)$$

The Raj estimator $\hat{Y}_{\text{RAJ}}$ possesses the following properties:

Theorem 4.4.7

(i) $E(\hat{Y}_{\text{RAJ}}) = Y$.

(ii) $\text{Var}(\hat{Y}_{\text{RAJ}}) \le \sum_{i\in U} p_i\left(\frac{y_i}{p_i} - Y\right)^2 = \text{Var}(\hat{Y}_{\text{HH}})$.

(iii) An unbiased estimator of $\text{Var}(\hat{Y}_{\text{RAJ}})$ is

$$\hat{\text{Var}}(\hat{Y}_{\text{RAJ}}) = \frac{1}{n(n-1)}\{t(r) - \hat{Y}_{\text{RAJ}}\}^2$$

Theorem 4.4.7 indicates that Raj's estimator $\hat{Y}_{\text{RAJ}}$ is more efficient than the Hansen-Hurwitz estimator $\hat{Y}_{\text{HH}}$. Detailed derivations of Theorems 4.4.5 and 4.4.7 are given by Arnab (2017).

4.4.5.2 Murthy's (1957) Estimator

Raj's estimator $\hat{Y}_{\text{RAJ}}$ depends on the order of selection of units in the sample. Raj's estimator based on the order sample $(i, \to j)$ where the unit i is selected in the first draw and the unit j is selected in the second draw is given by

$$\hat{Y}_{\text{RAJ}}(i,j) = \frac{1}{2}\left[\frac{y_i}{p_i}(1+p_i) + \frac{y_j}{p_j}(1-p_i)\right] \tag{4.4.13}$$

The estimator $\hat{Y}_{\text{RAJ}}(i,j)$ is different from the estimator

$$\hat{Y}_{\text{RAJ}}(j,i) = \frac{1}{2}\left[\frac{y_j}{p_j}(1+p_j) + \frac{y_i}{p_i}(1-p_j)\right] \tag{4.4.14}$$

based on the ordered sample $(j, \to i)$ where the unit j is selected in the first draw and the unit i is selected in the second draw.

Murthy (1957) symmetrized Raj's (1956) estimators and proposed the following alternative estimator:

$$\begin{aligned}\hat{Y}_{\text{MUR}} &= \frac{\hat{Y}_{\text{RAJ}}(i,j)\times p(i,j) + \hat{Y}_{\text{RAJ}}(j,i)\times p(j,i)}{p(i,j)+p(j,i)}\\ &= \frac{\frac{y_i}{p_i}(1-p_j) + \frac{y_j}{p_j}(1-p_j)}{2-p_i-p_j}\end{aligned} \tag{4.1.15}$$

The symmetrized Raj's estimator is known as Murthy's estimator $\hat{Y}_{\text{MUR}}$. Murthy's estimator $\hat{Y}_{\text{MUR}}$ is unbiased for the total Y and possesses lower variance than both Raj's estimators $\hat{Y}_{\text{RAJ}}(i,j)$ and $\hat{Y}_{\text{RAJ}}(j,i)$.

We denote $s = (j_1, j_2, .., j_n)$ with $j_1 < j_2, ... < j_n \in (i_1, i_2, \ldots, i_n)$ as the unordered sample derived from an ordered sample $(i_1, \to i_2, .., \to i_n)$ by arranging the labels $i_1, .., i_n$ in an ascending order. For example, the unordered sample $s = (2, 4, 6)$ can be derived from any of the following six ordered samples: $(2, 4, 6), (2, 6, 4), (4, 2, 6), (4, 6, 2), (6, 2, 4), (6, 4, 2)$.

Let s be an unordered sample obtained from ordered samples selected by the PPSWOR method and $p(s|i)$ be the conditional probability of selecting the unordered sample s given that the unit i is selected at the first draw. Then Murthy's estimator is obtained as

$$\hat{Y}_{\text{MUR}} = \frac{1}{p(s)} \sum_{i \in s} p(s|i) y_i \tag{4.1.16}$$

Let $p(s|i, j)$ be the conditional probability for the selection of s given that the units i and $j(\neq i)$ have been selected in the first two draws in any order.

Theorem 4.4.8

(i) $\hat{Y}_{\text{MUR}}$ is unbiased for Y.

(ii) Variance of $\hat{Y}_{\text{MUR}}$ is

$$\text{Var}(\hat{Y}_{\text{MUR}}) = \frac{1}{2} \sum_{i \neq} \sum_{j \in U} \beta_{ij} p_i p_j \left(\frac{y_i}{p_i} - \frac{y_j}{p_j} \right)^2$$

where $\beta_{ij} = 1 - \sum_{s \supset i, j} \frac{p(s|i)p(s|j)}{p(s)}$

(iii) An unbiased estimator of $\text{Var}(\hat{Y}_{\text{MUR}})$ is

$$\widehat{\text{Var}}(\hat{Y}_{\text{MUR}}) = \frac{1}{2\{p(s)\}^2} \sum_{i \neq} \sum_{j \in s} \{p(s|i, j)p(s) - p(s|i)p(s|j)\} p_i p_j \left(\frac{y_i}{p_i} - \frac{y_j}{p_j} \right)^2$$

Pathak and Sukla (1966) showed that $p(s|i, j)p(s) - p(s|i)p(s|j) \geq 0$. Hence, the variance estimator $\widehat{\text{Var}}(\hat{Y}_{\text{MUR}})$ is always nonnegative.

Remark 4.4.2 The Raj estimator $\hat{Y}_{\text{RAJ}}$ is inadmissible, while Murthy's estimator $\hat{Y}_{\text{MUR}}$ is admissible.

4.4.6 Inclusion Probability Proportional to Size (IPPS) Sampling Design

The variance of the Horvitz-Thompson estimator $\hat{Y}_{\text{HT}} = \sum_{i \in s} \frac{y_i}{\pi_i}$ based on a fixed effective sample of size n design becomes equal to zero if one chooses π_i proportional to y_i for every $i(\in U)$. But such a choice is not possible, as the values of y_i s are unknown before the survey. Consider the situation where auxiliary information x

that is known, positive, and approximately proportional to the study variable y is available prior to a survey. Under such a situation, we may assume that the variance of $\hat{Y}_{\mathrm{HT}}$ will be small if π_i's are chosen proportional to x_i's. A sampling scheme for which the inclusion probability π_i is proportional to x_i for all $i \in U$ is known as an IPPS(or π ps) sampling scheme. For an IPPS sampling scheme with a fixed effective sample size n, we have

$$\pi_i = n\frac{x_i}{X} = np_i \tag{4.6.1}$$

where $X = \sum_{i\in U} x_i$ and $p_i = x_i/X$.

Since $0 \leq \pi_i \leq 1$, for an IPPS sampling design, conditions (i) $x_i \geq 0$ and (ii) $x_i \leq X/n\ \forall i \in U$ hold. Various authors, including Narain (1951), Brewer (1963), Fellegi (1963), Rao (1965), Durbin (1967), and Hanurav (1967), proposed IPPS sampling designs of size $n = 2$ that are simple to implement. But, the IPPS sampling for the general sample size $n(> 2)$ is not so simple, and the expressions of the second-order inclusion probabilities π_{ij} are generally very complex. We will describe below a few important IPPS sampling schemes that are used in practice.

4.4.6.1 PPS Systematic Sampling Scheme

The PPS systematic sampling scheme was proposed by Hartley and Rao (1962). The sample scheme is very popular because it is easy to implement. For this sampling scheme, we first compute the cumulative totals

$$C_j = n\sum_{i=1}^{j} p_i \text{ for } j = 1, .., N \text{ with } C_0 = 0 \tag{4.6.2}$$

Then, select a random sample, say, r, known as random start from the uniform distribution $U(0, 1)$. The random start r selects the sample s whose label j satisfies

$$C_{j-1} \leq r + k < C_{j+1};\ j = 0, .., n-1 \tag{4.6.3}$$

For the sampling scheme inclusion probability of the ith unit is $\pi_i = C_i - C_{i-1} = np_i$.

The expression of π_{ij} to the $O(N^{-4})$ was provided by Hartley and Rao (1962) assuming the labels of the units were attached at random, p_i is of order $O(N^{-1})$, N is large, and n is relatively small to N as follows:

$$\pi_{ij} = n(n-1)p_ip_j\left[\begin{array}{l} 1 + \left\{(p_i + p_j) - \sum_j p_j^2\right\} \\ + \left\{2\left(p_i^2 + p_j^2\right) - 2\sum_j p_j^3 + 2p_ip_j - 3(p_i + p_j)\sum_j p_j^2 + 3\left(\sum_j p_j^2\right)^2\right\} \end{array}\right] \tag{4.6.4}$$

The expressions of the variance of $\hat{Y}_{\text{HT}}$ is given by Hartley and Rao (1962) to the order $O(N^{-2})$ as

$$\begin{aligned}\text{Var}\left(\hat{Y}_{\text{HT}}\right) =& \frac{1}{n}\left[\sum_i p_i z_i^2 - (n-1)\sum_i p_i^2 z_i^2\right] \\ &- \frac{n-1}{n}\left[2\sum_i p_i^3 z_i^2 - \left(\sum_i p_i^2\right)\left(\sum_i p_i^2 z_i^2\right) - 2\left(\sum_i p_i^2 z_i\right)^2\right] \\ =& \text{V}_{\text{GK}}\end{aligned} \quad (4.6.5)$$

where $z_i = \frac{y_i}{p_i} - Y$.

The expression of V_{GK} correct to $O(N^{-1})$ is

$$\text{V}_{\text{GK}} \cong \frac{1}{n}\sum_i p_i[1-(n-1)p_i]\left(\frac{y_i}{p_i} - Y\right)^2 \quad (4.6.6)$$

(vide Sukhatme et al. (1984)). Expression (4.6.6) indicates that the variance of the Horvitz-Thompson estimator based on the PPS systematic sampling scheme provides a smaller variance than the variance of the Hansen-Hurwitz estimator of the same sample size.

An approximate unbiased estimator of V_{GK} is

$$\hat{\text{V}}_{\text{GK}} = \frac{1}{n^2(n-1)}\frac{1}{2}\left[\sum_{i \neq j}\sum_{\in s}\left\{1 - n(p_i + p_j) + n\left(\sum_{i \in U} p_i^2\right)\right\}\left(\frac{y_i}{p_i} - \frac{y_j}{p_j}\right)^2\right] \quad (4.6.7)$$

4.4.6.2 Sampford's Sampling Scheme

In Sampford's (1967) sampling scheme, on the first draw, a unit, say, i is selected with probability $p_i(=x_i/X)$. The remaining $(n-1)$ units are selected from the whole population, U, with replacement with probability proportional to $\lambda_j = p_j/(1-p_i)$ for the $j(=1,..,N)$th unit. The selected n units are treated as a sample if all the n units are distinct. If all the units are not distinct, the entire procedure is repeated. The repetition is continued till all the selected units are distinct. For this sampling scheme $\pi_i = np_i$ and $\pi_i\pi_j - \pi_{ij} \geq 0$. The expression of π_{ij} is very complex. The variance of $\hat{Y}_{\text{HT}}$ correct to $O(N^{-2})$ was provided by Asok and Sukhatme (1976) as follows:

$$\text{Var}(\hat{Y}_{\text{HT}}) = \frac{1}{n}\left(\sum_i p_i z_i^2 - (n-1)\sum_i p_i^2 z_i^2\right)$$

$$-\frac{n-1}{n}\left\{2\sum_i p_i^3 z_i^2 - \left(\sum_i p_i^2\right)\left(\sum_i p_i^2 z_i^2\right) + (n-2)\left(\sum_i p_i^2 z_i\right)^2\right\}$$
$$= \mathrm{V}_{\mathrm{SAM}} \tag{4.6.8}$$

Asok and Sukhatme (1976) showed correct to $O(N^{-1})$

$$\mathrm{V}_{\mathrm{GK}} = \mathrm{V}_{\mathrm{SAM}} = \frac{1}{n}\left[\sum_i p_i z_i^2 - (n-1)\sum_i p_i^2 z_i^2\right]$$
$$\leq \mathrm{Var}(\hat{Y}_{\mathrm{HH}}) \tag{4.6.9}$$

For further details, readers are referred to Arnab (2017).

4.4.6.3 Poisson (Bernoulli) Sampling Scheme

In a Poisson sampling scheme, a sample is selected by performing N Bernoulli trials with a success probability $\pi_i = np_i$ for the ith trial. If the ith trial results in "success", then the $i(\in U)$th unit is selected in the sample. So, for the Poisson sampling scheme, all the units in a sample are distinct, but the sample size is not fixed. The inclusion probabilities for the ith, and ith and $j(\neq i)$th units are $\pi_i = np_i$ and $\pi_{ij} = \pi_i \pi_j = n^2 p_i p_j$. The variance of $\hat{Y}_{\mathrm{HT}}$ and its unbiased estimator are as follows:

$$\mathrm{Var}(\hat{Y}_{\mathrm{HT}}) = \sum_{i\in U} y_i^2\left(\frac{1}{\pi_i} - 1\right) \tag{4.6.10}$$

and

$$\hat{\mathrm{Var}}(\hat{Y}_{\mathrm{HT}}) = \sum_{i\in s} \frac{y_i^2}{\pi_i}\left(\frac{1}{\pi_i} - 1\right) \tag{4.6.11}$$

4.4.7 *Rao-Hartley-Cochran Sampling Scheme*

In the Rao et al. (1962) sampling scheme, at first, the entire population U of size N is divided at random into n (=sample size) disjoint groups so that the jth group G_j contains a predetermined number of N_j units with $\sum_j N_j = N$. Then in the second step, one unit is selected from each of the groups so formed with a probability proportional to its measure of size. For example, from the jth group, the kth unit is selected with a probability $x_k / \sum_{t \in G_j} x_t = p_k / \sum_{t \in G_j} p_t$. Let a sample $s = (j_1, \rightarrow, ..., j_k \rightarrow, .. \rightarrow, j_n)$ of size n be a sample selected by the Rao-Hartley-Cochran sampling scheme, where the unit j_k is selected from the group G_k. Then, we have the following results:

Theorem 4.4.9

(i) $\hat{Y}_{\text{RHC}} = \sum_{j_k \in s} \frac{y_{j_k}}{p_{j_k}} P_{j_k}$ is an unbiased estimator of Y

where $P_{j_k} = \sum_{j_t \in G_k} p_{j_t}$.

(ii) The variance of $\hat{Y}_{\text{RHC}}$ is

$$\text{Var}(\hat{Y}_{\text{RHC}}) = \frac{\sum_{i \in U} N_i^2 - N}{N(N-1)} \sum_{i \in U} p_i \left(\frac{y_i}{p_i} - Y \right)^2$$

(iii) An unbiased estimator of the variance of $\hat{Y}_{\text{RHC}}$ is

$$\hat{\text{Var}}(\hat{Y}_{\text{RHC}}) = \frac{\sum_{i \in U} N_i^2 - N}{N^2 - \sum_{i \in U} N_i^2} \sum_{j_k \in s} P_{j_k} \left(\frac{y_{j_k}}{p_{j_k}} - \hat{Y}_{\text{RHC}} \right)^2$$

The expression $\sum_{i \in U} N_i^2$ attains a minimum when $N_i = N/n$ for $i = 1, .., n$. Thus, if N/n is an integer, one should choose $N_i = N/n$ for $i = 1, .., n$ to make $\text{Var}(\hat{Y}_{\text{RHC}})$ a minimum. In this case, Theorem 4.4.9. reduces to

Theorem 4.4.10 For $N_i = N/n$, an integer, $i = 1, .., N$,

(i) $\hat{Y}_{\text{RHC}} = \sum_{j_k \in s} \frac{y_{j_k}}{p_{j_k}} P_{j_k}$ is an unbiased estimator of Y.

(ii) The variance of $\hat{Y}_{\text{RHC}}$ is

$$\text{Var}(\hat{Y}_{\text{RHC}}) = \frac{N - n}{n(N-1)} \sum_{i \in U} p_i \left(\frac{y_i}{p_i} - Y \right)^2$$

(iii) An unbiased estimator of the variance of $\hat{Y}_{\text{RHC}}$ is

$$\hat{\text{Var}}(\hat{Y}_{\text{RHC}}) = \frac{N - n}{N(n-1)} \sum_{j_k \in s} P_{j_k} \left(\frac{y_{j_k}}{p_{j_k}} - \hat{Y}_{\text{RHC}} \right)^2$$

Theorem 4.4.10 shows that the Rao-Hartley-Cochran estimator $\hat{Y}_{\text{RHC}}$ is more efficient than the Hansen-Hurwitz estimator $\hat{Y}_{\text{HH}}$ of the same size. The Rao-Hartley-Cochran sampling scheme is very useful because it is very simple to execute, possesses elegant expressions of the variance, and its unbiased estimator. The main drawback of the Rao-Hartley-Cochran strategy is that the estimator $\hat{Y}_{\text{RHC}}$ is inadmissible, and the standard Rao-Blackwellization technique does not yield any improved estimator that is easy to handle. For further details, readers are referred to Arnab (2017).

4.4.8 Probability Proportional to Aggregate Size (PPAS) Sampling Scheme

Lahiri (1951), Midzuno (1952), and Sen (1953) independently proposed the PPAS sampling scheme, which is also known as the Lahiri-Midzuno-Sen (LMS) sampling scheme. In this sampling scheme, on the first draw, a unit is selected with a probability proportional to its measure of size. The remaining $n-1$ units are selected from those that were not selected in the first draw by the SRSWOR method. For this sampling scheme, the probability of selecting of an ordered sample $\vec{s} = (i_1, \rightarrow i_2, \rightarrow, .., \rightarrow i_n)$ is

$$p(i_1, \rightarrow i_2, \rightarrow, .., \rightarrow i_n) = \frac{x_{i_1}}{X} \times \frac{1}{N-1} \times .... \times \frac{1}{N-n+1}$$

and the probability of selecting an unordered sample $s = (j_1, j_2, , .., j_n)$ with $j_1 < j_2 <, .., < j_n$ is

$$p(s) = \sum_{i \in s} \frac{x_i}{X} / \binom{N-1}{n-1} \tag{4.6.12}$$

Inclusion probabilities of the first two orders for this sampling scheme are $\pi_i = \frac{n-1}{N-1} + \frac{N-n}{N-1} p_i$ and $\pi_{ij} = \frac{n-1}{N-1}\left[\frac{n-2}{N-2} + \frac{N-n}{N-2}(p_i + p_j)\right]$. Now, writing $x_s = \sum_{i \in s} x_i$, $y_s = \sum_{i \in s} y_i$, $T_i = \frac{X}{\binom{N-1}{n-1}} \sum_{s \supset i} \frac{1}{x_s}$, and $T_{ij} = \frac{X}{\binom{N-1}{n-1}} \sum_{s \supset i,j} \frac{1}{x_s}$, one obtains the following results:

Theorem 4.4.11

(i) $\hat{Y}_{\text{LMS}} = \frac{y_s}{x_s} X$ is an unbiased estimator of Y.

(ii) The variance of $\hat{Y}_{\text{LMS}}$ is

$$\text{Var}(\hat{Y}_{\text{LMS}}) = \sum_{i \in U} y_i^2 \left(\frac{1}{T_i} - 1\right) + \sum_{i \neq} \sum_{i \in U} y_i y_j \left(\frac{1}{T_{ij}} - 1\right)$$

(iii) Two unbiased estimators of $\text{Var}(\hat{Y}_{\text{LMS}})$ are given by

$$\widehat{\text{Var}}_1(\hat{Y}_{\text{LMS}}) = \sum_{i \in s} \frac{y_i^2}{\pi_i}\left(\frac{1}{T_i} - 1\right) + \sum_{i \neq} \sum_{i \in s} \frac{y_i y_j}{\pi_{ij}}\left(\frac{1}{T_{ij}} - 1\right) \text{ and}$$

$$\widehat{\text{Var}}_2(\hat{Y}_{\text{LMS}}) = \hat{Y}_{\text{LMS}}^2 - \frac{X}{x_s}\left[\sum_{i \in s} y_i^2 + \frac{N-1}{n-1} \sum_{i \neq} \sum_{j \in s} y_i y_j\right]$$

4.5 Inference Under Superpopulation Model

So far, we have considered a fixed population setup where y_i is a fixed value of the study variable y for the ith unit of the population $U = (1, \ldots, N)$ and the vector $\boldsymbol{y} = (y_1, \ldots, y_N)$ is a fixed point in the N-dimensional Euclidean space R_N. In the superpopulation approach, we assume that the vector $\boldsymbol{y}$ is a realization of a random vector $\boldsymbol{Y} = (Y_1, .., Y_N)$ that possesses certain distribution ξ. The distribution ξ is called a superpopulation model or simply model. The distribution ξ is often indexed by certain unknown parameter θ which belongs to a known parametric space Ω_θ. In this section, we will consider concepts of design-based, model-based, and model-design-based inferences. Optimal strategies for such approaches will be identified.

4.5.1 Design-Based Approach

In a design-based approach, the population vector $\boldsymbol{y}$ is fixed. In order to estimate a parametric function, e.g., the population total Y, a sample s of suitable size n is selected from the population by using a sampling design p. Information relating to the study variables y_i s $(i \in s)$ are obtained from the selected sample s. Inference relating to the population parameter is obtained by establishing a linkage between the observed y_i's $(i \in s)$ and unobserved y_i's $(i \notin s)$ through the sampling design p. Here, expectation E_p denotes the average over all possible samples.

The estimator $t = t(s, \boldsymbol{y})$ is said to be a design-unbiased or p-unbiased if

$$E_p(t) = \sum_s t(s, \boldsymbol{y})p(s) = Y \quad \forall \boldsymbol{y} \in R_N \tag{4.5.1}$$

The class of p-unbiased estimators for Y is denoted by C_p. We have seen in Sect. 4.3 that the best estimator (minimum variance unbiased estimator) for the population total does not exist. There also does not exist any optimum strategy (combination of sampling design and estimator) that minimizes the variance

$$\text{Var}(t) = E_p(t - Y)^2 = \sum_s [t(s, y) - Y]^2 p(s) \quad \forall \boldsymbol{y} \in R_N \tag{4.5.2}$$

4.5.2 Model-Based Approach

In a model-based approach, the inference is based on the chosen superpopulation model ξ. The sampling design plays a little role in it. In the presence of a superpopulation model ξ, the population total Y is a random variable, and hence any estimator

t of Y will be called a predictor of Y. The estimator $t = t(s, y)$ is said to be a model-unbiased (ξ-unbiased) for the population total Y if

$$E_{\xi}(t) = E_{\xi}(Y) \quad \forall \theta \in \Omega_{\theta} \tag{4.5.3}$$

where $E_{\xi}(.)$ denotes expectation with respect to the model ξ. The class of ξ-unbiased predictor of Y will be denoted by C_{ξ}. In a model-based approach, we try to control the magnitude of

$$E_p E_{\xi}(t - Y)^2 = E_{\xi} E_p(t - Y)^2 \quad \forall \theta \in \Omega_{\theta} \tag{4.5.4}$$

Here we assume that the commutative of the operators $E_p E_{\xi} = E_{\xi} E_p$ is valid. This commutative property holds at least for a non-informative sampling design where $p(s)$ is independent of y_i's $i \in U$.

4.5.3 Model-Design-Based Approach

The model-design-based approach is also known as model-assisted or model-dependent approach. In this approach, both the sampling design p and the superpopulation model ξ are used in making inferences. This approach is effective as it protects against model misspecification. The predictor t will be called the model-design ($p\xi$) unbiased predictor for the total Y if

$$E_p E_{\xi}(t) = E_{\xi} E_p(Y) \quad \forall \theta \in \Omega_{\theta} \tag{4.5.5}$$

The class of $p\xi$-unbiased predictor of Y will be denoted by $C_{p\xi}$. The class $C_{p\xi}$ contains both the classes C_p and C_{ξ}.

4.6 Optimal Strategies Under Superpopulation Models

A combination of a sampling design p and a predictor t is called a sampling strategy. A strategy will be denoted by the duplet $h = (p, t)$. The class of strategies will be denoted by H. The strategy $h_0 = (p_0, t_0) \in$ H is said to be optimal in the class of strategies H if

$$E_{p_0} E_{\xi}(t_0 - Y)^2 \leq E_p E_{\xi}(t - Y)^2 \quad \forall \theta \in \Omega_{\theta},\ h(\neq h_0) \in \mathrm{H} \tag{4.6.1}$$

and $E_{p_0} E_{\xi}(t_0 - Y)^2 < E_p E_{\xi}(t - Y)^2$ for some $\theta \in \Omega_{\theta}$.

4.6.1 Optimal Strategies Under Design Unbiased Estimator

In this section, we will find optimal p-unbiased predictors under various superpopulation models.

Model $M1$: In this model, y_i's are independently distributed with mean $E_\xi(y_i) = \mu_i$ and variance $V_\xi(y_i) = \sigma_i^2$ where μ_i is known and σ_i^2 is positive but unknown.

Let t be a $p-$unbiased predictor of Y belonging to the class C_p. Then, one obtains from Godambe and Joshi (1965) the following theorem.

Theorem 4.6.1 For any sampling design p with inclusion probability $\pi_i > 0 \quad \forall i \in U$, the predictor

$$t_{\text{GP}} = \sum_{i\in s} \frac{y_i - \mu_i}{\pi_i} + \sum_{i\in U} \mu_i \tag{4.6.2}$$

is optimal in the class C_p in the sense

$$E_\xi E_p(t - Y)^2 \geq \sum_{i\in U} \left(\frac{1}{\pi_i} - 1\right)^2 \sigma_i^2 = E_\xi E_p(t_{GP} - Y)^2 \quad \forall t \in C_p \tag{4.6.3}$$

The predictor t_{GP} is called the **generalized difference estimator** (Basu, 1971). In case $\mu_i = \beta x_i$ where x_i is a known but β is unknown. In this case, (4.6.2) is reduced to

$$t_{\text{GEG}} = \sum_{i\in s} \frac{y_i}{\pi_i} + \hat{\beta}\left(\sum_{i\in s} \frac{x_i}{\pi_i} - \sum_{i\in U} x_i\right)$$

where $\hat{\beta}$ is an unbiased estimator of β.

The estimator t_{GEG} is called **generalized regression predictor**.

Let P_n be the class of fixed effective size of n (FESn) design such that $\sum_{i\in U} \pi_i = n$. The optimum value of π_i that minimizes $\sum_{i\in U} \left(\frac{1}{\pi_i} - 1\right)^2 \sigma_i^2$ subject to $\sum_{i\in U} \pi_i = n$ is $\pi_{i0} = n\sigma_i / \left(\sum_{i\in U} \sigma_i\right)$. Let $p_{n,\sigma} = p(n, \pi_{i0})$ be a FES(n) sampling design satisfying

(i) $\pi_{i0} = n\sigma_i / \left(\sum_{i\in U} \sigma_i\right)$ and
(ii) $\sum_{i\in s} \frac{y_i}{\pi_{i0}} = \sum_{i\in U} \mu_i$, then $t_{\text{GP0}} = \hat{Y}_{\text{HT0}} = \sum_{i\in s} \frac{y_i}{\pi_{i0}}$ satisfies

$$E_\xi E_p(t - Y)^2 \geq \frac{1}{n}\left(\sum_{i\in U} \sigma_i\right)^2 - \sum_{i\in U} \sigma_i = E_\xi E_p(\hat{Y}_{\text{HT0}} - Y)^2 \, \forall p \in P_n,\ t \in C_p \tag{4.6.4}$$

Equation (4.6.4) yields

Theorem 4.6.2 Under model $M1$, the strategy $h_1 = (p_{n,\sigma}, \hat{Y}_{HT0})$ is the optimal in the class of strategies $H_1 = (p, t); p \in P_n$ and $t \in C_p$.

Model $M11$: Model $M1$ with $\mu_i = \beta x_i$ and $\sigma_i^2 = \sigma^2 x_i^2$ where x_i is the value of the auxiliary variable x for the ith unit, which is assumed to be known and positive; β and σ^2 are positive but unknown constants.

Let us define $p_{nx}(\in P_n)$ be a sampling design with $\pi_{i0} = nx_i/X$. In this case, Theorem 4.6.2 reduces to

Theorem 4.6.3 Under the model $M11$, the strategy $h_2 = (p_{nx}, \hat{Y}_{HTx})$ is the optimal in the class of strategies $H_1 = (p, t); p \in P_n$ and $t \in C_p$, i.e.,

$$E_\xi E_p(t - Y)^2 \geq \sigma^2 \sum_{i \in U} \left(\frac{X^2}{n} - \sum_{i \in U} x_i^2 \right) = E_\xi E_{p_{nx}} (\hat{Y}_{HTx} - Y)^2 \quad \forall p \in P_n,\ t \in C_p \tag{4.6.5}$$

Model $M12$: Model $M11$ with $x_i = 1$ for every $i \in U$.

For the model $M12$, π_{i0} and $\hat{Y}_{HTx}$ reduce to $\pi_0 = n/N$ and $N\overline{y}_s$, respectively, where $\overline{y}_s = \frac{1}{n} \sum_{i \in s} y_i$. Let $p_0(\in P_n)$ be the sampling design for which $\pi_i = \pi_0 = n/N$. In this case, the strategy $h_0 = (p_0, N\overline{y}_s)$ becomes optimal.

Theorem 4.6.4 Under the model $M12$, the strategy $h_0 = (p_0, N\overline{y}_s)$ is optimal in the class of strategies $H_1 = (p, t); p \in P_n$ and $t \in C_p$, i.e.,

$$E_\xi E_p(t - Y)^2 \geq \frac{N(N-n)}{n} \sigma^2 = E_\xi E_{p_0} (N\overline{y}_s - Y)^2 \quad \forall p \in P_n, t \in C_p \tag{4.6.6}$$

4.6.2 *Optimal Strategies Under Model-Design-Unbiased Predictor*

Here we will consider the following equi-correlated model proposed by Cassel et al. (1976).

$$\textit{Model } M2{:}\ E_\xi(y_i) = \beta x_i,\ V_\xi(y_i) = \sigma^2 x_i^2 \text{ and } C_\xi(y_i, y_j) = \rho x_i x_j \text{ for } i \neq j \tag{4.6.7}$$

The class of linear $p\xi$-unbiased predictor $C_{lp\xi}$ consists of estimators

$$\tilde{t} = a_s + \sum_{i \in s} w_{si} y_i \tag{4.6.8}$$

where the constants a_s and w_{si} satisfy the $p\xi$-unbiasedness condition $E_p E_\xi(\tilde{t}) = E_\xi(Y)$, i.e.,

$$\text{(i)} \sum_s a_s p(s) = 0 \text{ and (ii)} \sum_s a_s \left(\sum_{i \in s} b_{si} x_i \right) = X \tag{4.6.9}$$

Minimizing $E_\xi\left[E_p(\tilde{t} - \beta X)^2\right]$ subject to (4.6.9), Cassel et al. (1976) derived the optimal strategy in the class of strategies $H_2 = (p, \tilde{t});\ p \in P_n, \tilde{t} \in C_{lp\xi}$ as follows:

Theorem 4.6.5 Under the model $M2$, the strategy $h_0 = (p, \hat{Y}_{\mathrm{HTx}})$ is optimal for estimating βX in the class of strategies $H_2 = (p, \tilde{t}); p \in P_n$ and $\tilde{t} \in C_{lp\xi}$, i.e.,

$$E_\xi E_p(\tilde{t} - \beta X)^2 \geq \sigma^2 X^2\{1 + (n-1)\rho\}/n = E_\xi E_p(\hat{Y}_{\mathrm{HTx}} - \beta X)^2\ \forall p \in P_n,\ \tilde{t} \in C_{lp\xi} \tag{4.6.10}$$

4.6.3 Optimal Strategies Under Model-Unbiased Predictors

The class of linear model-unbiased (ξ-unbiased) predictors $C_{l\xi}$ of the total Y consists of the predictors:

$$\hat{Y}_{l\xi} = w_s + \sum_{i \in s} w_{si} y_i \tag{4.6.11}$$

satisfying

$$E_\xi(\hat{Y}_{l\xi}) = E_\xi(Y) \text{ i.e.} w_s + \beta \sum_{i \in s} w_{si} x_i = \beta X \tag{4.6.12}$$

The predictor $\hat{Y}_{0l\xi}$ will be said to be optimum in the class $C_{l\xi}$ if

$$E_\xi\left[\sum_s (\hat{Y}_{0l\xi} - Y)^2 p(s)\right] \leq E_\xi\left[\sum_s (\hat{Y}_{l\xi} - Y)^2 p(s)\right] \tag{4.6.13}$$
$$\forall\ \hat{Y}_{l\xi}(\neq \hat{Y}_{0l\xi}) \in C_{l\xi} \text{ and } \theta \in \Omega$$

and

$$E_\xi\left[\sum_s (\hat{Y}_{0l\xi} - Y)^2 p(s)\right] < E_\xi\left[\sum_s (\hat{Y}_{l\xi} - Y)^2 p(s)\right]$$
$$\forall\ \hat{Y}_{l\xi}(\neq \hat{Y}_{0l\xi}) \in C_{l\xi} \text{ at least one } \theta \in \Omega_\theta$$

Consider the superpopulation model:

Model $M3$: $E_\xi(y_i) = \beta x_i$, $V_\xi(y_i) = \sigma^2 v(x_i)$ and $Cov_\xi(y_i, y_j) = 0$ for $i \neq j \in U$ (4.6.14)

where β and $\sigma^2(> 0)$ are unknown constants, x_i is a known positive auxiliary variable associated with the ith unit of the population, and $v(x_i)(> 0)$ is a known function of x_i.

Under the model M_3,

$$\begin{aligned} E_\xi(\hat{Y}_{l\xi} - Y)^2 &= V_\xi(\hat{Y}_{l\xi} - Y) + \left[E_\xi(\hat{Y}_{l\xi} - Y)\right]^2 \\ &= V_\xi(\hat{Y}_{l\xi} - Y), \text{ using (4.6.2)} \\ &= V_\xi(U_s - Y_{\bar{s}}) \end{aligned} \tag{4.6.15}$$

where $\hat{Y}_{l\xi} = U_s + Y_s$, $Y_s = \sum_{i\in s} y_i$, and $Y_{\bar{s}} = \sum_{i\notin s} y_i$.

Now noting $E_\xi(U_s) = \beta X_{\bar{s}}$ with $X_{\bar{s}} = \sum_{i\notin s} x_i$, one finds the best linear ξ-unbiased predictor of $\beta X_{\bar{s}}$ is $\hat{\beta}_{\text{BLUE}} X_{\bar{s}}$ where

$$\hat{\beta}_{\text{BLUE}} = \frac{\sum_{i\in s} \frac{y_i x_i}{v(x_i)}}{\sum_{i\in s} \frac{x_i^2}{v(x_i)}} \tag{4.6.16}$$

and hence the best linear ξ-unbiased predictor of βX_s is

$$\hat{Y}_{0l\xi} = Y_s + \hat{\beta}_{\text{BLUE}} X_{\bar{s}} \tag{4.6.17}$$

The above discussions are summarized as follows:

Theorem 4.6.6 For a given sampling design p and model $M3$:

$$E_\xi\left[\sum_s (\hat{Y}_{l\xi} - Y)^2\, p(s)\right] \geq E_\xi\left[\sum_s (\hat{\beta}_{\text{BLUE}} X_{\bar{s}} - Y_{\bar{s}})^2 p(s)\right] \tag{4.6.18}$$
$$\forall\ \hat{Y}_{l\xi} \in C_{l\xi} \text{ and } \theta \in \Omega_\theta$$

Let Ψ_n be the collection of samples of effective sample size n and P_n be the class of FED(n),then

$$\begin{aligned} E_\xi\left[\sum_s (\hat{\beta}_{\text{BLUE}} X_{\bar{s}} - Y_{\bar{s}})^2 p(s)\right] &= \sum_{s\in\Psi_n} V_\xi(\hat{\beta}_{\text{BLUE}} X_{\bar{s}} - Y_{\bar{s}}) p(s) \\ &= \sum_{s\in\Psi_n} \left[X_{\bar{s}}^2\, V_\xi(\hat{\beta}_{\text{BLUE}}) + V_\xi(Y_{\bar{s}})\right] p(s) \\ &= \sigma^2 \sum_{s\in\Psi_n} \left[\sum_{i\in\bar{s}} v(x_i) + X_{\bar{s}}^2 \left(\sum_{i\in s} \frac{x_i^2}{v(x_i)}\right)^{-1}\right] p(s) \end{aligned}$$

Let $s_0(\in \Psi_n)$ be such that

$$\operatorname*{Min}_{s\in\psi_n}\left\{\sum_{i\in\bar{s}} v(x_i) + X_{\bar{s}}^2\left(\sum_{i\in s}\frac{x_i^2}{v(x_i)}\right)^{-1}\right\} = \sum_{i\in\bar{s}_0} v(x_i) + X_{\bar{s}_0}^2\left(\sum_{i\in s_0}\frac{x_i^2}{v(x_i)}\right)^{-1} \quad (4.6.19)$$

Let $\tilde{p}_0$ be the purposive sampling design, which selects the sample s_0 with probability $p(s_0) = 1$, then $\tilde{h}_0 = (\tilde{p}_0, \hat{Y}_{00\xi})$ with $\hat{Y}_{00\xi} = Y_{s_0} + \hat{\beta}_{\text{BLUE}} X_{\bar{s}}$ becomes the optimum strategy in the class of strategies $H_3 = (p, \hat{Y}_{l\xi}); p \in P_n$ and $\hat{Y}_{l\xi} \in C_{l\xi u}$. Hence, we get the following theorem:

Theorem 4.6.7 Under the model $M3$, the strategy $\tilde{h}_0 = (\tilde{p}_0, \hat{Y}_{00\xi})$ is the optimum in the class of strategies $H_3 = (p, \hat{Y}_{l\xi})$,

i.e., $E_\xi\left[\sum_s (\hat{Y}_{00\xi} - Y)^2 p_0(s)\right] \le E_\xi\left[\sum_s (\hat{Y}_{l\xi} - Y)^2 p(s)\right] \le \forall \hat{Y}_{l\xi} (\neq \hat{Y}_{00\xi}) \in C_{l\xi}, p \in P_n$ and $\theta \in \Omega_\theta$

Particular Cases

Model: $M31$: $E_\xi(y_i) = \beta x_i$, $V_\xi(y_i) = \sigma^2 x_i$ and $Cov_\xi(y_i, y_j) = 0$ for $i \neq j \in U$.

Substituting $v(x_i) = \sigma^2 x_i$ in (4.6.17), one finds $\hat{\beta}_{\text{BLUE}} = \frac{\sum_{i\in s} y_i}{\sum_{i\in s} x_i}$ and the optimal predictor $\hat{Y}_{0l\xi}$ reduces to the well-known ratio predictor

$$\hat{Y}_R = Y_s + \frac{\sum_{i\in s} y_i}{\sum_{i\in s} x_i} X_{\bar{s}} = \frac{\sum_{i\in s} y_i}{\sum_{i\in s} x_i} X. \quad (4.6.20)$$

Let $s_{\max}$ be a sample of n distinct units for which $\sum_{i\in s} x_i$ is maximum. Then the expression

$$E_\xi\left[\sum_s (\hat{Y}_R - Y)^2 p(s)\right] = \sigma^2 X \sum_s \left(\frac{X}{X_s} - 1\right) p(s) \quad (4.6.21)$$

attains a minimum for the purposive sampling design $p_{\max}$ which selects the sample $s_{\max}$ with probability one. Thus, $h_{\max} = (p_{\max}, \hat{Y}_R)$ is the optimum strategy in the class of strategies H_3. A detailed review of the optimal strategies for estimating population characteristics is provided by Royall (1970), Cassel et al. (1977), and Arnab (2017).

4.7 Comparison of a Few Specific Strategies

In this section, performances of the Hansen-Hurwitz, Horvitz-Thompson, Rao-Hartley-Cochran, and Lahiri-Midzuno-Sen strategies are compared under the following superpopulation model:

$$\text{Model } M: E_\xi(y_i) = \beta x_i,\ V_\xi(y_i) = \sigma^2 v(x_i) \text{ and } C_\xi(y_i, y_j) = 0 \text{ for } i \neq j \in U \tag{4.7.1}$$

The expected variances of the Hansen-Hurwitz estimator $\hat{Y}_{\text{HH}} = \frac{1}{n}\sum_{r=1}^{n}\frac{y_{(r)}}{p_{(r)}}$ based on the PPSWR sampling scheme, the Horvitz-Thompson estimator $\hat{Y}_{\text{HT}} = \sum_{i\in s}\frac{y_i}{\pi_i}$ based on an IPPS sampling scheme, Rao-Hartley-Cochran estimator based on Rao-Hartley-Cochran sampling, and the Lahiri-Midzuno-Sen estimator $\hat{Y}_{\text{LMS}} = \frac{y_s}{x_s}X$ based on Lahiri-Midzuno-Sen sampling under the superpopulation model (4.7.1) are obtained as follows (for details, see Chaudhuri and Arnab (1979), and Arnab (2017)):

$$\begin{aligned} E_\xi\left[\text{Var}(\hat{Y}_{\text{HH}})\right] &= \frac{1}{n}E_\xi\left(\sum_{i\in U}\frac{y_i^2}{p_i} - Y^2\right) \\ &= \frac{\sigma^2}{n}\sum_{i\in U}\frac{v(x_i)}{x_i}(X - x_i) \\ &= Q_1(\text{say}) \end{aligned} \tag{4.7.2}$$

$$E_\xi\left[\text{Var}(\hat{Y}_{\text{HT}})\right] = E_\xi\left[\sum_{i\in U} y_i^2\left(\frac{1}{\pi_i} - 1\right) + \sum_{i\neq}\sum_{j\in U} y_i y_j\left(\frac{\pi_{ij}}{\pi_i\pi_j} - 1\right)\right]$$

For an IPPS sampling design $\pi_i = np_i = nx_i/X$, the above expression reduces to

$$\begin{aligned} E_\xi\left[\text{Var}(\hat{Y}_{\text{HT}})\right] &= \sigma^2\sum_{i\in U}\frac{v(x_i)}{x_i}\left(\frac{X}{n} - x_i\right) \\ &= Q_2(\text{say}) \end{aligned} \tag{4.7.3}$$

$$\begin{aligned} E_\xi\left[\text{Var}(\hat{Y}_{\text{RHC}})\right] &= \frac{N-n}{n(N-1)}E_\xi\left(\sum_{i\in U}\frac{y_i^2}{p_i} - Y^2\right) \\ &= \sigma^2\frac{N-n}{n(N-1)}\sum_{i\in U}\frac{v(x_i)}{x_i}(X - x_i) \\ &= Q_3(\text{say}) \end{aligned} \tag{4.7.4}$$

$$\begin{aligned} E_\xi\left[\text{Var}(\hat{Y}_{\text{LMS}})\right] &= E_\xi\left(\sum_{i\in U} T_i y_i^2 + \sum_{i\neq}\sum_{j\in U} T_{ij} y_i y_j - Y^2\right) \\ &= \sigma^2\sum_{i\in U}\frac{v(x_i)}{x_i}(T_i x_i - x_i) \\ &= Q_4(\text{say}) \end{aligned} \tag{4.7.5}$$

From expressions (4.7.2) to (4.7.5), the following results were obtained. For details, readers are referred to Chaudhuri and Arnab (1979) and Arnab (2017).

Theorem 4.7.1 Under the superpopulation model (4.7.1), the following inequality holds:

$E_\xi\left[\text{Var}(\hat{Y}_{\text{HT}})\right] \leq E_\xi\left[\text{Var}(\hat{Y}_{\text{RHC}})\right] \leq E_\xi\left[\text{Var}(\hat{Y}_{\text{LMS}})\right]$ if $v(x_i)/x_i$ is an increasing function of x_i.

$E_\xi\left[\text{Var}(\hat{Y}_{\text{HT}})\right] = E_\xi\left[\text{Var}(\hat{Y}_{\text{RHC}})\right] = E_\xi\left[\text{Var}(\hat{Y}_{\text{LMS}})\right]$ if $v(x_i)/x_i$ is a constant.

$E_\xi\left[\text{Var}(\hat{Y}_{\text{HT}})\right] \geq E_\xi\left[\text{Var}(\hat{Y}_{\text{RHC}})\right] \geq E_\xi\left[\text{Var}(\hat{Y}_{\text{LMS}})\right]$ if $v(x_i)/x_i$ is a decreasing function of x_i.

If $v(x_i) = \sigma^2 x_i^g$, Theorem 4.7.1 reduces to

Theorem 4.7.2 Under the superpopulation model (4.7.1), the following inequality holds:

$E_\xi\left[\text{Var}(\hat{Y}_{\text{HT}})\right] \lesseqgtr E_\xi\left[\text{Var}(\hat{Y}_{\text{RHC}})\right] \lesseqgtr E_\xi\left[\text{Var}(\hat{Y}_{\text{LMS}})\right]$ according to $g \lesseqgtr 1$.

4.8 Discussions

This chapter presents definitions of the basic terminologies such as finite population, sample, sample size, sampling design, sampling scheme, parameter, parameter space, estimators, and sampling strategies, among others. The relationship between sampling designs and sampling schemes is discussed. Inferences drawn from sampling from finite populations differ significantly from those drawn from sampling from infinite populations. Three types of inferences from finite populations, such as design-based, model-based, and model-design-based inferences, are proposed in the literature. In design-based inference, the parameter vector $\boldsymbol{y} = (y_1, .., y_N)$ is a fixed point in the N-dimensional Euclidean space. Godambe (1955) proved that the best estimator (MVUE) does not exist in the class of linear unbiased estimators. Hanurav (1966) modified Godambe's result further: the best estimator exists if the sampling design is uni-cluster; otherwise, it does not. Basu (1971) proved that the best estimator does not exist in the entire class of unbiased estimators. In model-based inference, the population vector $\boldsymbol{y}$ is not fixed. It is a random vector: a sample from an infinite population ξ whose distribution depends on a vector of an unknown parameter, say, θ. Such a population is termed a "superpopulation". Under such a superpopulation, the best estimator exists in some situations, and it is often based on a purposive sampling scheme. The model-design-based inference is a hybrid of model-based and design-based inferences. Under model-design-based inference, the best estimator exists in specific situations.

References

Arnab, R. (2017). *Survey sampling theory and applications*. Academic Press.

Asok, C., & Sukhatme, B. V. (1976). On Sampford's procedure of unequal probability sampling without replacement. *Journal of the American Statistical Association, 71*, 912–918.

Basu, D. (1958). On sampling with and without replacement. *Sankhyā, 20*, 287–294s.

Basu, D. (1971). An essay on the logical foundations of survey sampling, Part 1 (with discussion). In V. P. Godambe & D. A. Sprott (Eds.), *Foundations of statistical inference* (pp. 203–242). Holt, Rinehart and Winston.

Brewer, K. R. W. (1963). A model of systematic sampling schemes of unequal probabilities. *Australian Journal of Statistics, 5*, 5–13.

Cassel, C. M., Särndal, C. E., & Wretman, J. H. (1976). Some results on generalized difference estimation and generalized regression estimation for finite populations. *Biometrika, 63*, 615–620.

Cassel, C. M., Särndal, C. E., & Wretman, J. H. (1977). *Foundations of inference in survey sampling*. Wiley.

Chaudhuri, A., & Arnab, R. (1979). On the relative efficiencies of the sampling strategies under a superpopulation model. *Sankhyā, C, 41*, 40–53.

Durbin, J. (1967). Designs of multi-stage survey for estimation of sampling error. *Applied-Statistics, 16*, 152–164.

Fellegi, I. P. (1963). Sampling with varying probabilities without replacement rotation and non-rotation samples. *Journal of the American Statistical Association, 58*, 183–201.

Feller, W. (1957). *An introduction to probability theory and its applications* (Vol. 1). Wiley.

Godambe, V. P. (1955). A unified theory of sampling from finite populations. *Journal of the Royal Statistical Society, Series B, 17*, 269–278.

Godambe, V. P., & Joshi, V. M. (1965). Admissibility and Bayes estimation in sampling finite population I. *Annals of Mathematical Statistics, 36*, 1707–1722.

Hansen, M. H., & Hurwitz, W. N. (1943). On the theory of sampling from finite populations. *Annals of Mathematical Statistics, 14*, 333–362.

Hanurav, T. V. (1966). Some aspects of unified sampling theory. *Sankhyā, A, 28*, 175–204.

Hanurav, T. V. (1967). Optimum utilization of auxiliary information: PPS sampling of two units from a stratum. *Journal of the Royal Statistical Society, Series B, 29*, 374–391.

Hartley, H. O., & Rao, J. N. K. (1962). Sampling with unequal probabilities and without replacement. *Annals of Mathematical Statistics, 33*, 350–374.

Horvitz, D. G., & Thompson, D. J. (1952). A generalization of sampling without replacement from a finite population. *Journal of the American Statistical Association, 47*, 663–685.

Lahiri, D. B. (1951). A method of sample selection for providing unbiased ratio estimates. *Bulletin of the International Statistical Institute, 33*, 133–140.

Midzuno, H. (1952). On the sampling system with probabilities proportionate to sum of sizes. *Annals of the Institute of Statistical Mathematics, 3*, 99–107.

Murthy, M. N. (1957). Ordered and unordered estimators in sampling without replacement. *Sankhyā, 18*, 379–390.

Narain, R. D. (1951). On sampling without replacement with varying probabilities. *Journal of the Indian Society of Agricultural Statistics, 3*, 169–174.

Pathak, P. K. (1961). On the evaluation of moments of distinct units in a sample. *Sankhyā, A, 23*, 415–420.

Raj, D. (1956). Some estimators in sampling with varying probabilities without replacement. *Journal of the American Statistical Association, 51*, 269–284.

Rao, J. N. K. (1965). On two sample schemes of unequal probability sampling without replacement. *Journal of the Indian Society of Agricultural Statistics, 3*, 169–174.

Rao, J. N. K., Hartley, H. O., & Cochran, W. G. (1962). On a simple procedure of unequal probability sampling without replacement. *Journal of the Royal Statistical Society, Series B, 24*, 482–491.

Royall, R. M. (1970). On finite population sampling theory under certain linear regression models. *Biometrika, 57*, 377–387.

Sampford, M. R. (1967). On sampling without replacement with unequal probability selection. *Biometrika, 67*, 639–650.
Sen, A. R. (1953). On estimate of variance in sampling with varying probabilities *Journal of the Indian Society of Agricultural Statistics, 5*, 119–127.
Sukhatme, P. V., Sukhatme, B. V., Sukhatme, S., & Asok, C. (1984). *Sampling theory of surveys with applications.* Iowa state University Press.
Yates, F., & Gurndy, P. M. (1953). Selection without replacement from within strata with probability proportional to size. *Journal of the Royal Statistical Society, Series B, 15*, 253–261.

Chapter 5
Randomized Response Techniques for Complex Survey Designs

Abstract The general method of estimating the population total or mean for the various sampling strategies under randomized response techniques is proposed under a unified setup. The performance of the proposed strategies is thoroughly compared. Methods of estimating the nonnegative mean squares or variances of the estimators of the means or totals are proposed. In addition, some of the important RR techniques, such as Dalenius and Vitale (1974), Liu et al. (1975), Gjestavang and Singh's (2006) RR model, and Quatember's (2009) RR model, have been discussed in detail.

Keywords Linear homogeneous unbiased estimators · Mean-square estimation · Probability proportional to aggregate size (PPAS) · Probability proportional to size with replacement (PPSWR) · Probability proportional to size without replacement (PPSWOR) · Sampling strategies · Sufficient statistic · Variance estimation

5.1 Introduction

Consider a finite population $U = (1, .., i, .., N)$ of N identifiable units from which a sample s of size n is selected with probability $p(s)$ using a sampling design p. Let $\pi_i = \sum_{s \supset i} p(s)$ and $\pi_{ij} = \sum_{s \supset i,j} p(s)$ be the inclusion probabilities for the i th, and i and $j(\neq i)$ th units of the population, which are assumed to be positive. Further, let y_i be the value of a sensitive characteristic y for the ith unit of the population. The population total and the mean of y are denoted by $Y = \sum_{i \in U} y_i$ and $\overline{Y} = \sum_{i \in U} y_i/N$, respectively. In the case, y is a qualitative or categorical variable, we define $y_i = 1$ if the ith unit possesses the sensitive attribute, say, A and $y_i = 0$ otherwise. The proportion of units in the population that possesses the sensitive characteristic A is $\pi_A = \overline{Y}$. Our objective is to estimate $\overline{Y}$ or π_A by using a sample s selected from the population U. Because y is a sensitive variable, no direct response to y is available from the respondents. Hence, a suitable randomized response device is employed to obtain responses from each of the selected respondents in the sample independently. Let z_i be the randomized response obtained from the ith respondent. We assume that $r_i = g(z_i)$, a known function of z_i that satisfy the following model proposed by Arnab (1990):

R. Arnab, *Indirect Methods of Data Collection and Analysis from Surveys*, Indian Statistical Institute Series, https://doi.org/10.1007/978-981-97-6005-3_5

$$E_R(r_i) = y_i,\ V_R(r_i) = \phi_i \text{ and } C_R(r_i, r_j) = 0 \text{ for } i \neq j \tag{5.1.1}$$

Here r_i is an unbiased estimator of y_i and ϕ_i is a known function of y_i alone. We have shown in Sect. 3.9 that most of the RR techniques involving qualitative and quantitative characteristics satisfy the model (5.1.1). A list of such RR techniques is given in Table 3.9.1.

The methods of estimating the population total (or mean) for various sampling strategies are described in detail under the general RR model. Comparison of various strategies based on the SRSWR and varying probability sampling designs (VPPS) are proposed. Nonnegative methods of estimation of mean squares (MSEs) and variances are proposed for general sampling strategies. Some of the important RR techniques, such as Dalenius and Vitale (1974), Liu et al. (1975), Gjestavang and Singh's (2006) RR model, and Quatember's (2009) RR model, have been extensively discussed.

5.2 Estimation of Total and Proportion

In this section, we will demonstrate how one can modify the estimators for the direct response (DR) surveys described in Sect. 4.2.3 for the RR surveys to estimate parameters of interest. The linear homogeneous unbiased estimator of the population total Y in a DR survey is defined in Sect. 4.2.7 as follows:

$$t(s, y) = \sum_{i \in s} w_{si} y_i \tag{5.2.1}$$

where $\sum_{i \in s}$ denotes the sum over the distinct units in s and w_{si}'s are the known weights that are free from y_i or r_i s. The weights w_{si} s satisfy the unbiasedness condition:

$$\sum_{s \supset i} w_{si} p(s) = 1 \tag{5.2.2}$$

The variance of $t(s, y)$ and its unbiased estimator $\widehat{\text{Var}}[t(s, y)]$ were presented in Theorem 4.2.1. as follows:

$$\text{Var}[t(s, y)] = \sum_{i \in U} (\alpha_i - 1) y_i^2 + \sum_{i \neq} \sum_{j \in U} (\alpha_{ij} - 1) y_i y_j \tag{5.2.3}$$

and

$$\widehat{\text{Var}}[t(s, y)] = \sum_{i \in s} c_{si} y_i^2 + \sum_{i \neq} \sum_{j \in s} c_{sij} y_i y_j \tag{5.2.4}$$

where

$$\alpha_i = \sum_{s\supset i} w_{si}^2 p(s),\ \ \alpha_{ij} = \sum_{s\supset i,j} w_{si} w_{sj} p(s),$$
$$\sum_{s\supset i} c_{si} p(s) = \alpha_i - 1 \text{ and } \sum_{s\supset i,j} c_{si} p(s) = \alpha_{ij} - 1 \tag{5.2.5}$$

Since r_i is an unbiased estimator of y_i, we propose an unbiased estimator of Y for the RR survey by replacing y_i with its unbiased estimator r_i in $t(s, y)$ The resulting estimator is given by

$$t(s, r) = t(s, y)|_{y_i = r_i} = \sum_{i\in s} w_{si} r_i \tag{5.2.6}$$

Theorem 5.2.1

(i) $t(s, r)$ is an unbiased estimator Y.
(ii) The variance of $t(s, r)$ is

$$\begin{aligned}\text{Var}[t(s, r)] &= \text{Var}[t(s, y)]|_{y_i = r_i} + \sum_{i\in U} \alpha_i \phi_i \\ &= \left[\sum_{i\in U} (\alpha_i - 1) y_i^2 + \sum_{i\neq} \sum_{j\in U} (\alpha_{ij} - 1) y_i y_j\right] + \sum_{i\in U} \alpha_i \phi_i.\end{aligned} \tag{5.2.7}$$

(iii) An unbiased estimator of $\text{Var}[t(s, r)]$ is

$$\begin{aligned}\hat{\text{Var}}[t(s, r)] &= \hat{\text{Var}}[t(s, y)]|_{y_i = r_i} + \sum_{i\in s} w_{si} \hat{\phi}_i \\ &= \left[\sum_{i\in s} c_{si} r_i^2 + \sum_{i\neq} \sum_{j\in s} c_{sij} r_i r_j\right] + \sum_{i\in s} w_{si} \hat{\phi}_i\end{aligned} \tag{5.2.8}$$

where $\hat{\phi}_i$ is an unbiased estimator of ϕ_i, i.e., $E_R(\hat{\phi}_i) = \phi_i$.

Proof

(i)
$$\begin{aligned}E[t(s, r)] &= E_p\left[E_r\left(\sum_{i\in s} w_{si} r_i\right)\right] \\ &= E_p\left[\sum_{i\in s} w_{si} E_R(r_i)\right] \\ &= E_p[t(s, y)] \\ &= Y \text{ (noting } \sum_{s\supset i} w_{si} p(s) = 1)\end{aligned}$$

$$\text{(ii)} \quad \operatorname{Var}[t(s,r)] = V_p[E_R\{t(s,r)\}] + E_p[V_R\{t(s,r)\}] \tag{5.2.9}$$

Now noting,

$$\begin{aligned} V_p[E_R\{t(s,r)\}] &= V_p[t(s,y)] \\ &= \sum_{i\in U}(\alpha_i - 1)y_i^2 + \sum_{i\neq}\sum_{j\in U}(\alpha_{ij}-1)y_i y_j \quad \text{(using 5.2.3)} \end{aligned} \tag{5.2.10}$$

and

$$\begin{aligned} \text{(iii)}\ E_p[V_R\{t(s,r)\}] &= E_p\left[\sum_{i\in s} w_{si}^2\phi_i\right] \\ &= \sum_{i\in U}\phi_i \sum_{s\supset i} w_{si}^2 p(s) = \sum_{i\in U}\alpha_i\phi_i (\text{noting}\ \alpha_i = \sum_{s\supset i} w_{si}^2 p(s)) \end{aligned} \tag{5.2.11}$$

Equations (5.2.9), (5.2.10), and (5.2.11) yield the result.

$$\begin{aligned} E[\widehat{\operatorname{Var}}\{t(s,r)\}] &= E_p[E_R[\widehat{\operatorname{Var}}\{t(s,r)\}]] \\ &= E_p\left[\sum_{i\in s} c_{si}E_R(r_i^2) + \sum_{i\neq}\sum_{j\in s} c_{sij}E_R(r_i r_j)\right] \\ &\quad + E_p\left[\sum_{i\in s} w_{si}E_R(\hat{\phi}_i)\right] \\ &= E_p\left[\sum_{i\in s} c_{si}(y_i^2 + \phi_i) + \sum_{i\neq}\sum_{j\in s} c_{sij}y_i y_j\right] \\ &\quad + E_p\left[\sum_{i\in U} w_{si}\phi_i\right] \\ &= \left[\sum_{i\in U}(y_i^2+\phi_i)\sum_{s\supset i} c_{si}p(s) + \sum_{i\neq}\sum_{j\in U} y_i y_j \sum_{s\supset i,j} c_{sij}p(s)\right] \\ &\quad + E_p\left[\sum_{i\in U}\phi_i \sum_{s\supset i} w_{si}p(s)\right] \\ &= \operatorname{Var}[t(s,r)] \end{aligned}$$

(using $\sum_{s\supset i} c_{si}p(s) = \alpha_i - 1$, $\sum_{s\supset i,j} c_{sij}p(s) = \alpha_{ij} - 1$, and $\sum_{s\supset i} w_{si}p(s) = 1$)

Estimation of Proportion

For estimating population proportion $\pi_A = Y/N$, one needs to modify Theorem 5.2.1 by remembering $y_i = 1$ for $i \in A$ and $y_i = 0$ for $i \notin A$. The modified formula for the proportion is obtained straight from Theorem 5.2.1 as follows:

Theorem 5.2.2

(i) $\hat{\pi}_A = \frac{1}{N} t(s, r) = \frac{1}{N} \sum_{i \in s} w_{si} r_i$ is an unbiased estimator Y.
(ii) The variance of $\hat{\pi}_A$ is

$$\text{Var}[\hat{\pi}_A] = \frac{1}{N^2}\left[\sum_{i \in A}(\alpha_i - 1)y_i + \sum_{i \neq}\sum_{j \in A}(\alpha_{ij} - 1)y_i y_j\right] + \frac{1}{N^2}\left(\sum_{i \in U}\alpha_i \phi_i\right).$$

(iii) An unbiased estimator of $\text{Var}[\hat{\pi}_A]$ is

$$\widehat{\text{Var}}[\hat{\pi}_A] = \frac{1}{N^2}\left[\sum_{i \in s} c_{si} r_i^2 + \sum_{i \neq}\sum_{j \in s} c_{sij} r_i r_j\right] + \frac{1}{N^2}\sum_{i \in s} w_{si}\hat{\phi}_i$$

Remark 5.2.1 Chaudhuri (2011) proposed an alternative expression of variance of $\text{Var}[\hat{\pi}_A]$ as

$$\widehat{\text{Var}}[\hat{\pi}_A]_1 = \frac{1}{N^2}\left[\sum_{i \in s}(\alpha_i - 1)\frac{r_i}{\pi_i} + \sum_{i \neq}\sum_{j \in s}(\alpha_{ij} - 1)\frac{r_i r_j}{\pi_{ij}}\right] + \frac{1}{N^2}\left(\sum_{i \in s}\frac{\alpha_i \hat{\phi}_i}{\pi_i}\right). \tag{5.2.13}$$

It should be noted that $\hat{\text{Var}}[\hat{\pi}_A] = \hat{\text{Var}}[\hat{\pi}_A]_1$ whenever $\hat{\phi}_i = r_i(r_i - 1)$.

5.3 Estimation of the Total, Variance, and Unbiased Estimator of the Variance for Different Sampling Strategies

5.3.1 Arbitrary Sampling Design

Substituting $w_{si} = 1/\pi_i$ in Eq. (5.2.6), we obtain the estimator of the total Y for the Horvitz-Thompson estimator as

$$\hat{Y}_{\text{HT}}(r) = \sum_{i \in s}\frac{r_i}{\pi_i} \tag{5.3.1}$$

The estimator $\hat{Y}_{\mathrm{HT}}(r)$ is unbiased for Y as

$$\begin{aligned} E\left[\hat{Y}_{\mathrm{HT}}(r)\right] &= E_p\left[\sum_{i\in s}\frac{E_R(r_i)}{\pi_i}\right] \\ &= E_p\left[\sum_{i\in s}\frac{y_i}{\pi_i}\right] \\ &= Y \end{aligned} \tag{5.3.2}$$

The variance of $\hat{Y}_{\mathrm{HT}}(r)$ is

$$\begin{aligned} \mathrm{V}_{\mathrm{HT}}(r) &= \mathrm{Var}\left[\hat{Y}_{\mathrm{HT}}(r)\right] \\ &= V_p\left[E_R\left(\sum_{i\in s}\frac{r_i}{\pi_i}\right)\right] + E_p\left[V_R\left(\sum_{i\in s}\frac{r_i}{\pi_i}\right)\right] \\ &= V_p\left(\sum_{i\in s}\frac{y_i}{\pi_i}\right) + E_p\left(\sum_{i\in s}\frac{\phi_i}{\pi_i^2}\right) \end{aligned} \tag{5.3.3}$$

Now noting $E_p\left(\sum_{i\in s}\frac{\phi_i}{\pi_i^2}\right) = \sum_{i\in U}\frac{\phi_i}{\pi_i}$ and using Eq. (4.4.3), one obtains

$$\mathrm{V}_{\mathrm{HT}}(r) = \left[\sum_{i\in U}\frac{1-\pi_i}{\pi_i}y_i^2 + \sum_{i\neq}\sum_{j\in U}\frac{\pi_{ij}-\pi_i\pi_j}{\pi_i\pi_j}y_iy_j\right] + \left(\sum_{i\in U}\frac{\phi_i}{\pi_i}\right) \tag{5.3.4}$$

Substituting $w_{si} = 1/\pi_i$, $c_{si} = \frac{\alpha_i - 1}{\pi_i}$ and $c_{si} = \frac{\alpha_{ij}-1}{\pi_{ij}}$ in $\hat{\mathrm{Var}}[t(s,r)]$ of Theorem 5.2.1, one finds

$$\hat{\mathrm{V}}_{\mathrm{HT}}(r) = \left[\sum_{i\in s}\frac{1-\pi_i}{\pi_i^2}r_i^2 + \sum_{i\neq}\sum_{j\in s}\frac{\pi_{ij}-\pi_i\pi_j}{\pi_i\pi_j\pi_{ij}}r_ir_j\right] + \left(\sum_{i\in s}\frac{\hat{\phi}_i}{\pi_i}\right) \tag{5.3.5}$$

For a fixed effective sample of size n design, the variance of $\hat{\mathrm{V}}_{\mathrm{HT}}(r)$ and its unbiased estimator are as follows:

$$\mathrm{V}_{\mathrm{HT}}(r) = \sum_{i\neq}\sum_{j\in U}(\pi_i\pi_j - \pi_{ij})\left(\frac{y_i}{\pi_i} - \frac{y_j}{\pi_j}\right)^2 + \left(\sum_{i\in U}\frac{\phi_i}{\pi_i}\right) \tag{5.3.6}$$

$$\hat{\mathrm{V}}_{\mathrm{YG}}(r) = \sum_{i\neq}\sum_{j\in s}\frac{\pi_i\pi_j - \pi_{ij}}{\pi_{ij}}\left(\frac{r_i}{\pi_i} - \frac{r_j}{\pi_j}\right)^2 + \left(\sum_{i\in s}\frac{\hat{\phi}_i}{\pi_i}\right) \tag{5.3.7}$$

5.3.2 SRSWOR Sampling Design

For the SRSWOR, $\pi_i = n/N$ and $\pi_{ij} = n(n-1)/\{N(N-1)\}$. In this case, expressions $\hat{Y}_{HT}(r)$, $V_{HT}(r)$, and $\hat{V}_{HT}(r) = \widehat{V}_{YG}(r)$ reduce to

$$\begin{aligned}\hat{Y}_s(r) &= N\bar{r}_s,\\ V\big[\hat{Y}_s(r)\big] &= N^2\left(\frac{1}{n}-\frac{1}{N}\right)S_y^2 + \frac{N}{n}\sum_{i\in U}\phi_i\\ \text{and } \hat{V}\big[\hat{Y}_s(r)\big] &= N^2\left(\frac{1}{n}-\frac{1}{N}\right)s_r^2 + \frac{N}{n}\sum_{i\in s}\hat{\phi}_i\end{aligned} \tag{5.3.8}$$

where $\bar{r}_s = \sum_{i\in s} r_i/n$, $(N-1)S_y^2 = \sum_{i\in U}(y_i - \overline{Y})^2$, and $(n-1)s_r^2 = \sum_{i\in s}(r_i - \bar{r}_s)^2$.

5.3.3 PPSWOR Sampling Design

Murthy's estimators for the total Y, variance and unbiased estimator of variance for the DR surveys based on the PPSWOR sampling scheme (described in Sect. 4.4.5) are given in Theorem 4.4.8 as follows:

$$\hat{Y}_{MUR} = \sum_{i\in s}\frac{p(s|i)}{p(s)}y_i,\ \mathrm{Var}(\hat{Y}_{MUR}) = \sum_{i\neq}\sum_{j\in U}\beta_{ij}p_ip_j\left(\frac{y_i}{p_i}-\frac{y_j}{p_j}\right)^2$$

and

$$\widehat{\mathrm{Var}}(\hat{Y}_{MUR}) = \frac{1}{2\{p(s)\}^2}\sum_{i\neq}\sum_{j\in s}\{p(s|i,j)p(s) - p(s|i)p(s|j)\}p_ip_j\left(\frac{y_i}{p_i}-\frac{y_j}{p_j}\right)^2$$

Using Theorem 5.2.1, Murthy's estimator for the total, the variance of the estimator, and an unbiased estimator of the variance for the RR survey are obtained as follows:

$$\hat{Y}_{MUR}(r) = \hat{Y}_{MUR}\Big|_{y_i=r_i} = \sum_{i\in s}\frac{p(s|i)}{p(s)}r_i, \tag{5.3.9}$$

$$\begin{aligned}\mathrm{Var}\big[\hat{Y}_{MUR}(r)\big] &= \mathrm{Var}\big[\hat{Y}_{MUR}\big] + \sum_{i\in U}\alpha_i^*\phi_i \,with\, \alpha_i^* = \sum_{s\supset i}\frac{\{p(s|i)\}^2}{p(s)}\\ &= \sum_{i\neq}\sum_{j\in U}\beta_{ij}p_ip_j\left(\frac{y_i}{p_i}-\frac{y_j}{p_j}\right)^2 + \sum_{i\in U}\alpha_i^*\phi_i\end{aligned} \tag{5.3.10}$$

and

$$
\begin{aligned}
\widehat{\text{Var}}\left[\hat{Y}_{\text{MUR}}(r)\right] &= \text{Var}\left[\hat{Y}_{\text{MUR}}\right]\Big|_{y_i=r_i} + \sum_{i\in s}\alpha_i^*\hat{\phi}_i \\
&= \frac{1}{2\{p(s)\}^2}\sum_{i\neq}\sum_{j\in s}\{p(s|i,j)p(s) - p(s|i)p(s|j)\}p_ip_j\left(\frac{r_i}{p_i} - \frac{r_j}{p_j}\right)^2 \\
&\quad + \sum_{i\in s}\frac{p(s|i)}{p(s)}\hat{\phi}_i
\end{aligned}
\tag{5.3.11}
$$

5.3.4 Rao-Hartley-Cochran Sampling Scheme

For the Rao-Hartley-Cochran (RHC) sampling scheme with N/n an integer (defined in Sect. 4.4.7), the estimator for the total, its variance, and an unbiased estimator of the variance for the DR surveys are obtained from Theorem 4.4.10 as follows:

$$
\hat{Y}_{\text{RHC}} = \sum_{j_k\in s}\frac{y_{j_k}}{p_{j_k}}P_{j_k},\ \text{Var}(\hat{Y}_{\text{RHC}}) = \frac{N-n}{n(N-1)}\sum_{i\in U}p_i\left(\frac{y_i}{p_i} - Y\right)^2
$$

and

$$
\widehat{\text{Var}}(\hat{Y}_{\text{RHC}}) = \frac{N-n}{N(n-1)}\sum_{j_k\in s}P_{j_k}\left(\frac{y_{j_k}}{p_{j_k}} - \hat{Y}_{\text{RHC}}\right)^2
$$

Now noting $w_{si} = P_{j_k}/p_{j_k}$, one finds from Theorem 5.2.1, the expression of the unbiased estimator of the total Y as

$$
\hat{Y}_{\text{RHC}}(r) = \hat{Y}_{\text{RHC}}\Big|_{y_{j_k}=r_{j_k}} = \sum_{j_k\in s}\frac{r_{j_k}}{p_{j_k}}P_{j_k}
\tag{5.3.12}
$$

where r_{j_k} is the RR obtained from the j_kth unit of the group G_k that was formed for the selection of sample by the RHC sampling scheme.

Although in the RHC sampling scheme the expression of $w_{si} = P_{j_k}/p_{j_k}$ is simple, the expression of $p(s)$ is not so. Hence, we compute the expression of variance as follows:

$$
\begin{aligned}
\text{Var}\left[\hat{Y}_{\text{RHC}}(r)\right] &= V\left[E_R\{\hat{Y}_{\text{RHC}}(r)\}\right] + E\left[V_R\{\hat{Y}_{\text{RHC}}(r)\}\right] \\
&= V(\hat{Y}_{\text{RHC}}) + E\left[\sum_{j_k\in s}\left(\frac{P_{j_k}}{p_{j_k}}\right)^2\phi_{j_k}\right]
\end{aligned}
\tag{5.3.13}
$$

Now noting $\mathrm{Var}\left[\hat{Y}_{\mathrm{RHC}}\right] = \frac{N-n}{n(N-1)} \sum_{i\in U} p_i \left(\frac{y_i}{p_i} - Y\right)^2$ and $E\left[\sum_{j_k\in s} \left(\frac{P_{j_k}}{p_{j_k}}\right)^2 \phi_{j_k}\right] = \sum_{i\in U} a_i \phi_i$ with $a_i = \frac{1}{n(N-1)}\left[N(n-1) + \frac{N-n}{p_i}\right]$ (see Arnab (2017, page 525–526)), one finds

$$\mathrm{Var}\left[\hat{Y}_{\mathrm{RHC}}(r)\right] = \frac{N-n}{n(N-1)} \sum_{i\in U} p_i \left(\frac{y_i}{p_i} - Y\right)^2 + \sum_{i\in U} a_i \phi_i \tag{5.3.14}$$

Finally using (5.2.8), the expression of the unbiased estimator $\widehat{\mathrm{Var}}\left[\hat{Y}_{\mathrm{RHC}}(r)\right]$ is obtained as follows:

$$\widehat{\mathrm{Var}}\left[\hat{Y}_{\mathrm{RHC}}(r)\right] = \frac{N-n}{N(n-1)} \sum_{j_k\in s} P_{j_k} \left(\frac{r_{j_k}}{p_{j_k}} - \hat{Y}_{\mathrm{RHC}}(r)\right)^2 + \sum_{j_k\in s} \frac{P_{j_k}}{p_{j_k}} \hat{\phi}_{j_k} \tag{5.3.15}$$

Remark 5.3.1 It is important to note that the variance expression

$$\mathrm{Var}\left[\hat{Y}_{\mathrm{RHC}}(r)\right] = \frac{N-n}{n(N-1)} \sum_{i\in U} p_i \left(\frac{y_i}{p_i} - Y\right)^2 + \sum_{i\in U} \phi_i \left(\frac{1}{P_i} - 1\right)$$

provided by Chaudhuri and Christofides (2013, page 40) is incorrect.

5.3.5 PPAS Sampling Scheme

The expressions of the unbiased estimator of the total, variance, and unbiased estimator of the variance for the PPAS sampling scheme proposed by Lahiri (1951), Midzuno (1952), and Sen (1953) for the DR survey are given in Theorem 4.4.11 as follows:

$$\hat{Y}_{\mathrm{LMS}} = \frac{y_s}{x_s} X, \quad \mathrm{Var}(\hat{Y}_{\mathrm{LMS}}) = \sum_{i\in U} \left(\frac{1}{T_i} - 1\right) y_i^2 + \sum\sum_{i\neq j\in U} \left(\frac{1}{T_{ij}} - 1\right) y_i y_j$$

and

$$\widehat{\mathrm{Var}}(\hat{Y}_{\mathrm{LMS}}) = \sum_{i\in s} \left(\frac{1}{T_i} - 1\right) \frac{y_i^2}{\pi_i} + \sum\sum_{i\neq j\in s} \left(\frac{1}{T_{ij}} - 1\right) \frac{y_i y_j}{\pi_{ij}}.$$

Details of the sampling scheme and notations are given in Sect. 4.4.3.

Now noting $w_{si} = X/x_s$, one finds the expression for the estimator of the total, and unbiased estimator of the variance for the RR surveys using Theorem 5.2.1 as follows:

$$\hat{Y}_{\text{LMS}}(r) = \hat{Y}_{\text{LMS}}(r)\big|_{y_i=r_i} = \frac{r_s}{x_s}X \text{ with } r_s = \sum_{i\in s} r_i$$

$$\begin{aligned}\text{Var}\big[\hat{Y}_{\text{LMS}}(r)\big] &= \text{Var}(\hat{Y}_{\text{LMS}})\big|_{y_i=r_i} + \sum_{i\in U}\overline{\alpha}_i\phi_i \\ &= \sum_{i\in U}\left(\frac{1}{T_i}-1\right)y_i^2 + \sum\sum_{i\neq j\in U}\left(\frac{1}{T_{ij}}-1\right)y_iy_j + \sum_{i\in U}\overline{\alpha}_i\phi_i \end{aligned} \quad (5.3.16)$$

where $\overline{\alpha}_i = \sum_{s\supset i}\left(\frac{X}{x_s}\right)^2 p(s) = \sum_{s\supset i}\frac{X}{x_s}\Big/\binom{N-1}{n-1}$

$$\begin{aligned}\hat{\text{Var}}\big[\hat{Y}_{\text{LMS}}(r)\big] &= \hat{\text{Var}}(\hat{Y}_{\text{LMS}})\big|_{y_i=r_i} + \sum_{i\in s}\frac{X}{x_s}\hat{\phi}_i \\ &= \sum_{i\in s}\left(\frac{1}{T_i}-1\right)\frac{r_i^2}{\pi_i} + \sum\sum_{i\neq j\in s}\left(\frac{1}{T_{ij}}-1\right)\frac{r_ir_j}{\pi_{ij}} + \sum_{i\in s}\frac{X}{x_s}\hat{\phi}_i \end{aligned} \quad (5.3.17)$$

5.3.6 PPSWR Sampling Scheme

For the PPSWR sampling scheme, a respondent may be selected more than once in a sample. Hence, we assume that if a respondent is selected $n_i(\geq 1)$ times, the respondent performs RR trials n_i times independently and provides n_i randomized responses independently. We denote $r_{(j)}$ as the randomize response obtained from the respondent that selected at the $j(=1, .., n)$th draw with probability $p_{(j)}$, i.e., if the jth draw produces $i(=1, ..N)$th unit, then $r_{(j)} = r_i$, $p_{(j)} = p_i$, and $\phi_{(j)} = \phi_i$. The Hansen-Hurwitz estimator for the total Y for the PPSWR sampling scheme is

$$\hat{Y}_{\text{HH}}(r) = \frac{1}{n}\sum_{j=1}^{n}\frac{r_{(j)}}{p_{(j)}} \quad (5.3.18)$$

The expectation, variance, and unbiased estimator of the variance of $\hat{Y}_{\text{HH}}(r)$ are obtained as follows:

$$\begin{aligned}E\big[\hat{Y}_{\text{HH}}(r)\big] &= E_p\big[E_R\{\hat{Y}_{\text{HH}}(r)\}\big] \\ &= E_p\left(\frac{1}{n}\sum_{j=1}^{n}\frac{y_{(j)}}{p_{(j)}}\right) \\ &= Y,\end{aligned}$$

$$
\begin{aligned}
\operatorname{Var}\left[\hat{Y}_{\mathrm{HH}}(r)\right] &= V_p\left[E_R\{\hat{Y}_{\mathrm{HH}}(r)\}\right] + E_p\left[V_R\{\hat{Y}_{\mathrm{HH}}(r)\}\right] \\
&= V_p\left(\frac{1}{n}\sum_{j=1}^{n}\frac{y_{(j)}}{p_{(j)}}\right) + E_p\left(\frac{1}{n^2}\sum_{j=1}^{n}\frac{\phi_{(j)}}{p_{(j)}^2}\right) \\
&= \frac{1}{n}\sum_{i\in U} p_i\left(\frac{y_i}{p_i} - Y\right)^2 + \frac{1}{n}\sum_{i\in U}\frac{\phi_i}{p_i}
\end{aligned} \tag{5.3.19}
$$

An unbiased estimator of variance is

$$
\widehat{\operatorname{Var}}\left[\hat{Y}_{\mathrm{HH}}(r)\right] = \frac{1}{n(n-1)}\sum_{j=1}^{n}\left(\frac{r_{(j)}}{p_{(j)}} - \hat{Y}_{\mathrm{HH}}(r)\right)^2 \tag{5.3.20}
$$

as $E\{\widehat{\operatorname{Var}}[\hat{Y}_{\mathrm{HH}}(r)]\} = \operatorname{Var}\left[\hat{Y}_{\mathrm{HH}}(r)\right]$.

5.3.7 SRSWR Sampling Scheme

The PPSWR sampling scheme reduces to SRSWR if $p_i = 1/N$ for all $i \in U$. Hence substituting $p_i = 1/N$ in the expressions of $\hat{Y}_{\mathrm{HH}}(r)$, $\operatorname{Var}\left[\hat{Y}_{\mathrm{HH}}(r)\right]$, and $\widehat{\operatorname{Var}}\left[\hat{Y}_{\mathrm{HH}}(r)\right]$, one obtains unbiased estimator of the total Y, its variance and unbiased estimator of the variance for the SRSWR sampling scheme as follows:

$$
\hat{Y}_{\mathrm{WR}}(r) = N\,\overline{r} \quad \text{with } \overline{r} = \frac{1}{n}\sum_{j=1}^{n} r_{(j)} \tag{5.3.21}
$$

$$
\operatorname{Var}\left[\hat{Y}_{\mathrm{WR}}(r)\right] = N^2\frac{\sigma_y^2 + \overline{\phi}}{n} \tag{5.3.22}
$$

and

$$
\widehat{\operatorname{Var}}\left[\hat{Y}_{\mathrm{WR}}(r)\right] = N^2\frac{\overline{s}_r^2}{n} \tag{5.3.23}
$$

where $\sigma_y^2 = \frac{1}{N}\sum\limits_{i\in U}(y_i - \overline{Y})^2$, $\overline{\phi} = \frac{1}{N}\sum\limits_{i\in U}\phi_i$, and $\overline{s}_r^2 = \sum\limits_{j=1}^{n}(r_{(j)} - \overline{r})^2/(n-1)$.

Remark 5.3.2 Chaudhuri and Pal (2008), and Chaudhuri (2011) proposed a series of unbiased estimators of $\operatorname{Var}(T_1)$ and $\operatorname{Var}(T_2)$ which are highly complex and difficult to use. On the other hand, the proposed unbiased estimators

$$
\widehat{\operatorname{Var}}(T_1) = \left(\frac{1}{\nu_s} - \frac{1}{N}\right)s_{\overline{r}}^2 + \frac{1}{N}\left(\frac{1}{\nu_s}\sum_{i\in s}\frac{\hat{\phi}_i}{n_{i(s)}}\right) \text{ and}
$$

$$\widehat{\text{Var}}(T_2) == \left(\frac{1}{\nu_s} - \frac{1}{N}\right) s_{r*}^2 + \left(\frac{1}{N}\frac{1}{\nu_s}\sum_{i \in s} \hat{\phi}_i\right)$$

presented here are simple and easy to handle.

Remark 5.3.3 It is important to note that for the PPSWR and SRSWR sampling designs, estimation of variance of the estimator of the total Y (or mean), unbiased estimator of ϕ_i is not required. But estimator of ϕ_i, i.e., $\hat{\phi}_i$ is required for estimating variance for the without replacement sampling designs.

5.3.8 A Tabular Presentation

The following table shows expressions of unbiased estimators of the population total, variances, and unbiased estimators of the variances for various sampling strategies (Table 5.1).

5.4 Comparison of Strategies

In a SRSWR sampling scheme, a unit can be repeated more than once in a sample. For the DR surveys, Basu (1958), Raj and Khamis (1958), and Chikkagoudar (1966), among others, compared the performances of the sample mean based on all the units, including repetition, with the sample mean based on the set of distinct units only. They found that the variance of the sample mean based on distinct units is smaller than that of the sample mean based on all the units. Basu (1958) showed that the data based on the set of distinct units provides a sufficient statistic. Hence, the result follows from the application of the Rao-Blackwellization technique.

It is also well known that for the DR surveys, the Horvitz-Thompson estimator based on Sampford's π ps sampling scheme, the RHC estimator based on the RHC sampling scheme, and Murthy's estimator based on the PPSWOR sampling scheme perform better than the Hansen-Hurwitz estimator based on the PPSWR sampling scheme for the same sample size. In this section, we will show that some of the results that hold for DR surveys may not hold for RR surveys.

In fact, Arnab (1995) proved the following results that do not hold for the DR surveys but hold for the RR surveys: (i) admissibility of the sample mean based on all units in the class of linear unbiased estimators under the SRSWR sampling scheme; (ii) admissibility of the Hansen-Hurwitz estimator in the class of linear unbiased estimators under the PPSWR sampling scheme; and (iii) the Horvitz-Thompson estimator is not the minimum variance unbiased estimator of the population total under the with replacement (WR) noncensus uni-cluster sampling design (UCSD).

Table 5.1 Estimators of the total, variances, and estimators of variances for different sampling strategies

Sampling designs	Unbiased estimators	Variance	Unbiased estimator of variance
Arbitrary	$\hat{Y}_{\mathrm{HT}}(r) = \sum_{i\in s}\frac{r_i}{\pi_i}$	$\sum_{i\in U}\frac{1-\pi_i}{\pi_i}y_i^2 + \sum_{i\neq}\sum_{j\in U}\frac{\pi_{ij}-\pi_i\pi_j}{\pi_i\pi_j}y_iy_j + \left(\sum_{i\in U}\frac{\phi_i}{\pi_i}\right)$	$\left[\sum_{i\in s}\frac{1-\pi_i}{\pi_i^2}r_i^2 + \sum_{i\neq}\sum_{j\in s}\frac{\pi_{ij}-\pi_i\pi_j}{\pi_i\pi_j\pi_{ij}}r_ir_j\right] + \left(\sum_{i\in s}\frac{\hat{\phi}_i}{\pi_i}\right)$
SRSWR	$\hat{Y}_{\mathrm{WR}}(r) = N\bar{r}$	$N^2\frac{\sigma_y^2+\bar{\phi}}{n}$	$N^2\frac{s_r^2}{n}$
SRSWOR	$\hat{Y}_s(r) = N\bar{r}_s$	$N^2\left(\frac{1}{n}-\frac{1}{N}\right)S_y^2 + \frac{N^2\bar{\phi}}{n}$	$N^2\left(\frac{1}{n}-\frac{1}{N}\right)s_r^2 + \frac{N}{n}\sum_{i\in s}\hat{\phi}_i$
PPSWR	$\hat{Y}_{\mathrm{HH}}(r) = \frac{1}{n}\sum_{j=1}^{n}\frac{r_{(j)}}{p_{(j)}}$	$\frac{1}{n}\sum_{i\in U}p_i\left(\frac{y_i}{p_i}-Y\right)^2 + \frac{1}{n}\sum_{i\in U}\frac{\phi_i}{p_i}$	$\frac{1}{n(n-1)}\sum_{j=1}^{n}\left(\frac{r_{(j)}}{p_{(j)}} - \hat{Y}_{\mathrm{HH}}(r)\right)^2$
PPSWOR	$\hat{Y}_{\mathrm{MUR}}(r) = \sum_{i\in s}\frac{p(s\|i)}{p(s)}r_i$	$\sum_{i\neq}\sum_{j\in U}\beta_{ij}p_ip_j\left(\frac{y_i}{p_i}-\frac{y_j}{p_j}\right)^2 + \sum_{i\in U}\alpha_i^*\phi_i$	$\frac{1}{2\{p(s)\}^2}\sum_{i\neq}\sum_{j\in s}q_{ij}p_ip_j\left(\frac{r_i}{p_i}-\frac{r_j}{p_j}\right)^2 + \sum_{i\in s}\frac{p(s\|i)}{p(s)}\hat{\phi}_i$; $q_{ij} = \{p(s\|i,j)p(s) - p(s\|i)p(s\|j)\}$

(continued)

Table 5.1 (continued)

Sampling designs	Unbiased estimators	Variance	Unbiased estimator of variance
RHC	$\hat{Y}_{\mathrm{RHC}}(r) = \sum_{j_k \in s} \frac{r_{j_k}}{p_{j_k}} P_{j_k}$	$\frac{N-n}{n(N-1)} \sum_{i \in U} p_i \left(\frac{y_i}{p_i} - Y\right)^2 + \sum_{i \in U} a_i \phi_i$	$\frac{N-n}{N(n-1)} \sum_{j_k \in s} P_{j_k} \left(\frac{r_{j_k}}{p_{j_k}} - \hat{Y}_{\mathrm{RHC}}(r)\right)^2 + \sum_{j_k \in s} \frac{P_{j_k}}{p_{j_k}} \hat{\phi}_{j_k}$
LMS	$\hat{Y}_{\mathrm{LMS}}(r) = \frac{r_s}{x_s} X$	$\sum_{i \in U} \left(\frac{1}{T_i} - 1\right) y_i^2 + \sum_{i \neq} \sum_{j \in U} \left(\frac{1}{T_{ij}} - 1\right) y_i y_j + \sum_{i \in U} \overline{\alpha}_i \phi_i$	$\sum_{i \in s} \left(\frac{1}{T_i} - 1\right) \frac{r_i^2}{\pi_i} + \sum_{i \neq} \sum_{j \in s} \left(\frac{1}{T_{ij}} - 1\right) \frac{r_i r_j}{\pi_{ij}} + \sum_{i \in s} \frac{X}{x_s} \hat{\phi}_i$

5.4.1 Strategies Based on SRSWR Sampling Scheme

The estimator of the population mean $\overline{Y}$ under RR model (5.1.1) is

$$\hat{\overline{Y}}_{\mathrm{WR}}(r) = \frac{\hat{Y}_{\mathrm{WR}}(r)}{N} = \frac{1}{n}\sum_{j=1}^{n} r_{(j)} = \overline{r}_{\mathrm{WR}}$$

where $r_{(j)}$ is the revised RR obtained from the respondent that selected at jth draw.

In an SWSWR, a unit may appear more than once in a sample. Let $n_i(s)$ be the number of times ith unit appears in the sample s, and each time the respondent performs the RR trial independently. Let r_{ij} be the revised response obtained by the ith respondent at the jth trial ($j = 1, \ldots n_i(s); n_i(s) \geq 1$). Then,

$$\overline{r}_{\mathrm{WR}} = \frac{1}{n}\sum_{i\in s} n_i(s)\overline{r}_i \tag{5.4.1}$$

where $\overline{r}_i = \frac{1}{n}\sum_{j=1}^{n_i(s)} r_{ij}$ and $\sum_{i\in s}$ is the sum of distinct units in s.

Denoting ν_s as the number of distinct units in the sample s, Arnab (1999) proposed the following alternative estimator for the mean $\overline{Y}$ as

$$T_1 = \frac{1}{\nu_s}\sum_{i\in s} \overline{r}_i \tag{5.4.2}$$

Arnab (1999) also proposed an alternative strategy where a respondent supplies only one RR even if he or she is selected more than once. Let r_i be the revised RR obtained from the ith respondent. Arnab's (1999) estimator of $\overline{Y}$ is

$$T_2 = \frac{1}{\nu_s}\sum_{i\in s} r_i \tag{5.4.3}$$

Mangat et al. (1995) proposed an alternative estimator of $\overline{Y}$ as

$$T_3 = \frac{1}{\nu_0}\sum_{i\in s_0} r_i \tag{5.4.4}$$

where s_0 is an SRSWOR sample of size $\nu_0 = E_p(\nu_s)$, the expected number of distinct units ν_s of an SRSWR sample of size n.

The variances (Var) and unbiased estimators of variances ($\widehat{\mathrm{Var}}$) of the above estimators are given in the following theorems:

Theorem 5.4.1

(i) $E(\overline{r}_{\mathrm{WR}}) = \overline{Y}$,

(ii) $\mathrm{Var}(\overline{r}_{\mathrm{WR}}) = \frac{\sigma_y^2 + \overline{\phi}}{n}$,
(iii) $\hat{\mathrm{Var}}(\overline{r}_{\mathrm{WR}}) = \frac{1}{n} s_r^2$,

where $s_r^2 = \frac{1}{n-1} \sum_{r=1}^{n} (r_{(j)} - \overline{r}_{\mathrm{WR}})^2$ and $\overline{\phi} = \frac{1}{N} \sum_{i \in U} \phi_i$.

Proof Follows straight from the results of Sect. 5.3.6.

Theorem 5.4.2

(i) $E(T_1) = \overline{Y}$,
(ii) $\mathrm{Var}(T_1) = \left[E_p\left(\frac{1}{\nu_s}\right) - \frac{1}{N}\right] S_y^2 + E_p\left(\frac{1}{\nu_s^2} \sum_{i \in s} \frac{\phi_i}{n_i(s)}\right)$,
(iii) $\hat{\mathrm{Var}}(T_1) = \left(\frac{1}{\nu_s} - \frac{1}{N}\right) s_{\overline{r}}^2 + \frac{1}{N}\left(\frac{1}{\nu_s} \sum_{i \in s} \frac{\hat{\phi}_i}{n_{i(s)}}\right)$,

where $s_{\overline{r}}^2 = \frac{1}{\nu_s - 1} \sum_{i \in s} (\overline{r}_i - T_1)^2$.

Proof

(i)
$$E(T_1) = E_p\left[\frac{1}{\nu_s} \sum_{i \in s} E_R(\overline{r}_i)\right]$$
$$= E_p\left(\frac{1}{\nu_s} \sum_{i \in s} y_i\right),$$
$$= \overline{Y}$$

(ii)
$$\mathrm{Var}(T_1) = V_p\left(\frac{1}{\nu_s} \sum_{i \in s} y_i\right) + E_p\left(\frac{1}{\nu_s^2} \sum_{i \in s} \frac{\phi_i}{n_i(s)}\right),$$
$$= V_p(\overline{y}_{\nu_s}) + E_p\left(\frac{1}{\nu_s^2} \sum_{i \in s} \frac{\phi_i}{n_i(s)}\right)$$

(where $\overline{y}_{\nu_s} = \frac{1}{\nu_s} \sum_{i \in s} y_i$ = sample mean of distinct units of Sin DR survey)

$$= \left[E_p\left(\frac{1}{\nu_s}\right) - \frac{1}{N}\right] S_y^2 + E_p\left(\frac{1}{\nu_s^2} \sum_{i \in s} \frac{\phi_i}{n_i(s)}\right)$$

(iii)
$$E[\hat{\mathrm{Var}}(T_1)] = E_p\left[\left(\frac{1}{\nu_s} - \frac{1}{N}\right) E_R(s_{\overline{r}}^2)\right] + E_P\left(\frac{1}{N} \frac{1}{\nu_s} \sum_{i \in s} \frac{E_R(\hat{\phi}_i)}{n_{i(s)}}\right).$$

Now noting $E_R(\hat{\phi}_i) = \phi_i$,

$$E_R(s_{\overline{r}}^2) = \frac{1}{\nu_s - 1} E_R\left(\sum_{i \in s} \overline{r}_i^2 - \nu_s T_1^2\right)$$

$$= \frac{1}{\nu_s - 1}\left[\sum_{i\in s}\left(y_i^2 + \frac{\phi_i}{n_{i(s)}}\right) - \nu_s\left(\overline{y}_s^2 + \frac{1}{\nu_s^2}\sum_{i\in s}\frac{\phi_i}{n_{i(s)}}\right)\right]$$
$$= \frac{1}{\nu_s - 1}\sum_{i\in s}(y_i - \overline{y}_s)^2 + \frac{1}{\nu_s}\sum_{i\in s}\frac{\phi_i}{n_{i(s)}}$$

and

$$E_p\left\{E_R\left(\frac{1}{\nu_s} - \frac{1}{N}\right)\overline{s}_r^2\right\} = \left[E_p\left(\frac{1}{\nu_s}\right) - \frac{1}{N}\right]S_y^2 + E_p\left(\frac{1}{\nu_s^2}\sum_{i\in s}\frac{\phi_i}{n_{i(s)}}\right)$$
$$- \frac{1}{N}E_p\left(\frac{1}{\nu_s}\sum_{i\in s}\frac{\phi_i}{n_{i(s)}}\right)$$
$$= \text{Var}(T_1) - \frac{1}{N}E_p\left(\frac{1}{\nu_s}\sum_{i\in s}\frac{\phi_i}{n_{i(s)}}\right),$$

we can verify (iii).

Theorem 5.4.3

(i) $E(T_2) = \overline{Y}$,

(ii) $\text{Var}(T_2) = \left[E_p\left(\frac{1}{\nu_s}\right) - \frac{1}{N}\right]S_y^2 + \overline{\phi}E_p\left(\frac{1}{\nu_s}\right)$,

(iii) $\widehat{\text{Var}}(T_2) = \left(\frac{1}{\nu_s} - \frac{1}{N}\right)s_{r*}^2 + \left(\frac{1}{N}\frac{1}{\nu_s}\sum_{i\in s}\hat{\phi}_i\right)$,

where $s_{r*}^2 = \frac{1}{\nu_s - 1}\sum_{i\in s}(r_i - T_2)^2$.

Proof

(i)
$$E(T_2) = E_p\left[\frac{1}{\nu_s}\sum_{i\in s}E_R(r_i)\right]$$
$$= E_p\left(\frac{1}{\nu_s}\sum_{i\in s}y_i\right),$$
$$= \overline{Y}$$

(ii)
$$\text{Var}(T_2) = V_p\left(\frac{1}{\nu_s}\sum_{i\in s}E_R(r_i)\right) + E_p\left(\frac{1}{\nu_s^2}\sum_{i\in s}V_R(r_i)\right)$$
$$= V_p(\overline{y}_{\nu_s}) + \overline{\phi}E_p\left(\frac{1}{\nu_s}\right),$$
$$= \left[E_p\left(\frac{1}{\nu_s}\right) - \frac{1}{N}\right]S_y^2 + \overline{\phi}E_p\left(\frac{1}{\nu_s}\right)$$

(iii) Noting $E\left(\frac{1}{N}\frac{1}{\nu_s}\sum_{i\in s}\hat{\phi}_i\right) = E_P\left(\frac{1}{N}\frac{1}{\nu_s}\sum_{i\in s}\phi_i\right) = \frac{\overline{\phi}}{N}$.

and

$$E\left[\left(\frac{1}{\nu_s}-\frac{1}{N}\right)s_{r*}^2\right]=E_p\left[\left(\frac{1}{\nu_s}-\frac{1}{N}\right)\frac{1}{\nu_s-1}E_R\left(\sum_{i\in s}r_i^2-\nu_sT_2^2\right)\right]$$
$$=E_p\left[\left(\frac{1}{\nu_s}-\frac{1}{N}\right)\frac{1}{\nu_s-1}\sum_{i\in s}(y_i-\overline{y}_s)^2+\left(\frac{1}{\nu_s}-\frac{1}{N}\right)\frac{1}{\nu_s}\sum_{i\in s}\phi_i\right]$$
$$=\left[E_p\left(\frac{1}{\nu_s}\right)-\frac{1}{N}\right]S_y^2+E_p\left(\frac{1}{\nu_s^2}\sum_{i\in s}\phi_i\right)-E_p\frac{1}{\nu_s}\frac{1}{N}\sum_{i\in s}\phi_i$$
$$=\mathrm{Var}(T_2)-\frac{1}{N}\overline{\phi},$$

we verify part (iii) of the theorem.

Theorem 5.4.4

(i) $E(T_3)=\overline{Y}$,

(ii) $\mathrm{Var}(T_3)==\left(\frac{1}{\nu_0}-\frac{1}{N}\right)S_y^2+\frac{1}{\nu_0}\overline{\phi}$,

(iii) $\hat{\mathrm{Var}}(T_3)=\left(\frac{1}{\nu_0}-\frac{1}{N}\right)s_0^2+\frac{1}{\nu_0}\sum_{i\in s}\hat{\phi}_i$,

where $s_0^2=\frac{1}{n_0-1}\sum_{i\in s_0}(r_i-T_3)^2$.

Proof Since s_0 is an SRSWOR sample of size n_0, the theorem follows from the results derived in Sect. 5.3.2.

5.4.1.1 Comparison of Strategies

(i) T_2 versus T_1,T_3 and $\overline{r}_{\mathrm{WR}}$.

Theorems 5.4.2 and 5.4.3 yield

$$\mathrm{Var}(T_2)-\mathrm{Var}(T_1)=E_p\left[\frac{\overline{\phi}}{\nu_s}-\frac{1}{\nu_s^2}\sum_{i\in s}\frac{\phi_i}{n_i(s)}\right]$$
$$\geq E_p\left[\frac{\overline{\phi}}{\nu_s}-\frac{1}{\nu_s^2}\sum_{i\in s}\phi_i\right]\text{ since } n_i(s)\geq 1$$
$$=E_p\left(\frac{\overline{\phi}}{\nu_s}-\frac{\overline{\phi}}{\nu_s}\right)$$
$$=0 \tag{5.4.5}$$

$$\begin{aligned}\mathrm{Var}(T_2) - \mathrm{Var}(T_3) &= \left[E_p\left(\frac{1}{\nu_s}\right) - \frac{1}{\nu_0}\right]S_y^2 + \overline{\phi}\left[E_p\left(\frac{1}{\nu_s}\right) - \frac{1}{\nu_0}\right]\\ &= (S_y^2 + \overline{\phi})E_p\left(\frac{1}{\nu_s} - \frac{1}{\nu_0}\right)\\ &\geq 0 \text{ as } E_p\left(\frac{1}{\nu_s}\right) \geq \frac{1}{E_p(\nu_s)} = \frac{1}{\nu_0}\end{aligned} \quad (5.4.6)$$

$$\begin{aligned}\mathrm{Var}(\overline{r}_{\mathrm{WR}}) - \mathrm{Var}(T_2) &= \left[\left\{\frac{\sigma_y^2}{n} - V_p(\overline{y}_{\nu_s})\right\} + \overline{\phi}\left\{\frac{1}{n} - E_p\left(\frac{1}{\nu_s}\right)\right\}\right]\\ &= \left[\left\{\frac{1}{n} - E_p\left(\frac{1}{\nu_s} - \frac{1}{N}\right)\frac{N}{N-1}\right\}\sigma_y^2 - \overline{\phi}\left\{E_p\left(\frac{1}{\nu_s}\right) - \frac{1}{n}\right\}\right]\end{aligned} \quad (5.4.7)$$

It is well known from Basu (1958) that $\sigma_y^2/n - V_p(\overline{y}_{\nu_s})$, the first component of (5.4.7) is positive while the second component is negative as $n > \nu_s$. Hence, the magnitude of the difference will depend on the values of σ_y^2 and $\overline{\phi}$. Hence,

$$\mathrm{Var}(\overline{r}_{\mathrm{WR}}) \begin{matrix}>\\=\\<\end{matrix} \mathrm{Var}(T_2) \text{ according as } \frac{\sigma_y^2}{\overline{\phi}} \begin{matrix}>\\=\\<\end{matrix} \left\{E_p\left(\frac{1}{\nu_s}\right) - \frac{1}{n}\right\}\left\{\frac{1}{n} - E_p\left(\frac{1}{\nu_s} - \frac{1}{N}\right)\frac{N}{N-1}\right\}^{-1} \quad (5.4.8)$$

(ii) $\overline{r}_{\mathrm{WR}}$ versus T_1 and T_3

$$\mathrm{Var}(\overline{r}_{\mathrm{WR}}) - \mathrm{Var}(T_1) = \left[\frac{\sigma_y^2}{n} - V_p(\overline{y}_{\nu_s})\right] + \left[\frac{\overline{\phi}}{n} - E_p\left(\frac{1}{\nu_s^2}\sum_{i\in s}\frac{\phi_i}{n_i(s)}\right)\right] \quad (5.4.9)$$

The first component of (5.4.9) is positive but the second component is negative if all ϕ_i's are equal to $\overline{\phi}$. Hence, the sign of the difference will depend on the magnitudes of σ_y^2 and $\overline{\phi}$. Hence,

$$\mathrm{Var}(\overline{r}_{\mathrm{WR}}) \begin{matrix}>\\=\\<\end{matrix} \mathrm{Var}(T_2) \text{ according as } \left\{\frac{\sigma_y^2}{n} - V_p(\overline{y}_{\nu_s})\right\} \begin{matrix}>\\=\\<\end{matrix} E_p\left(\frac{1}{\nu_s^2}\sum_{i\in s}\frac{\phi_i}{n_i(s)}\right) - \frac{\overline{\phi}}{n} \quad (5.4.10)$$

$$\mathrm{Var}(\overline{r}_{\mathrm{WR}}) - \mathrm{Var}(T_3) = \left[\frac{\sigma_y^2}{n} - \left(\frac{1}{\nu_0} - \frac{1}{N}\right)S_y^2\right] + \left(\frac{1}{n} - \frac{1}{\nu_0}\right)\overline{\phi} \quad (5.4.11)$$

Here also, the first component of (5.4.11) is positive and the second component is negative. Hence,

$$\mathrm{Var}(\overline{r}_{\mathrm{WR}}) \mathop{\gtreqless} \mathrm{Var}(T_3) \text{ according as } \left\{\frac{\sigma_y^2}{n} - V_p(\overline{y}_{\nu_s})\right\} \gtreqless \left(\frac{1}{\nu_0} - \frac{1}{n}\right)\overline{\phi} \tag{5.4.12}$$

For Warner's model, $y_i = 1(0)$ if ith unit does (does not) belong to the sensitive group A and $\phi_i = \frac{P(1-P)}{(2P-1)^2} = \phi_{\mathrm{W}}$. In this case, Mangat et al. (1995) obtained sufficient conditions for superiority and inferiority of T_3 over $\overline{r}_{\mathrm{WR}}$ are

$$\begin{aligned} &\sigma_y^2 > n(N-1)(6N+n-1)/[N\{6Nn-12-n(n-1)\}] \\ \text{and}\quad &\sigma_y^2 < n(N-1)\{359N+60(n-1)\}/[N\{360Nn-720N-60n(n-1)\}] \end{aligned} \tag{5.4.13}$$

From the above discussions, it reveals that none of the proposed estimators $\overline{r}_{\mathrm{WR}}$, T_1, T_2, and T_3 performs best in all the situations. Estimator T_2 is found inadmissible as it is dominated by the estimators T_1 and T_3.

Remark 5.4.1 Chaudhuri (2011) proposed the following two additional Horvitz-Thompson estimators for the mean $\overline{Y}$ under Warner's model with $\phi_i = \phi_{\mathrm{W}}$.

$$T_4 = \frac{1}{N\pi}\sum_{i\in s} r_i = \frac{1}{\nu}\sum_{i\in s} r_i \text{ and } T_5 = \frac{1}{N\pi}\sum_{i\in s} \overline{r}_i = \frac{1}{\nu}\sum_{i\in s} \overline{r}_i \tag{5.4.14}$$

where $\pi_i = \pi$ is the inclusion probability of the ith unit that satisfies $\sum_{i\in U}\pi_i = N\pi = E_p(\nu_s) = \nu$.

Both the estimators T_4 and T_5 are unbiased for $\overline{Y}$. But the variance of T_5 is lower than that of T_4 as

$$\begin{aligned} \mathrm{Var}(T_4) - \mathrm{Var}(T_5) &= E_p[V_R(T_4) - V_R(T_5)] \\ &= \frac{\phi_{\mathrm{W}}}{\nu^2} E_p\left(\nu_s - \sum_{i\in s}\frac{1}{n_i(s)}\right) \\ &\geq 0 \text{ as } n_i(s) \geq 1 \end{aligned} \tag{5.4.15}$$

The expressions of the variances of T_4 and T_5, and their unbiased estimators are obtained directly from the expressions as follows:

$$\mathrm{Var}(T_4) = Q + \frac{\phi_W}{\nu},\ \widehat{\mathrm{Var}}(T_4) = \hat{Q} + \left(\frac{\phi_{\mathrm{W}}\nu_s}{N\nu}\right) \tag{5.4.16}$$

$$\mathrm{Var}(T_5) = Q + \frac{\phi_W}{\nu^2} E_p\left(\sum_{i\in s}\frac{1}{n_{i(s)}}\right) \text{ and } \widehat{\mathrm{Var}}(T_5) = \hat{Q} + \left(\frac{\phi_{\mathrm{W}}}{N\nu}\sum_{i\in s}\frac{1}{n_i(s)}\right) \tag{5.4.17}$$

where $Q = \frac{1}{N^2}\left[\left(\frac{1}{\pi} - 1\right)\sum_{i\in U} y_i^2 + \left(\frac{\pi*}{\pi^2} - 1\right)\right)\sum_{i\neq}\sum_{j\in U} y_i y_j\right]$,

$\hat{Q} = \frac{1}{N^2}\left[\frac{1-\pi}{\pi^2}\sum_{i\in s}\bar{r}_i^2 + \frac{\pi*-\pi^2}{\pi^2\pi*}\sum_{i\neq}\sum_{j\in s}\bar{r}_i\bar{r}_j\right]$ and $\pi* =$ joint inclusion probability of any two units $= 1 - 2\left(1 - \frac{1}{N}\right)^n + \left(1 - \frac{2}{N}\right)^n$.

5.4.2 *Strategies Based on Varying Probability Sampling Scheme*

In this section, we will compare strategies based on various unequal probability sampling schemes under the RR model (5.1.1).

5.4.2.1 Horvitz-Thompson Estimator Versus Hansen-Hurwitz Estimator

The variance of the Horvitz-Thompson estimator for the population total Y based on an IPPS (or πps) sampling design of a fixed sample size n under the RR model (5.1.1) is given by

$$\begin{aligned}\mathrm{Var}\left[\hat{Y}_{\mathrm{HT}}(r)\right] &= \mathrm{Var}\left(\sum_{i\in s}\frac{r_i}{\pi_i}\right)\\ &= \frac{1}{2}\sum_{i\neq}\sum_{j\in U}(\pi_i\pi_j - \pi_{ij})\left(\frac{y_i}{\pi_i} - \frac{y_j}{\pi_j}\right)^2 + \sum_{i\in U}\frac{\phi_i}{\pi_i}\end{aligned} \tag{5.4.18}$$

The variance of the Hansen-Hurwitz estimator based on a PPSWR sampling scheme with normed size measure p_i for the ith unit under the RR model (5.1.1) is

$$\begin{aligned}\mathrm{Var}\left[\hat{Y}_{\mathrm{HH}}(r)\right] &= \mathrm{Var}\left(\frac{1}{n}\sum_{j=1}^{n}\frac{r_{(j)}}{p_{(j)}}\right)\\ &= \frac{1}{n}\sum_{i\in U}p_i\left(\frac{y_i}{p_i} - Y\right)^2 + \frac{1}{n}\sum_{i\in U}\frac{\phi_i}{p_i}\end{aligned} \tag{5.4.19}$$

Now noting the difference $\mathrm{Var}\left[\hat{Y}_{\mathrm{HH}}(r)\right] - \mathrm{Var}\left[\hat{Y}_{\mathrm{HT}}(r)\right] \geq 0$ for Sampford's (1967) (see Eq. (4.6.9)) and Brewer's (1975) sampling schemes (vide Sengupta (1986)), we conclude that for the RR model (5.1.1), the Horvitz-Thompson estimator based on Sampford's and Brewer's (1975) IPPS sampling schemes is more efficient than the

Hansen-Hurwitz estimator based on a PPSWR sampling scheme of the same sample size.

5.4.2.2 Rao-Hartley-Cochran Estimator Versus Hansen-Hurwitz Estimator

The variance of the RHC estimator (Theorem 4.4.10) based on a RR model when N/n is an integer is

$$\mathrm{Var}\left[\hat{Y}_{\mathrm{RHC}}(r)\right] = \frac{N-n}{N(n-1)}\sum_{i\in U} p_i\left(\frac{y_i}{p_i} - Y\right)^2 + \sum_{i\in U} a_i\phi_i \tag{5.4.20}$$

Equations (5.4.19) and (5.4.20) yield

$$\begin{aligned}\mathrm{Var}\left[\hat{Y}_{\mathrm{HH}}(r)\right]-\mathrm{Var}\left[\hat{Y}_{\mathrm{RHC}}(r)\right] &= \frac{n-1}{n(N-1)}\sum_{i\in U} p_i\left(\frac{y_i}{p_i} - Y\right)^2 \\ &+\frac{n-1}{n(N-1)}\sum_{i\in U}\phi_i\left(\frac{1}{p_i} - N\right)\end{aligned} \tag{5.4.21}$$

The first component of the above expression is positive, but the second component may not always be positive. Hence, we cannot conclude the superiority of the RHC estimator over the HH estimator. However, for the class of RR models where $\phi_i = \phi_{\mathrm{W}}$, is a constant free from y-values, which includes Warner's model, one finds

$$\begin{aligned}\mathrm{Var}\left[\hat{Y}_{\mathrm{HH}}(r)\right]-\mathrm{Var}\left[\hat{Y}_{\mathrm{RHC}}(r)\right] &= \frac{n-1}{n(N-1)}\sum_{i\in U} p_i\left(\frac{y_i}{p_i} - Y\right)^2 \\ &+\frac{n-1}{n(N-1)}\phi_{\mathrm{W}}\left(\sum_{i\in U}\frac{1}{p_i} - N^2\right) \geq 0\end{aligned} \tag{5.4.22}$$

as $\sum\limits_{i\in U}\frac{1}{p_i} \geq N^2$.

5.4.2.3 Murthy's Estimator Versus Hansen-Hurwitz Estimator

Difference between the variances of Hansen-Hurwitz and Murthy's estimator is

$$\mathrm{Var}\left[\hat{Y}_{\mathrm{HH}}(r)\right]-\mathrm{Var}\left[\hat{Y}_{\mathrm{MUR}}(r)\right] = \sum_{i\neq j}\sum_{\in U}\left(\frac{1}{n} - \beta_{ij}\right)p_ip_j\left(\frac{y_i}{p_i} - \frac{y_j}{p_j}\right)^2$$

$$+\sum_{i\in U}\phi_i\left[\frac{1}{np_i}-\frac{\{p(s|i)\}^2}{p(s)}\right] \tag{5.4.23}$$

Here also, the first component is positive, but the second component is not. For the RR model with $\phi_i = \phi_0$, the second component becomes positive (Sengupta, 2017). Hence, Murthy's estimator performs better than the Hansen-Hurwitz estimator if $\phi_i = \phi_0$.

5.5 Nonnegative Mean Square and Variance Estimation

Estimation of mean-square error (MSE) for the case of a biased estimator and variance for an unbiased estimator is essential for deriving the precision of an estimator, deriving confidence intervals, testing hypotheses of parameters, and determining the optimal sample size of a survey design, among others. Although the MSE or variance of an estimator is a nonnegative quantity, unbiased estimators may not be. If the variance of an estimator becomes negative, it cannot be used for interval estimation or testing hypotheses. In this section, expressions for MSEs and variances of a few unbiased estimators based on RR surveys are obtained, and their nonnegativity properties are studied.

5.5.1 DR Surveys

Consider the linear homogeneous estimator of the total Y as

$$e(s,y)=\sum_{i\in s}b_{si}y_i \tag{5.5.1}$$

where b_{si} s are known constants free of y_i's and $b_{si}=0$ if $i\notin s$.

The MSE of $e(s,y)$ is defined as

$$\begin{aligned}
\mathrm{M}[e(s,y)] &= E\left[\sum_{i\in s}b_{si}y_i - Y\right]^2\\
&= E\left[\sum_{i\in U}b_{si}I_{si}y_i - Y\right]^2;\ I_{si}=\begin{cases}1 \text{ if } i\in s\\ 0 \text{ if } i\notin s\end{cases}\\
&= E\left[\sum_{i\in U}(b_{si}I_{si}-1)y_i\right]^2
\end{aligned}$$

$$= \sum_{i \in U} \gamma_{ii} y_i^2 + \sum_{i \neq} \sum_{j \in U} \gamma_{ij} y_i y_j \tag{5.5.2}$$

where

$$\gamma_{ii} = E_p(b_{si} I_{si} - 1)^2 = \sum_{s \supset i} b_{si}^2 p(s) - 2 \sum_{s \supset i} b_{si} p(s) + 1 \tag{5.5.3}$$

and

$$\gamma_{ij} = E_p(b_{si} I_{si} - 1)(b_{si} I_{sj} - 1) = \sum_{s \supset i,j} b_{si} b_{sj} p(s) - \sum_{s \supset i} b_{si} p(s) - \sum_{s \supset i} b_{sj} p(s) + 1 \tag{5.5.4}$$

Let us state the results derived by Rao and Vijayan (1977), which are also given by Arnab (2017), as follows:

Theorem 5.5.1 If the mean-square error $\mathrm{M}[e(s, y)]$ becomes zero for $y_i / w_i = k =$ a constant where w_i s are known nonzero constant free from y_i s for $\forall\, i \in U$. Then,

(i) The $\mathrm{M}[e(s, y)]$ can be expressed as

$$\mathrm{M}[e(s, y)] = -\frac{1}{2} \sum_{i \neq} \sum_{j \in U} w_i w_j \gamma_{ij} \left(\frac{y_i}{w_i} - \frac{y_j}{w_j} \right)^2$$

(ii) A nonnegative quadratic unbiased estimator of $\mathrm{M}[e(s, y)]$ is necessarily of the form

$$\widehat{\mathrm{M}}[e(s, y)] = -\frac{1}{2} \sum_{i \neq} \sum_{j \in s} w_i w_j \gamma_{sij} \left(\frac{y_i}{w_i} - \frac{y_j}{w_j} \right)^2 \tag{5.5.5}$$

where γ_{sij} are constants satisfying

$$E(\gamma_{sij}) = \sum_{s} \gamma_{sij} p(s) = \gamma_{ij} \tag{5.5.6}$$

One can choose γ_{sij} in various ways. However, one of the popular choices of γ_{sij} is γ_{ij} / π_{ij}.

Example 5.5.1 The Hansen-Hurwitz estimator for the total Y under DR survey is

$$\hat{Y}_{\mathrm{HH}} = \frac{1}{n} \sum_{i \in s} n_i(s) \frac{y_i}{p_i}$$

where $n_i(s)$ is the number of times the ith unit appeared in s. The estimator $\hat{Y}_{\text{HH}}$ is unbiased for the total Y and it becomes a constant k when $y_i = kp_i \,\forall\, i \in U$. Hence, the MSE (=Variance) is zero when $y_i = kp_i$ So, we write

$$\text{Var}(\hat{Y}_{\text{HH}}) = -\frac{1}{2}\sum_{i\neq}\sum_{j\in U} \gamma_{ij}p_ip_j\left(\frac{y_i}{p_i} - \frac{y_j}{p_j}\right)^2 \tag{5.5.7}$$

where

$$\begin{aligned}\gamma_{ij} &= E\left(\frac{n_i(s)}{np_i} - 1\right)\left(\frac{n_j(s)}{np_j} - 1\right) = \frac{E\{n_i(s)n_j(s)\}}{n^2p_ip_j} - \frac{E\{n_i(s)\}}{np_i} - \frac{E\{n_j(s)\}}{np_j} + 1\\ &= \frac{-np_ip_j + n^2p_ip_j}{n^2p_ip_j} - 1 = -\frac{1}{n}\end{aligned} \tag{5.5.8}$$

Equation (5.5.8) yields

$$\text{Var}(\hat{Y}_{\text{HH}}) = \frac{1}{2n}\sum_{i\neq}\sum_{j\in U} p_ip_j\left(\frac{y_i}{p_i} - \frac{y_j}{p_j}\right)^2.$$

Now writing $\gamma_{sij} = \gamma_{ij}\frac{n_i(s)n_j(s)}{E\{n_i(s)n_j(s)\}} = -\frac{n_i(s)n_j(s)}{n^2(n-1)p_ip_j}$, an unbiased estimator of $\text{Var}(\hat{Y}_{\text{HH}})$ is obtained as

$$\begin{aligned}\widehat{\text{Var}}(\hat{Y}_{\text{HH}}) &= \frac{1}{2n^2(n-1)}\sum_{i\neq}\sum_{j\in s} n_i(s)n_j(s)\left(\frac{y_i}{p_i} - \frac{y_j}{p_j}\right)^2\\ &= \frac{1}{n(n-1)}\sum_{i\in s} n_j(s)\left(\frac{y_i}{p_i} - \hat{Y}_{\text{HH}}\right)^2\end{aligned} \tag{5.5.9}$$

Clearly the above estimator is nonnegative.

Example 5.5.2 The Horvitz-Thompson estimator for the total Y is

$$\hat{Y}_{\text{HT}} = \sum_{i\in s}\frac{y_i}{\pi_i} \tag{5.5.10}$$

where π_i is the inclusion probability of the ith unit, which is assumed to be positive for all $i \in U$.

The estimator $\hat{Y}_{\text{HT}}$ is unbiased of Y for any sampling design. For a fixed effective size n sampling design, $\hat{Y}_{\text{HT}}$ becomes equal to nk (=constant) if $y_i = k\pi_i$ and its variance or MSE becomes equal to zero. Hence, in this case, the variance of $\hat{Y}_{\text{HT}}$ can be written as

$$\text{Var}(\hat{Y}_{\text{HT}}) = -\frac{1}{2}\sum_{i\neq}\sum_{j\in U}\gamma_{ij}\pi_i\pi_j\left(\frac{y_i}{\pi_i}-\frac{y_j}{\pi_j}\right)^2 \tag{5.5.11}$$

where $\gamma_{ij} = \frac{1}{\pi_i\pi_j}\sum_{s\supset i,j} p(s) - \frac{1}{\pi_i}\sum_{s\supset i} p(s) - \frac{1}{\pi_j}\sum_{s\supset j} p(s) + 1 = \frac{\pi_{ij}}{\pi_i\pi_j} - 1$

In this case, a nonnegative unbiased estimator should be of the form

$$\widehat{\text{Var}}(\hat{Y}_{\text{HT}}) = -\frac{1}{2}\sum_{i\neq}\sum_{j\in s}\gamma_{sij}\pi_i\pi_j\left(\frac{y_i}{\pi_i}-\frac{y_j}{\pi_j}\right)^2 \tag{5.5.12}$$

where γ_{sij} satisfies $E_p(\gamma_{sij}) = \gamma_{ij}$.

For the choice $\gamma_{sij} = \gamma_{ij}/\pi_{ij}$, $\widehat{\text{Var}}(\hat{Y}_{\text{HT}})$ reduces to the Yates and Grundy's (1953) variance estimator

$$\widehat{\text{Var}}_{YG}(\hat{Y}_{\text{HT}}) = \frac{1}{2}\sum_{i\neq}\sum_{j\in s}\frac{\pi_i\pi_j-\pi_{ij}}{\pi_{ij}}\left(\frac{y_i}{\pi_i}-\frac{y_j}{\pi_j}\right)^2$$

Example 5.5.3 The ratio estimator for Y based on a SRSWOR sample s of size n is

$$\hat{Y}_R = \frac{y_s}{x_s}X \tag{5.5.13}$$

where $x_s = \sum_{i\in s} x_i$, $y_s = \sum_{i\in s} y_i$ and $X = \sum_{i\in U} x_i$.

The estimator $\hat{Y}_R$ becomes a constant kX if $y_i = kx_i\ \forall\, i \in U$. Hence, the MSE of $\hat{Y}_R$ can be expressed as

$$\text{MSE}(\hat{Y}_R) = -\frac{1}{2}\sum_{i\neq}\sum_{j\in U}\gamma_{ij}\left(\frac{y_i}{x_i}-\frac{y_j}{x_j}\right)^2 \tag{5.5.14}$$

Now noting $b_{si} = \frac{X}{x_s}I_{si}$ with $I_{si} = \begin{cases}1 \text{ for } i\in s\\ 0 \text{ for } i\notin s\end{cases}$, one finds

where

$$\begin{aligned}\gamma_{ij} &= E\left(\frac{X}{x_s}I_{si}-1\right)\left(\frac{X}{x_s}I_{sj}-1\right) = X^2E\left(\frac{I_{si}I_{sj}}{x_s^2}\right) - XE\left(\frac{I_{si}}{x_s}\right) - XE\left(\frac{I_{sj}}{x_s}\right) + 1\\ &= \left[X^2\sum_{s\supset i,j}\frac{1}{x_s^2} - X\sum_{s\supset i}\frac{1}{x_s} - X\sum_{s\supset j}\frac{1}{x_s} + \binom{N}{n}\right]\Big/\binom{N}{n}\end{aligned}$$

Noting the second-order inclusion probabilities for a SRSWOR sampling design is $\pi_{ij} = \frac{n(n-1)}{N(N-1)}$, an unbiased estimator of $\text{MSE}(\hat{Y}_R)$ is

$$\begin{aligned}\widehat{\text{MSE}}(\hat{Y}_R) &= -\frac{1}{2}\sum_{i\neq}\sum_{j\in s}\frac{\gamma_{ij}}{\pi_{ij}}\left(\frac{y_i}{x_i}-\frac{y_j}{x_j}\right)^2\\ &= -\frac{N(N-1)}{2n(n-1)}\sum_{i\neq}\sum_{j\in s}\gamma_{ij}\left(\frac{y_i}{x_i}-\frac{y_j}{x_j}\right)^2 \qquad (5.5.15)\end{aligned}$$

5.5.2 RR Surveys

Let a sample s of size n be selected from a population with probability $p(s)$ using a sampling design p. Each of the selected units in the sample was provided with a randomized response independently using a suitable RR device. Let r_i be the revised RR obtained from the ith respondent which satisfy the model (5.1.1):

$$E_R(r_i) = y_i,\ V_R(r_i) = \phi_i \text{ and } C_R(r_i, r_j) = 0 \text{ for } i \neq j$$

A linear estimator of the population total Y for the RR survey is obtained from the estimator $e(s, y)$ of DR survey by replacing y_i with r_i and it is given as

$$e(s, r) = e(s, r)|_{y_i=r_i} = \sum_{i\in s} b_{si}r_i \qquad (5.5.16)$$

where b_{si}'s are constants free from y_i and $r_i's$, and $b_{si} = 0$ for $i \notin s$.

The mean-square error of $e(s, r)$ is

$$\begin{aligned}\text{M}[e(s, r)] &= E[e(s, r) - Y]^2\\ &= E_pE_R\left[\sum_{i\in s} b_{si}r_i - Y\right]^2\\ &= E_p\left[\left\{E_R\sum_{i\in U}(b_{si}I_{si}r_i - y_i)\right\}^2 + V_R\left(\sum_{i\in U} b_{si}I_{si}r_i - Y\right)\right]\\ &= E_p\left[\sum_{i\in U}(b_{si}I_{si} - 1)y_i\right]^2 + E_p\left(\sum_{i\in U} b_{si}^2I_{si}\phi_i\right)\\ &= \text{M}[e(s, y)] + \sum_{i\in U}\alpha_i\phi_i \qquad (5.5.17)\end{aligned}$$

where $\text{M}[e(s, y)]$ is given in (5.5.2) and

$$\alpha_i = \sum_{s} b_{si}^2p(s)I_{si} = \sum_{s\supset i} b_{si}^2p(s) \qquad (5.5.18)$$

Consider the situation where the mean-square error $\mathrm{M}[e(s, y)]$ becomes zero for $y_i/w_i = k =$ a constant and w_i's are known nonzero constants free from y_i's for $\forall i \in U$. In this case, a possible nonnegative estimator of the $\mathrm{M}[e(s, y)]$ can be obtained by using Theorem 5.5.1 as

$$\widehat{\mathrm{M}}[e(s, y)] = -\frac{1}{2}\sum_{i\neq}\sum_{j\in s} w_i w_j \gamma_{sij}\left(\frac{y_i}{w_i} - \frac{y_j}{w_j}\right)^2 \tag{5.5.19}$$

Now noting $E_R\left[\left(\frac{r_i}{w_i} - \frac{r_j}{w_j}\right)^2 - \left(\frac{\hat{\phi}_i}{w_i^2} + \frac{\hat{\phi}_j}{w_j^2}\right)\right] = \left(\frac{y_i}{w_i} - \frac{y_j}{w_j}\right)^2$ when $E_R(\hat{\phi}_i) = \phi_i$, Chaudhuri and Stenger (2005), Chaudhuri (2011), and Adhikary (2015) obtained an unbiased estimator of $\mathrm{M}[e(s, y)]$ for the RR survey as

$$\widehat{\widehat{\mathrm{M}}}[e(s, y)] = -\frac{1}{2}\sum_{i\neq}\sum_{j\in s} w_i w_j \gamma_{sij}\left\{\left(\frac{r_i}{w_i} - \frac{r_j}{w_j}\right)^2 - \left(\frac{\hat{\phi}_i}{w_i^2} + \frac{\hat{\phi}_j}{w_j^2}\right)\right\} \tag{5.5.20}$$

Finally, from (5.5.18) and (5.5.20), they proposed an unbiased estimator of $\mathrm{M}[e(s, r)]$ as

$$\widehat{\mathrm{M}}[e(s, r)] = \widehat{\widehat{\mathrm{M}}}[e(s, r)] + \sum_{i\in s} b_{si}^2 \hat{\phi}_i \tag{5.5.21}$$

The estimator (5.5.20) suffers from the drawback that it can take negative values even if the estimator $\widehat{\mathrm{M}}[e(s, y)]$ for the DR survey is positive. To illustrate this demerit, consider the following example.

Example 5.5.4 Horvitz-Thompson estimator for the RR survey is

$$\hat{Y}_{\mathrm{HT}}(r) = \sum_{i\in s}\frac{r_i}{\pi_i} \tag{5.5.22}$$

The estimator $\hat{Y}_{\mathrm{HT}}(r)$ is unbiased for Y and has the variance

$$\mathrm{Var}\left[\hat{Y}_{\mathrm{HT}}(r)\right] = \mathrm{Var}\left(\sum_{i\in s}\frac{y_i}{\pi_i}\right) + \sum_{i\in U}\frac{\phi_i}{\pi_i} \tag{5.5.23}$$

From Example 5.5.2, a nonnegative estimator of $\mathrm{Var}\left(\sum_{i\in s}\frac{y_i}{\pi_i}\right)$ for a fixed effective size sampling design satisfying $\pi_i\pi_j - \pi_{ij} \geq 0$ for $i \neq j \in U$ was obtained as

$$\widehat{\mathrm{Var}}_{YG}(\hat{Y}_{\mathrm{HT}}) = \frac{1}{2}\sum_{i\neq}\sum_{j\in s}\frac{\pi_i\pi_j - \pi_{ij}}{\pi_{ij}}\left(\frac{y_i}{\pi_i} - \frac{y_j}{\pi_j}\right)^2$$

Equation (5.5.20) yields an unbiased estimator of $\mathrm{M}\left[\sum_{i\in s}\frac{r_i}{\pi_i}\right]$ as

$$\hat{\mathrm{M}}\left[\sum_{i\in s}\frac{r_i}{\pi_i}\right]=\widehat{\mathrm{M}}\left[\sum_{i\in s}\frac{y_i}{\pi_i}\right]+\sum_{i\in s}\frac{\hat{\phi}_i}{\pi_i^2}$$
$$=\frac{1}{2}\sum_{i\neq}\sum_{j\in s}\frac{\pi_i\pi_j-\pi_{ij}}{\pi_{ij}}\left\{\left(\frac{r_i}{\pi_i}-\frac{r_j}{\pi_j}\right)^2-\left(\frac{\hat{\phi}_i}{\pi_i^2}+\frac{\hat{\phi}_j}{\pi_j^2}\right)\right\}+\sum_{i\in s}\frac{\hat{\phi}_i}{\pi_i^2} \tag{5.5.24}$$

Consider the situation when a sample $s=(1,2)$ contains units 1, 2 with $r_1=\pi_1$, $r_2=\pi_2$ and $\hat{\phi}_i=\hat{\phi}_2=\phi_0$. In this case $\hat{\mathrm{M}}\left[\sum_{i\in s}\frac{r_i}{\pi_i}\right]=-\phi_0\left(\frac{1}{\pi_1^2}+\frac{1}{\pi_2^2}\right)\frac{\pi_1\pi_2}{\pi_{12}}<0$.

The proposed alternative unbiased estimator of $\mathrm{M}[e(s,r)]$ is given as follows:

$$\widehat{\mathrm{M}}_0[e(s,r)]=\widehat{\mathrm{M}}[e(s,y)]\Big|_{y_i=r_i}+\sum_{i\in s}\varphi_i\frac{\hat{\phi}_i}{\pi_i} \tag{5.5.25}$$

where $\pi_i(>0)$ is the inclusion probability of the ith unit and

$$\varphi_i=2\sum_{s\supset i}b_{si}p(s)-1 \tag{5.5.26}$$

Theorem 5.5.2

(i) $\widehat{\mathrm{M}}_0[e(s,r)]=\widehat{\mathrm{M}}[e(s,y)]\Big|_{y_i=r_i}+\sum_{i\in s}\varphi_i\frac{\hat{\phi}_i}{\pi_i}$ is unbiased for $\mathrm{M}[e(s,r)]$

where $E_R(\hat{\phi}_i)=\phi_i$ and $\varphi_i=2\sum_{s\supset i}b_{si}p(s)-1$.

(ii) $\hat{\mathrm{M}}_0[e(s,r)]$ is nonnegative if $\widehat{\mathrm{M}}[e(s,y)]\geq 0$ and $\varphi_i\geq 0\,\forall\, i\in U$.

Proof

(i) Since $\widehat{\mathrm{M}}[e(s,y)]$ is unbiased for $\mathrm{M}[e(s,r)]$, we have

$$E_p\{\widehat{\mathrm{M}}[e(s,y)]\}=\mathrm{M}[e(s,y)]$$
$$=\sum_{i\in U}\gamma_{ii}y_i^2+\sum_{i\neq}\sum_{j\in U}\gamma_{ij}y_iy_j \tag{5.5.27}$$

Hence,

$$E_p\{\widehat{\mathrm{M}}[e(s,y)]\}\Big|_{y_i=r_i}=\sum_{i\in U}\gamma_{ii}r_i^2+\sum_{i\neq}\sum_{j\in U}\gamma_{ij}r_ir_j \tag{5.5.28}$$

and

$$
\begin{aligned}
E\{\hat{M}_0[e(s,r)]\} &= E_p E_R\big[\{\hat{M}_0[e(s,r)]\}\big] + E_p \sum_{i\in s} \varphi_i E_R(\hat{\phi}_i)/\pi_i \\
&= E_R\big[E_p\{\hat{M}_0[e(s,r)]\}\big] + E_p\left(\sum_{i\in s} \varphi_i \phi_i/\pi_i\right) \\
&= E_R\left[\sum_{i\in U} \gamma_{ii} r_i^2 + \sum_{i\neq}\sum_{j\in U} \gamma_{ij} r_i r_j\right] + \sum_{i\in U} \varphi_i \phi_i \\
&\quad \text{(using (5.2.28))} \\
&= \left[\sum_{i\in U} \gamma_{ii}(y_i^2 + \phi_i) + \sum_{i\neq}\sum_{j\in U} \gamma_{ij} y_i y_j\right] + \sum_{i\in U} \varphi_i \phi_i \\
&= \mathrm{M}[e(s,y)] + \sum_{i\in U} \phi_i(\gamma_{ii} + \varphi_i) \\
&= \mathrm{M}[e(s,y)] + \sum_{i\in U} \alpha_i \phi_i \text{ [using (5.5.3), (5.5.18) and (5.2.26)]} \\
&= \mathrm{M}[e(s,y)]
\end{aligned}
$$

(ii) If $\widehat{\mathrm{M}}[e(s,y)] \geq 0 \; \forall \; -\infty < y_i < \infty, \; i \in U$ then $\widehat{\mathrm{M}}[e(s,r)] \geq 0$ for $\forall \; -\infty < r_i < \infty, \; i \in U$ and hence $\hat{\mathrm{M}}_0[e(s,r)] \geq 0$ if $\varphi_i \geq 0 \forall i \in U$.

In case $e(s,y)$ is unbiased for Y, we have $\sum_{s\supset i} b_{si} p(s) = 1 \, \forall \, i \in U$, $\gamma_{ii} = \alpha_i - 1$ and $\gamma_{ij} = \alpha_{ij} - 1$ $\alpha_{ij} = \sum_s p(s) \sum_{s\supset i,j} b_{si} b_{sj}$. In this case, we have $\mathrm{M}[e(s,y)]$ and $\mathrm{M}[e(s,r)]$ reduce to

$$
\mathrm{M}[e(s,y)] = \mathrm{Var}[e(s,y)] = \sum_{i\in U} (\alpha_i - 1) y_i^2 + \sum_{i\neq}\sum_{j\in U} (\alpha_{ij} - 1) y_i y_j
$$

and

$$
\mathrm{M}[e(s,r)] = \mathrm{Var}[e(s,r)] = \sum_{i\in U} (\alpha_i - 1) y_i^2 + \sum_{i\neq}\sum_{j\in U} (\alpha_{ij} - 1) y_i y_j + \sum_{i\in U} \alpha_i \phi_i \tag{5.5.29}
$$

Denoting $\widehat{\mathrm{M}}[e(s,y)] = \widehat{\mathrm{Var}}[e(s,y)]$ as an unbiased estimator of $\mathrm{Var}[e(s,y)]$, we have

Theorem 5.5.3

(i) $\hat{\mathrm{Var}}[e(s,r)] = \hat{\mathrm{Var}}[e(s,y)]\big|_{y_i=r_i} + \sum_{i\in s} \frac{\hat{\phi}_i}{\pi_i}$ is unbiased for $\mathrm{Var}[e(s,r)]$.
(ii) $\mathrm{Var}[e(s,r)]$ is nonnegative if $\hat{\mathrm{Var}}[e(s,y)] \geq 0$.

Example 5.5.5 The Horvitz-Thompson estimator $\hat{Y}_{\mathrm{HT}} = \sum_{i\in s} \frac{y_i}{\pi_i}$ based on a fixed sample size n design for the DR survey is unbiased for the total Y. The variance of

$\hat{Y}_{\text{HT}}$ and its unbiased estimators are

$$\text{Var}(\hat{Y}_{\text{HT}}) = \frac{1}{2}\sum_{i \neq}\sum_{j \in U} (\pi_i \pi_j - \pi_{ij})\left(\frac{y_i}{\pi_i} - \frac{y_j}{\pi_j}\right)^2$$

and

$$\widehat{\text{Var}}(\hat{Y}_{\text{HT}}) = \frac{1}{2}\sum_{i \neq}\sum_{j \in s} \frac{(\pi_i \pi_j - \pi_{ij})}{\pi_{ij}}\left(\frac{y_i}{\pi_i} - \frac{y_j}{\pi_j}\right)^2$$

Using Theorem 5.5.3, the Horvitz-Thompson estimator, its variance, and its unbiased estimators for the RR surveys are obtained as follows:

$$\hat{Y}_{\text{HT}}(r) = \sum_{i \in s} \frac{r_i}{\pi_i}, \quad \text{Var}[\hat{Y}_{\text{HT}}(r)] = \frac{1}{2}\sum_{i \neq}\sum_{j \in U} (\pi_i \pi_j - \pi_{ij})\left(\frac{y_i}{\pi_i} - \frac{y_j}{\pi_j}\right)^2 + \sum_{i \in U} \frac{\phi_i}{\pi_i}$$

and

$$\begin{aligned} \widehat{\text{Var}}[\hat{Y}_{\text{HT}}(r)] &= \widehat{\text{Var}}(\hat{Y}_{\text{HT}})\Big|_{y_i = r_i} + \sum_{i \in s} \frac{\hat{\phi}_i}{\pi_i} \\ &= \frac{1}{2}\sum_{i \neq}\sum_{j \in s} \frac{(\pi_i \pi_j - \pi_{ij})}{\pi_{ij}}\left(\frac{r_i}{\pi_i} - \frac{r_j}{\pi_j}\right)^2 + \sum_{i \in s} \frac{\hat{\phi}_i}{\pi_i} \end{aligned} \tag{5.5.30}$$

The estimator $\widehat{\text{Var}}[\hat{Y}_{\text{HT}}(r)]$ is nonnegative if $\widehat{\text{Var}}(\hat{Y}_{\text{HT}})$ and $\hat{\phi}_i$ are nonnegative.

Example 5.5.6 For the DR survey, Murthy's estimator (see Sect. 4.4.5.2) for the total Y under the PPSWOR sampling design is

$$\hat{Y}_{\text{MUR}} = \sum_{i \in s} \frac{p(s|i)}{p(s)} y_i \tag{5.5.31}$$

By applying Theorem 5.5.3, the expressions of the unbiased estimator of Y, its variance, and the unbiased estimator of the variance for the RR survey can be obtained. The expressions are already given in (5.3.9), (5.3.10), and (5.3.11). The proposed unbiased estimator (5.3.11) is always nonnegative as the variance for the DR survey $\widehat{\text{Var}}(\hat{Y}_{\text{MUR}})$ is nonnegative.

5.6 Few Important RR Techniques

5.6.1 *Christofides' RR Technique*

In Christofides's (2003) RR technique, each of the sampled respondents is asked to pick a card at random from a box bearing numbers $1, .., L$ with proportions $p_1, .., p_L (p_i > 0, \sum_i p_i = 1)$. If a respondent picks a card bearing the number $x (= 1, .., L)$, then the respondent should report $z = x$ if he or she does not possess the sensitive characteristic A. On the other hand, if the respondent possesses the characteristic A, the respondent should report $z = L + 1 - x$ as his or her randomized response. For example, let $L = 8$ and a respondent pick a card bearing the number $x = 4$. Then the respondent should report 4 if the respondent does not possess the characteristic A. Otherwise, if the respondent possesses the characteristic A, he or she should report 5 as his or her response.

Let z_i be the RR obtained from the respondent $i (\in U)$ selected in the sample s. Then,

$$z_i = \begin{cases} j(1 - y_i) & \text{with probability } p_j \\ (L + 1 - j) y_i & \text{with probability } p_j \end{cases} ; j = 1, .., L \tag{5.6.1}$$

where $y_i = \begin{cases} 0, & if\ i \notin A \\ 1, & if\ i \in A \end{cases}$.

Equation (5.6.1) yields

$$\begin{aligned} E_R(z_i) &= (1 - y_i) \left(\sum_j j p_j \right) + y_i \sum_j (L + 1 - j) p_j \\ &= \{(L + 1) - 2\mu\} y_i + \mu \\ &\text{where } \mu = \sum_j j p_j \end{aligned} \tag{5.6.2}$$

$$\begin{aligned} E_R(z_i^2) &= (1 - y_i) \left(\sum_j j^2 p_j \right) + y_i \sum_j (L + 1 - j)^2 p_j \\ &= (\sigma^2 + \mu^2)(1 - y_i) + \{(L + 1)^2 + (\sigma^2 + \mu^2) - 2(L + 1)\mu\} y_i \\ &(\text{where } \sigma^2 = \sum_j j^2 p_j - \mu^2) \\ &= (\sigma^2 + \mu^2) + (L + 1)\{(L + 1) - 2\mu\} y_i \end{aligned}$$

and

$$V_R(z_i) = \sigma^2 \tag{5.6.3}$$

Let us define

$$r_i = \frac{z_i - \mu}{L+1-2\mu} \text{ with } L+1 \neq 2\mu \tag{5.6.4}$$

Equations (5.6.2) and (5.6.3) yield the following model:

$$E_R(r_i) = y_i,\ V_R(r_i) = \frac{\sigma^2}{(L+1-2u)^2} = \phi_i \text{ and } \mathrm{Cov}(r_i, r_j) = 0 \text{ for } i \neq j \tag{5.6.5}$$

The estimators for the population proportion $\pi = \sum_{i \in U} y_i/N$, variance of the estimator, and unbiased estimator of the variance for various sampling can be obtained from Table 5.1.

5.6.2 *Dalenius and Vitale's RR Technique*

Dalenius and Vitale (1974) provided a RR technique to estimate the mean μ of a discrete sensitive variable x such as the number of abortions, the number of students who cheated in examinations, etc. Suppose that the random variable x takes the values 1, 2,.., t with unknown probabilities $p_1, \ldots, p_t$; $\sum_j p_j = 1$. Each participant in the sample is asked to choose a number at random from 1, 2,…, t using a suitable randomizing device such as a ticket with numbers 1, 2,…, t in equal proportion, a random number table, a spinner, or other similar devices. The respondent should report "yes" if the selected number is less than his or her true value. Otherwise, the respondent should answer "no". Let us denote $z_i = 1$ if the ith respondent answers "yes" and $z_i = 0$ if the ith respondent answers "no". Then,

$$\begin{aligned} \Pr ob(z_i = 1) &= \theta \\ &= \text{Probability of getting an answer "Yes"} \\ &= \frac{1}{t}\sum_{j=1}^{t} p_j j = \mu/t \end{aligned} \tag{5.6.6}$$

5.6.2.1 SRSWR Sampling Design

Let a sample s of size n be selected by the SRSWR method and $\hat{\theta}$ be the proportion of "yes" answers in the sample. Then one can easily verify the following result.

Theorem 5.6.1

(i) $\hat{\mu} = t\hat{\theta}$ is an unbiased estimator of μ.
(ii) The variance of $\hat{\mu}$ is

$$\mathrm{Var}(\hat{\mu}) = t^2\theta(1-\theta)/n.$$

(iii) $\hat{\text{Var}}(\hat{\mu}) = t^2\hat{\theta}(1-\hat{\theta})/n$ is an unbiased estimator of $\text{Var}(\hat{\mu})$.

5.6.2.2 Arbitrary Sampling Design

Chaudhuri (2002, 2011), and Chaudhuri and Christofides (2013) explored Dalenius and Vitale's (1974) RR technique for complex survey designs for estimating the population mean. Arnab (2023) has shown that the entire derivations of the estimator of the population mean, variance, and unbiased estimator of the variance were incorrect. The corrected expressions of the mean, variance, and the unbiased estimator of the variance are presented as follows:

Let a sample s of size n be selected from a population with inclusion probabilities π_i for the ith unit and $\pi_{ij}(>0)$ for the ith and jth unit ($i \neq j \in U$). Equation (5.6.6) yields

$$E_R(z_i) = \theta = \frac{\mu}{t} \text{ and } V_R(z_i) = \theta(1-\theta) \tag{5.6.7}$$

Then Eq. (5.6.7) yields Horvitz-Thompson analog estimator of μ as

$$\hat{\mu}_{\text{HT}} = \frac{t}{N}\sum_{i \in s}\frac{z_i}{\pi_i} \tag{5.6.8}$$

The estimator $\hat{\mu}_{\text{HT}}$ has the following properties.

Theorem 5.6.2

(i) $\hat{\mu}_{\text{HT}} = \frac{t}{N}\sum\limits_{i \in s}\frac{z_i}{\pi_i}$ is unbiased for μ.

(ii) The variance of $\hat{\mu}_{\text{HT}}$ is

$$\text{Var}(\hat{\mu}_{\text{HT}}) = \frac{t^2}{N^2}\left[\theta^2\left\{\sum_{i \in U}\left(\frac{1}{\pi_i}-1\right)+\sum\sum_{i \neq j \in U}\left(\frac{\pi_{ij}}{\pi_i\pi_j}-1\right)\right\}+\theta(1-\theta)\sum_{i \in U}\frac{1}{\pi_i}\right]$$

(iii) An unbiased estimator of $\text{Var}(\hat{\mu}_{\text{HT}})$ is

$$\hat{\text{Var}}(\hat{\mu}_{\text{HT}}) = \frac{t^2}{N^2}\left[\frac{\overline{z}(n\overline{z}-1)}{n-1}\left\{\sum_{i \in s}\frac{1-\pi_i}{\pi_i^2}z_i+\sum\sum_{i \neq j \in s}\frac{\pi_{ij}-\pi_i\pi_j}{\pi_i\pi_j\pi_{ij}}z_iz_j\right\}+\frac{\overline{z}(1-\overline{z})}{n-1}\sum_{i \in s}\frac{1}{\pi_i^2}\right]$$

where $\overline{z} = \frac{1}{n}\sum\limits_{i \in s}z_i$.

Proof

(i) $$E(\hat{\mu}_{\mathrm{HT}}) = \frac{t}{N}E_p\left[\sum_{i\in s}\frac{E_R(z_i)}{\pi_i}\right] = \frac{t\theta}{N}E_p\left[\sum_{i\in s}\frac{1}{\pi_i}\right] = \frac{t\theta}{N}N = \mu$$

(ii) $$\begin{aligned}\mathrm{Var}(\hat{\mu}_{\mathrm{HT}}) &= V_p[E_R(\hat{\mu}_{\mathrm{HT}})] + E_p[V_R(\hat{\mu}_{\mathrm{HT}})] \\ &= \frac{t^2}{N^2}\left[V_p\left(\sum_{i\in s}\frac{\theta}{\pi_i}\right) + E_p\left\{\sum_{i\in s}\frac{\theta(1-\theta)}{\pi_i^2}\right\}\right] \\ &= \frac{t^2}{N^2}\left[\begin{array}{l}\theta^2\left\{\sum_{i\in U}\frac{1-\pi_i}{\pi_i} + \sum\sum_{i\neq j\in U}\frac{\pi_{ij}-\pi_i\pi_j}{\pi_i\pi_j}\right\} \\ +\theta(1-\theta)\left(\sum_{i\in U}\frac{1}{\pi_i}\right)\end{array}\right]\end{aligned}$$

(iii) $$E[\hat{\mathrm{Var}}(\hat{\mu}_{\mathrm{HT}})] = \frac{t^2}{N^2}E_p\left[\begin{array}{l}\frac{E_R\{z(nz-1)\}}{n-1}\left\{\sum_{i\in s}\frac{1-\pi_i}{\pi_i^2} + \sum\sum_{i\neq j\in s}\frac{\pi_{ij}-\pi_i\pi_j}{\pi_i\pi_j\pi_{ij}}\right\} \\ +\frac{E_R\{z(1-z)\}}{n-1}\sum_{i\in s}\frac{1}{\pi_i^2}\end{array}\right]$$

Now noting $E_R\left\{\frac{z(nz-1)}{n-1}\right\} = E_R\left(\frac{nz^2-z}{n-1}\right) = \theta^2$ and $E_R\left\{\frac{z(1-z)}{n-1}\right\} = \theta(1-\theta)$, one finds

$$\begin{aligned}E[\hat{\mathrm{Var}}(\hat{\mu}_{\mathrm{HT}})] &= \frac{t^2}{N^2}E_p\left[\theta^2\left\{\sum_{i\in s}\frac{1-\pi_i}{\pi_i^2} + \sum\sum_{i\neq j\in s}\frac{\pi_{ij}-\pi_i\pi_j}{\pi_i\pi_j\pi_{ij}}\right\} + \theta(1-\theta)\sum_{i\in s}\frac{1}{\pi_i}\right] \\ &= \mathrm{Var}(\hat{\mu}_{\mathrm{HT}})\end{aligned}$$

5.6.3 Liu, Chow, and Mosley's RR Technique

Liu et al. (1975) proposed a RR technique that can be used for both discrete qualitative and quantitative sensitive characteristics. However, their method was limited to the SRSWR method only. Chaudhuri (2002, 2011), and Chaudhuri and Christofides (2013) tried to extend Liu et al. (1975) RR technique for complex survey designs.

However, their entire exercises for derivations of the estimators of the population mean, variance, and unbiased estimator of the variance were incorrect. The corrected expressions of the estimators of the total, variance, and unbiased estimators of the variances under complex survey designs are presented here.

5.6.3.1 Randomization Device

Let a population be divided into T mutually exclusive and exhaustive groups, with at least a few of the groups being identified as sensitive. The proportion of persons belonging to the tth group is θ_t, $(t = 1, .., T)$ and $\sum_{t=1}^{T} \theta_t = 1$. Liu et al. (1975) randomization device consists of a flask made of transparent material with a long, narrow neck, so that only one ball can pass through it. T different colored balls are put in the flask. Each color represents a particular group: Color-1 represents group 1, Color-2 represents group 2,...., Color-T represents group T. The number of balls bearing color t is $m_t(\geq 1)$ and $m_1 \neq m_2 \neq, ..., m_T$. The total number of balls in the flask is $\sum_{t=1}^{T} m_t = m$. The respondents are asked to close the top edge of the flask with a stopper, shake it thoroughly, turn it upside down, and allow all the balls to fall into the neck. The neck has m locations marked from the bottom with numbers 1, 2,..., m. The respondents are requested to report at which location the first ball from the bottom of the neck identifies his or her color (group). For example, if a respondent belongs to Group 1, he or she should report where the first ball of color-1 has fallen.

5.6.3.2 Estimation of Proportion

Let a sample s of size n be selected from a population $U = (1, .., i, .., N)$ with probability $p(s)$ by using a general sampling design $p(s)$. Let π_i and $\pi_{ij}(> 0)$ denote the inclusion probabilities of the ith, and ith and $j(\neq i)$th unit of U.

Then,

$P_{t1} =$ Probability [a respondent belonging to group t reports 1] $= \frac{m_t}{m}$,

$P_{t2} =$ Probability [a respondent belonging to group t reports 2] $= \left(1 - \frac{m_t}{m}\right)\frac{m_t}{m-1}$.

And, in general,

$P_{tj} =$ Probability [a respondent belonging to group t reports j]

$$= \frac{(m-j)!(m-m_t)!m_t}{m!(m-m_t+1-j)!}; j = 1, 2, .., M \tag{5.6.9}$$

where $M = \underset{1 \leq t \leq T}{\text{Max}}(m - m_t + 1)$.

For details, readers are referred to Liu et al. (1975).

The probability of obtaining a response "j" from a respondent $i(\in U)$ depends on the unit i and it is given by

$$\lambda_i(j) = \sum_{t=1}^{T} P_{tj} y_i(t) \tag{5.6.10}$$

where $y_i(t) = \begin{cases} 1, & \text{if the } i\text{th person belong to the category } t \\ 0, & \text{otherwise} \end{cases}$; $i = 1, .., N$; $t = 1, .., T$

The total number and proportion of persons belonging to the category t are

$$\sum_{i \in U} y_i(t) = N_t \text{ and } \theta_t = \frac{N_t}{N}, \tag{5.6.11}$$

respectively.

Let,

$$z_i(j) = \begin{cases} 1, & \text{if the } i\text{th person provides response } j \\ 0, & \text{otherwise} \end{cases} \quad j = 1, .., M;\ i \in U$$

Then,

$$E_R[z_i(j)] = \sum_{t=1}^{T} P_{tj} y_i(t) = \lambda_i(j) \tag{5.6.12}$$

$$V_R[z_i(j)] = \sum_{t=1}^{T} P_{tj} y_i(t) - \left(\sum_{t=1}^{T} P_{tj} y_i(t) \right)^2 = \sum_{t=1}^{T} P_{tj}(1 - P_{tj}) y_i(t) = \delta_i(j, j)$$

and

$$\begin{aligned} \text{Cov}[z_i(j), z_i(k)] &= -\left(\sum_{t=1}^{T} P_{tj} y_i(t) \right)\left(\sum_{t=1}^{T} P_{tk} y_i(t) \right) \\ &= -\sum_{t=1}^{T} P_{tj} P_{tk} y_i(t) = \delta_i(j, k) \text{ for } j \neq k \end{aligned}$$

Further, noting, $\sum_{t=1}^{T} y_i(t) = 1$, we obtain the following linear model:

$$\boldsymbol{Z}_i = \boldsymbol{P}\boldsymbol{Y}_i + \boldsymbol{e}_i \tag{5.6.13}$$

where

$$\boldsymbol{Z}_i = \begin{pmatrix} z_i(1) - P_{T1} \\ \cdot \\ z_i(j) - P_{Tj} \\ \cdot \\ z_i(M) - P_{TM} \end{pmatrix},$$

$$\boldsymbol{Y}_i = \begin{pmatrix} y_i(1) \\ \cdot \\ y_i(t) \\ \cdot \\ y_i(T-1) \end{pmatrix},$$

$$\boldsymbol{P} = \begin{pmatrix} P_{11} - P_{T1} & \cdot & P_{t1} - P_{T1} & \cdot\, P_{T-11} - P_{T1} \\ \cdot & \cdot & \cdot & \cdot\;\cdot \\ P_{1j} - P_{Tj} & \cdot & P_{tj} - P_{Tj} & \cdot\, P_{T-1j} - P_{Tj} \\ \cdot & \cdot & \cdot & \cdot \quad \cdot \\ P_{1M} - P_{TM} & \cdot & P_{tM} - P_{TM} & \cdot\, P_{T-1M} - P_{TM} \end{pmatrix},$$

$$\boldsymbol{e}_i = \begin{pmatrix} e_i(1) \\ \cdot \\ e_i(j) \\ \cdot \\ e_i(M) \end{pmatrix}$$

with

$E_R(\boldsymbol{e}_i) = \boldsymbol{0}$ and the variance-covariance matrix of $V_R(\boldsymbol{e}_i)$ is $\boldsymbol{D}_i = ((\delta_i(jk)))$.

Assuming $\boldsymbol{P}$ and $\boldsymbol{D}_i$ are of full rank, $\boldsymbol{Y}_i$ can be estimated by the weighted least squares method as

$$\hat{\boldsymbol{Y}}_i = (\boldsymbol{P}'\boldsymbol{D}_i^{-1}\boldsymbol{P})^{-1}\boldsymbol{P}'\boldsymbol{Z}_i \tag{5.6.14}$$

and $y_i(T)$ can be estimated by

$$\hat{y}_i(T) = 1 - \sum_{t=1}^{T-1} \hat{y}_i(t) \tag{5.6.15}$$

The variance and covariance matrix of $\hat{\boldsymbol{Y}}_i$ is

$$\text{Var}(\hat{\boldsymbol{Y}}_i) = (\boldsymbol{P}'\boldsymbol{D}^{-1}\boldsymbol{P})^{-1} = \boldsymbol{\psi}_i = ((\psi_i(t, t'))) \tag{5.6.16}$$

The Horvitz and Thompson (1952)-type estimator of θ_t, the proportion of persons belonging to the category $t(= 1, .., T)$ is

$$\hat{\theta}_{\mathrm{HT}}^{t} = \frac{1}{N} \sum_{i \in s} \frac{\hat{y}_i(t)}{\pi_i} \tag{5.6.17}$$

where $\hat{y}_i(t)$ is obtained from Eq. (5.6.15).

The mean and variance of $\hat{\theta}_{\mathrm{HT}}^{t}$ are given in the following theorem without derivation.

Theorem 5.6.3

(i) $\hat{\theta}_{\mathrm{HT}}^{t}$ is an unbiased estimator of θ_t

(ii) The variance of $\hat{\theta}_{\mathrm{HT}}^{t}$ is

$$\begin{aligned}\mathrm{Var}\left(\hat{\theta}_{\mathrm{HT}}^{t}\right) &= \frac{1}{N^2}\left[\left\{\sum_{i \in U}\left(\frac{1}{\pi_i}-1\right)y_i^2(t) - \sum_{i \neq}\sum_{j \in U}\left(\frac{\pi_{ij}}{\pi_i\pi_j}-1\right)y_i(t)y_j(t)\right\} + \sum_{i \in U}\frac{\psi_i(t,t)}{\pi_i}\right] \\ &= \frac{1}{N^2}\left[\left\{\sum_{i \in U_t}\left(\frac{1}{\pi_i}-1\right) + \sum_{i \neq}\sum_{j \in U_t}\left(\frac{\pi_{it}}{\pi_i\pi_j}-1\right)\right\} + \sum_{i \in U_t}\frac{\psi_i(t,t)}{\pi_i}\right].\end{aligned}$$

It is difficult to estimate $\psi_i(t,t)$ based on a single randomized response $\hat{y}_i(t)$, as pointed out by Chaudhuri and Christofides (2013). In order to estimate $\psi_i(t,t)$, $q(\geq 2)$ independent randomized responses $\mathbf{Z}_{iq}$, $r = 1, \ldots, Q$ are obtained from each of the selected respondents, yielding Q independent estimates of $y_i(t)$ as $\hat{y}_{i1}(t), \ldots, \hat{y}_{iQ}(t)$.

Denoting $\hat{\bar{y}}_i(t) = \frac{1}{Q}\sum_{q=1}^{Q}\hat{y}_{iq}(t)$, one obtains

$E_R\left(\hat{\bar{y}}_i(t)\right) = y_i(t)$, $V_R\left(\hat{\bar{y}}_i(t)\right) = \frac{V_R\{\hat{y}_q(t)\}}{Q} = \psi_i(t,t)$ and an unbiased estimator of $\psi_i(t,t)$ as

$$\hat{\psi}_i(t,t) = \frac{1}{Q(Q-1)}\sum_{q=1}^{Q}\{\hat{y}_{iq}(t) - \hat{\bar{y}}_i(t)\}^2 \tag{5.6.18}$$

From Eq. (5.6.18) and Theorem 5.6.3, one can derive the following result. The proof is omitted because it is simple.

Theorem 5.6.4

(i) $\hat{\hat{\theta}}_{\mathrm{HT}}^{t} = \frac{1}{N}\sum_{i \in s}\frac{\hat{\bar{y}}_i(t)}{\pi_i}$ is an unbiased estimator of θ_t.

(ii) The variance of $\hat{\hat{\theta}}_{\mathrm{HT}}^{t}$ is

$$\mathrm{Var}\left(\hat{\hat{\theta}}_{\mathrm{HT}}^{t}\right) = \frac{1}{N^2}\left[\left\{\sum_{i \in U}\left(\frac{1}{\pi_i}-1\right)y_i^2(t) + \sum_{i \neq}\sum_{i' \in U}\left(\frac{\pi_{ii'}}{\pi_i\pi_{i'}}-1\right)y_i(t)y_{i'}(t)\right\} + \sum_{i \in U}\frac{\psi_i(t,t)}{\pi_i}\right].$$

(iii) An unbiased estimator of $\mathrm{Var}(\hat{\hat{\theta}}_{\mathrm{HT}}^{t})$ is

$$\widehat{\mathrm{Var}}\left(\hat{\hat{\theta}}_{\mathrm{HT}}^{t}\right) = \frac{1}{N^2}\left[\left\{\sum_{i \in s}\left(\frac{1}{\pi_i}-1\right)\frac{\hat{\bar{y}}_i^2(t)}{\pi_i} + \sum_{i \neq}\sum_{i' \in s}\left(\frac{\pi_{ii'}}{\pi_i\pi_{i'}}-1\right)\frac{\hat{y}_i(t)\hat{y}_{i'}(t)}{\pi_{ii'}}\right\} + \sum_{i \in s}\frac{\psi_i(t,t)}{\pi_i}\right].$$

5.6.4 RR Model of Gjestavang and Singh (2006)

In RR model of Gjestavang and Singh (2006), respondents belonging to the sensitive groups A are asked to perform a randomization device R_1 and respondents belonging to the non-sensitive group $\overline{A}$ are asked to perform a randomization device R_2. The RR device R_1: two positive real numbers α_1 and β_1 are chosen and directed to the selected respondent to choose a random sample S_1 from some distribution (the distribution need not be known). The selected respondents are then asked to report $1+\beta_1 S_1$ and $1-\beta_1 S_1$ with probability $p_{\text{GS}} = \alpha_1/(\alpha_1+\beta_1)$ and $1-p_{\text{GS}}$, respectively. The RR device R_2: two positive numbers α_2 and β_2 are chosen, and each of the selected respondents is asked to choose a random sample S_2 as in R_1. Then the selected respondent is directed to report a scrambled response $\beta_2 S_2$ and $-\alpha_2 S_2$ with probability $T_{\text{GS}} = \beta_2/(\alpha_2+\beta_2)$ and $1 - T_{\text{GS}}$, respectively.

Let z_i be the RR reported by the ith respondent in the sample s. Then,

$$z_i = \begin{cases} (1+\beta_1 S_1)y_i \text{ with probability } p_{\text{GS}} \\ (1-\alpha_1 S_1)y_i \text{ with probability } 1-p_{\text{GS}} \\ \beta_2 S_2(1-y_i) \text{ with probability } T_{\text{GS}} \\ -\alpha_2 S_2(1-y_i) \text{ with probability } 1-T_{GS} \end{cases} \tag{5.6.19}$$

where $y_i = 1(0)$ if the ith unit of the population U belongs to the group $A(\overline{A})$.

Writing $E(S_i) = \theta_i$ and $V(S_i) = \gamma_i^2$, one finds

$$\begin{aligned} E_R(z_i) =& (1+\beta_1\theta_1)p_{\text{GS}}y_i + (1-\alpha_1\theta_1)(1-p_{\text{GS}})y_i \\ &+ \beta_2\theta_2(1-y_i)T_{\text{GS}} - \alpha_2\theta_2(1-y_i)(1-T_{\text{GS}}) \\ =& y_i \end{aligned} \tag{5.6.20}$$

$$\begin{aligned} V_R(z_i) =& \{(1+2\beta_1\theta_1) + \beta_1^2(\gamma_1^2+\theta_1^2)\}p_{\text{GS}}y_i + \{(1-2\alpha_1\theta_1) \\ &+ \alpha_1^2(\theta_1^2+\gamma_1^2)\}(1-p_{\text{GS}})y_i + \{\beta_2^2(\gamma_2^2+\theta_2^2)\}(1-y_i)T_{\text{GS}} \\ &+ \{\alpha_2^2(\theta_2^2+\gamma_2^2)\}(1-y_i)(1-T_{\text{GS}}) - y_i^2 \\ =& \left[\{1+\alpha_1\beta_1\}(\gamma_1^2+\theta_1^2)\right]y_i + \left[\alpha_2\beta_2(\gamma_2^2+\theta_2^2)\right](1-y_i) - y_i \\ =& \left[(\alpha_1\beta_1)(\gamma_1^2+\theta_1^2)\right]y_i + \left[\alpha_2\beta_2(\gamma_2^2+\theta_2^2)\right](1-y_i) \\ =& V_R(r_i) = \phi_i \end{aligned} \tag{5.6.21}$$

Now assuming $\frac{\alpha_2\beta_2}{\alpha_1\beta_1} = \frac{\gamma_1^2+\theta_1^2}{\gamma_2^2+\theta_2^2}$ and $\gamma_2^2+\theta_2^2 = 1$, Gjestvang and Singh obtained

$$V_R(z_i) = V_R(r_i) = \alpha_2\beta_2 = \phi_{\text{GS}} \tag{5.6.22}$$

Finally, from (5.6.20), (5.6.21), and (5.6.22) and noting $r_i = z_i$, one finds the following model:

$$E_R(r_i) = y_i,\ V_R(r_i) = \alpha_2\beta_2(\gamma_2^2 + \theta_2^2) \text{ and } C_R(r_i, r_j) = 0 \text{ for } i \neq j \tag{5.6.23}$$

From the model (5.6.23), one finds expressions of the unbiased estimator of the proportion π, its variance, and an unbiased estimator of the variance for various sampling strategies by using Table 5.1.

For the SRSWR sampling of size n, Gjestavang and Singh (2006) proposed an unbiased estimator of the proportion π as

$$\hat{\pi}_{SG} = \frac{1}{n}\sum_{i=1}^{n} r_{(j)} \tag{5.6.24}$$

where $z_{(j)} = r_{(j)}$ is the RR obtained from the respondent selected at the jth draw $j = 1, .., n$.

The variance and unbiased estimators of the variance are obtained as

$$\text{Var}(\hat{\pi}_{SG}) = \frac{1}{n}[\pi(1-\pi) + \alpha_2\beta_2] \text{ and } \hat{\text{Var}}(\hat{\pi}_{SG}) = \frac{1}{n(n-1)}\sum_{j=1}^{n}(z_{(j)} - \hat{\pi}_{SG})^2 \tag{5.6.25}$$

Remark 5.6.1 The model of Gjestavang and Singh (2006) reduces to models of Warner (1967), Mangat and Singh (1990), and Mangat (1994), respectively, if

(i) $(1 + \beta_1\theta_1)p_{GS} + (1 - \alpha_1\theta_1)(1 - p_{GS}) = P_W$ and $\beta_2\theta_2 T_{GS} - \alpha_2\theta_2(1 - T_{GS}) = 1 - P_W$.
(ii) $(1 + \beta_1\theta_1)p_{GS} + (1 - \alpha_1\theta_1)(1 - p_{GS}) = (1 - P_W)(1 - T_0)$ and $\beta_2\theta_2 T_{GS} - \alpha_2\theta_2(1 - T_{GS}) = 1 - (1 - P_W)(1 - T_0)$.
(iii) $(1+\beta_1\theta_1)p_{GS} + (1-\alpha_1\theta_1)(1-p_{GS}) = 1$ and $(1+\beta_1\theta_1)p_{GS} + (1-\alpha_1\theta_1)(1-p_{GS}) = 1$.

Remark 5.6.2 The strategy of Gjestavang and Singh (2006) will be more, equal, or less efficient than Warner's strategy (1967) according to

$$\begin{aligned}
\text{Var}(\hat{\pi}_{SG}) &= \frac{1}{n}[\pi(1-\pi) + \alpha_2\beta_2]\ <, \\
= \text{ or } > \text{Var}(\hat{\pi}_W) &= \frac{1}{n}\left[\pi(1-\pi) + \frac{P_w(1-P_w)}{(2P_w - 1)^2}\right] \\
\text{i.e. if } \alpha_2\beta_2\ <, = &\text{ or} > \frac{P_W(1-P_W)}{(2P_W - 1)^2}
\end{aligned}$$

Remark 5.6.3 Gjestavang and Singh (2006) claimed that their proposed method is preferable to the existing methods because, in their method, the distributions of the scrambled variables need not be known. This claim, however, is not true. From the expression of variance (5.6.20), one can find that the variance of the response z_i can be very large unless one chooses the means and variances of the scrambled variables carefully.

5.6.5 RR Model of Antoch et al. (2022)

Antoch et al. (2022) proposed the following alternative RR method for quantitative sensitive variables. The method is very simple to implement, and, at the same time, protect respondent's privacy. Let y_i be the value of the sensitive variable y which is positive, and M be its maximum value which is known. In this method, the ith $(i \in U)$ respondent selected in the sample s is asked to draw a random sample x_i from a uniform distribution (0, M) and should report

$$z_i = \begin{cases} 0 \text{ if } x_i > y_i \\ 1 \text{ if } x_i \leq y_i \end{cases} \tag{5.6.26}$$

as his or her response.

In this method,

$$E_R(z_i) = P(z_i = 1) = \frac{y_i}{M} \text{ and } V_R(z_i) = \frac{y_i(M - y_i)}{M^2}$$

In this case, the revised response $r_i = Mz_i$ satisfies the model:

$$E_R(r_i) = y_i,\ V_R(r_i) = y_i(M - y_i) = \phi_{Ai} \text{ and } C_R(r_i, r_j) = 0 \text{ for } i \neq j \in U \tag{5.6.27}$$

The Horvitz-Thompson estimator for the total $Y = \sum_{i \in U} y_i$ is given by

$$\hat{Y}_{\mathrm{HT}} = \sum_{i \in s} \frac{r_i}{\pi_i} = M \sum_{i \in s} \frac{z_i}{\pi_i}$$

For an SRSWOR sampling, $\hat{Y}_{\mathrm{HT}}$ reduces to $\hat{Y}_{\mathrm{Wo}} = N\frac{1}{n}\sum_{i \in s} r_i = N\frac{M}{n}\sum_{i \in s} z_i$. The variance of $\hat{Y}_{\mathrm{Wo}}$ is given by

$$\mathrm{Var}(\hat{Y}_{Wo}) = N^2\left[\left(\frac{1}{n} - \frac{1}{N}\right)S_y^2 + \frac{1}{nN}\sum_{i \in U} y_i(M - y_i)\right]$$

and its variance and an unbiased estimator of the variance come out, respectively, as follows:

Theorem 5.6.5

(i) $\hat{Y}_{\mathrm{HT}} = \sum_{i \in s} \frac{r_i}{\pi_i}$ is an unbiased estimator of Y where $\pi_i(> 0)$ is the inclusion probability of the ith unit.

(ii) The variance of $\hat{Y}_{\mathrm{HT}}$ is

$$\mathrm{Var}(\hat{Y}_{\mathrm{HT}}) = \frac{1}{2}\sum_{i \neq}\sum_{j \in U}(\pi_i\pi_j - \pi_{ij})\left(\frac{y_i}{\pi_i} - \frac{y_j}{\pi_j}\right)^2 + \sum_{i \in U}\frac{\phi_{A}}{\pi}$$

where $\pi_{ij}(i \neq j)$ is the inclusion probability of the ith unit.

(iii) An unbiased estimator of $\mathrm{Var}(\hat{Y}_{\mathrm{HT}})$ is

$$\hat{\mathrm{Var}}(\hat{Y}_{\mathrm{HT}}) = \frac{1}{2}\sum_{i \neq}\sum_{j \in s}\frac{\pi_i\pi_j - \pi_{ij}}{\pi_{ij}}\left(\frac{r_i}{\pi_i} - \frac{r_j}{\pi_j}\right)^2 + \sum_{i \in s}\frac{\hat{\phi}_{A}}{\pi_i}.$$

Remark 5.6.4 If both the upper and lower bounds of the sensitive variable y, m and M are known, respondents are asked to select a sample x from a uniform population (m, M). The ith respondent should report

$$z_i = \begin{cases} 0 \text{ if } x_i > y_i \\ 1 \text{ if } m \leq x_i \leq y_i \end{cases}$$

as his or her RR. In this case,

$E_R(z_i) = P(z_i = 1) = \frac{y_i - m}{M - m}$, $V_R(z_i) = \frac{y_i - m}{M - m}\left(1 - \frac{y_i - m}{M - m}\right)$, and the revised RR $r_i = \frac{(M - m)z_i + m}{M - m}$ follows the model

$E_R(r_i) = y_i$, $V_R(r_i) = \frac{y_i - m}{M - m}\left(1 - \frac{y_i - m}{M - m}\right)$.

5.6.6 *Quatember's RR Model*

Quatember (2009) proposed an RR technique by combining the questions of unrelated question models of Warner (1965), Greenberg et al. (1971), and forced RR model of Folsom et al. (1973). In Quatember's model, a respondent is asked to answer "yes" or "no" to one of the following five questions with a certain probability:

Q1: "Are you member of the sensitive group A?" with probability p_1.
Q2: "Are you a member of the non-sensitive group $\overline{A}$?" with probability p_2.
Q3: "Are you not a member of the innocuous group B?" with probability p_3.
Q4: Report "yes" with probability p_4.
Q5: Report "no" with probability $p_5(= 1 - p_1 - p_2 - p_3 - p_4)$.

5.6.6.1 Estimation of the Population Proportion π

Let a sample of size n be selected from a population by the SRSWR method, and n_Q be the number of respondents in the sample who answered "yes". The probability of obtaining a "yes" answer from a respondent selected at random is

$$\lambda_Q = \pi p_1 + (1 - \pi)p_2 + \pi_B p_3 + p_4 \tag{5.6.28}$$

Let $\widehat{\lambda}_Q = n_Q/n$ be the proportion of "yes" answers in the sample. Assuming π_B, the proportion of individuals in the population belonging to the innocuous group B is known, the following theorem is obtained.

Theorem 5.6.6

(i) $\hat{\pi}_Q = \frac{\hat{\lambda}_Q - (\pi_B p_3 + p_2 + p_4)}{p_1 - p_2}$; $p_1 \neq p_2$ is an unbiased estimator of π.
(ii) The variance of $\hat{\pi}_Q$ is

$$\text{Var}(\hat{\pi}_Q) = \frac{\lambda_Q(1 - \lambda_Q)}{n(p_1 - p_2)^2}.$$

(iii) An unbiased estimator of the variance of $\text{Var}(\hat{\pi}_Q)$ is

$$\widehat{\text{Var}}(\hat{\pi}_Q) = \frac{\hat{\lambda}_Q(1 - \hat{\lambda}_Q)}{(n - 1)(p_1 - p_2)^2}.$$

5.6.6.2 General Sampling Design

Let us define

$$x_i = \begin{cases} 1, \text{ if the ith respondent } \in \text{ B} \\ 0, \text{ if the ith respondent } \notin \text{ B} \end{cases}, \; y_i = \begin{cases} 1, \text{ if the ith respondent } \in \text{ A} \\ 0, \text{ if the ith respondent } \notin \text{ A} \end{cases}$$

and

$$z_i = \begin{cases} 1, \text{ if the ith respondent answers "yes"} \\ 0, \text{ if the ith respondent answers "no"} \end{cases} \tag{5.6.29}$$

Then,

$$\begin{aligned} E_R(z_i) &= p_1 y_i + p_2(1 - y_i) + p_3 x_i + p_4 \\ &= (p_1 - p_2)y_i + p_2 + p_3\pi_x + p_3(x_i - \pi_B) + p_4 \\ &= \alpha y_i + \beta + p_3(x_i - \pi_B); \end{aligned} \tag{5.6.30}$$

where $\alpha = (p_1 - p_2)$, $\beta = p_2 + p_3\pi_B + p_4$ and $\pi_B = \sum\limits_{i \in U} x_i/N$.

$$
\begin{aligned}
V_R(z_i) &= \alpha y_i + \beta + p_3(x_i - \pi_B) - \{\alpha y_i + \beta + p_3(x_i - \pi_B)\}^2 \\
&= \alpha[(1-\alpha) - 2\beta]y_i + p_3(x_i - \pi_B) - p_3^2(x_i - \pi_B)^2 + 2\beta p_3(x_i - \pi_B) \\
&\quad - 2\alpha p_3(x_i - \pi_B)y_i + \beta(1-\beta) \\
&= \phi_{Qi}
\end{aligned}
\tag{5.6.31}
$$

Equations (5.6.30) and (5.6.31) show that the revised RR

$$
r_i = \frac{z_i - \beta}{\alpha} \tag{5.6.32}
$$

satisfies the model

$$
E_R(r_i) = y_i + \frac{p_3}{\alpha}(x_i - \pi_B),\ V_R(r_i) = \frac{\phi_{Qi}}{\alpha^2} \text{ and } Cov(r_i, r_j) = 0 \, for\ i \neq j \tag{5.6.33}
$$

The expressions for the estimator of the proportion π, the variance of the estimator, and the unbiased estimator of the variance for the various strategies can be obtained from Table 5.1. In particular, the Horvitz-Thompson estimator for the proportion π is given by

$$
\hat{\pi}_{\text{HTQ}} = \frac{1}{N}\sum_{i \in s} \frac{r_i}{\pi_i} \tag{5.6.34}
$$

where π_i is the inclusion probability of the ith unit of the population U. Denoting π_{ij} the inclusion probability for the ith and $j(\neq i)$th units, the following results are obtained.

Theorem 5.6.7

(i) $\hat{\pi}_{\text{HTQ}}$ is an unbiased estimator of π.
(ii) For a fixed effective size sampling design, the variance of $\hat{\pi}_{\text{HTQ}}$ is

$$
\text{Var}(\hat{\pi}_{\text{HTQ}}) = \frac{1}{N^2}\left[\left\{V_{\text{HT}yy} + \left(\frac{p_3}{\alpha}\right)^2 V_{\text{HT}xx} + 2\frac{p_3}{\alpha}V_{\text{HT}xy}\right\} + \frac{1}{\alpha^2}\sum_{i \in U}\frac{\phi_{Qi}}{\pi_i}\right]
$$

where $V_{\text{HT}uv} = \frac{1}{2}\sum\sum_{i \neq j \in U}(\pi_i\pi_j - \pi_{ij})\left(\frac{u_i}{\pi_i} - \frac{u_j}{\pi_j}\right)\left(\frac{v_i}{\pi_i} - \frac{v_j}{\pi_j}\right)$; $u, v = x, y$.

(iii) An unbiased estimator of $\text{Var}(\hat{\pi}_{\text{HTQ}})$ is

$$
\widehat{\text{Var}}(\hat{\pi}_{\text{HTQ}}) = \frac{1}{N^2}\left[\sum_{i \neq}\sum_{j \in s}\frac{\pi_i\pi_j - \pi_{ij}}{\pi_{ij}}\left(\frac{r_i}{\pi_i} - \frac{r_j}{\pi_j}\right)^2 + \sum_{i \in s}\frac{\hat{\phi}_{Qi}}{\pi_i}\right]
$$

where $\hat{\phi}_{Qi}$ is an unbiased estimator of ϕ_{Qi}.

Proof

(i)
$$
\begin{aligned}
E(\hat{\pi}_{\mathrm{HTQ}}) &= E_p[E_R(\hat{\pi}_{\mathrm{HTQ}})] \\
&= E_p\left[\frac{1}{N}\sum_{i\in s}\left(\frac{y_i}{\pi_i} + \frac{p_3}{\alpha}\frac{(x_i-\pi_B)}{\pi_i}\right)\right] \\
&= \frac{1}{N}\sum_{i\in U} y_i + \frac{p_3}{\alpha}\frac{1}{N}\sum_{i\in U}(x_i-\pi_B) \\
&= \pi \left(\text{since} \sum_{i\in U}(x_i-\pi_B)=0\right)
\end{aligned}
$$

(ii) The variance of $\hat{\pi}_{\mathrm{HTQ}}$ is

$$
\begin{aligned}
\mathrm{Var}(\hat{\pi}_{\mathrm{HTQ}}) &= V_p[E_R(\hat{\pi}_{\mathrm{HTQ}})] + E_p[V_R(\hat{\pi}_{\mathrm{HTQ}})] \\
&= V_p\left[\frac{1}{N}\sum_{i\in s}\frac{1}{\pi_i}\left\{y_i + \frac{p_3}{\alpha}(x_i-\pi_B)\right\}\right] + \frac{1}{N^2\alpha^2}E_p\left(\sum_{i\in s}\frac{\phi_{Qi}}{\pi_i^2}\right) \\
&= \frac{1}{N^2}\left[\left\{V_{\mathrm{HTyy}} + \left(\frac{p_3}{\alpha}\right)^2 V_{\mathrm{HTxx}} + 2\frac{p_3}{\alpha}V_{\mathrm{HTxy}}\right\} + \frac{1}{\alpha^2}\sum_{i\in U}\frac{\phi_{Qi}}{\pi_i}\right].
\end{aligned}
$$

(iii)
$$
\begin{aligned}
E[\hat{\mathrm{Var}}(\hat{\pi}_{\mathrm{HTQ}})] &= E_p[E_R\{\hat{\mathrm{Var}}(\hat{\pi}_{\mathrm{HTQ}})\}] \\
&= \mathrm{Var}(\hat{\pi}_{\mathrm{HTQ}})
\end{aligned}
$$

Remark 5.6.5 It is difficult to obtain $\hat{\phi}_{Qi}$, an unbiased estimator of ϕ_{Qi} from a single response z_i because x_i is unknown. In this case, a common practice is to take two responses z_{i1} and z_{i2} from each of the selected respondents independently. From the responses z_{i1} and z_{i2} thus obtained, an unbiased estimator of y_i is obtained as $\overline{r}_i = (r_{i1} + r_{i2})/2$ where $r_{i1} = (z_{i1} - \beta)/\alpha$ and $r_{i2} = (z_{i2} - \beta)/\alpha$. In this case, the Horvitz-Thompson estimator of π and an unbiased estimator of $V_R(\overline{r}_i) = \phi_i$ are obtained as $\frac{1}{N}\sum_{i\in s}\frac{\overline{r}_i}{\pi_i}$ and $\hat{\phi}_i = (r_{i1} - r_{i2})^2/2$, respectively.

Remark 5.6.6 The expression of $P(z_i = 1) = E_R(z_i) = p_1 y_i + p_2(1-y_i) + p_3\pi_B + p_4$ given by Chaudhuri and Christofides (2013, page 154) is incorrect as it is free of x_i. Hence, the expressions of the variance $V_p(\hat{\theta})$ (Eq. (7.4), page 154) and its unbiased estimators $\hat{V}_p(\hat{\theta})$ are also incorrect.

5.7 Discussions

The most general RR method was proposed by Arnab (1990). Almost all the RR models available in the literature can be obtained from the proposed model as special cases. The general method of estimation of the population total or mean for the DR (direct response) surveys can be extended to the RR surveys. Similarly, the method of nonnegative estimation for the DR survey can also be extended to the RR survey. Some of the important RR techniques, e.g., Christofides (2003), Dalenius and Vitale (1974), Liu et al. (1975), Quatember (2009), and Antoch et al.'s (2022) models, have been discussed in detail.

References

Adhikary, A. K. (2015). Mean square estimation in randomized response surveys. *Model Assisted Statistics and Applications, 10*, 397–409.

Antoch, J., Mola, F., & Vozar, O. (2022). New randomized response technique for estimating the population total of a quantitative variable. *Statistika, 102*(22), 205–227. https://doi.org/10.54694/stat.2022.11

Arnab, R. (1990). On commutativity of design and model expectations in randomized response surveys. *Communications in Statistics-Theory and Methods, 19*, 3751–3757.

Arnab, R. (1995). On admissibility and optimality of sampling strategies in randomized response surveys. *Sankhyā, Ser B, 37*, 385–390.

Arnab, R. (1999). On use of distinct respondents in RR surveys. *Biometrical Journal, 41*, 507–513.

Arnab, R. (2017). *Survey sampling theory and applications*. Academic Press.

Arnab, R. (2023). Dalenius and Vitale randomized response technique for complex survey designs. *Communications in Statistics-Theory and Methods, 52*, 684–692.

Basu, D. (1958). On sampling with and without replacement. *Sankhyā, 20*, 287–294.

Brewer, K. R. W. (1975). A simple procedure for sampling npswor. *Australian Journal of Statistics, 17*(3), 166–172.

Chaudhuri, A. (2002). Estimating sensitive proportions from randomized responses in unequal probability sampling. *Calcutta Statistical Association Bulletin, 52*, 315–322.

Chaudhuri, A. (2011). *Randomized response and indirect questioning techniques in surveys*. Chapman & Hall, CRC Press, Taylor & Francis Group.

Chaudhuri, A., & Christofides, T. (2013). *Indirect questioning in sample surveys*. Springer.

Chaudhuri, A., & Pal, S. (2008). Estimating sensitive proportions from Warner's randomized responses in alternative ways restricting to only distinct units sampled. *Metrika, 68*, 147–156.

Chaudhuri, A., & Stanger, H. (2005). *Survey sampling: theory and methods* (2nd ed.). Chappman and Hall, CRC Press, Taylor, and Francis Group.

Chikkagoudar, M. S. (1966). A note on inverse sampling scheme with equal probabilities. *Sankhyā, A, 28*, 93–96.

Christofied, T. C. (2003). A generalized randomized response technique. *Metrika 57*, 195–200.

Dalenius, T., & Vitale, R. A. (1974). *Anew RR design for estimating the mean of a distribution*. Technical Report 78, Brown University, Providence, RI.

Folsom, R. E., Greenberg, B. G., Horvitz, D. G., & Abernathy, J. R. (1973). The two alternative questions randomized response model for human surveys. *Journal of the American Statistical Association, 68*, 525–530.

Gjestavang, C. R., & Singh, S. (2006). A new randomized response model. *Journal of the Royal Statistical Society, Series B, 68*, 523–530.

Greenberg, B. G., Kuebler, R. R., Abernathy, J. R., & Horvitz, D. G. (1971). Application of the randomized 21 response technique in obtaining quantitative data. *Journal of the American Statistical Association., 66*, 243–250.

Horvitz, D. G., & Thompson, D. J. (1952). A generalization of sampling without replacement from a finite population. *Journal of the American Statistical Association, 47*, 663–685.

Lahiri, D. B. (1951). A method of sample selection for providing unbiased ratio estimates. *Bulletin of the International Statistical Institute, 33*, 133–140.

Liu, P. T., Chow, L. P., & Mosley, W. H. (1975). Use of randomized response technique with a new randomizing device. *Journal of the American Statistical Association, 70*, 329–332.
Mangat, N. S. (1994). An improved Randomized response strategy. *Journal of the Royal Statistical Society, Series B, 56*, 93–95.
Mangat, N. S., & Singh, R. (1990). An alternative randomized response procedure. *Biometrika, 77*, 349–442.
Mangat, N. S., Singh, R., Singh, S., Bellhouse, D. R., & Kashani, H.B. (1995). On efficiency of estimator using distinct respondents in randomized response survey. *Survey Methodology, 21*, 21–23.
Midzuno, H. (1952). On the sampling system with probabilities proportionate to sum of sizes. *Annals of the Institute of Statistical Mathematics, 3*, 99–107.
Quatember, A. (2009). A standardization of randomized response strategies. *Survey Methodology, 35*, 143–152.
Raj, D., & Khamis, S. H. (1958). Some remarks on sampling with replacement. *Annals of the Mathematics and Statistics, 29*, 550–557.
Rao, J. N. K., & Vijayan, K. (1977). On estimating variance in sampling with probability proportional to aggregate size. *Journal of the American Statistical Association, 72*, 579–584.
Sampford, M. R. (1967). On sampling without replacement with unequal probability selection. *Biometrika, 67*, 639–650.
Sen, A. R. (1953). On estimate of variance in sampling with varying probabilities. *Journal of the Indian Society of Agricultural Statistics, 5*, 119–127.
Sengupta, S. (1986). A comparison between PPSWR and Brewer's πPSWOR procedures. *Calcutta Statistical Association Bulletin, 35*, 207–220.
Sengupta, S. (2017). On the comparison of Warner's and unrelated question randomized response plans for estimating sensitive finite population proportions. *Communications in Statistics-Theory and Methods, 46*, 2370–2374.
Warner, S. L. (1965). Randomized response: A survey technique for eliminating evasive answer bias. *Journal of the American Statistical Association, 60*, 63–69.
Yates, F., & Grundy, P. M. (1953). Selection without replacement from within strata with probability proportional to size. *Journal of the Royal Statistical Society, Series B, 15*, 235–261.

Chapter 6
RR Techniques with Repeated Trials

Abstract In this chapter, RR techniques involving more than one response from each of the respondents are considered. The most general model for multiple randomized responses, along with the general method of estimation, is proposed in detail. Details of Franklins' (1989a, 1989b) RR model are provided. Applications of Franklin's model to various sampling designs are also presented. In particular, uses of the binomial, negative binomial, geometric, hypergeometric, negative hypergeometric, and Polya's urn processes in Franklin's model are proposed. Kuk (1990) model and methods of improvement over Kuk's model by Arnab (2018) are also presented. A balanced RR model for improving cooperation from respondents is proposed.

Keywords Balanced RR model · Binomial distribution · Geometric distribution · Hypergeometric distribution · Negative binomial distribution · Polya's urn processes

6.1 Introduction

We have discussed several important RR models in previous chapters, including Warner (1965), Greenberg et al. (1969), forced RR model of Boruch (1972), and model of Mangat and Singh (1990), where each respondent provided only one response. It is well known that the estimates based on the RR surveys are less efficient than the corresponding DR surveys. The simplest way to improve the efficiency of the estimates of the parameters of interest is to collect several additional responses from each respondent by allotting a little or no extra cost. For example, Warner's RR model can be improved if more than one response, say, $k(> 1)$ responses, is obtained from each of the respondents by drawing k cards independently.

In this section, we consider a few important RR models that include Franklin (1989a), Kuk (1990), Singh and Grewal (2013), and Hussain et al. (2014, 2017), among others. The very general method of the RR model proposed by Arnab et al. (2016) and methods of estimation of parameters from the model are presented. All the existing methods for multiple RR models may be considered special cases of

R. Arnab, *Indirect Methods of Data Collection and Analysis from Surveys*, Indian Statistical Institute Series, https://doi.org/10.1007/978-981-97-6005-3_6

model of Arnab et al. (2017). For Franklin, Kuk, Singh and Grewal, among other models, respondents from the sensitive group should use different RR techniques than those from the non-sensitive group. As a result, respondents may feel that there might be some trap behind the proposed technique that might reveal their personal information. To improve cooperation from the respondents, a balanced RR model (BRR) is proposed where respondents are given the option to choose one of the RR models from a set of several alternatives with a given probability.

6.2 Multiple RR Model

In a multiple RR model, each of the selected respondents in the sample provides more than one randomized response. For instance, the ith respondent produces $k(\geq 1)$ independent randomized responses $z_{i1}, .., z_{ij}, .., z_{ik}$ using RR models $R_1, .., R_j, .., R_k$, respectively. For example, R_1 may be model of Warner (1965), R_2 be model of Horvitz et al. (1967), R_3 be model of Greenberg et al. (1969), and so on. Suppose that the ith respondent provided a response z_{ij} by using Warner's (1965) model $R_j (j = 1, .., k)$ with the parameter $P_j (\neq 1/2)$. In this case, $z_{ij} = 1(0)$ if the ith respondent answered "yes" ("no"). Details of the Warner model are described in Sect. 1.2. The means, variances, and covariances of z_{ij} s under the model R_j are given by

$$\begin{aligned} E_R(z_{ij}) &= P_j y_i + (1 - P_j)(1 - y_i), \\ V_R(z_{ij}) &= P_j(1 - P_j) \text{ and } C_R(z_{ij}, z_{i'j'}) = 0 \text{ for } (i, j) \neq (i', j') \end{aligned} \tag{6.2.1}$$

where $y_i = 1$, if the ith respondent possesses a certain sensitive attribute, say A, and $y_i = 0$, otherwise (Fig. 6.1).

The model (6.2.1) yields an unbiased estimator of y_i as $r_{ij} = \frac{z_{ij} - (1 - P_j)}{2P_j - 1}$. The unbiased estimator r_{ij} will be called a revised RR.

Let us consider a more general case, where the revised response r_{ij} obtained from the RR z_{ij} that was provided by the ith respondent in the jth trial satisfies the general model (3.9.1) proposed by Arnab (1990) as follows:

$$E_R(r_{ij}) = y_i, \ V_R(r_{ij}) = \phi_{ij} \text{ and } C_R(r_{ij}, r_{i'j'}) = 0 \text{ for } (i, j) \neq (i', j') \tag{6.2.2}$$

From the revised responses r_{ij} s, an estimate of y_i is obtained as

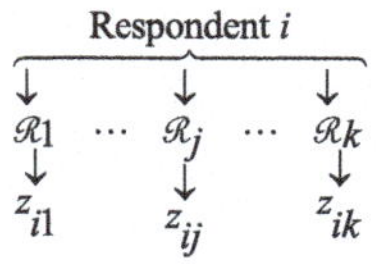

Fig. 6.1 Multiple randomized responses

$$\overline{r}_{wi} = \sum_{j=1}^{k} w_{ij} r_{ij} \tag{6.2.3}$$

where w_{ij} s are suitably chosen weights. The optimal choice of w_{ij} that minimizes the variance of $V_R(\overline{r}_{wi})$ is $w_{0i} = \frac{1}{\phi_{ij}} / \sum_{j=1}^{k} \frac{1}{\phi_{ij}}$. The weight $w_{ij} = w_{0i}$ yields the optimum unbiased estimator of y_i as

$$\overline{r}_{Ci} = \sum_{j=1}^{k} w_{i0} r_{ij} \tag{6.2.4}$$

The reduced response $\overline{r}_{0i}$ satisfies the model:

$$E_R(\overline{r}_{0i}) = y_i,\ V_R(\overline{r}_{0i}) = \psi_{0i} = \left(\sum_{j=1}^{k} \frac{1}{\phi_{ij}} \right)^{-1} \text{ and } C_R(\overline{r}_{0i}, \overline{r}_{0j}) = 0 \text{ for } i \neq j \tag{6.2.5}$$

6.2.1 *Estimation of Proportion π*

6.2.1.1 A General Sampling Design

Consider the general sampling design, where a sample s of a fixed effective size (FES) n is selected from a population U with probability $p(s)$ using a sampling design p. The inclusion probabilities of the ith, and ith and jth$(j \neq i)$ units are denoted by π_i, and $\pi_{ij}(> 0)$, respectively. We denote $y_i = 1(0)$ if the ith unit of the population U possesses (does not possess) the sensitive attribute A. The proportion of individuals in the population that possesses the attribute A is $\pi = \sum_{i \in U} y_i / N$.

The Horvitz-Thompson (1952)-type estimator for π is

$$\hat{\pi}_{\text{HT}} = \frac{1}{N} \sum_{i \in s} \frac{\overline{r}_{0i}}{\pi_i} \tag{6.2.6}$$

where $i \in s$ denotes the sum over the distinct units in s.

The properties of $\hat{\pi}_{\text{HT}}$ are as follows:

Theorem 6.2.1

(i) $\hat{\pi}_{\text{HT}}$ is an unbiased estimator of π
(ii) The variance of $\hat{\pi}_{\text{HT}}$ is

$$\text{Var}(\hat{\pi}_{\text{HT}}) = \frac{1}{N^2} \left[\sum_{i<} \sum_{j \in U} (\pi_i \pi_j - \pi_{ij}) \left(\frac{y_i}{\pi_i} - \frac{y_i}{\pi_j} \right)^2 + \sum_{i \in U} \frac{\psi_{0i}}{\pi_i} \right]$$

where $\psi_{0i} = \left(\sum_{j=1}^{k} \frac{1}{\phi_{ij}}\right)^{-1}$.

(iii) An unbiased estimator of $\mathrm{Var}(\hat{\pi}_{\mathrm{HT}})$ is

$$\widehat{\mathrm{Var}}(\hat{\pi}_{\mathrm{HT}}) = \frac{1}{N^2}\left[\sum_{i<}\sum_{j\in s} \frac{\pi_i\pi_j - \pi_{ij}}{\pi_{ij}}\left(\frac{\bar{r}_{0i}}{\pi_i} - \frac{\bar{r}_{0j}}{\pi_j}\right)^2 + \sum_{i\in s}\frac{\hat{\psi}_{0i}}{\pi_i}\right]$$

where $\hat{\psi}_{0i} = \bar{r}_{0i}(\bar{r}_{0i} - 1)$ is an unbiased estimator of ψ_{0i}.

Proof

(i)
$$\begin{aligned} E(\hat{\pi}_{\mathrm{HT}}) &= E_p\left[\frac{1}{N}\sum_{i\in s}\frac{E_R(\bar{r}_{0i})}{\pi_i}\right] \\ &= \frac{1}{N}\sum_{i\in U} y_i \\ &= \pi \end{aligned}$$

(ii)
$$\begin{aligned} \mathrm{Var}(\hat{\pi}_{\mathrm{HT}}) &= V_p\left[E_R\left(\frac{1}{N}\sum_{i\in s}\frac{\bar{r}_{0i}}{\pi_i}\right)\right] + E_p\left[V_R\left(\frac{1}{N}\sum_{i\in s}\frac{\bar{r}_{0i}}{\pi_i}\right)\right] \\ &= \frac{1}{N^2}\left[V_p\left(\sum_{i\in s}\frac{y_i}{\pi_i}\right) + E_p\left(\sum_{i\in s}\frac{\psi_i}{\pi_i^2}\right)\right] \\ &= \frac{1}{N^2}\left[\sum_{i<}\sum_{j\in U}(\pi_i\pi_j - \pi_{ij})\left(\frac{y_i}{\pi_i} - \frac{y_i}{\pi_j}\right)^2 + \sum_{i\in U}\frac{\psi_{i0}}{\pi_i}\right] \end{aligned}$$

(iii)
$$\begin{aligned} E\left[\widehat{\mathrm{Var}}(\hat{\pi}_{\mathrm{HT}})\right] &= \frac{1}{N^2}E_p\left[\sum_{i<}\sum_{j\in s}\frac{\pi_i\pi_j - \pi_{ij}}{\pi_{ij}}E_R\left(\frac{r_{0i}}{\pi_i} - \frac{r_{0j}}{\pi_j}\right)^2 + \sum_{i\in s}\frac{E_R(\hat{\psi}_{i0})}{\pi_i}\right] \\ &= \frac{1}{N^2}E_p\left[\sum_{i<}\sum_{j\in s}(\pi_i\pi_j - \pi_{ij})\left\{\begin{array}{c}\left(\frac{y_i}{\pi_i} - \frac{y_j}{\pi_j}\right)^2 \\ +\left(\frac{\psi_{0i}}{\pi_i^2} + \frac{\psi_{0j}}{\pi_j^2}\right)\end{array}\right\} + \sum_{i\in s}\frac{\psi_{i0}}{\pi_i}\right] \end{aligned}$$

Now using the relation $\sum_{j(\neq i)}\pi_{ij} = (n-1)\pi_i$, one obtains

$$E[\widehat{\mathrm{Var}}(\hat{\pi}_{\mathrm{HT}})] = \mathrm{Var}(\hat{\pi}_{\mathrm{HT}}).$$

For the SRSWOR design, $\hat{\pi}_{\mathrm{HT}}$ reduces to $\hat{\pi}_{\mathrm{WO}} = \bar{r}_{0s} = \frac{1}{n}\sum_{i\in s} r_{0i}$. In this case, the following results are obtained as a corollary of Theorem 6.2.1.

Theorem 6.2.2 For the SRSWOR sampling,

(i) $\bar{r}_{0s}$ is an unbiased estimator of π.

(ii) The variance of $\overline{r}_{0s}$ is

$$\text{Var}(\overline{r}_{0s}) = \frac{N-n}{n}\frac{\pi(1-\pi)}{N-1} + \frac{1}{n}\overline{\psi}_0$$

where $\overline{\psi}_0 = \frac{1}{N}\sum_{i\in U}\psi_{0i}$.

(iii) An unbiased estimator of $\text{Var}(\overline{r}_{0s})$ is

$$\widehat{\text{Var}}(\overline{r}_{0s}) = \frac{N-n}{Nn}s_r^2 + \frac{\hat{\overline{\psi}}_0}{n}$$

where $(n-1)s_r^2 = \sum_{i\in s}(r_i - \overline{r}_{0s})^2$ and $\hat{\overline{\psi}}_0 = \frac{1}{n}\sum_{i\in s}\hat{\psi}_{0i}$.

6.2.1.2 SRSWR Sampling

Consider the situation where a sample s of size n is selected by the SRSWR method. In this case, a unit can be repeated more than once in a sample. Denoting $\sum_{i=1}^{n}$ as the sum over all the n units of the sample s with repetition, one obtains

Theorem 6.2.3

(i) $\overline{r}_{0\text{w}} = \frac{1}{n}\sum_{i=1}^{n}\overline{r}_{0i}$ is an unbiased estimator of π.
(ii) The variance of $\overline{r}_{0\text{w}}$ is $\text{Var}(\overline{r}_{0\text{w}}) = \frac{\pi(1-\pi)}{n} + \frac{\overline{\psi}_0}{n}$.
(iii) An unbiased estimator of $\text{Var}(\overline{r}_{0\text{w}})$ is $\widehat{\text{Var}}(\overline{r}_{0\text{w}}) = \frac{1}{n(n-1)}\sum_{i=1}^{n}(\overline{r}_{0i} - \overline{r}_{0\text{w}})^2$.

6.2.2 *Special Case: Warner's Model*

Arnab et al. (2017) considered a situation where a sample s of n respondents was selected by the SRSWR method. Each of the selected respondents was asked to perform k RR trials independently using Warner's model. In particular, for the jth randomized device $R_j(j = 1, .., k)$, respondents were asked to select one card at random from a pack containing two types of cards bearing the statements "I belong to the sensitive group A" with proportion $P_j(\neq 1/2)$ and "I do not belong to the sensitive group A" with proportion $1 - P_j$, respectively. Respondents were requested to report "yes" if the statement matched their status; otherwise, they should report "no". Let z_{ij} be the RR obtained from the ith respondent at jth trial (Fig. 6.2).

Define

$$z_{ij} = \begin{cases} 1, \text{ if the ith respondent answers "yes" at jth trial} \\ 0, \text{ if the ith respondent answers "no" at jth trial} \end{cases} \quad i \in U;\ j = 1, .., k \tag{6.2.7}$$

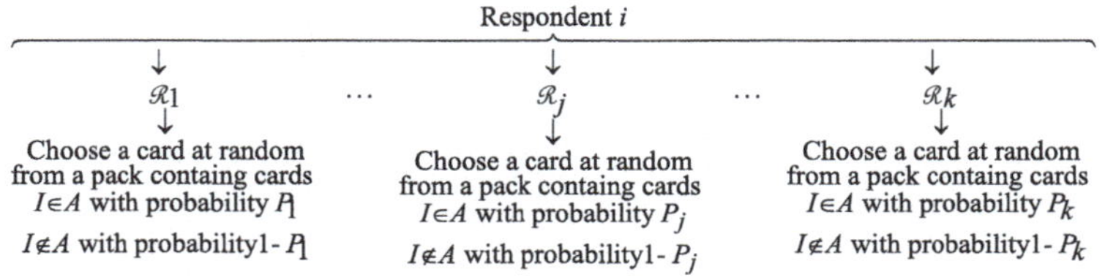

Fig. 6.2 Multiple responses with Warner's trials

and

$$y_i = \begin{cases} 1, & \text{if } i \in A \\ 0, & \text{if } i \notin A \end{cases} ; i \in U \tag{6.2.8}$$

Responses thus obtained can be presented as follows (Table. 6.1):
Equations (6.2.7) and (6.2.8) yield

$$\begin{aligned} E_R(z_{ij}) &= y_i P_j + (1 - y_i)(1 - P_j); \\ V_R(z_{ij}) &= P_j(1 - P_j) \text{ and } C_R(z_{ij}, z_{i'j'}) = 0 \text{ for } (i, j) \neq (i', j') \end{aligned} \tag{6.2.9}$$

Now writing

$$r_{ij} = \frac{z_{ij} - (1 - P_j)}{2P_j - 1}$$

the following reduced RR model is obtained:

$$\begin{aligned} E_R(r_{ij}) &= y_i, \ V_R(r_{ij}) = \phi_{ij} = \frac{P_j(1 - P_j)}{(2P_j - 1)^2} = \phi_{0j}, \text{ and} \\ C_R(r_{ij}, r_{i'j'}) &= 0 \text{ for } (i, j) \neq (i', j') \end{aligned} \tag{6.2.10}$$

Table 6.1 Multiple responses from RR trials

Respondent	RR from trials					Mean
	1	…	j	…	k	
1	z_{11}	…	z_{1j}	…	z_{1k}	$\bar{z}_{1\cdot}$
…	…	…	…	…	…	…
i	z_{i1}	…	z_{ij}	…	z_{ik}	$\bar{z}_{i\cdot}$
…	…	…	…	…	…	…
n	z_{n1}	…	z_{nj}	…	z_{nk}	$\bar{z}_{n\cdot}$
Mean	$\bar{z}_{\cdot 1}$	…	$\bar{z}_{\cdot j}$	…	$\bar{z}_{\cdot k}$	$\bar{z}_{\cdot\cdot}$

Using the model (6.2.10), Arnab et al. (2017) proposed an unbiased estimator of π as

$$\hat{\pi}_{\text{AS}} = \frac{1}{n}\sum_{i=1}^{n}\hat{\lambda}_{0i} \tag{6.2.11}$$

where $\hat{\lambda}_{0i} = \left(\sum_{j=1}^{k} 1/\phi_{0j}\right)^{-1} \sum_{j=1}^{k} \frac{r_{ij}}{\phi_{0j}} = \left[\sum_{j=1}^{k} \frac{(2P_j-1)^2}{P_j(1-P_j)}\right]^{-1} \sum_{j=1}^{k} \frac{(2P_j-1)^2 r_{ij}}{P_j(1-P_j)}$.

Expressions of the variance of $\hat{\pi}_{\text{AS}}$ and its unbiased estimators are obtained as follows:

$$\text{Var}(\hat{\pi}_{\text{AS}}) = \frac{\pi(1-\pi)}{n} + \frac{\phi_{\text{AS}}}{n} \tag{6.2.12}$$

and

$$\widehat{\text{Var}}(\hat{\pi}_{\text{AS}}) = \frac{1}{n(n-1)}\sum_{i=1}^{n}\left(\hat{\lambda}_{0i} - \hat{\pi}_{\text{AS}}\right)^2$$

where

$$\phi_{\text{AS}} = \left[\sum_{j=1}^{k}\frac{(2P_j-1)^2}{P_j(1-P_j)}\right]^{-1}$$

In case, $P_j = P$ for $j = 1, .., k$, the variance of $\hat{\pi}_{\text{AS}}$ reduces to

$$\text{Var}(\hat{\pi}_{\text{AS}}) = \frac{\pi(1-\pi)}{n} + \frac{P(1-P)}{nk(2P-1)^2} \tag{6.2.13}$$

The expression (6.2.13) is the variance of Warner's (1965) technique where each of the respondents repeated the trial $k(\geq 1)$ times independently.

6.2.2.1 Warner Model with $k = 2$

Odumade and Singh (2009) considered RR technique of Arnab and Shangodoyin (2015) with $k = 2$. In this technique, each of the selected respondents performed two independent trials using the Warner model. Responses from each of the selected respondents were collected in the form (X, Y), where X is the response "yes" or "no" originated from the trial R_1 and Y was the response from the trial R_2. The cell frequencies n_{ij} and their probabilities θ_{ij} (in brackets) are summarized in Table 6.2.

In the above table, probabilities of getting response "yes, yes", "yes, no", "no, yes", and "no, no" from a respondent selected at random are

Table 6.2 Cell frequencies n_{ij} (probabilities θ_{ij})

		Response from R_2		
		Yes	No	Total
Response from R_1	Yes	$n_{11}(\theta_{11})$	$n_{10}(\theta_{10})$	$n_{1\cdot}$
	No	$n_{01}(\theta_{01})$	$n_{00}(\theta_{00})$	$n_{0\cdot}$
Total		$n_{\cdot 1}$	$n_{\cdot 0}$	n

$$\begin{aligned}&\theta_{11}=\pi P_1P_2+(1-\pi)(1-P_1)(1-P_2);\ \theta_{10}=\pi P_1(1-P_2)+(1-\pi)(1-P_1)P_2\\&\theta_{01}=\pi(1-P_1)P_2+(1-\pi)P_1(1-P_2);\ \theta_{00}=\pi(1-P_1)(1-P_2)+(1-\pi)P_1P_2\end{aligned}\tag{6.2.14}$$

Odumade and Singh (2009) proposed an unbiased estimator of π by minimizing least square distance

$$D=\frac{1}{2}\sum_{i=0}^{1}\sum_{j=0}^{1}(\theta_{ij}-\hat{\theta}_{ij})^2;\ \hat{\theta}_{ij}=\frac{n_{ij}}{n},j=0,1\tag{6.2.15}$$

with respect to θ_{ij} as

$$\hat{\pi}_{\mathrm{OS}}=\frac{1}{2}+\frac{(P_1+P_2-1)(n_{11}-n_{00})+(P_1-P_2)(n_{10}-n_{00})}{2n\left[(P_1+P_2-1)^2+(P_1-P_2)^2\right]}\tag{6.2.16}$$

The variance of $\hat{\pi}_{\mathrm{OS}}$ is

$$\begin{aligned}&\mathrm{Var}(\hat{\pi}_{\mathrm{OS}})=\\&\frac{(P_1+P_2-1)^2\{P_1P_2+(1-P_1)(1-P_2)\}+(P_1-P_2)^2\{P_2(1-P_1)+P_1(1-P_2)\}}{4n\left[(P_1+P_2-1)^2+(P_1-P_2)^2\right]^2}\\&-\frac{(2\pi-1)^2}{4n}\end{aligned}\tag{6.2.17}$$

6.2.2.2 Comparison with Arnab and Shangodoyin's Estimator

For $k=2$, Arnab et al. (2017) showed that

$$\begin{aligned}\mathrm{Var}(\hat{\pi}_{\mathrm{OS}})-\mathrm{Var}(\hat{\pi}_{\mathrm{AS}})=&\frac{(2P_1-1)^4\phi_{01}+(2P_2-1)^4\phi_{02}}{n\left[(2P_1-1)^2+(2P_2-1)^2\right]^2}-\frac{\phi_{01}\phi_{02}}{n(\phi_{01}+\phi_{02})}\ \text{with}\\ \phi_{0j}=&\frac{P_j(1-P_j)}{(2P_j-1)^2};j=1,2\end{aligned}$$

$$= \frac{1}{n}\left[\frac{P_1(1-P_1) - P_2(1-P_2)}{(2P_1-1)^2 + (2P_2-1)^2}\right]^2 \frac{1}{\phi_{01} + \phi_{02}} \geq 0$$

Thus, estimator $\hat{\pi}_{\text{AS}}$ of Arnab et al. (2017) is more efficient than the estimator $\hat{\pi}_{\text{OS}}$ of Odumade and Singh (2009).

Arnab et al. (2012) extended Odumade and Singh's (2009) technique for complex survey designs. Arnab and Shangodoyin (2015) proposed alternative estimators based on the maximum likelihood method that yield better estimators than those proposed by Odumade and Singh (2009) and Arnab et al. (2012). Jayraj et al. (2018) used the estimating function method that was proposed by Godambe and Kale (1991) for estimating π based on a sample selected by the SRSWR method. The proposed estimator is

$$\hat{\pi}_{\text{J}} = \frac{(P_1+P_2-1)\left[P_1P_2\hat{\theta}_{11} - (1-P_1)(1-P_2)\hat{\theta}_{00}\right] + (P_1-P_2)\left[P_1(1-P_2)\hat{\theta}_{10} - P_2(1-P_1)\hat{\theta}_{01}\right]}{(P_1+P_2-1)^2(\hat{\theta}_{11}+\hat{\theta}_{00}) + (P_1-P_2)^2(\hat{\theta}_{10}+\hat{\theta}_{01})} \tag{6.2.18}$$

The estimator $\hat{\pi}_{\text{J}}$ is biased and the expression of the MSE (mean-square error) is highly complex. Jayraj et al. (2018) compared the relative efficiency of $\hat{\pi}_{\text{J}}$ with respect to $\hat{\pi}_{\text{OS}}$ numerically and found that $\hat{\pi}_{\text{J}}$ is more efficient than $\hat{\pi}_{\text{OS}}$. However, no theoretical result was obtained.

6.3 Franklin's Model

Franklin (1989a, 1989b) proposed an RR technique for estimating π, the proportion of persons belonging to a certain sensitive group A. Franklin's technique was extended by several researchers, including Kuk (1990), Singh and Grewal (2013), Hussain et al. (2014, 2017), Chaudhuri and Christofides (2013), and Singh and Sedory (2013), among others.

In Franklins' (1989a, 1989b) RR technique, a sample s of n respondents was selected by the SRSWR method. If the selected respondent labeled $i(\in U)$ in the sample s belonged to the sensitive group A, then, he or she was directed to draw $k(\geq 1)$ independent samples g_{ij} from the distribution g_j, $j = 1, .., k$. Otherwise, if the respondent belonged to the non-sensitive category $\overline{A}$ (complementary to A), then he or she was asked to draw k independent samples h_{ij} from the distribution h_j, $j = 1, .., k$. The samples were drawn by using a suitable randomizing device such as a deck of cards, a spinner, a random number table, a computer, etc. The confidentiality of the respondent was maintained as the whole experiment was performed in the absence of the interviewer. The mean and variance of $g_j(h_j)$ were known as $\mu_{gj}(\mu_{hj})$ and $\sigma^2_{gj}(\sigma^2_{hj})$, respectively (Fig. 6.3).

Let z_{ij} be the response obtained from the $j(= 1, .., k)$th trial of the $i(\in U)$th respondent, then z_{ij} can be expressed as

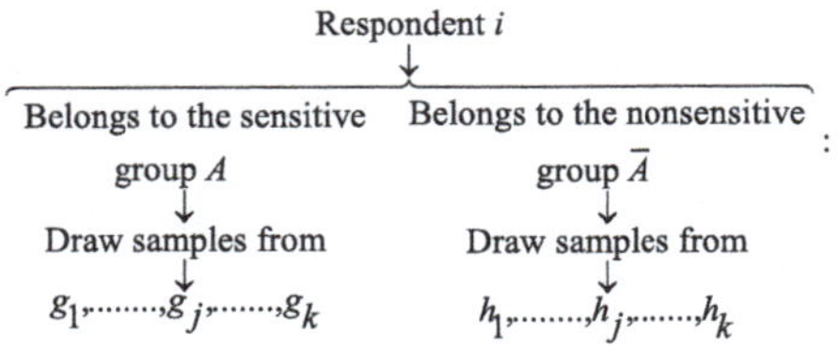

Fig. 6.3 Franklin's model

Table. 6.3 Responses from franklin's RR model

	Trials					
Respondents	1	...	j	...	k	Row average
1	z_{11}	...	z_{1j}		z_{1k}	$\bar{z}_{1\cdot}$
...	...	...	...	...	...	...
i	z_{i1}	...	z_{ij}		z_{ik}	$\bar{z}_{i\cdot}$
...	...	...	...	...	...	...
n	z_{n1}	...	z_{nj}		z_{nk}	$\bar{z}_{n\cdot}$
Column average	$\bar{z}_{\cdot 1}$	...	$\bar{z}_{\cdot j}$		$\bar{z}_{\cdot k}$	

$$z_{ij} = y_i g_{ij} + (1 - y_i)h_{ij};\ i(\in U), j = 1, .., k \tag{6.3.1}$$

Responses z_{ij} s follow the model

$$E_R(z_{ij}) = y_i(\mu_{gj} - \mu_{hj}) + \mu_{hj}$$

and

$$\begin{aligned} V_R(z_{ij}) &= y_i(\mu_{gj}^2 + \sigma_{gj}^2) + (1 - y_i)(\mu_{hj}^2 + \sigma_{hj}^2) - \left\{y_i\mu_{gj} + (1 - y_i)\mu_{hj}\right\}^2 \\ &= y_i\sigma_{gj}^2 + (1 - y_i)\sigma_{hj}^2 \end{aligned} \tag{6.3.2}$$

Responses z_{ij} s so obtained can be arranged as follows (Table. 6.3).

6.3.1 *Franklin's Estimators*

Franklin (1989a) proposed the following two estimators of the population proportion π.

6.3.1.1 Estimator Based on Row Averages

Equation (6.3.2) yields

$$E_R(\bar{z}_{i\cdot}) = \frac{1}{k}\sum_{j=1}^{k}\left\{y_i\left(\mu_{gj} - \mu_{hj}\right) + \mu_{hj}\right\} = y_i\left(\overline{\mu}_g - \overline{\mu}_h\right) + \overline{\mu}_h \tag{6.3.3}$$

and

$$\begin{aligned} V_R(\bar{z}_{i\cdot}) &= \frac{1}{k^2}\sum_{j=1}^{k}\left\{y_i\sigma_{gj}^2 + (1-y_i)\sigma_{hj}^2\right\} \\ &= \frac{1}{k}\left\{y_i\overline{\sigma}_g^2 + (1-y_i)\overline{\sigma}_h^2\right\} \\ &= \frac{1}{k}\left\{y_i\left(\overline{\sigma}_g^2 - \overline{\sigma}_h^2\right) + \overline{\sigma}_h^2\right\} \end{aligned} \tag{6.3.4}$$

where $\overline{\mu}_g = \frac{1}{k}\sum_{j=1}^{k}\mu_{gj}$, $\overline{\mu}_h = \frac{1}{k}\sum_{j=1}^{k}\mu_{hj}$, $\overline{\sigma}_g^2 = \frac{1}{k}\sum_{j=1}^{k}\overline{\sigma}_{gj}^2$, and $\overline{\sigma}_h^2 = \frac{1}{k}\sum_{j=1}^{k}\overline{\sigma}_{hj}^2$.

Franklin's estimator of π based on the row averages is given by

$$\hat{\pi}_{\text{F1}} = \frac{1}{n}\sum_{i=1}^{n}\frac{\bar{z}_{i\cdot} - \overline{\mu}_h}{\overline{\mu}_g - \overline{\mu}_h},\ \overline{\mu}_g \neq \overline{\mu}_h \tag{6.3.5}$$

The mean and variance of $\hat{\pi}_{\text{F1}}$ are given by

$$\begin{aligned} E(\hat{\pi}_{\text{F1}}) &= E\left[\frac{1}{n}\sum_{i=1}^{n}E_R\left(\frac{\bar{z}_{i\cdot} - \overline{\mu}_h}{\overline{\mu}_g - \overline{\mu}_h}\right)\right] \\ &= \frac{1}{N}\sum_{i\in U}y_i \\ &= \pi \end{aligned} \tag{6.3.6}$$

and

$$\begin{aligned} \text{Var}(\hat{\pi}_{\text{F1}}) &= E\left[\frac{1}{n^2}\sum_{i=1}^{n}V_R\left(\frac{\bar{z}_{i\cdot} - \overline{\mu}_h}{\overline{\mu}_g - \overline{\mu}_h}\right)\right] + V\left[\frac{1}{n}\sum_{i=1}^{n}E_R\left(\frac{\bar{z}_{i\cdot} - \overline{\mu}_h}{\overline{\mu}_g - \overline{\mu}_h}\right)\right] \\ &= E\left[\frac{1}{kn^2}\sum_{i=1}^{n}\frac{y_i\left(\overline{\sigma}_g^2 - \overline{\sigma}_h^2\right) + \overline{\sigma}_h^2}{\left(\overline{\mu}_g - \overline{\mu}_h\right)^2}\right] + V\left(\frac{1}{n}\sum_{i=1}^{n}y_i\right) \\ &= \frac{\pi(1-\pi)}{n} + \frac{\pi\left(\overline{\sigma}_g^2 - \overline{\sigma}_h^2\right) + \overline{\sigma}_h^2}{nk\left(\overline{\mu}_g - \overline{\mu}_h\right)^2} \end{aligned} \tag{6.3.7}$$

Noting that $\overline{r}_{\text{F1}i} = \frac{\bar{z}_{i\cdot} - \overline{\mu}_h}{\overline{\mu}_g - \overline{\mu}_h}$, $i = 1, .., n$ are independently and identically distributed (iid) random variables, one finds an unbiased estimator of $\text{Var}(\hat{\pi}_{\text{F1}})$ as

$$\widehat{\text{Var}}(\hat{\pi}_{\text{F1}}) = \frac{1}{n(n-1)} \sum_{i=1}^{n} (\overline{r}_{\text{F1}i} - \hat{\pi}_{\text{F1}})^2 \tag{6.3.8}$$

6.3.1.2 Estimator Based on Column Averages

Franklins' estimator based on the column averages is

$$\hat{\pi}_{\text{F2}} = \sum_{j=1}^{k} w_j \frac{\overline{z}_{\cdot j} - \mu_{hj}}{\mu_{gj} - \mu_{hj}}, \mu_{gj} \neq \mu_{hj} \tag{6.3.9}$$

where w_j s (>0) are suitably chosen weights that satisfy $\sum_{j=1}^{k} w_j = 1$.

Now noting,

$$\begin{aligned} E_R(\overline{z}_{\cdot j}) &= \frac{1}{n} \sum_{i=1}^{n} \{y_i \mu_{gj} + (1 - y_i)\mu_{hj}\} \\ &= (\mu_{gj} - \mu_{hj}) \left(\frac{1}{n} \sum_{i=1}^{n} y_i \right) + \mu_{hj} \end{aligned}$$

and

$$\begin{aligned} V_R(\overline{z}_{\cdot j}) &= \frac{1}{n^2} \sum_{i=1}^{n} \left\{ y_i \sigma_{gj}^2 + (1 - y_i)\sigma_{hj}^2 \right\} \\ &= \left(\sigma_{gj}^2 - \sigma_{hj}^2 \right) \frac{1}{n^2} \sum_{i=1}^{n} y_i + \frac{\sigma_{hj}^2}{n} \end{aligned}$$

one finds

$$\begin{aligned} E(\hat{\pi}_{\text{F2}}) &= E\left[\sum_{j=1}^{k} w_j E_R \left(\frac{\overline{z}_{\cdot j} - \mu_{hj}}{\mu_{gj} - \mu_{hj}} \right) \right] \\ &= E\left[\sum_{j=1}^{k} w_j \left(\frac{1}{n} \sum_{i=1}^{n} y_i \right) \right] \\ &= E\left(\frac{1}{n} \sum_{i=1}^{n} y_i \right) \\ &= \pi \end{aligned} \tag{6.3.10}$$

and

$$\begin{aligned}\operatorname{Var}(\hat{\pi}_{F2}) &= V_p[E_R(\hat{\pi}_{F2})] + E_p[V_R(\hat{\pi}_{F2})]\\ &= V_p\left[\frac{1}{n}\left(\sum_{i=1}^{n} y_i\right)\right] + E_p\sum_{j=1}^{k} w_j^2 V_R\left(\frac{\bar{z}_{\cdot j} - \mu_{hj}}{\mu_{gj} - \mu_{hj}}\right)\\ &= \frac{\pi(1-\pi)}{n} + \frac{1}{n}\sum_{j=1}^{k} w_j^2\left\{\frac{\left(\sigma_{gj}^2 - \sigma_{hj}^2\right)\pi + \sigma_{hj}^2}{\left(\mu_{gj} - \mu_{hj}\right)^2}\right\}\end{aligned} \tag{6.3.11}$$

Choices of w_j s

Since π is unknown, Franklin proposed the optimum values of w_j as

$$w_{0j} = \frac{D_j}{D}; D_j = |\mu_{gj} - \mu_{hj}| and \sum_{1}^{k} D_j$$

The optimum Franklin's estimator based on the column averages is

$$\hat{\pi}_{\mathrm{Fw}}^{0} = \sum_{j=1}^{k} \frac{D_j}{D}\frac{\bar{z}_{\cdot j} - \mu_{hj}}{\mu_{gj} - \mu_{hj}} \tag{6.3.12}$$

The proposed optimum estimator $\hat{\pi}_{\mathrm{Fw}}^{0}$ is unbiased of π and the variance of $\hat{\pi}_{\mathrm{FW}}^{0}$ is

$$\begin{aligned}\operatorname{Var}\left(\hat{\pi}_{\mathrm{Fw}}^{0}\right) &= \frac{\pi(1-\pi)}{n} + \frac{1}{nD^2}\sum_{j=1}^{k}\left\{\left(\sigma_{gj}^2 - \sigma_{hj}^2\right)\pi + \sigma_{hj}^2\right\}\\ &= \frac{\pi(1-\pi)}{n} + \frac{k\left[\pi\left(\bar{\sigma}_g^2 - \bar{\sigma}_h^2\right) + \bar{\sigma}_h^2\right]}{n\left(\sum_{j=1}^{k}\left|\mu_{gj} - \mu_{hj}\right|\right)^2}\end{aligned} \tag{6.3.13}$$

To estimate $\operatorname{Var}\left(\hat{\pi}_{\mathrm{Fw}}^{0}\right)$, one can express $\hat{\pi}_{\mathrm{Fw}}^{0}$ as

$$\begin{aligned}\hat{\pi}_{\mathrm{Fw}}^{0} &= \sum_{j=1}^{k} w_{0j}\frac{\bar{z}_{\cdot j} - \mu_{hj}}{\mu_{gj} - \mu_{hj}}\\ &= \frac{1}{n}\sum_{i=1}^{n}\sum_{j=1}^{k} w_{0j}\frac{z_{ij} - \mu_{hj}}{\mu_{gj} - \mu_{hj}}\\ &= \frac{1}{n}\sum_{i=1}^{n}\bar{r}_{\mathrm{Fw}i} \text{ with } \bar{r}_{\mathrm{Fw}i} = \sum_{j=1}^{k} w_{0j}\frac{z_{ij} - \mu_{hj}}{\mu_{gj} - \mu_{hj}}\end{aligned}$$

Since $\bar{r}_{\mathrm{Fw}i}$ s are iid, an unbiased estimator of the variance Var $(\hat{\pi}_{\mathrm{FW}}^{0}$ is

$$\widehat{\mathrm{Var}}(\hat{\pi}^0_{\mathrm{Fw}}) = \frac{1}{n(n-1)} \sum_{i=1}^{n} (\bar{r}_{\mathrm{FW}i} - \hat{\pi}^0_{\mathrm{FW}})^2 \tag{6.3.14}$$

6.3.1.3 Comparison Between the Row and Column Estimators

(i) The estimator $\hat{\pi}_{\mathrm{F1}}$ is defined if the condition $\overline{\mu}_g \neq \overline{\mu}_h$ is satisfied, while $\hat{\pi}^0_{\mathrm{Fw}}$ is defined when $\mu_{gj} \neq \mu_{hj}$ for every $j = 1, .., k$. Hence, $\hat{\pi}_{\mathrm{F1}}$ is less restrictive than $\hat{\pi}^0_{\mathrm{Fw}}$ from the application point of view. (ii) The variance of the estimator $\hat{\pi}_{\mathrm{F1}}(\hat{\pi}_{\mathrm{F2}})$ decreases with the increase of the difference between $\overline{\mu}_g$ and $\overline{\mu}_h$ (μ_{gj} and μ_{hj}), but, in this case, protection of the privacy of the respondent diminishes. (iii) The column average estimator $\hat{\pi}^0_{\mathrm{Fw}}$ is more efficient than the row average estimator $\hat{\pi}_{\mathrm{F1}}$ since the difference

$$\mathrm{Var}(\hat{\pi}_{\mathrm{F1}}) - \mathrm{Var}(\hat{\pi}^0_{\mathrm{FW}}) = \frac{1}{n}\left\{\pi\left(\overline{\sigma}^2_g - \overline{\sigma}^2_h\right) + \overline{\sigma}^2_h\right\}\left[\frac{1}{k(\overline{\mu}_g - \overline{\mu}_h)^2} - \frac{k}{\left(\sum_{j=1}^{k}\left|\mu_{gj} - \mu_{hj}\right|\right)^2}\right]$$

$$\geq 0$$

as

$$(\overline{\mu}_g - \overline{\mu}_h)^2 = \left|\frac{1}{k}\sum_{j=1}^{k}(\mu_{gj} - \mu_{hj})\right|^2 \leq \frac{1}{k^2}\left[\sum_{j=1}^{k}\left|\mu_{gj} - \mu_{hj}\right|\right]^2.$$

The difference $\mathrm{Var}(\hat{\pi}_{\mathrm{F1}}) - \mathrm{Var}(\hat{\pi}^0_{\mathrm{FW}}) = 0$ if $\mu_{gj} = \overline{\mu}_g$ and $\mu_{hj} = \overline{\mu}_h$ for $j = 1, .., k$.

6.3.2 *General Sampling Designs*

Consider the situation where a sample s of size n is selected by a general sampling design with π_i and π_{ij} are the inclusion probabilities for the ith, ith, and $j(\neq i)$th units. In this case, Eq. (6.3.1) yields the following two working models based on the k observations from a given respondent $i(\in U)$.

$$E_R(\overline{r}_{\mathrm{F1}i}) = y_i,\ V_R(\overline{r}_{\mathrm{F1}i}) = \overline{\phi}_{\mathrm{F1}i} \text{ and } \mathrm{Cov}(\overline{r}_{\mathrm{F1}i}, \overline{r}_{\mathrm{F1}j}) = 0 \text{ for } i \neq j \tag{6.3.15}$$

and

$$E_R(\overline{r}_{\mathrm{F2}i}) = y_i,\ V_R(\overline{r}_{\mathrm{F2}i}) = \overline{\phi}_{\mathrm{F2}i} \text{ and } \mathrm{Cov}(\overline{r}_{\mathrm{F2}i}, \overline{r}_{\mathrm{F2}j}) = 0 \text{ for } i \neq j \tag{6.3.16}$$

where

$$\overline{r}_{F1i} = \frac{\overline{z}_i - \overline{\mu}_h}{\overline{\mu}_g - \overline{\mu}_h}; \overline{\mu}_g \neq \overline{\mu}_h,\ \overline{r}_{F2i} = \frac{1}{k}\sum_{j=1}^{k}\frac{z_{ij} - \mu_{hj}}{\mu_{gj} - \mu_{hj}}; \mu_{gj} \neq \mu_{hj},$$

$$\phi_{F1i} = \frac{1}{k^2}\sum_{j=1}^{k}\frac{(\sigma_{gj}^2 - \sigma_{hj}^2)y_i + \sigma_{hj}^2}{(\overline{\mu}_g - \overline{\mu}_h)^2} \text{ and } \phi_{F2i} = \frac{1}{k^2}\sum_{j=1}^{k}\frac{(\sigma_{gj}^2 - \sigma_{hj}^2)y_i + \sigma_{hj}^2}{(\mu_{gj} - \mu_{hj})^2}$$

Models (6.3.15) yield two obvious unbiased estimators of ϕ_{F1i} as

$$\hat{\phi}_{F1i}(1) = \overline{r}_{F1i}(\overline{r}_{F1i} - 1) \text{ and } \hat{\phi}_{F1i}(2) = \frac{1}{k^2}\sum_{j=1}^{k}\frac{(\sigma_{gj}^2 - \sigma_{hj}^2)\overline{r}_{F1i} - \sigma_{hj}^2}{(\overline{\mu}_g - \overline{\mu}_h)^2}$$

Similarly, model (6.3.16) provides two unbiased estimators of $\hat{\phi}_{F2i}$ as

$$\hat{\phi}_{F2i}(1) = \overline{r}_{F2i}(\overline{r}_{F1i} - 1) \text{ and } \hat{\phi}_{F2i}(2) = \frac{1}{k^2}\sum_{j=1}^{k}\frac{(\sigma_{gj}^2 - \sigma_{hj}^2)\overline{r}_{F2i} + \sigma_{hj}^2}{(\mu_{gj} - \mu_{hj})^2} \quad (6.3.17)$$

It is difficult to compare the performances of the models (6.3.15) and (6.3.16) in general. However, one finds that the model (6.3.15) is less restrictive than that of (6.3.16), as (6.3.16) demands $\mu_{gj} \neq \mu_{hj}$ for every $j = 1, ..k$ while (6.3.15) demands $\overline{\mu}_g \neq \overline{\mu}_h$ only.

From the models (6.3.15) and (6.3.16), one finds two Horvitz-Thompson (1952)-type (HT-type) unbiased estimators of π for the general sampling design as

$$\hat{\pi}_{HT}^{(j)} = \frac{1}{N}\sum_{i\in s}\frac{\overline{r}_{Fji}}{\pi_i}, \ j = 1, 2 \quad (6.3.18)$$

The properties of $\hat{\pi}_{HT}^{(j)}$ are obtained from Theorem 6.2.1 as follows

Theorem 6.3.1 For $j = 1, 2$,

(i) $\hat{\pi}_{HT}^{(j)}$ is an unbiased estimator of π.
(ii) The variance of $\hat{\pi}_{HT}^{(j)}$ is

$$\mathrm{Var}\left(\hat{\pi}_{HT}^{(j)}\right) = \frac{1}{N^2}\left[\sum_{i<}\sum_{k\in U}(\pi_i\pi_k - \pi_{ik})\left(\frac{y_i}{\pi_i} - \frac{y_k}{\pi_k}\right)^2 + \sum_{i\in U}\frac{\phi_{Fji}}{\pi_i}\right].$$

(iii) Two unbiased estimators of $\mathrm{Var}\left(\hat{\pi}_{HT}^{(j)}\right)$ are

$$\widehat{\mathrm{Var}}\left(\hat{\pi}_{HT}^{(j)}\right)_t = \frac{1}{N^2}\left[\sum_{i<}\sum_{k\in s}\left(\frac{\pi_i\pi_k - \pi_{ik}}{\pi_{ik}}\right)\left(\frac{\overline{r}_{Fji}}{\pi_i} - \frac{\overline{r}_{Fjk}}{\pi_k}\right)^2 + \sum_{i\in s}\frac{\hat{\phi}_{Fji}(t)}{\pi_i}\right], t = 1, 2.$$

6.4 Extensions of Franklin's Model

6.4.1 Kuk's Model

Kuk (1990) applied the Binomial distribution to Franklin's model. Kuk's randomization device is comprised of two packs of cards, pack-1 and pack-2, each of which contains black and red cards. The proportions of black cards in pack-1 and pack-2 are, respectively, P_1 and $P_2(\neq P_1)$. Unnoticed by the interviewer, each respondent in the sample was requested to draw a fixed number of $k(\geq 1)$ cards, with replacement from pack-1 (pack-2) if he or she belonged to the sensitive group A(non-sensitive group $\overline{A}$). The respondents were directed to report the number of black cards drawn as their RR responses. Let z_i be the RR obtained from the ith respondent. Then,

$$z_i = y_i z_{1i} + (1 - y_i) z_{2i} \tag{6.4.1}$$

where $z_{1i}(z_{2i})$ denotes the total number of black cards drawn by ith the respondent from pack-1 (pack-2) and $y_i = 1(0)$ if the ith respondent belongs to the group $A(\overline{A})$. Since z_{1i} and z_{2i} follow binomial $B(k, P_1)$ and $B(k, P_2)$ distributions, one finds $E_R(z_i) = k\left[y_i P_1 + (1 - y_i)P_2\right]$, $V_R(z_i) = k\left[y_i P_1(1 - P_1) + (1 - y_i)P_2(1 - P_2)\right]$ and $C_R(z_i, z_j) = 0$ for $i \neq j$.

Further denoting,

$$r_i = \frac{\hat{\lambda}_i - P_2}{P_1 - P_2} \tag{6.4.2}$$

with $\hat{\lambda}_i = \frac{z_i}{k}$ as the proportion of black cards drawn by the ith respondent, the following reduced model is obtained:

$$\begin{aligned} E_R(r_i) &= y_i, \\ V_R(r_i) &= \frac{y_i(P_1 - P_2)(1 - P_1 - P_2) + P_2(1 - P_2)}{k(P_1 - P_2)^2} = \phi_{Ki} \text{ and} \\ C_R(r_i, r_j) &= 0 \text{ for } i \neq j \end{aligned} \tag{6.4.3}$$

An unbiased estimator of ϕ_{Ki} is $\hat{\phi}_{Ki} = \frac{r_i(P_1 - P_2)(1 - P_1 - P_2) + P_2(1 - P_2)}{k(P_1 - P_2)^2}$.

6.4.1.1 Estimation of Proportion π

SRSWR Sampling

Kuk (1990) selected a sample of n respondents by the SRSWR method and proposed an unbiased estimator of π as

$$\hat{\pi}_{\mathrm{K}} = \frac{1}{n}\sum_{i=1}^{n} r_i = \frac{\overline{\lambda} - P_2}{P_1 - P_2} \text{ where } \overline{\lambda} = \frac{1}{n}\sum_{i=1}^{n} \hat{\lambda}_i \tag{6.4.4}$$

Noting,

$$E(\hat{\pi}_{\mathrm{K}}) = E_p\left[\frac{1}{n}\sum_{i=1}^{n} E_R(r_i)\right] = E_p\left(\frac{1}{n}\sum_{i=1}^{n} y_i\right) = \pi, \tag{6.4.5}$$

$$\begin{aligned}\mathrm{Var}(\hat{\pi}_{\mathrm{K}}) &= V_p\left[\frac{1}{n}\sum_{i=1}^{n} y_i\right] + E_p\left[\frac{1}{n^2}\sum_{i=1}^{n} \phi_{\mathrm{K}i}\right] \\ &= \frac{\pi(1-\pi)}{n} + \frac{1}{nN}\sum_{i\in U} \phi_{\mathrm{K}i} \\ &= \frac{\pi(1-\pi)}{n} + \frac{\overline{\phi}_{\mathrm{K}}}{n}\end{aligned} \tag{6.4.6}$$

(where $\overline{\phi}_{\mathrm{K}} = \frac{1}{N}\sum_{i\in U} \phi_{\mathrm{K}i} = \frac{\pi P_1(1-P_1)+(1-\pi)P_2(1-P_2)}{k(P_1-P_2)^2}$) and remembering r_i, $i = 1, .., n$ are independent and identically distributed random variables, the following theorem is obtained.

Theorem 6.4.1

(i) $\hat{\pi}_{\mathrm{K}}$ is an unbiased estimator of π.

(ii) The variance of $\hat{\pi}_{\mathrm{K}}$ is

$$\mathrm{Var}(\hat{\pi}_{\mathrm{K}}) = \frac{\pi(1-\pi)}{n} + \frac{\overline{\phi}_{\mathrm{K}}}{n}.$$

(iii) An unbiased estimator of $\mathrm{Var}(\hat{\pi}_{\mathrm{K}})$ is

$$\widehat{\mathrm{Var}}(\hat{\pi}_{\mathrm{K}}) = \frac{1}{n(n-1)}\sum_{i=1}^{n} (r_i - \hat{\pi}_{\mathrm{K}})^2.$$

Remark 6.4.1 The variance $\mathrm{Var}(\hat{\pi}_{\mathrm{K}})$ decreases with an increase in the number of trials k. Hence, the efficiency of Kuk's model increases with the number of trials. Kuk's (1990) model reduces to Warner's (1965) model if $P_1 = P_2 = P$ and $k = 1$.

General Sampling Design

For the general sampling design described in Sect. 6.4.1, the HT-type estimator for the proportion π is given by

$$\hat{\pi}_{\text{HTK}} = \frac{1}{N}\sum_{i\in s}\frac{r_i}{\pi_i} \tag{6.4.7}$$

The properties of $\hat{\pi}_{\text{HTK}}$.are obtained from Theorem 6.4.1 as follows:

Theorem 6.4.2

(i) $\hat{\pi}_{\text{HTK}}$ is an unbiased estimator of π.
(ii) The variance of $\hat{\pi}_{\text{HTK}}$ is

$$\begin{aligned}\text{Var}(\hat{\pi}_{\text{HTK}}) &= \frac{1}{N^2}\left[\sum_{i\in U}\left(\frac{1}{\pi_i}-1\right)y_i + \sum\sum_{i\neq j\in U}\left(\frac{\pi_{ij}}{\pi_i\pi_j}-1\right)y_iy_j + \sum_{i\in U}\frac{\phi_{Ki}}{\pi_i}\right]\\ &= \frac{1}{N^2}\left[\sum_{i\in A}\left(\frac{1}{\pi_i}-1\right) + \sum\sum_{i\neq j\in A}\left(\frac{\pi_{ij}}{\pi_i\pi_j}-1\right) + \sum_{i\in U}\frac{\phi_{Ki}}{\pi_i}\right].\end{aligned}$$

(iii) An unbiased estimator of $\text{Var}(\hat{\pi}_{\text{HTK}})$ is

$$\widehat{\text{Var}}(\hat{\pi}_{\text{HTK}}) = \frac{1}{N^2}\left[\sum_{i\in s}\left(\frac{1}{\pi_i}-1\right)\frac{r_i^2}{\pi_i} + \sum\sum_{i\neq j\in s}\left(\frac{\pi_{ij}}{\pi_i\pi_j}-1\right)\frac{r_ir_j}{\pi_{ij}} + \sum_{i\in s}\frac{\hat{\phi}_{Ki}}{\pi_i}\right].$$

SRSWOR Sampling

For the simple random sampling without replacement (SRSWOR) of size n, $\pi_i = \frac{n}{N}$ and $\pi_{ij} = \frac{n(n-1)}{N(N-1)}$. In this case, $\hat{\pi}_{\text{HTK}} = \overline{r}_s = \frac{1}{n}\sum_{i\in s} r_i$ is unbiased for π, $\text{Var}(\overline{r}_s) = \frac{N-n}{n(N-1)}\pi(1-\pi) + \frac{\overline{\phi}_K}{n}$ and an unbiased estimator of $\text{Var}(\overline{r}_s)$ is $\widehat{\text{Var}}(\overline{r}_s) = \frac{N-n}{nN}s_r^2 + \frac{1}{nN}\sum_{i\in s}\hat{\phi}_{Ki}$ where $(n-1)s_r^2 = \sum_{i\in s}(r_i-\overline{r}_s)^2$.

Remark 6.4.2 The estimators of π, their variances, and their unbiased estimators for various strategies can be obtained from Table 5.3.1.

6.4.2 Model of Arnab (2021)

Arnab (2018) proposed an alternative RR model that is more efficient than model of Kuk (1990). Arnab's randomization device comprises two sets of cards, each of which consists of k packs of cards of two colors: black and red, as in Kuk's model. The proportions of the black and red cards of the $j(=1, ..., k)$th pack in set-1 are $P_{1j}(\neq 1/2)$ and $1-P_{1j}$ while the proportions of the black and red cards of the jth pack in set-2 are $P_{2j}(\neq 1/2)$ and $1-P_{2j}$. Each of the selected respondents in the

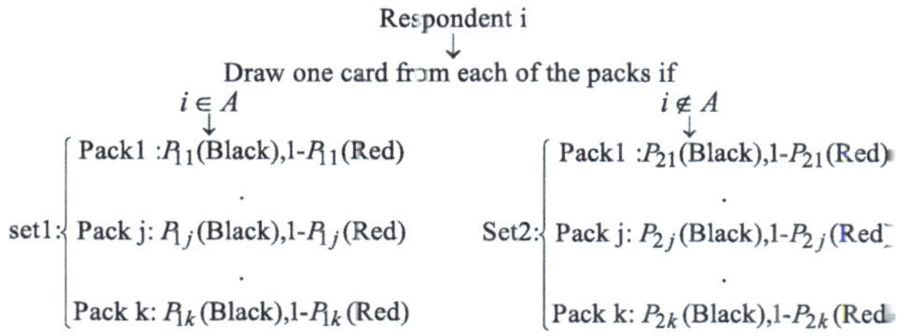

Fig. 6.4 Model of Arnab (2021)

sample s was asked to draw one card at random from each of the k packs of set-1 if he or she was a member of the sensitive group A. The respondents belonging to the non-sensitive group $\overline{A}$, were asked to draw one card at random from each of the k packs of set-2 independently. Respondents should report z, the total number of black cards drawn as their response. The entire experiment was conducted in the respondents' absence to protect their privacy. Define (Fig. 6.4)

$$z_i(j) = \begin{cases} 1 \text{ if ith respondents draws a black card from the jth pack} \\ 0 \text{ if ith respondents draws a red card from the jth pack} \end{cases} \quad j = 1, .., k \tag{6.4.8}$$

Then,

$$\text{Prob}\ (z_i(j) = 1) = \begin{cases} P_{1j} \text{ if } i \in A \\ P_{2j} \text{ if } i \notin A \end{cases} \tag{6.4.9}$$

Now noting $y_i = \begin{cases} 1 \text{ if } i \in A \\ 0 \text{ if } i \notin A \end{cases}$, one finds

$$\begin{aligned} E_R(z_i(j)) &= y_i P_{1j} + (1 - y_i) P_{2j}, \\ V_R(z_i(j)) &= y_i P_{1j} + (1 - y_i) P_{2j} - \{y_i P_{1jj} + (1 - y_i) P_{2j}\}^2 \\ &= y_i P_{1j}(1 - P_{1j}) + (1 - y_i) P_{2j}(1 - P_{2j}) \end{aligned} \tag{6.4.10}$$

Denoting $z_i = \sum_{j=1}^{k} z_i(j)$, the total number of black cards drawn by the ith respondent, one obtains from (6.4.10)

$$\begin{aligned} E_R(z_i) &= k\{(\bar{P}_1 - \bar{P}_2) y_i + \bar{P}_2\} \text{ and} \\ V_R(z_i) &= y_i \sum_{j=1}^{k} (P_{1j} - P_{2j})(1 - P_{1j} - P_{2j}) + \sum_{j=1}^{k} P_{2j}(1 - P_{2j}) \end{aligned} \tag{6.4.11}$$

where $\overline{P}_1 = \frac{1}{k} \sum_{j=1}^{k} P_{1j}$ and $\overline{P}_2 = \frac{1}{k} \sum_{j=1}^{k} P_{2j}$.

Now denoting

$$r_i = \frac{(z_i/k) - \overline{P}_2}{\overline{P}_1 - \overline{P}_2}, \tag{6.4.12}$$

the following working model is obtained:

$$E_R(r_i) = y_i,\ V_R(r_i) = \phi_{Ai} \text{ and } C_R(z_i, z_j) = 0 \text{ for } i \neq j \tag{6.4.13}$$

where

$$\phi_{Ai} = \frac{y_i \sum_{j=1}^{k} (P_{1j} - P_{2j})(1 - P_{1j} - P_{2j}) + \sum_{j=1}^{k} P_{2j}(1 - P_{2j})}{k^2(\overline{P}_1 - \overline{P}_2)^2} \tag{6.4.14}$$

Equation (6.4.14) yields an unbiased estimator of ϕ_{Ai} as follows:

$$\hat{\phi}_{Ai} = \frac{r_i \sum_{j=1}^{k} (P_{1j} - P_{2j})(1 - P_{1j} - P_{2j}) + \sum_{j=1}^{k} P_{2j}(1 - P_{2j})}{k^2(\overline{P}_1 - \overline{P}_2)^2} \tag{6.4.15}$$

Using (5.3.1), one can find the expressions of the unbiased estimators of π, variance of the estimator and the estimators of variances for different sampling strategies based on the model (6.4.13).

Theorem 6.4.3 RR model of Arnab (2018) is more efficient than model of Kuk (1990) if $\overline{P}_1 = P_1$ and $\overline{P}_2 = P_2$.

Proof For a given sampling design, model of Arnab (2018) is more efficient than the model of Kuk (1990) as the difference

$$\begin{aligned}
\phi_{Ki} - \phi_{Ai} &= \frac{y_i(P_1 - P_2)(1 - P_1 - P_2) + P_2(1 - P_2)}{k(P_1 - P_2)^2} \\
&\quad - \frac{y_i \sum_{j=1}^{k} (P_{1j} - P_{2j})\left(1 - P_{1j} - P_{2j}\right) + \sum_{j=1}^{k} P_{2j}\left(1 - P_{2j}\right)}{k^2(P_1 - P_2)^2} \\
&= \frac{y_i\left[\frac{1}{k}\sum\limits_{j=1}^{k}\left(P_{1j} - P_{2j}\right)^2 - (P_1 - P_2)^2\right] + \frac{1}{k}\left[\sum_{j=1}^{k} P_{2j}^2 - P_2^2\right]}{k(P_1 - P_2)^2} \\
&\geq 0
\end{aligned}$$

6.4.3 Singh and Grewal's Model

Singh and Grewal (2013) applied the Geometric distribution to the Franklin model. In the proposed model, each respondent is provided with two packs of cards, as in model of Kuk (1990). Pack-1 consists of P_{G1} proportion of cards bearing the statement "I belong to the sensitive group A" and $1 - P_{G1}$ proportion of cards written with the statement "I do not belong to the sensitive group A". In pack-2, $P_{G2}(\neq P_{G1})$ proportion of cards written with the statement "I do not belong to the sensitive group A" and $1 - P_{G2}$ proportion of cards with the statement "I belong to the sensitive group A".

If a respondent belongs to the sensitive group A, he or she is instructed to draw cards one by one with replacement from the pack-1 until he or she draws a card bearing the statement "I belong to the sensitive group A" for the first time. On the other hand, if a respondent belongs to the non-sensitive group $\overline{A}$, he or she is asked to draw cards one by one from the pack-2 until he or she draws a card bearing the statement "I do not belong to the sensitive group A". The respondent is asked to report the total number of cards z drawn as his or her response.

Let z_i be the RR reported by the $i(\in U)$th respondent and y_i as in (6.2.8). Then,

$$z_i = z_{i1}y_i + z_{i2}(1 - y_i) \tag{6.4.16}$$

where z_{i1} and z_{i2} are RR obtained from the pack-1 and pack-2, respectively.

Noting z_{i1} and z_{i2} follow Geometric distributions with parameters P_{G1} and P_{G2}, one finds that

$$\begin{aligned} E_R(z_i) &= \frac{y_i}{P_{G1}} + \frac{(1 - y_i)}{P_{G2}}, \\ V_R(z_i) &= \frac{1 - P_{G1}}{P_{G1}^2} y_i + \frac{1 - P_{G2}}{P_{G2}^2}(1 - y_i) \text{ and} \\ C_R(z_i, z_j) &= 0 \text{ for } i \neq j \end{aligned}$$

On writing $r_i = \frac{(z_i P_{G2} - 1)P_{G1}}{P_{G2} - P_{G1}}$, one obtains the following reduced RR model:

$$\begin{aligned} E_R(r_i) &= y_i, \\ V_R(r_i) &= \frac{P_{G2}^2(1 - P_{G1})y_i + P_{G1}^2(1 - P_{G2})(1 - y_i)}{(P_{G2} - P_{G1})^2} = \phi_{SG} \text{ and} \\ C_R\left(r_i, r_j\right) &= 0 \text{ for } i \neq j \end{aligned} \tag{6.4.17}$$

6.4.3.1 Estimation of Proportion π

Singh and Grewal (2013) considered the SRSWR sampling of size n. They proposed unbiased estimator of π is

$$\hat{\pi}_{\mathrm{SG}} = \frac{1}{n}\sum_{i=1}^{n} r_i = \frac{1}{P_{\mathrm{G2}} - P_{\mathrm{G1}}}\left(\frac{P_{\mathrm{G2}}P_{\mathrm{G1}}}{n}\sum_{i=1}^{n} z_i - P_{\mathrm{G1}}\right),\ P_{\mathrm{G1}} \neq P_{\mathrm{G2}} \tag{6.4.18}$$

The properties of $\hat{\pi}_{\mathrm{SG}}$ are obtained from Theorem 6.4.1 as follows:

Theorem 6.4.4

(i) $\hat{\pi}_{\mathrm{SG}}$ is an unbiased estimator of π.
(ii) The variance of $\hat{\pi}_{\mathrm{SG}}$ is

$$\mathrm{Var}(\hat{\pi}_{\mathrm{SG}}) = \frac{\pi(1-\pi)}{n} + \frac{\overline{\phi}_{\mathrm{SG}}}{n}$$

where $\overline{\phi}_{\mathrm{SG}} = \frac{1}{N}\sum_{i\in U}\phi_{\mathrm{SG}i} = \frac{P_{\mathrm{G2}}^2(1-P_{\mathrm{G1}})\pi + P_{\mathrm{G1}}^2(1-P_{\mathrm{G2}})(1-\pi)}{(P_{\mathrm{G2}}-P_{\mathrm{G1}})^2}$.

(iii) An unbiased estimator of $\mathrm{Var}(\hat{\pi}_{\mathrm{SG}})$ is

$$\widehat{\mathrm{Var}}(\hat{\pi}_{\mathrm{SG}}) = \frac{1}{n(n-1)}\sum_{i=1}^{n}(r_i - \hat{\pi}_{\mathrm{SG}})^2.$$

Estimators of proportion of π and their properties for the general sampling design can be derived directly from Theorem 6.4.2.

Remark 6.3.1 Chaudhuri and Christofied (2013) extended model of Singh and Grewal (2013) for RR models of Kuk (1990) and Greenberg et al. (1969).

6.4.4 *Model A of Hussain et al.*

Hussain et al. (2014) used the Negative Binomial distribution in the Franklin model. The proposed model is like model of Sing and Grewal (2013) except that the respondent selects cards one by one with replacement until he or she receives a pre-specified number $c(\geq 1)$ cards of his or her status. The respondents should report the total number of cards drawn as their response. Let z_i be the RR provided by the ith respondent. Then z_i can be expressed as

$$z_i = z_{i1}y_i + z_{i2}(1-y_i) \tag{6.4.19}$$

where z_{i1} and z_{i2} follow the Negative Binomial distribution with parameters (c, P_{G1}) and (c, P_{G2}), respectively.

On noting that $E_R(z_{i1}) = \frac{c}{P_{G1}}$, $E_R(z_{i2}) = \frac{c}{P_{G2}}$, $V_R(z_{i1}) = \frac{c(1-P_G)}{P_{G1}^2}$, $V_R(z_{i2}) = \frac{c(1-P_G)}{P_{G2}^2}$ and writing

$$r_i = \frac{(z_i P_{G2} - c)P_{G1}}{c(P_{G2} - P_{G1})}, \; P_{G1} \neq P_{G2},$$

the following RR model is obtained:

$$\begin{aligned} E_R(r_i) &= y_i, \\ V_R(r_i) &= \frac{P_{G2}^2(1 - P_{G1})y_i + P_{G1}^2(1 - P_{G2})(1 - y_i)}{c(P_{G2} - P_{G1})^2} = \phi_{HAi} \text{ and} \\ C_R(r_i, r_j) &= 0 \text{ for } i \neq j \end{aligned} \tag{6.4.20}$$

The estimators of the population proportion π for various sampling designs for the model (6.4.20) can be obtained using Table 5.3.1. For the SRSWR sample of size n, Hussain et al. (2014) derived the following results:

Theorem 6.4.5

(i) $\hat{\pi}_{HA} = \frac{1}{n}\sum_{i=1}^{n} r_i = \frac{1}{c(P_{G2}-P_{G1})}\left(\frac{P_{G2}P_{G1}}{n}\sum_{i=1}^{n} z_i - cP_{G1}\right)$ is an unbiased estimator of π.

(ii) The variance of $\hat{\pi}_{HA}$ is

$$\text{Var}(\hat{\pi}_{HA}) = \frac{\pi(1-\pi)}{n} + \frac{\overline{\phi}_{HA}}{n}$$

where $\overline{\phi}_{HA} = \frac{1}{N}\sum_{i \in U} \phi_{Hi} = \frac{P_{G2}^2(1-P_{G1})\pi - P_{G1}^2(1-P_{G2})(1-\pi)}{c(P_{G2}-P_{G1})^2}$.

(iii) An unbiased estimator of $\text{Var}(\hat{\pi}_{HA})$ is

$$\widehat{\text{Var}}(\hat{\pi}_{HA}) = \frac{1}{n(n-1)}\sum_{i=1}^{n}(r_i - \hat{\pi}_{HA})^2.$$

6.4.5 *Model B of Hossain et al.*

Hussain et al. (2017) applied the generalized geometric distribution to the Franklin (1989b) model. The model is similar to models of Singh and Grewal (2013) and Hussain et al. (2014). The difference is that an individual belonging to the group $A(\overline{A})$ is instructed to report the number of cards drawn to produce c_1 (c_2) consecutive cards of his or her actual status for the first time as their responses. In this case, the RR obtained from the ith respondent can be written as

$$z_i = z_{i1}y_i + z_{i2}(1 - y_i) \tag{6.4.21}$$

where z_{i1} and z_{i2} follow Geometric distribution of order $c_1(GD_{c_1})$ and $c_2(GD_{c_2})$, respectively.

Noting

$E_R(z_{ij}) = \mu_{\mathrm{HB}j} = \frac{1-P_{\mathrm{Gj}}^{c_j}}{\left(1-P_{\mathrm{Gj}}\right)P_{\mathrm{Gj}}^{c_j}}, V_R(z_{ij}) = \sigma^2_{\mathrm{HB}j} = \left[\frac{1-(2c_j+1)\left(1-P_{\mathrm{Gj}}\right)P_{\mathrm{Gj}}^{c_j}-P_{\mathrm{Gj}}^{2c_j+1}}{\left(1-P_{\mathrm{Gj}}\right)^2 P_{\mathrm{Gj}}^{2c_j}}\right]$ for $j = 1, 2$ and writing $r_i = \frac{z_i - \mu_{\mathrm{HB2}}}{\mu_{\mathrm{HB1}} - \mu_{\mathrm{HB2}}}$, $\mu_{\mathrm{HB1}} \neq \mu_{\mathrm{HB2}}$, the following reduced model is obtained:

$$E_R(r_i) = y_i, V_R(r_i) = \sigma^2_{\mathrm{HB1}}y_i + \sigma^2_{\mathrm{HB2}}(1 - y_i) = \phi_{\mathrm{HB}i} \text{ and } C_R(r_i, r_j) = 0 \text{ for } i \neq j \tag{6.4.22}$$

From the model (6.4.22), one can easily find unbiased estimators of π, variances and their unbiased estimators for various strategies using Table 5.3.1. For the SRSWR sampling, Hussain et al. (2017) obtained the following results:

Theorem 6.4.6

(i) $\hat{\pi}_{\mathrm{HB}} = \frac{1}{n}\sum_{i=1}^n r_i = \frac{1}{(\mu_{\mathrm{HB1}} - \mu_{\mathrm{HB2}})}\left(\frac{1}{n}\sum_{i=1}^n z_i - \mu_{\mathrm{HB2}}\right)$ is an unbiased estimator of π.

(ii) The variance of $\hat{\pi}_{\mathrm{HB}}$ is

$$\mathrm{Var}(\hat{\pi}_{\mathrm{HB}}) = \frac{\pi(1-\pi)}{n} + \frac{\overline{\phi}_{\mathrm{HB}}}{n}$$

where $\overline{\phi}_{\mathrm{HB}} = \frac{1}{N}\sum_{i \in U}\phi_{\mathrm{HB}i} = \frac{\sigma^2_{\mathrm{HB1}}\pi + \sigma^2_{\mathrm{HB2}}(1-\pi)}{(\mu_{\mathrm{HB1}} - \mu_{\mathrm{HB2}})^2}$.

(iii) An unbiased estimator of $\mathrm{Var}(\hat{\pi}_{\mathrm{HB}})$ is

$$\widehat{\mathrm{Var}}(\hat{\pi}_{\mathrm{HB}}) = \frac{1}{n(n-1)}\sum_{i=1}^n (r_i - \hat{\pi}_{\mathrm{HB}})^2.$$

6.4.6 Model of Al-Sobhi et al.

Al-Sobhi et al. (2014) proposed two RR models by applying Polya's urn process to Franklin's model.

6.4.6.1 Model A

The randomization device for Model A is comprised of two decks of cards: deck-1 and deck-2, each of which contains two types of cards, e.g., black and red. The deck-1 contains a_1 black and b_1 red cards, while the deck-2 contains a_2 black and b_2 red cards. Each respondent belonging to the sensitive group A (non-sensitive group $\overline{A}$) is asked to draw $k_1\,(k_2)$ cards one by one from deck-1 (deck-2). For each draw, he or she is required to replace the card drawn and add $c_1\,(c_2)$ cards of the same color. The respondent is requested to submit the number of black cards drawn as his or her randomized response.

Let $Z_i = Z_{i1}(Z_{i2})$ be the RR reported by the $i(\in U)$th respondent if he or she belonged to $A(\overline{A})$. In this case, one can express

$$Z_i = Z_{i1}y_i + Z_{i2}(1 - y_i) \tag{6.4.23}$$

Now noting,

$$\Pr ob(Z_{1i} = z_1) = \binom{k_1}{z_1}\frac{a_1^{(c_1,z_1)}b_1^{(c_1,k_1-z_1)}}{(a_1+b_1)^{(c_1,k_1)}},\ z_1 = 0, 1, .., k_1, \tag{6.4.24}$$

$$\Pr ob(Z_{2i} = z_2) = \binom{k_2}{z_2}\frac{a_2^{(c_2,z_2)}b_1^{(c_2,k_2-z_2)}}{(a_2+b_2)^{(c_2,k_2)}},\ z_2 = 0, 1, .., k_2, \tag{6.4.25}$$

and

$$r^{(s,j)} = r(r+s)(r+2s)....[r+(j-1)s],$$

one finds

$$\begin{aligned} E_R(Z_{1i}) = \mu_{A1} &= \frac{k_1a_1}{a_1+b_1},\ E_R(Z_{2i}) = \mu_{A2} = \frac{k_2a_2}{a_2+b_2}, \\ V_R(Z_{1i}) = \sigma_{A1}^2 &= \frac{k_1a_1b_1}{(a_1+b_1)^2}\left[1+(k_1-1)\frac{c_1}{a_1+b_1+c_1}\right] \\ V_R(Z_{2i}) = \sigma_{A2}^2 &= \frac{k_2a_2b_2}{(a_2+b_2)^2}\left[1+(k_2-1)\frac{c_2}{a_2+b_2+c_2}\right] \end{aligned} \tag{6.4.26}$$

Further, writing

$$r_i = \frac{Z_{i1} - \mu_{A2}}{\mu_{A1} - \mu_{A2}};\ \mu_{A1} \neq \mu_{A2} \tag{6.4.27}$$

the following reduced model is obtained:

$$E_R(r_i) = y_i,\ V_R(r_i) = \frac{\sigma_{A1}^2 y_i + \sigma_{A2}^2(1-y_i)}{(\mu_{A1}-\mu_{A2})^2} = \phi_{ALi},\ C_R(r_i, r_j) = 0 \text{ for } i \neq j \tag{6.4.28}$$

Estimation of Proportion π

Using Table 5.3.1, one obtains estimators of π, variances and their unbiased estimators for various sampling strategies. Al-Sobhi et al. (2014) derived the following results for the SRSWR sampling design of size n.

Theorem 6.4.7 For a SRSWR sampling,

(i) $\hat{\pi}_{\text{AL}} = \frac{1}{n}\sum_{i=1}^{n} r_i = \frac{\overline{Z}_{\text{AL}} - \mu_{A2}}{\mu_{A1} - \mu_{A2}}$ is an unbiased estimate of π

where $\overline{Z}_{\text{AL}} = \frac{1}{n}\sum_{i=1}^{n} Z_i$.

(ii) The variance of $\hat{\pi}_{\text{AL}}$ is

$$\text{Var}(\hat{\pi}_{\text{AL}}) = \frac{\pi(1-\pi)}{n} + \frac{\pi\sigma_{A1}^2 + (1-\pi)\sigma_{A2}^2}{n(\mu_{A2} - \mu_{A1})^2}.$$

(iii) An unbiased estimator of $\text{Var}(\hat{\pi}_{\text{AL}})$ is

$$\widehat{\text{Var}}(\hat{\pi}_{\text{AL}}) = \frac{1}{n}\sum_{i=1}^{n} (r_i - \hat{\pi}_{\text{AL}})^2.$$

Particular Cases

(i) For $c_1 = c_2 = 0$ and $k_1 = k_2 = c$, Model A reduces to model of Kuk (1990).
(ii) Dihidar (2016) considered Model A with $c_1 = c_2 = -1$ and $k_1 = k_2 = c$. In this case, Z_{1i} and Z_{2i} become samples from the Hypergeometric distributions with the parameters $(a_1 + b_1, a_1, c)$ and $(a_2 + b_2, a_2, c)$. Dihidar (2016) presented estimators of the proportion, variances, and unbiased estimators of the variances of various sampling strategies.

6.4.6.2 Model B

Model B is similar to Model A. Here also, two decks of cards, deck-1 and deck-2, containing a_1 black and b_1 red cards, and a_2 black and b_2 red cards are taken as in Model A. Each respondent belonging to the sensitive (non-sensitive) group A ($\overline{A}$) is requested to use deck-1 (deck-2) and asked to draw cards one by one until a fixed number $r_1 (r_2)$ of black cards are drawn. On each draw, the respondent replaces the card drawn and adds $c_1 = 1 (c_2 = 1)$ cards of the same color. The respondent is requested to submit the total number of cards drawn as his or her response. Let $Z_i = Z_{i1} (Z_{i2})$ be the RR reported by the ith respondent belonging to the group $A(\overline{A})$. So, we write

$$Z_i = Z_{i1}y_i + Z_{i2}(1 - y_i) \tag{6.4.29}$$

In this case,

$$\Pr ob(Z_{i1} = z_1) = \binom{z_1 - 1}{r_1 - 1}\frac{a_1^{(1,r_1)}b_1^{(1,z_1-r_1)}}{(a_1 + b_1)^{(1+z_1)}},\ z_1 = r_1, r_1 + 1, \ldots\infty$$

and

$$\Pr ob(Z_{i1} = z_2) = \binom{z_2 - 1}{r_1 - 1}\frac{a_2^{(1,r_2)}b_1^{(1,z_2-r_2)}}{(a_1 + b_1)^{(1+z_2)}},\ z_2 = r_1, r_1 + 1, \ldots\infty \tag{6.4.30}$$

On writing,

$$E_R(Z_{i1}) = \mu_{B1} = \frac{r_1(a_1 + b_1 - 1)}{a_1 - 1},\ E_R(Z_{i2}) = \mu_{B2} = \frac{r_2(a_2 + b_2 - 1)}{a_2 - 1},$$

$$V_R(Z_{i1}) = \sigma_{B1}^2 = \frac{r_1(a_1 + b_1 - 1)}{(a_1 - 1)(a_1 - 2)}[b_1 + r_1(a_1 + b_1 - 2)] - \left\{\frac{r_1(a_1 + b_1 - 1)}{a_1 - 1}\right\}^2,$$

$$V_R(Z_{i2}) = \sigma_{B2}^2 = \frac{r_2(a_2 + b_2 - 1)}{(a_2 - 1)(a_2 - 2)}[b_2 + r_2(a_2 + b_2 - 2)] - \left\{\frac{r_2(a_2 + b_2 - 1)}{a_2 - 1}\right\}^2$$

and

$$r_i = \frac{z_i - \mu_{B2}}{\mu_{B1} - \mu_{B2}};\ \mu_{B1} \neq \mu_{B2}, \tag{6.4.31}$$

the following reduced model is obtained:

$$E_R(r_i) = y_i,\ V_R(r_i) = \frac{y_i\sigma_{B1}^2 + (1 - y_i)\sigma_{B2}^2}{(\mu_{B1} - \mu_{B2})^2} \text{ and } C_R(r_i, r_j) = 0 \text{ for } i \neq j \tag{6.4.32}$$

The model (6.4.32) is similar to the model (6.4.28). Hence, we derive similar results to those obtained in Theorem 6.4.7 for model A.

Particular Cases

(i) Al-Sobhi et al. (2014) considered the SRSWR sampling design of size n and randomizing device (Polya's urn process) with $c_1 = c_2 = 1$. They obtained an unbiased estimator of π and its variance as

$$\hat{\pi}_{BL} = \frac{\overline{Z}_{BL} - \mu_{B2}}{\mu_{B1} - \mu_{B2}} \tag{6.4.33}$$

and

$$\mathrm{Var}(\hat{\pi}_{\mathrm{BL}}) = \frac{\pi(1-\pi)}{n} + \frac{\pi\sigma_{\mathrm{B1}}^2 + (1-\pi)\sigma_{\mathrm{B2}}^2}{n(\mu_{\mathrm{B1}} - \mu_{\mathrm{B2}})^2} \tag{6.4.34}$$

where

$$\overline{Z}_{\mathrm{BL}} = \frac{1}{n}\sum_{i=1}^{n} Z_i.$$

(ii) Singh and Grewal (2013) considered the situation where $c_1 = c_2 = 0$ and $r_1 = r_2 = 1$. In this case, Z_i is a mixture of two geometric distributions with parameters $p_1 = \frac{a_1}{a_1+b_1}$ and $p_2 = \frac{a_2}{a_2+b_2}$. Details are given in Sect. 6.4.3.

(iii) Hussain et al. (2014) considered the situation where $c_1 = c_2 = 0$. In this case, Z_i is a mixture of two negative binomial distributions with parameters $\left(r_1, p_1 = \frac{a_1}{a_1+b_1}\right)$ and $\left(r_2, p_2 = \frac{a_2}{a_2+b_2}\right)$. Details are given in Sect. 6.4.4.

(iv) Singh and Sedory (2013) considered the SRSWR sampling design and the Polya's urn process with $c_1 = c_2 = -1$ as randomizing device. In this case, Z_i is a mixture of two negative hypergeometric distributions with parameters $(a_1 + b_1, a_1, r_1)$ and $(a_2 + b_2, a_2, r_2)$.

(v) Dihidar (2016) extended results of Singh and Sedory (2013) for the general sampling strategies.

(vi) Su et al. (2015) extended model of Franklin (1989a), where responses become a sample from the Bernoulli population.

6.5 Balanced RR Model (BRR)

In Franklin's (1989a, 1989b) RR technique, respondents belonging to the sensitive group A are directed to perform a RR technique that is quite different from that of the non-sensitive group $\overline{A}$. In such a procedure, respondents may feel that there might be a trap in requiring a particular method for one group and another for others. As a result, spontaneous cooperation, the primary purpose of the RR technique, may be violated to a certain extent. In the proposed balanced randomized response (BRR) model, respondents are allowed to choose one from a set of several RR techniques with a certain probability.

6.5.1 A General BRR Model

In the proposed BRR model, the respondents are initially given the option to choose one of the $k(\geq 1)$ randomization devices from the devices $R_1, .., R_k$ with probabilities $W_1, .., W_k$; $(W_j \geq 0, \sum_j W_j = 1)$. In the second stage, respondents should provide

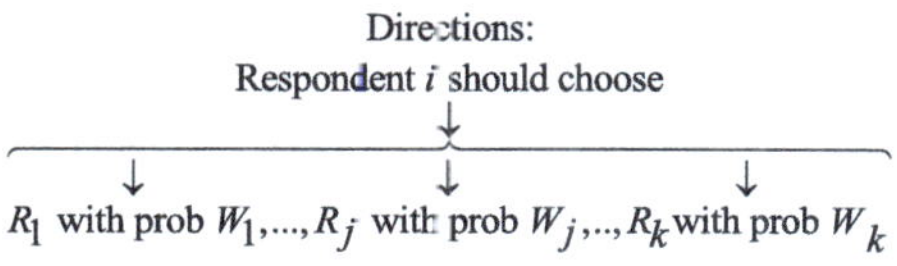

Fig. 6.5 BRR model

RR using the chosen device. Let $z_i(j)$ be the RR provided by the ith respondent while performing the device R_j. Assume that the responses $z_i(j)$ s follow the model:

$$E_{R_j}\{z_i(j)\} = \mu_{ij},\ V_{R_j}\{z_i(j)\} = \phi_i(j) \text{ and } C_{R_j}\{z_i(j), z_{i'}(j')\} = 0 \text{ for } (i,j) \neq (i',j') \tag{6.5.1}$$

where E_{R_j}, V_{R_j}, and C_{R_j} denote the expectation, variance, and covariance with respect to the model R_j (Fig. 6.5).

Let z_i be the RR supplied by a respondent without disclosing to the interviewer which randomization method was used. Then we have

$$\begin{aligned} E_R(z_i) &= E_{\mathcal{R}}\left[E_R\left(z_i|R_j\right)\right] \\ &= E_{\mathcal{R}}\left(\mu_{ij}\right) \\ &= \sum_j W_j\mu_{ij} = \overline{\mu}_i \end{aligned}$$

and

$$\begin{aligned} V_R(z_i) &= E_R\left[V\left(z_i|R_j\right)\right] + V_R\left[E\left(z_i|R_j\right)\right] \\ &= E_R(\phi_i(j)) + V_R\left(\mu_{ij}\right) \\ &= \sum_j W_j\phi_i(j) + \sum_j W_j\left(\mu_{ij} - \overline{\mu}_i\right)^2 \end{aligned}$$

Consider the situation where

$$\mu_{ij} = \alpha_j y_i + \beta_j,\ \phi_i(j) = \gamma_j y_i + \delta_j \text{ and } y_i = \begin{cases} 1 \text{ if } i \in A \\ 0 \text{ if } i \notin A \end{cases} \tag{6.5.2}$$

In this case, Eq. (6.4.1) yields

$$E_R(z_i) = y_i\overline{\alpha}_{\mathrm{w}} + \overline{\beta}_{\mathrm{w}},\ V_R(z_i) = Q_0 + Q_1 y_i = \phi_{\mathrm{w}i} \text{ and } C_R(z_i, z_j) = 0 \text{ for } i \neq j$$

where $Q_0 = \overline{\delta}_{\mathrm{w}} + \sum_j W_j(\beta_j - \overline{\beta}_{\mathrm{w}})^2$, $Q_1 = \overline{\gamma}_{\mathrm{w}} + \sum_j W_j(c_j - \overline{\alpha}_{\mathrm{w}})^2 + 2\sum_j W_j(\alpha_j - \overline{\alpha}_{\mathrm{w}})(\beta_j - \overline{\beta}_{\mathrm{w}})$, $\overline{\alpha}_{\mathrm{w}} = \sum_j \alpha_j W_j$, $\overline{\beta}_{\mathrm{w}} = \sum_j \beta_j W_j$, $\overline{\gamma}_{\mathrm{w}} = \sum_j \gamma_j W_j$, and $\overline{\delta}_{\mathrm{w}} = \sum_j \delta_j W_j$.

Finally writing

$$r_i = \frac{z_i - \overline{\beta}_{\mathrm{w}}}{\overline{\alpha}_{\mathrm{w}}}, \tag{6.5.3}$$

the following working model is obtained:

$$E_R(r_i) = y_i,\ V_R(r_i) = \psi_i \text{ and } C_R\big(r_i, r_j\big) = 0 \text{ for } i \neq j \tag{6.5.4}$$

where $\psi_i = \phi_{\mathrm{w}i}/\overline{\alpha}_{\mathrm{w}}^2$.

The model (6.5.4) leads to two obvious unbiased estimators of ψ_i as $\hat{\psi}_{i1} = Q_0 + Q_1 r_i$ and $\hat{\psi}_{i2} = r_i(r_i - 1)$.

The Horvitz-Thompson (1952)-type estimator for the population proportion π is obtained as

$$\hat{\pi}_{\mathrm{BHT}} = \frac{1}{N}\sum_{i\in s}\frac{r_i}{\pi_i} \tag{6.5.5}$$

The properties of $\hat{\pi}_{\mathrm{BHT}}$ are stated in the following theorem.

Theorem 6.5.1

(i) $\hat{\pi}_{\mathrm{BHT}}$ is unbiased for π.

(ii) $\mathrm{Var}(\hat{\pi}_{\mathrm{BHT}}) = \frac{1}{N^2}\left[\sum\sum_{i<j\in U}(\pi_i\pi_j - \pi_{ij})\left(\frac{y_i}{\pi_i} - \frac{y_j}{\pi_i}\right)^2 + \sum_{i\in U}\frac{\psi_i}{\pi_i}\right].$

(iii) Two unbiased estimators of $\mathrm{Var}(\hat{\pi}_{\mathrm{BHT}})$ are

$$\widehat{\mathrm{Var}}(\hat{\pi}_{\mathrm{BHT}})_j = \frac{1}{N^2}\left[\sum\sum_{i<j\in U}\frac{(\pi_i\pi_j - \pi_{ij})}{\pi_{ij}}\left(\frac{r_i}{\pi_i} - \frac{r_j}{\pi_i}\right)^2 + \sum_{i\in U}\frac{\hat{\psi}_{ij}}{\pi_i}\right];\ j = 1, 2.$$

For a SRSWR of size n, the proposed estimator of π is

$$\hat{\pi}_{\mathrm{BW}} = \frac{1}{n}\sum_{i=1}^{n} r_i \tag{6.5.6}$$

Theorem 6.5.2

(i) $\hat{\pi}_{\mathrm{BW}}$ is unbiased for π.

(ii) $\mathrm{Var}(\hat{\pi}_{\mathrm{BW}}) = \frac{1}{n}\left[\pi(1-\pi) + \frac{1}{N}\sum_{i\in U}\psi_i\right].$

(iii) $\widehat{\mathrm{Var}}(\hat{\pi}_{\mathrm{BW}}) = \frac{1}{n(n-1)}\sum_{i=1}^{n}(r_i - \hat{\pi}_{\mathrm{BW}})^2.$

6.5.2 BRR with Warner's Model

Consider the situation where the randomization device R_1 comprises drawing of a card at random from a pack of two types of cards bearing statements "I belong to the sensitive group A" and "I belong to the non-sensitive group $\overline{A}$" with proportions P and $1 - P$ while for R_2 respondents are asked to draw a card at random from a pack bearing statements "I belong to the sensitive group A" and "I belong to the non-sensitive group $\overline{A}$" with proportions $1 - T$ and T, respectively. Respondent should answer "yes" ("no") if the card drawn matches (does not match) with the respondent's status. Let $z_{ij} = 1(0)$ if the ith respondent answers "yes" ("no"). In this case,

$W_1 = W$, $W_2 = 1 - W$, $\mu_{i1} = y_iP + (1 - y_i)(1 - P)$, $\phi_i(1) = P(1 - P)$, $\mu_{i2} = y_i(1 - T) + (1 - y_i)T$, $\phi_{i2} = T(1 - T)$,$\alpha_1 = (2P - 1)$, $\alpha_2 = (1 - 2T)$, $\beta_1 = (1 - P)$, $\beta_2 = T$, $\gamma_1 = \gamma_2 = \overline{\gamma}_{\text{w}} = 0$, $\delta_1 = P(1 - P)$, $\delta_2 = T(1 - T)$, $\overline{\alpha}_{\text{w}} = 2W(P + T - 1) + (1 - 2T)$, $\overline{\beta}_{\text{w}} = W(1 - P - T) + T$, $\overline{\delta}_{\text{w}} = WP(1 - P) + (1 - W)T(1 - T)$, $Q_0 = WP(1 - P) + (1 - W)T(1 - T) + W(1 - W)(1 - P - T)^2$, and $Q_1 = 0$.

$$r_i = \frac{z_i - [W(1 - P - T) + T]}{[2W(P + T - 1) + (1 - 2T)]} \tag{6.5.7}$$

For SRSWR sampling, an unbiased estimator for π is

$$\hat{\pi}_{\text{BWW}} = \frac{\overline{z}_{\text{w}} - [W(1 - P - T) + T]}{[2W(P + T - 1) + (1 - 2T)]};\ \overline{z}_{\text{w}} = \frac{1}{n}\sum_{i=1}^{n} z \tag{6.5.8}$$

The variance and unbiased estimators of the variance are obtained as

$$\begin{aligned}\text{Var}(\hat{\pi}_{\text{BWW}}) &= \frac{\pi(1 - \pi)}{n} \\ &\quad + \frac{1}{n}\frac{WP(1 - P) + (1 - W)T(1 - T) + W(1 - W)(1 - P - T)^2}{[2W(P + T - 1) + (1 - 2T)]^2}\end{aligned} \tag{6.5.9}$$

and

$$\widehat{\text{Var}}(\hat{\pi}_{\text{BWW}}) = \frac{1}{n(n - 1)}\sum_{i=1}^{n}(r_i - \hat{\pi}_{\text{BWW}})^2 \tag{6.5.10}$$

Remark 6.5.1. Yennum et al. (2019) considered two-stage procedures using Warner's (1965), negative binomial and negative hypergeometric distributions under the SRSWR sampling design, but no explicit expressions of the variances were obtained.

6.6 Discussions

According to Kuk (1990), the main advantage of Franklin's model is that respondents do not have to answer "yes" or "no" to sensitive and embarrassing questions like "Do you belong to sensitive group A?" The respondents should report the number of balls they have drawn. So, in this model, one may expect enhanced cooperation from the respondents. Apart from that, repeated use of the same procedure several times may yield a more precise estimate without increasing the sampling cost. In Franklin's model, the variance of the estimator decreases with the absolute differences $|\mu_{ij} - \mu_{2j}|$. However, to preserve the respondent's privacy, the difference $|\mu_{ij} - \mu_{2j}|$ should be kept at a certain level. For a fixed number of trials, the application of the hypergeometric distribution will certainly produce more efficient estimators than the Binomial distribution. It is known that model of Arnab (2018) provides a better estimator than model of Kuk (1990) in all situations. In model of Warner (1965), respondents supply only one response, while in Kuk and hypergeometric models, respondents supply a fixed number of responses. But in the geometric, negative binomial, and generalized geometric distributions, the number of responses is not fixed: it is a random variable. Several researchers, including Hussain et al. (2017), showed the superiority of their proposed model over models of Kuk (1990), Singh and Grewal (2013), and Hussain et al. (2014) numerically. But such comparisons are not valid as all the RR techniques are not based on the same number of trials. It can be easily verified that none of the RR models described above is optimal in all situations. Further details are given in Chap. 11. The administration of RR techniques involving geometric distribution, negative binomial, generalized geometric, negative hypergeometric distribution, and Polya's urn process to the respondents is not easy, as the investigators may face severe difficulty describing the methods to the respondents and the respondents may not feel comfortable executing such randomization devices. Hence, the methods may be prone to erroneous results due to a misunderstanding of the method on the respondent's part. Sometimes, respondents need to perform a large number of trials, which may cause less cooperation from the respondents, and the purpose of the RR model is defeated.

In Franklin's (1989a, 1989b) RR technique, respondents from the sensitive group are required to perform different RR techniques than those from the non-sensitive group. In this situation, respondents might think that there might be some trap revealing their personal information behind the proposed RR techniques. To remove such difficulties and boost respondents' confidence, the BRR has been proposed, where respondents are given the option to choose one of the several RR models with certain probabilities.

References

Al-sobhi, M. M., Hussain, Z., & Al-Zahrani, B. (2014). General randomized response techniques using Polya's urn process as a randomization device. *Plos One, 9*, e115612.

Arnab, R. (1990). On commutativity of design and model expectations in randomized response surveys. *Communications in Statistics-Theory and Methods, 19*, 3751–3757.

Arnab, R., Singh, S., & North, D. (2012). Use of two decks of cards in randomized response techniques for complex designs. *Communications in Statistics-Theory and Methods, 41*, 3198–3210.

Arnab, R., & Shangodoyin, D. K. (2015). Randomized response technique using maximum likelihood estimator. *Communications in Statistics-Theory and Methods, 44*, 3340–3349.

Arnab, R., Shangodoyin, D. K., & Dwivedi, V. K. (2016). Generalization of randomized response techniques for complex survey designs. *Journal of the Indian Society of Agricultural Statistics, 70*, 131–138.

Arnab, R., Shangodoyin, D. K., & Kgosi, P. M. (2017). Randomized response techniques with multiple responses. *Journal of the Indian Society of Agricultural Statistics, 71*, 201–205.

Arnab, R. (2018). An improved randomized response technique. *Journal of the Indian Society of Agricultural Statistics, 72*, 141–145.

Arnab, R., & Shangodoyin, D. K. (2020). Stratified randomized response model for multiple responses. *Journal of the Indian Society of Agricultural Statistics, 74*, 1–10.

Arnab, R. (2021). Randomized response trial using geometric distribution. *Journal of the Indian Society of Agricultural Statistics, 75*, 195–204.

Boruch, R. F. (1972). Relations among statistical methods for assuring confidentiality of social research data. *Social Science Research, 1*, 403–413.

Chaudhuri, A., & Christofied, T. (2013). *Indirect questioning in sample surveys*. Springer, New York.

Dihidar, K. (2016). Estimating sensitive population proportion by generating randomized response following direct and inverse hypergeometric distribution. In A. Chaudhuri, T. C., Christofides & C. R. Rao (Eds.), *Hand book of Statistics* (Vol. 34, pp. 443–469).

Franklin. (1989a). Randomized response sampling from dichotomous populations with continuous randomization. *Survey Methodology, 15*, 225–235.

Franklin, L. A. (1989b). A comparison of estimators for randomized response sampling with continuous distribution from dichotomous populations. *Communications in Statistics-Theory and Methods, 18*, 489–505.

Godambe, V. P., & Kale. (1991). Estimating functions: An overview. In V. P. Godambe (Ed.), *Estimating functions* (pp. 3–20). Oxford University Press.

Greenberg, B. D., Abul-Ela, A. L. A., Simmons, W. R., & Horvitz, D. G. (1969). The unrelated question randomized response model: Theoretical framework. *Journal of the American Statistical Association, 64*, 520–539.

Horvitz, D. G., Shah, B. V., Simmons, W. R. (1967). The unrelated question randomized response model. In *Proceedings of Social, Statistical section, American Statistical Association* (pp. 65–72).

Horvitz, D. G., & Thompson, D. J. (1952). A generalization of sampling without replacement from a finite population. *Journal of the American Statistical Association, 47*, 663–685.

Hussain, Z., Shabbir, J., Pervez, Z., Shah, S. F., & Khan, M. (2017). Generalized geometric distribution of order k: A flexible choice to randomized response. *Communications in Statistics-Simulation and Computation, 46*, 4708–4721.

Hussain, Z., Shabbir, J., Al-zahrani, B., & Khan, M. (2014). On using negative binomial distribution as a randomization device in sensitive surveys. *Communications in Statistics Simulation and Computation, 45*, 3584–3596.

Jayraj, A., Sedory, S. A., Singh, S., & Odumade, O. (2018). A new estimator of proportion with a linear function using data from two-decks randomized response model. *Communications in Statistics-Theory and Methods, 38*, 439–446.

Kuk, A. Y. (1990). Asking sensitive question indirectly. *Biometrika, 77*, 436–438.
Mangat, N. S., & Singh, R. (1990). An alternative randomized response procedure. *Biometrika, 77*, 349–442.
Odumade, O., & Singh, S. (2009). Efficient use of two decks of cards in randomized response sampling. *Communications in Statistics-Theory and Methods, 38*, 439–446.
Singh, S., & Sedory, S. A. (2013). A new randomization device for sensitive characteristics: An application to negative hypergeometric distribution. *Metron, 71*, 3–8.
Singh, S., & Grewal, I. S. (2013). Geometric distribution as a randomization device: Implemented to the Kuk's model. *International Journal of Complementary Mathematical Sciences, 8*, 243–248.
Su, S., Sedory, S. A., & Singh, S. (2015). Kuk's model adjusted for protection and efficiency. *Sociological Methods and Research, 44*, 534–551.
Warner, S. L. (1965). Randomize response: A survey technique for eliminating evasive answer bias. *Journal of the American Statistical Association, 60*, 63–69.
Yennum, N., Sedory, S. A., & Singh, S. (2019). Improved strategy to collect sensitive data by using geometric distribution as a randomization device. *Communications in Statistics—Theory and Methods, 48*, 5777–5795.

Chapter 7
Randomized Response with Multiple Attributes

Abstract Randomized response techniques for estimating proportions of more than one sensitive attribute are proposed. The categories (attributes) need not necessarily be disjoint. There might be some overlaps among the categories; an individual may belong to more than one category. For example, an individual may have both HIV infection (one category) and tuberculosis (another category). The works of Christofides (2005), Lee, Sedory, Singh, 2011, A magical talk: Estimating at least seven measures of qualitative variables from a single sample using randomized response technique. International Statistical Institute: Proc.58th World Statistical Congress, 2011, Dublin (session STS001), 1948–1957.; Lee et al., Statistical Probability Letters, 83:399–409, 2013, and Arnab, Communications in statistics-theory and methods 52:94–103, 2023 for estimating the proportions of multiple attributes have been discussed in detail. Methods of estimating the proportions of each of the categories are also proposed. Expressions of the estimators of variances and their unbiased estimators are presented in detail for the various sampling strategies and RR models.

Keywords Randomized response · Sensitive characteristics · Sampling strategies · Unbiased estimators

7.1 Introduction

Consider a finite population U where respondents may possess none or at least one of the $k(\geq 1)$ sensitive attributes $A_1, \ldots, A_k$. The attributes need not necessarily be disjoint. There might be some overlap among the attributes, so an individual may have more than one attribute. For example, an individual may have both HIV and tuberculosis infections. Our objective is to estimate the proportion of individuals belonging to each of the categories as well as their combinations, e.g., the prevalence of HIV, tuberculosis, and both HIV and tuberculosis, among others, in a single sample selected according to a sampling design. Randomized response models for the multiple distinct sensitive attributes where the attributes have no overlap

R. Arnab, *Indirect Methods of Data Collection and Analysis from Surveys*, Indian Statistical Institute Series, https://doi.org/10.1007/978-981-97-6005-3_7

were proposed by several authors, including Abul-Ela et al. (1967), Burke (1981, 1982, 1990), Chang and Liang (1996), Drane (1976), Eriksson (1973), Greenberg et al. (1989), Kim and Flueck (1978), Mukhopadhyaya (1980), Mukerjee (1981), and Thamane (1981). Some of the methods have been discussed in Arnab (2017). Recently, Barabesi et al. (2012), Christofides (2005), Hsieh et al. (2018), Lee et al. (2013), and Moshagen and Musch (2012) proposed RR models for two sensitive characteristics, while Barabesi et al. (2012) proposed a model for multiple sensitive attributes. Most of them are restricted to SRSWR sampling only. Arnab (2023) has proposed a more general RR model that is applicable to any sampling design. Most of the existing models can be derived from Arnab's (2023) model as special cases. Some of the important RR models are described in detail in this chapter.

7.2 Two Sensitive Attributes

7.2.1 *Christofieds Model*

Christofides (2005) considered the situation where an individual in a population U may possess one, two, or none of the sensitive attributes A_1 and A_2. The proportions of individuals in the population possessing the sensitive attributes A_1, A_2 and $A_1 \cap A_2$ will be denoted by $\pi_{(1)}$, $\pi_{(2)}$ and $\pi_{(12)}$, respectively. The conditional probabilities of individuals possessing the attribute A_1 among those who possess the attribute A_2 and those who do not possess the attribute A_2 will be denoted by $\pi_{(1|2)}$ and, $\pi_{(1|\bar{2})}$, respectively. In Christofides's (2005) RR technique, a sample of n units (individuals) is selected by the SRSWR method. Each of the individuals selected in the sample is provided with two randomization devices, one for each of the two characters A_1 and A_2. For the character $A_1(A_2)$, the ith respondent is requested to draw one number, say, $x_i(1)(x_i(2))$ from a set of integers $1, \ldots, L_1(1, \ldots, L_2)$ with probability $p_1, \ldots, p_{L1}$ $(q_1, \ldots, q_{L2})$, $p_j > 0$, $\sum_{j=1}^{L_1} p_j = 1,$, $(q_j > 0, \sum_{j=1}^{L_2} q_j = 1)$, respectively, and asked to report $z_i(1) = L_1 + 1 - x_i(1)$ $(z_i(2) = L_2 + 1 - x_i(2))$ if the person belongs to the sensitive group $A_1(A_2)$. Otherwise, if the respondent does not belong to the sensitive group $A_1(A_2)$, he or she should report $z_i(1) = x_i(1)(z_i(2) = x_i(2))$ as his or her randomized response. The means and variances of $x_i(1)$ and $x_i(2)$ are given by (Fig. 7.1)

$$E(x_i(1)) = \sum_{k=1}^{L1+1} kp_k = \mu_{c1};\ E(x_i(2)) = \sum_{k=1}^{L2+1} kq_k = \mu_{c2};$$

and

$$V(x_i(1)) = \sum_{k=1}^{L1} k^2 p_k - \mu_{C1}^2 = \sigma_{C1}^2 \text{ and } V(x_i(2)) = \sum_{k=1}^{L2} k^2 q_k - \mu_{C2}^2 = \sigma_{C2}^2 \quad (7.2.1)$$

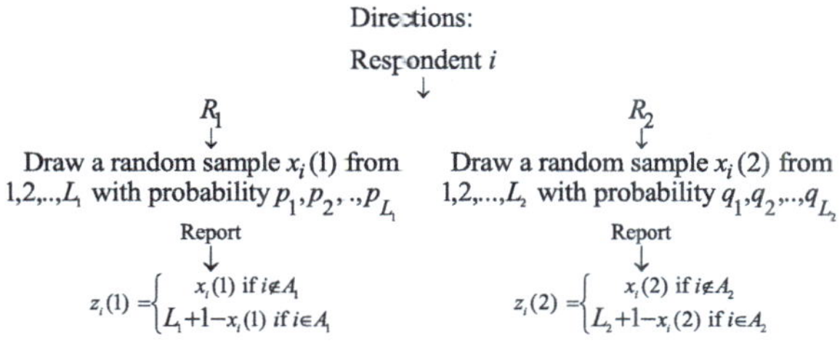

Fig. 7.1 Christofides (2005) model for two attributes

Christofides (2005) derived the following results:

Theorem 7.2.1 For $L_j + 1 - 2\mu_{C(j)} \neq 0, j = 1, 2,$

(i) $\hat{\pi}_{C(j)} = \frac{\bar{z}(j) - \mu_{C(j)}}{L_j + 1 - 2\mu_{C(j)}}$ is an unbiased estimator of $\pi_{(j)}$

where $\bar{z}(j) = \frac{1}{n}\sum_{i=1}^{n} z_i(j), j = 1, 2.$

(ii) The variance of $\hat{\pi}_{C(j)}$ is

$$\operatorname{Var}\left(\hat{\pi}_{C(j)}\right) = \frac{1}{n}\left[\pi_{(j)}\left(1 - \pi_{(j)}\right) + \phi_C(j)\right]$$

where $\phi_C(j) = \sigma_{Cj}^2 / D_{(j)}^2$ and $D_{(j)}^2 = \left(L_j + 1 - 2\mu_{Cj}\right)^2$.

(iii) $\hat{\pi}_{C(12)} = \frac{\bar{t} - \bar{z}(1)\mu_{C(2)} - \bar{z}(2)\mu_{C(1)} + \mu_{C(1)}\mu_{C(2)}}{D_{(1)}D_{(2)}}$ is an unbiased estimator of $\pi_{(12)}$

where $\bar{t} = \frac{1}{n}\sum_{i=1}^{n} z_i(1) z_i(2)$.

(iv) The variance of $\hat{\pi}_{C(12)}$ is

$$\operatorname{Var}\left(\hat{\pi}_{C(12)}\right) = \frac{1}{n}\left[\pi_{(12)}\left(1 - \pi_{(12)}\right) + \frac{\pi_{(1)} D_{(1)}^2 \sigma_{C2}^2 + \pi_{(2)} D_{(2)}^2 \sigma_{C1}^2 + \sigma_{C1}^2 \sigma_{C2}^2}{D_{(1)}^2 D_{(2)}^2}\right]$$

(v) An unbiased estimator of $\operatorname{Var}(\hat{\pi}_{C(12)})$ is

$$\widehat{\operatorname{Var}}\left(\hat{\pi}_{C(12)}\right) = \frac{1}{n-1}\left[\hat{\pi}_{C(12)}\left(1 - \hat{\pi}_{C(12)}\right) + \frac{\hat{\pi}_{C(1)} D_{(1)}^2 \sigma_{C2}^2 + \hat{\pi}_{C(2)} D_{(2)}^2 \sigma_{C1}^2 + \sigma_{C1}^2 \sigma_{C2}^2}{D_{(1)}^2 D_{(2)}^2}\right]$$

Christofides (2005) also provided the estimators of the conditional proportions of $\pi_{(1|2)}$ and $\pi_{(1|\bar{2})}$ as follows:

$$\hat{\pi}_{(1|2)} = \frac{\hat{\pi}_{(12)}}{\hat{\pi}_{(2)}} \text{ and } \hat{\pi}_{(1|\bar{2})} = \frac{\hat{\pi}_{(1)} - \hat{\pi}_{(12)}}{1 - \hat{\pi}_{(2)}} \tag{7.2.2}$$

Since the proposed estimators (7.2.2) are ratio estimators, they are not unbiased. The expressions of bias and mean-square errors were obtained by assuming the sample size n is large and using Cochran's (1977) results.

7.2.2 *Lee et al. Model*

Lee et al. (2011, 2013) proposed two RR models named "Simple Model" and "Crossed Model" for estimating population proportions and conditional probabilities for two sensitive attributes A_1 and A_2 based on a sample size n selected by the SRSWR method. The simple model comprises two independent (Warner's, 1965) trials with parameters, P_1 and P_2, while the crossed model is a modification of Warner's model. The models are described as follows.

7.2.2.1 Simple Model

Each of the respondents selected in the sample is directed to perform two (Warner's, 1965) trials, say, R_1 and R_2 independently. For the randomization device of $R_1(R_2)$, respondents are requested to select a card at random from a pack consisting of two types of cards bearing the statements "I possess the sensitive attribute $A_1(A_2)$" and "I do not possess the sensitive attribute $A_1(A_2)$" with proportions $P_1(P_2)$ and $1 - P_1(1 - P_2)$, respectively (Fig. 7.2).

The respondents should report "yes" if the card drawn matches their status; otherwise, report "no" as their response. Data was collected in the form of a 2×2 contingency table (X, Y), where $X(Y)$ denotes responses "yes" or "no" from the trail $R_1(R_2)$. Let the cell frequencies (yes, yes), (yes, no), (no, yes), and (no, no) and their probabilities be denoted by $n_{11}, n_{10}, n_{01}, n_{00}$ and, $\theta_{11}, \theta_{10}, \theta_{01}, \theta_{00}$, respectively (Fig. 7.3).

Simple model:

R_1:		R_2:
$I \in A_1$ with probability P_1		$I \in A_2$ with probability P_2
$I \notin A_1$ with probability $1-P_1$		$I \notin A_2$ with probability $1-P_2$

Fig. 7.2 Lee et al.'s (2011) Simple model

		Randomization R_2:	
Randomization R_1:	Responses:	yes	no
	yes	$n_{11}(\theta_{11})$	$n_{10}(\theta_{10})$
	no	$n_{01}(\theta_{01})$	$n_{00}(\theta_{00})$

Fig. 7.3 Cell frequencies (probabilities)

Let the proportions of individuals in the population that belong to the groups A_1, A_2 and $A_1 \cap A_2$ be $\pi_{(1)}, \pi_{(2)}$ and $\pi_{(12)}$.

The expressions of the cell probabilities θ_{ij}'s are computed as follows:

$$\begin{aligned}
\text{Prob}(yes, yes) = \theta_{11} &= (2P_1-1)(2P_2-1)\pi_{(1,2)} + (2P_1-1)(1-P_2)\pi_{(1)} \\
&\quad + (1-P_1)(2P_2-1)\pi_{(2)} + (1-P_1)(1-P_2) \\
\text{Prob}(yes, no) = \theta_{10} &= -(2P_1-1)(2P_2-1)\pi_{(1,2)} + (2P_1-1)P_2\pi_{(1)} \\
&\quad + (1-P_1)(2P_2-1)\pi_{(2)} + (1-P_1)P_2 \\
\text{Prob}(no, yes) = \theta_{01} &= -(2P_1-1)(2P_2-1)\pi_{(1,2)} + (2P_1-1)(1-P_2)\pi_{(1)} \\
&\quad + P_1(2P_2-1)\pi_{(2)} + P_1(1-P_2) \\
\text{Prob}(no, no) = \theta_{00} &= (2P_1-1)(2P_2-1)\pi_{(1,2)} + (2P_1-1)P_2\pi_{(1)} \\
&\quad + P_1(2P_2-1)\pi_{(2)} + P_1P_2 \qquad (7.2.3)
\end{aligned}$$

Denoting $\hat{\theta}_{ij} = n_{ij}/n$ and minimizing $D = \sum_{i=0}^{1}\sum_{j=0}^{1}\left(\theta_{ij} - \hat{\theta}_{ij}\right)^2$ with respect to $\pi_{(1)}, \pi_{(2)}$ and $\pi_{(1,2)}$, Lee et al. (2011) derived the following results:

Theorem 7.2.2 For, $P_1, P_2 \neq 1/2$,

(i) $\hat{\pi}_{(1)} = \frac{\hat{\theta}_{11}+\hat{\theta}_{10}-\hat{\theta}_{01}-\hat{\theta}_{00}+(2P_1-1)}{2(2P_1-1)}$, $\hat{\pi}_{(2)} = \frac{\hat{\theta}_{11}-\hat{\theta}_{10}+\hat{\theta}_{01}-\hat{\theta}_{00}+(2P_2-1)}{2(2P_2-1)}$ and

$$\hat{\pi}_{(12)} = \frac{(P_1+P_2)\hat{\theta}_{11} + (P_2-P_1)\hat{\theta}_{10} + (P_1-P_2)\hat{\theta}_{01} + (2-P_1-P_2)\hat{\theta}_{00} - P_2(1-P_1) - P_1(1-P_2)}{2(2P_1-1)(2P_2-1)}$$

are unbiased estimators of $\pi_{(1)}, \pi_{(2)}$ and $\pi_{(12)}$, respectively.

(ii) The variances of $\hat{\pi}_{(1)}, \hat{\pi}_{(2)}$ and $\hat{\pi}_{(12)}$ are

$$\text{Var}\left(\hat{\pi}_{(j)}\right) = \frac{1}{n}\left[\pi_{(j)}\left(1-\hat{\pi}_{(j)}\right) + \frac{P_j(1-P_j)}{(2P_j-1)^2}\right]; j = 1, 2$$

and

$$\text{Var}\left(\hat{\pi}_{(12)}\right) = \frac{1}{n}\left[\begin{array}{l}\pi_{(12)}\left(1-\hat{\pi}_{(12)}\right) \\ +\dfrac{(2P_1-1)^2P_2(1-P_2)\pi_{(1)} + P_1(1-P_1)(2P_2-1)^2\pi_{(2)} + P_1P_2(1-P_1)(1-P_2)}{(2P_1-1)^2(2P_2-1)^2}\end{array}\right]$$

(iii) $\hat{\pi}_{(1|2)} = \frac{\hat{\pi}_{(12)}}{\hat{\pi}_{(2)}}$ is a biased estimator of $\hat{\pi}_{(1|2)}$.

Approximate expressions for bias and mean-square error of $\hat{\pi}_{(1|2)}$ are.

$B\left(\hat{\pi}_{(1|2)}\right) = \pi_{(1|2)}\left[\frac{\text{Var}\hat{\pi}_{(2)}}{\pi_2^2} - \frac{\text{Cov}\left(\hat{\pi}_{(12)}, \hat{\pi}_{(2)}\right)}{\pi_{(12)}\pi_{(2)}}\right]$ and

$\text{MSE}\left(\hat{\pi}_{(1|2)}\right) = \pi_{(1|2)}^2\left[\frac{\text{Var}\left(\hat{\pi}_{(12)}\right)}{\pi_{12}^2} - 2\frac{\text{Cov}\left(\hat{\pi}_{(12)}, \hat{\pi}_{(2)}\right)}{\pi_{(12)}\pi_{(2)}} + \frac{\text{Var}\left(\hat{\pi}_{(2)}\right)}{\pi_2^2}\right].$

where $\text{Cov}\left(\hat{\pi}_{(12)}, \hat{\pi}_{(2)}\right) = \frac{1}{n}\left[\pi_{(12)}(1-\pi_2) + \frac{P_2(1-P_2)}{(2P_2-1)^2}\pi_{(1)}\right].$

7.2.2.2 Crossed Model

In the crossed model, each respondent is directed to perform two randomization devices, say, $\tilde{R}_1$ and $\tilde{R}_2$ independently. The randomization device $\tilde{R}_1(\tilde{R}_2)$ comprises selecting a card at random from a pack consisting of two types of cards bearing the statements "I possess the sensitive attribute $A_1(A_2)$" and "I do not possess the sensitive attribute $A_2(A_1)$" with proportions $P_1(P_2)$ and $1-P_1(1-P_2)$, respectively. Respondents were requested to report "yes" if the card drawn matched with their status, otherwise, they should report "no" as their response (Fig. 7.4).

The probabilities of obtaining responses (yes, yes), (yes, no), (no, yes) and (no, no) are given by:

$$\begin{aligned}
\tilde{\theta}_{11} &= \pi_{(12)}\{P_1P_2 + (1-P_1)(1-P_2)\} - \pi_{(1)}(1-P_1)(1-P_2) - \pi_{(2)} \\
&\quad (1-P_1)(1-P_2) + (1-P_1)(1-P_2) \\
\tilde{\theta}_{10} &= -\pi_{(12)}\{P_1P_2 + (1-P_1)(1-P_2)\} - \pi_{(1)}\{(1-P_1)P_2 - 1\} - \pi_{(2)} \\
&\quad (1-P_1)P_2 + (1-P_1)P_2 \\
\tilde{\theta}_{01} &= -\pi_{(12)}\{P_1P_2 + (1-P_1)(1-P_2)\} - \pi_{(1)}P_1(1-P_2) - \pi_{(2)} \\
&\quad \{P_1(1-P_2) - 1\} + P_1(1-P_2) \\
\tilde{\theta}_{00} &= \pi_{(12)}\{P_1P_2 + (1-P_1)(1-P_2)\} - \pi_{(1)}P_1P_2 - \pi_{(2)}P_1P_2 + P_1P_2
\end{aligned}$$

Responses from the four categories and their probabilities (in brackets) are presented in the following Table 7.1.

On minimizing $D = \sum_{i=0}^{1}\sum_{j=0}^{1}(\tilde{\theta}_{ij} - \hat{\tilde{\theta}}_{ij})^2$ with respect to $\pi_{(1)}$,$\pi_{(2)}$ and $\pi_{(1,2)}$, Lee et al. (2013) derived the unbiased estimators of the proportions $\pi_{(1)}$,$\pi_{(2)}$, $\pi_{(1,2)}$ and conditional probabilities $\pi_{(1|2)}$, $\pi_{(1|\bar{2})}$, etc., as in the simple model. The derived results are as follows:

Crossed model:

$\tilde{R}_1$:		$\tilde{R}_2$:
$I \in A_1$ with probability P_1		$I \in A_2$ with probability P_2
$I \notin A_2$ with probability $1-P_1$		$I \notin A_1$ with probability $1-P_2$

Fig. 7.4 Lee et al.'s (2013) crossed model

Table 7.1 Cell frequencies (probabilities)

Randomization $\tilde{R}_1$		Randomization $\tilde{R}_2$	
	Responses:	Yes	No
	Yes	$\tilde{n}_{11}(\tilde{\theta}_{11})$	$\tilde{n}_{10}(\tilde{\theta}_{10})$
	No	$\tilde{n}_{01}(\tilde{\theta}_{01})$	$\tilde{n}_{00}(\tilde{\theta}_{00})$

Theorem 7.2.3 For $P_1, P_2 \neq 1/2$,

(i) $\hat{\pi}_{(1)} = \frac{1}{2} + \frac{(P_2-P_1+1)\left(\hat{\hat{\theta}}_{11}-\hat{\hat{\theta}}_{00}\right)+(P_1+P_2-1)\left(\hat{\hat{\theta}}_{10}-\hat{\hat{\theta}}_{01}\right)}{2(P_1+P_2-1)}$,

(ii) $\hat{\pi}_{(2)} = \frac{1}{2} + \frac{(P_1-P_2+1)\left(\hat{\hat{\theta}}_{11}-\hat{\hat{\theta}}_{00}\right)+(P_1+P_2-1)\left(\hat{\hat{\theta}}_{01}-\hat{\hat{\theta}}_{10}\right)}{2(P_1+P_2-1)}$

and

$$\hat{\pi}_{(12)} = \frac{P_1P_2\hat{\hat{\theta}}_{11} - (1-P_1)(1-P_2)\hat{\hat{\theta}}_{00}}{(P_1+P_2-1)\{P_1P_2+(1-P_1)(1-P_2)\}}$$

are unbiased estimators of $\pi_{(1)}$,$\pi_{(2)}$ and $\pi_{(12)}$, respectively.

(ii) The variances of $\hat{\pi}_{(1)}$,$\hat{\pi}_{(2)}$ and $\hat{\pi}_{(12)}$ are

$$\operatorname{Var}\left(\hat{\pi}_{(1)}\right) = \frac{1}{n}\left[\begin{array}{l}\pi_{(1)}\left(1-\pi_{(1)}\right) \\ +\dfrac{P_2(1-P_1)\{P_1P_2+(1-P_1)(1-P_2)\}(1-\pi_{(1)}-\pi_{(2)}+2\pi_{(12)})}{(P_1+P_2-1)^2}\end{array}\right],$$

$$\operatorname{Var}\left(\hat{\pi}_{(2)}\right) = \frac{1}{n}\left[\begin{array}{l}\pi_{(2)}\left(1-\pi_{(2)}\right) \\ +\dfrac{P_1(1-P_2)\{P_1P_2+(1-P_1)(1-P_2)\}(1-\pi_{(1)}-\pi_{(2)}+2\pi_{(12)})}{(P_1+P_2-1)^2}\end{array}\right]$$

and

$$\operatorname{Var}(\hat{\pi}_{(12)}) = \frac{1}{n}\left[\begin{array}{l}-\pi_{(12)}^2+ \\ \dfrac{\left\{(P_1P_2)^2+(1-P_1)^2(1-P_2)^2\right\}\pi_{(12)}+P_1P_2(1-P_1)(1-P_2)(P_1+P_2-1)(1-\pi_{(1)}-\pi_{(2)})}{\{P_1P_2+(1-P_1)(1-P_2)\}(P_1+P_2-1)^2}\end{array}\right]$$

(iii) $\hat{\pi}_{(1|2)} = \frac{\hat{\pi}_{(12)}}{\hat{\pi}_{(2)}}$ is a biased estimator of $\hat{\pi}_{(1|2)}$.

Approximate expressions for bias and mean-square error are.

$B\left(\hat{\pi}_{(1|2)}\right) = \pi_{(1|2)}\left[\frac{\operatorname{Var}\left(\hat{\pi}_{(2)}\right)}{\pi_2^2} - \frac{\operatorname{Cov}\left(\hat{\pi}_{(12)},\hat{\pi}_{(2)}\right)}{\pi_{(12)}\pi_{(2)}}\right]$ and

$\operatorname{MSE}\left(\hat{\pi}_{(1|2)}\right) = \pi_{(1|2)}^2\left[\frac{\operatorname{Var}\left(\hat{\pi}_{(12)}\right)}{\pi_{12}^2} - 2\frac{\operatorname{Cov}\left(\hat{\pi}_{(12)},\hat{\pi}_{(2)}\right)}{\pi_{(12)}\pi_{(2)}} + \frac{\operatorname{Var}\left(\hat{\pi}_{(2)}\right)}{\pi_2^2}\right]$.

where

$$\operatorname{Cov}\left(\hat{\pi}_{(12)}, \hat{\pi}_{(2)}\right) = \frac{1}{n}\left[\pi_{(12)}(1-\hat{\pi}_{(2)}) + \frac{P_1(1-P_2)(P_2-P_1+1)}{(P_1+P_2-1)^2}\pi_{(12)} + \frac{P_1P_2(1-P_1)(1-P_2)(P_1-P_2+1)(1-\pi_{(1)}-\pi_{(2)})}{\{P_1P_2+(1-P_1)(1-P_2)\}(P_1+P_2-1)^2}\right]$$

Lee et al. (2013) compared the performances of the proposed two models numerically. However, they did not provide any theoretical comparison.

7.3 Models for Multiple Attributes

Arnab's General Model

Arnab (2023) proposed a more general RR model for multiple sensitive attributes. All the models for multiple attributes may be obtained as special cases of the proposed model. In this model, each individual selected in the sample is requested to perform k RR trials $R_1, R_2, \ldots..R_k$ for each of the k sensitive attributes $A_1, \ldots, A_k$ independently. The RR trails R_j s ($j = 1, \ldots, k$) need not be the same. For instance,R_1 may be Warner's (1965) model, R_2 maybe Christofides's (2003) model, and so on. Let $z_i(j)$ be the RR obtained by the ith respondent from the randomization device R_j and

$$z_i = (z_i(1), \ldots, z_i(j), \ldots, z_i(k)) \tag{7.3.1}$$

be the vector of responses provided by the ith respondent.

Further, let

$$y_i(j) = \begin{cases} 1, \text{ if the ith individual} \in A_j \\ 0, \text{ if the ith individual} \notin A_j \end{cases} ; j = 1, \ldots, k \tag{7.3.2}$$

Then,

$N_j = \sum_{i \in U} y_i(j) =$ total number of individuals in the group A_j.

$N_{jt} = \sum_{i \in U} y_i(j) y_i(t) =$ total number of individuals belonging to the group $A_j \cap A_t$.

$\pi_{(j)} =$ Proportion of individuals belong to $A_j = \sum_{i \in U} y_i(j)/N = N_j/N$.

$\pi_{(j,t)} =$ Proportion of individuals belong to A_j and $A_t = N_{jt}/N$.

$\pi_{(t|j)} = N_{jt}/N_j =$ Proportion of individuals that belong to the group A_t among those who belong to the group A_j,etc.

Let the RR $z_i(j)$ yields an unbiased estimator of $y_i(j)$ as $\hat{y}_i(j)$. We call $\hat{y}_i(j)$, the reduced response obtained from $z_i(j)$. The reduced response $\hat{y}_i(j)$ is assumed to follow the general model proposed by Arnab (1990):

$$E_R(\hat{y}_i(j)) = y_i(j),\ V_R(\hat{y}_i(j)) = \phi_i(j) \text{ and } C_R\big(\hat{y}_i(j), \hat{y}_{i'}(j')\big) = 0 \text{ for } (i,j) \neq (i',j') \tag{7.3.3}$$

An unbiased estimator of $\phi_i(j)$ will be denoted by $\hat{\phi}_i(j)$.

Example 7.3.1 In the event that $z_i(j)$ is obtained by Warner's model with the parameter P_j, we obtain

$E_R(z_i(j)) = E_R(z_i^2(j)) = y_i(j)P_j + (1 - y_i(j))(1 - P_j)$, $\hat{y}_i(j) = \frac{z_i(j)-(1-P_j)}{2P_j-1}$ and $\phi_i(j) = \frac{P_j(1-P_j)}{(2P_j-1)^2} = \hat{\phi}_i(j)$.

Estimation of Overlaps

Let us define the category

$$F = (j_1, j_2, \ldots, j_t) = A_{j_1} \cap A_{j_2} \cap \cdots \cap A_{j_t} \tag{7.3.4}$$

where $(j_1, j_2, \ldots, j_t) \subset (1, 2, \ldots, k)$.

On writing,

$$F_i = y_i(j_1) \times \cdots \times y_i(j_t) = \prod_{c=1}^{t} y_i(j_c), \tag{7.3.5}$$

one finds the proportion of individuals belonging to the group F in the population as

$$\pi_{(F)} = \frac{1}{N}\sum_{i \in U} F_i \tag{7.3.6}$$

Further, denoting

$$\hat{F}_i = \hat{y}_i(j_1) \times \cdots \times \hat{y}_i(j_t) = \prod_{c=1}^{t} \hat{y}_i(j_c) \tag{7.3.7}$$

we obtain

$$\begin{aligned}
E_R(\hat{F}_i) &= \prod_{c=1}^{t} E_R(\hat{y}_i(j_c)) = \prod_{c=1}^{t} y_i(j_c) = F_i, \\
V_R(\hat{F}_i) &= E_R(\hat{F}_i^2) - F_i^2 \\
&= \prod_{c=1}^{t} E_R(\hat{y}_i^2(j_c)) - F_i^2 \\
&= \prod_{c=1}^{t} \{\phi_i(j_c) + y_i(j_c)\} - F_i^2;\ as\, y_i(j_c) = 0\, or\, 1 \\
&= \phi_{Fi},\ say
\end{aligned} \tag{7.3.8}$$

From Eq. (7.3.8), an unbiased estimator of ϕ_{Fi} is obtained as

$$\hat{\phi}_{Fi} = \hat{F}_i(1 - \hat{F}_i) \tag{7.3.9}$$

$$\begin{aligned}\text{because, } E_R(\hat{\phi}_{Fi}) &= E_R\left[\hat{F}_i(\hat{F}_i - 1)\right] \\ &= V_R(\hat{F}_i) + F_i^2 - F_i \\ &= V_R(\hat{F}_i) \text{ as } F_i = 0, 1 \\ &= \phi_{Fi} \\ &\text{and } C_R(\hat{F}_i, \hat{F}_j) = 0 \quad for\, i \neq j\end{aligned}$$

Co-variance of Estimators of Overlaps

Consider another category G $= (q_1, q_{2,\ldots,} q_r) = A_{q_1} \cap A_{q_2} \cap \cdots \cap A_{q_r}$ where $(q_1, q_{2,\ldots,} q_r) \subset (1, \ldots, t)$. The proportion of individuals belongs to the category G in the population is

$$\pi_{(G)} = \frac{1}{N} \sum_{i \in U} G_i \tag{7.3.10}$$

where $G_i = y_i(q_1) \times \cdots \times y_i(q_r) = \prod_{d=1}^{r} y_i(q_d)$.

An unbiased estimator of G_i is

$$\hat{G}_i = \hat{y}_i(q_1) \times \cdots \times \hat{y}_i(q_r)$$

Co-variance between $\hat{F}_i$ and $\hat{G}_i$ with respect to the model R is given by

$$\begin{aligned}C_R\left[\hat{F}_i, \hat{G}_i\right] &= E_R\left[\left\{\prod_{j \in F \cap G} \hat{y}_i^2(j)\right\} \times \left\{\prod_{j \in F - G \cup G - F} \hat{y}_i(j)\right\}\right] - F_i G_i \\ &= \left[\prod_{j \in F \cap G} \{\phi_i(j) + y_i(j)\}\right] \times \left[\prod_{j \in F - G \cup G - F} y_i(j)\right] - F_i G_i \\ &= \left[\prod_{j \in F \cap G} \{\phi_i(j)\}\right] \times \left[\prod_{j \in F - G \cup G - F} y_i(j)\right] \\ &= \phi_{FGi}\end{aligned} \tag{7.3.11}$$

An unbiased estimator of ϕ_{FGi} is

$$\hat{\phi}_{FGi} = \left[\prod_{j \in F \cap G} \{\hat{\phi}_i(j)\}\right] \times \left[\prod_{j \in F - G \cup G - F} \hat{y}_i(j)\right] \tag{7.3.12}$$

Estimation of Proportion Variance and Co-variances of Overlaps

Consider the general sampling design, where a sample of effective size n is selected from a population U of N units with the inclusion probabilities π_i and π_{ij} for the ith and the pair i and $j(\neq i)$th units. In this case, the Horvitz-Thompson type estimators of $\pi_{(\mathrm{F})}$ and $\pi_{(G)}$ are given by

$$\hat{\pi}_{(\mathrm{F})\mathrm{HT}} = \frac{1}{N}\sum_{i\in s}\frac{\hat{F}_i}{\pi_i} \quad \text{and} \quad \hat{\pi}_{(\mathrm{G})\mathrm{HT}} = \frac{1}{N}\sum_{i\in s}\frac{\hat{G}_i}{\pi_i} \tag{7.3.13}$$

Properties of the estimators $\hat{\pi}_{(\mathrm{F})\mathrm{HT}}$ and $\hat{\pi}_{(\mathrm{G})\mathrm{HT}}$ are given as follows:

Theorem 7.3.1 For Q = G,H;

(i) $\hat{\pi}_{(\mathrm{Q})\mathrm{HT}}$ is unbiased estimators for $\pi_{(Q)}$
(ii) The variance of $\hat{\pi}_{(\mathrm{Q})\mathrm{HT}}$ is

$$\mathrm{Var}\left[\hat{\pi}_{(\mathrm{Q})\mathrm{HT}}\right] = \frac{1}{N^2}\left[\sum_{i\neq}\sum_{j\in U}(\pi_i\pi_j - \pi_{ij})\left(\frac{Q_i}{\pi_i} - \frac{Q_j}{\pi_j}\right)^2 + \sum_{i\in U}\frac{\phi_{Qi}}{\pi_i}\right];$$

(iii) An unbiased estimator of the variance of $\mathrm{Var}\left[\hat{\pi}_{(\mathrm{Q})\mathrm{HT}}\right]$ is

$$\hat{\mathrm{Var}}\left[\hat{\pi}_{(\mathrm{Q})\mathrm{HT}}\right] = \frac{1}{N^2}\left[\sum_{i\neq}\sum_{j\in s}\frac{\pi_i\pi_j - \pi_{ij}}{\pi_{ij}}\left(\frac{\hat{Q}_i}{\pi_i} - \frac{\hat{Q}_j}{\pi_j}\right)^2 + \sum_{\in}\frac{\hat{\phi}_{Qi}}{\pi_i}\right]$$

(iv) Co-variance of $\hat{\pi}_{(\mathrm{F})\mathrm{HT}}$ and $\hat{\pi}_{(\mathrm{G})\mathrm{HT}}$ is

$$\mathrm{Cov}\left[\hat{\pi}_{(\mathrm{F})\mathrm{HT}}, \hat{\pi}_{(\mathrm{G})\mathrm{HT}}\right] = \frac{1}{N^2}\left[\sum_{i<}\sum_{j\in U}(\pi_i\pi_j - \pi_{ij})\left(\frac{F_i}{\pi_i} - \frac{F_j}{\pi_j}\right)\left(\frac{G_i}{\pi_i} - \frac{G_j}{\pi_j}\right) + \sum_{i\in U}\frac{\phi_{\mathrm{FG}i}}{\pi_i}\right]$$

(v) An unbiased estimator of $\mathrm{Cov}\left[\hat{\pi}_{(\mathrm{F})\mathrm{HT}}, \hat{\pi}_{(\mathrm{G})\mathrm{HT}}\right]$ is

$$\hat{\mathrm{Cov}}\left[\hat{\pi}_{(\mathrm{F})\mathrm{HT}}, \hat{\pi}_{(\mathrm{G})\mathrm{HT}}\right] = \frac{1}{N^2}\left[\sum_{i<}\sum_{j\in s}\frac{\pi_i\pi_j - \pi_{ij}}{\pi_{ij}}\left(\frac{\hat{F}_i}{\pi_i} - \frac{\hat{F}_j}{\pi_j}\right)\left(\frac{\hat{G}_i}{\pi_i} - \frac{\hat{G}_j}{\pi_j}\right) + \sum_{i\in U}\frac{\hat{\phi}_{\mathrm{FG}i}}{\pi_i}\right]$$

Proof Let $E_p(E_R)$, $V_p(V_R)$ and $C_p(C_R)$ denote the expectation, variance, and co-variance with respect to the sampling design p(RR model R).

(i)
$$E\left[\hat{\pi}_{(\mathrm{Q})\mathrm{HT}}\right] = E_p\left[E_R\left\{\hat{\pi}_{(\mathrm{Q})\mathrm{HT}}\right\}\right]$$
$$= E_p\left[\frac{1}{N}\sum_{i\in s}\frac{Q_i}{\pi_i}\right]$$
$$= \pi_{(\mathrm{Q})}$$

(ii) $\mathrm{Var}\left[\hat{\pi}_{(\mathrm{Q})\mathrm{HT}}\right] = V_p\left[E_R\left(\hat{\pi}_{(\mathrm{Q})\mathrm{HT}}\right)\right] + E_p\left[V_R\left(\hat{\pi}_{(\mathrm{Q})\mathrm{HT}}\right)\right]$

$$= V_p\left[\frac{1}{N}\left\{\sum\nolimits_{i\in s}\frac{Q_i}{\pi_i}\right\}\right] + E_p\left[\frac{1}{N^2}\sum\nolimits_{i\in s}\frac{\phi_{\mathrm{Q}i}}{\pi_i^2}\right]$$

$$= \frac{1}{N^2}\left[\sum\nolimits_{i\neq}\sum\nolimits_{j\in U}(\pi_i\pi_j - \pi_{ij})\left(\frac{Q_i}{\pi_i} - \frac{Q_j}{\pi_j}\right)^2 + \sum\nolimits_{i\in U}\frac{\phi_{\mathrm{Q}i}}{\pi_i}\right]$$

(iii) $E\left\{\hat{\mathrm{Var}}\left[\hat{\pi}_{(\mathrm{Q})\mathrm{HT}}\right]\right\} = \frac{1}{N^2}$

$$E_P\left[\sum_{i\neq}\sum_{j\in s}\frac{\pi_i\pi_j - \pi_{ij}}{\pi_{ij}}\left\{\begin{array}{l}\left(\frac{Q_i}{\pi_i} - \frac{Q_j}{\pi_j}\right)^2\\ +\left(\frac{\phi_{\mathrm{Q}i}}{\pi_i^2} + \frac{\phi_{\mathrm{Q}j}}{\pi_j^2}\right)\end{array}\right\} + \sum_{i\in s}\frac{\phi_{\mathrm{Q}i}}{\pi_i}\right]$$

$$= \mathrm{Var}\left[\hat{\pi}_{(\mathrm{Q})\mathrm{HT}}\right] \text{ using } \sum_{j(\neq i)}\pi_{ij} = (n-1)\pi_i$$

(iv) $\mathrm{Cov}\left[\hat{\pi}_{(\mathrm{F})\mathrm{HT}}, \hat{\pi}_{(\mathrm{G})\mathrm{HT}}\right] = C_P\left[E_R\left(\hat{\pi}_{(\mathrm{F})\mathrm{HT}}\right), E_R\left(\hat{\pi}_{(\mathrm{G})\mathrm{HT}}\right)\right]$

$$+ E_P\left[C_R\left\{\hat{\pi}_{(\mathrm{G})\mathrm{HT}}, \hat{\pi}_{(\mathrm{G})\mathrm{HT}}\right\}\right]$$

$$= C_P\left[\frac{1}{N}\sum_{i\in s}\frac{F_i}{\pi_i}, \frac{1}{N}\sum_{i\in s}\frac{G_i}{\pi_i}\right]$$

$$+ E_P\left[\frac{1}{N^2}\sum_{i\in s}\frac{\phi_{\mathrm{FG}i}}{\pi_i^2}\right]$$

$$= \frac{1}{N^2}\left[\begin{array}{l}\sum_{i<}\sum_{j\in U}(\pi_i\pi_j - \pi_{ij})\\ \left(\frac{F_i}{\pi_i} - \frac{F_j}{\pi_j}\right)\left(\frac{G_i}{\pi_i} - \frac{G_j}{\pi_j}\right) + \sum_{i\in U}\frac{\phi_{\mathrm{FG}i}}{\pi_i}\end{array}\right]$$

(v) $E\left\{\hat{\mathrm{Cov}}\left[\hat{\pi}_{(\mathrm{F})\mathrm{HT}}, \hat{\pi}_{(\mathrm{G})\mathrm{HT}}\right]\right\}$

$$= \frac{1}{N^2}E_p\left[\begin{array}{l}\sum_{i<}\sum_{j\in s}\frac{\pi_i\pi_j - \pi_{ij}}{\pi_{ij}}E_R\left(\frac{\hat{F}_i}{\pi_i} - \frac{\hat{F}_j}{\pi_j}\right)\left(\frac{\hat{G}_i}{\pi_i} - \frac{\hat{G}_j}{\pi_j}\right)\\ +\sum_{i\in s}\frac{E_R\left(\hat{\phi}_{\mathrm{FG}i}\right)}{\pi_i}\end{array}\right]$$

$$= \frac{1}{N^2}E_p\left[\sum_{i<}\sum_{j\in s}\frac{\pi_i\pi_j - \pi_{ij}}{\pi_{ij}}\left\{\begin{array}{l}\left(\frac{F_i}{\pi_i} - \frac{F_j}{\pi_j}\right)\left(\frac{G_i}{\pi_i} - \frac{G_j}{\pi_j}\right)\\ +\left(\frac{\phi_{FGi}}{\pi_i^2} + \frac{\phi_{FGj}}{\pi_j^2}\right)\end{array}\right\} + \sum_{i\in s}\frac{\phi_{\mathrm{FG}i}}{\pi_i}\right]$$

$$= \mathrm{Cov}\left[\hat{\pi}_{(\mathrm{F})\mathrm{HT}}, \hat{\pi}_{(\mathrm{G})\mathrm{HT}}\right], \text{ using } \sum_{j\neq i}\pi_{ij} = (n-1)\pi_i$$

SRSWR Design

For the SRSWR, the proposed estimators of $\pi_{(F)}$ and $\pi_{(G)}$ are

$$\hat{\pi}_{(F)W} = \frac{1}{n}\sum_{i=1}^{n}\hat{F}_i \quad \text{and} \quad \hat{\pi}_{(G)W} = \frac{1}{n}\sum_{i=1}^{n}\hat{G}_i \tag{7.3.14}$$

where $\sum_{i=1}^{n}$ denotes the sum of all the selected n units with repetition.

Properties of the estimators $\hat{\pi}_{(F)W}$ and $\hat{\pi}_{(G)W}$ are as follows:

Theorem 7.3.2 For Q = G,H;

(i) $\hat{\pi}_{(Q)W}$ is an unbiased estimator of $\pi_{(Q)}$

(ii) The variance of $\hat{\pi}_{(Q)W}$ is

$$\mathrm{Var}\left[\hat{\pi}_{(Q)W}\right] = \frac{1}{n}\left[\pi_Q(1-\pi_Q) + \overline{\phi}_Q\right];\ \overline{\phi}_Q = \frac{1}{N}\sum_{i\in U} Q_i$$

(iii) An unbiased estimator of the variance of $\hat{\pi}_{(Q)W}$ is

$$\hat{\mathrm{Var}}\left[\hat{\pi}_{(Q)HT}\right] = \frac{1}{n(n-1)}\sum_{i=1}^{n}\left(\hat{Q}_i - \hat{\pi}_{(Q)W}\right)^2$$

(iv) Co-variance between $\hat{\pi}_{(F)W}$ and $\hat{\pi}_{(G)W}$ is

$$\mathrm{Cov}\left[\hat{\pi}_{(F)W}, \hat{\pi}_{(G)W}\right] = \frac{\pi_{FG}(1-\pi_{FG})}{n} + \frac{\overline{\phi}_{FG}}{n}$$

where $\pi_{FG} = \frac{1}{N}\sum_{i\in U} F_i G_i$ and $\overline{\phi}_{FG} = \frac{1}{N}\sum_{i\in U}\phi_{FGi}$

(v) An unbiased estimator of $\mathrm{Cov}\left[\hat{\pi}_{(F)W}, \hat{\pi}_{(G)W}\right]$ is

$$\hat{\mathrm{Cov}}\left[\hat{\pi}_{(F)W}, \hat{\pi}_{(G)W}\right] = \frac{1}{n(n-1)}\sum_{i=1}^{n}\left(\hat{F}_i - \hat{\pi}_{(F)W}\right)\left(\hat{G}_i - \hat{\pi}_{(G)W}\right)$$

Proof

(i) $$\begin{aligned} E\left[\hat{\pi}_{(Q)HT}\right] &= E_p\left[E_R\{\hat{\pi}_{(Q)W}\}\right] \\ &= E_p\left(\frac{1}{n}\sum\nolimits_{i=1}^{n} Q_i\right) \\ &= \pi_Q \end{aligned}$$

(ii) $\text{Var}\left[\hat{\pi}_{(\text{Q})\text{W}}\right] = V_p\left[E_R\{\hat{\pi}_{(\text{Q})\text{W}}\}\right] + E_p\left[V_R\{\hat{\pi}_{(\text{Q})\text{W}}\}\right]$

$$=V_p\left(\frac{1}{n}\sum\nolimits_{i=1}^{n} Q_i\right) + E_p\left(\frac{1}{n^2}\sum\nolimits_{i=1}^{n} \phi_{\text{Q}i}\right)$$

$$= \frac{1}{n}\left[\pi_Q(1-\pi_Q) + \overline{\phi}_{\text{Q}}\right]$$

(iii) Follows from the fact that $\hat{Q}_1, \ldots, \hat{Q}_n$ are iid random variables.

(iv) $\text{Cov}\left[\hat{\pi}_{(\text{F})\text{W}}, \hat{\pi}_{(\text{G})\text{W}}\right] = C_P\left[E_R\left(\hat{\pi}_{(\text{F})\text{W}}\right), E_R\left(\hat{\pi}_{(\text{G})\text{W}}\right)\right] + E_P\left[C_R\left\{\hat{\pi}_{(\text{F})\text{W}}, \hat{\pi}_{(\text{G})\text{W}}\right\}\right]$

$$= C_P\left(\frac{1}{n}\sum\nolimits_{i=1}^{n} F_i, \frac{1}{n}\sum\nolimits_{i=1}^{n} G_i\right) + E_P\left(\frac{1}{n^2}\sum\nolimits_{i=1}^{n} \phi_{\text{FG}i}\right)$$

$$= \frac{\pi_{\text{FG}}(1-\pi_{\text{FG}})}{n} + \frac{\overline{\phi}_{\text{FG}}}{n}$$

Estimation of Unions of Groups

Proportion of individuals belongs either to the category H = F ∪ G is

$$\pi_{(\text{H})} = P(\text{F} \cup \text{G}) \\ = \pi_{(\text{F})} + \pi_{(\text{G})} - \pi_{(\text{F}\cap\text{G})} \quad (7.3.15)$$

where $\pi_{(\text{G})} = \frac{1}{N}\sum_{i\in U} G_i$, $\pi_{(\text{F})} = \frac{1}{N}\sum_{i\in U} F_i$ and $\pi_{(\text{F}\cap\text{G})} = \frac{1}{N}\sum_{i\in U} F_i G_i$

An unbiased estimator of the union $\pi_{(\text{H})}$ is given by

$$\hat{\pi}_{(\text{H})} = \hat{\pi}_{(\text{F})} + \hat{\pi}_{(\text{G})} - \hat{\pi}_{(\text{F}\cap\text{G})} \quad (7.1.16)$$

Using Theorems 7.3.1. and 7.3.2, the following results may be derived.

Theorem 7.3.3

(i) $\hat{\pi}_{(\text{H})}$ is an unbiased estimator of $\pi_{(\text{H})}$

(ii) The variance of $\hat{\pi}_{(\text{H})}$ is

$$\text{Var}\left[\hat{\pi}_{(\text{H})}\right] = \text{Var}\left[\hat{\pi}_{(\text{F})}\right] + \text{Var}\left[\hat{\pi}_{(\text{G})}\right] + \text{Var}\left[\hat{\pi}_{(\text{FG})}\right] \\ + 2\text{Cov}\left[\hat{\pi}_{(\text{F})}, \hat{\pi}_{(\text{G})}\right] - 2\text{Cov}\left[\hat{\pi}_{(\text{F})}, \hat{\pi}_{(\text{FG})}\right] - 2\text{Cov}\left[\hat{\pi}_{(\text{G})}, \hat{\pi}_{(\text{FG})}\right]$$

(iii) An unbiased estimator of $\text{Var}\left[\hat{\pi}_{(\text{H})}\right]$ is

$$\hat{\text{Var}}\left[\hat{\pi}_{(\text{H})}\right] = \hat{\text{Var}}\left[\hat{\pi}_{(\text{F})}\right] + \hat{\text{Var}}\left[\hat{\pi}_{(\text{G})}\right] + \hat{\text{Var}}\left[\hat{\pi}_{(\text{FG})}\right] \\ + 2\hat{\text{Cov}}\left[\hat{\pi}_{(\text{F})}, \hat{\pi}_{(\text{G})}\right] - 2\hat{\text{Cov}}\left[\hat{\pi}_{(\text{F})}, \hat{\pi}_{(\text{FG})}\right] - 2\hat{\text{Cov}}\left[\hat{\pi}_{(\text{G})}, \hat{\pi}_{(\text{FG})}\right]$$

Expressions of $\hat{\pi}_{(\text{H})}$, $\text{Var}\left[\hat{\pi}_{(\text{H})}\right]$ and $\hat{\text{Var}}\left[\hat{\pi}_{(\text{H})}\right]$ for the different sampling strategies can be obtained from Theorems 7.3.1, 7.3.2, and 7.3.3.

Special Cases

Warner's Model with SRSWR

Consider the SRSWR sampling with Warner's model where $F = A_j$ and $G = A_k$. Proportions of individuals in the population belong to the groups A_j,A_k are, respectively,

$$\pi_{(j)}=\frac{1}{N}\sum_{i\in U} y_i(j) \text{ and } \pi_{(k)}=\frac{1}{N}\sum_{i\in U} y_i(k) \tag{7.3.17}$$

Unbiased estimator of $\pi_{(j)}$ and $\pi_{(k)}$ are

$$\hat{\pi}_{(j)}=\frac{1}{n}\sum_{i=1}^{n} \hat{y}_i(j) \text{ and } \hat{\pi}_{(k)}=\frac{1}{n}\sum_{i=1}^{n} \hat{y}_i(k) \tag{7.3.18}$$

where $\hat{y}_i(j) = \frac{z_i(j)-(1-P_j)}{2P_j-1}$ and $\hat{y}_i(k) = \frac{z_i(k)-(1-P_k)}{2P_k-1}$.

The proportion of individuals in the population that possess the attributes A_j or A_k, i.e., belong to the group $H = A_j \cup A_k$ is $\pi_H = \pi_{(j)} + \pi_{(k)} - \pi_{(jk)}$. An unbiased estimator of π_H is

$$\hat{\pi}_H = \hat{\pi}_{(j)} + \hat{\pi}_{(k)} - \hat{\pi}_{(jk)} \tag{7.3.19}$$

The variance of $\hat{\pi}_H$ is

$$\begin{aligned}\text{Var}(\hat{\pi}_H) &= \text{Var}(\hat{\pi}_{(j)}) + \text{Var}(\hat{\pi}_{(k)}) + \text{Var}(\hat{\pi}_{(j,k)}) \\ &\quad + 2\text{Cov}(\hat{\pi}_{(j)}, \hat{\pi}_{(k)}) - 2\text{Cov}(\hat{\pi}_{(j)}, \hat{\pi}_{(j,k)}) - 2\text{Cov}(\hat{\pi}_{(k)}, \hat{\pi}_{(j,k)})\end{aligned} \tag{7.3.20}$$

Now noting,
$\text{Var}(\hat{\pi}_{(j)}) = \frac{\pi_{(j)}(1-\pi_{(j)})}{n} + \frac{\phi_{(j)}}{n}$; $\text{Var}(\hat{\pi}_{(k)}) = \frac{\pi_{(k)}(1-\pi_{(k)})}{n} + \frac{\phi_{(k)}}{n}$; $\phi_{(t)} = \frac{P_t(1-Pt)}{(2P_t-1)^2}, t = j, k$;
$\text{Var}(\hat{\pi}_{(j,k)}) = \frac{1}{n}\left[\pi_{(jk)}(1 - \pi_{(jk)}) + \pi_{(j)}\phi_{(k)} + \pi_{(k)}\phi_{(j)} + \phi_{(j)}\phi_{(k)}\right]$,

$\text{Cov}(\hat{\pi}_{(j)},\hat{\pi}_{(k)}) = \frac{\pi_{(jk)}-\pi_{(j)}\pi_{(k)}}{n}$, $\text{Cov}(\hat{\pi}_{(j)}, \hat{\pi}_{(j,k)}) = \frac{1}{n}\left[\phi_{(j)}\pi_{(k)} + \pi_{(jk)}(1 - \pi_{(j)})\right]$ and $\text{Cov}(\hat{\pi}_{(k)}, \hat{\pi}_{(j,k)}) = \frac{1}{n}\left[\phi_{(k)}\pi_{(j)} + \pi_{(jk)}(1 - \pi_{(k)})\right]$, one finds,

$$\text{Var}(\hat{\pi}_H) = \frac{1}{n}\begin{bmatrix} \{\pi_{(j)}(1 - \pi_{(j)}) + \phi_{(j)}\} + \{\pi_{(k)}(1 - \pi_{(k)}) + \phi_{(k)}\}+ \\ \{\pi_{(jk)}(1 - \pi_{(jk)}) + \pi_{(j)}\phi_{(k)} + \pi_{(k)}\phi_{(j)} + \phi_{(j)}\phi_{(k)}\} \\ +2\{\pi_{(jk)} - \pi_{(j)}\pi_{(k)}\} - 2\{\phi_{(j)}\pi_{(k)} + \pi_{(jk)}(1 - \pi_{(j)})\} - \\ 2\{\phi_{(k)}\pi_{(j)} + \pi_{(jk)}(1 - \pi_{(k)})\} \end{bmatrix}$$

$$
= \frac{1}{n}\begin{bmatrix} \pi_{(j)}(1-\pi_{(j)}) + \pi_{(k)}(1-\pi_{(k)}) + \pi_{(jk)}(1-\pi_{(jk)}) + \\ 2\{\pi_{(jk)} - \pi_{(j)}\pi_{(k)}\} - 2\{\pi_{(jk)}(1-\pi_{(j)})\} \\ -2\{\pi_{(jk)}(1-\pi_{(k)})\} + \{\pi_{(j)}\phi_{(k)} + \pi_{(k)}\phi_{(j)} - 2\phi_{(k)}\pi_{(j)} \\ -2\phi_{(j)}\pi_{(k)}\} + \{\phi_{(j)} + \phi_{(k)} + \phi_{(j)}\phi_{(k)}\} \end{bmatrix}
$$

$$
= \frac{1}{n}\begin{bmatrix} \{\pi_{(j)}(1-\pi_{(j)}) + \pi_{(k)}(1-\pi_{(k)}) + \pi_{(jk)}(1-\pi_{(jk)}) \\ -2\pi_{(j)}\pi_{(k)} - 2\pi_{(jk)}(1-\pi_{(j)}-\pi_{(k)})\} \\ +\{\pi_{(j)}\phi_{(k)} + \pi_{(k)}\phi_{(j)} - 2\pi_{(j)}\phi_{(k)} - 2\pi_{(k)}\phi_{(j)}\} \\ +\{\phi_{(j)} + \phi_{(k)} + \phi_{(j)}\phi_{(k)}\} \end{bmatrix} \tag{7.3.21}
$$

Franklin's Model with SRSWR

Barabesi et al. (2012) proposed Franklin's (1989) RR model for estimating the proportion of individuals that have more than one sensitive characteristic. For the proposed Franklin's model, each of the selected respondents in the sample s draws a sample g_{ij} from the distribution g_j if he or she belongs to the group A_j. Otherwise, if the respondent does not belong to the group A_j, then he or she draws a random sample h_{ij} from the distribution h_j. The samples g_{ij} and $h_{ij}; j = 1, \ldots, k$ are selected independently. The means (variances) of g_j and h_j are known and are denoted by $\mu_{gj}\left(\sigma_{gj}^2\right)$, and $\mu_{hj}\left(\sigma_{hj}^2\right)$, respectively. As usual, the randomization stage is unobserved by the interviewer, to protect the respondent's privacy. In this case, the i th respondent reports a vector $z_i' = \left(z_{i1}, \ldots, z_{ij}, \ldots, z_{ik}\right)$ as his or her response (Fig. 7.5).

Let us denote

$$
\delta_i(j) = \begin{cases} 1 \text{ if } i \text{ is a member of } A_j \\ 0 \text{ if } i \text{ is not a member of } A_j \end{cases} ; j = 1, \ldots, k \tag{7.3.22}
$$

Then,

$$
z_{ij} = \delta_i(j) g_{ij} + (1 - \delta_i(j)) h_{ij} \tag{7.3.23}
$$

and Eq. (7.3.23) yields,

$$
E_R\left(z_{ij}\right) = \mu_{gj}\delta_i(j) + (1-\delta_i(j))\mu_{hj}, \quad V_R(z_{ij}) = \delta_i(j)\sigma_{gj}^2 + (1-\delta_i(j))\sigma_{hj}^2
$$
$$
\text{and } C_R\left(z_{ij}, z_{i'j'}\right) = 0 \text{ for } (i,j) \neq (i',j') \tag{7.3.24}
$$

Respondent i

$\downarrow$

$$
z_{i1} = \begin{cases} g_{1i} \text{ if } i \in A_1 \\ h_{1i} \text{ if } i \notin A_1 \end{cases} \cdots \quad z_{ij} = \begin{cases} g_{ji} \text{ if } i \in A_j \\ h_{ji} \text{ if } i \notin A_j \end{cases} \cdots \quad z_{ik} = \begin{cases} g_{ki} \text{ if } i \in A_k \\ h_{ki} \text{ if } i \notin A_k \end{cases}
$$

Fig. 7.5 Franklin's (1989) model

Furthermore, writing

$$\hat{\delta}_i(j) = \frac{z_{ij} - \mu_{hj}}{\mu_{gj} - \mu_{hj}} \text{ and assuming } \mu_{gj} \neq \mu_{hj} \text{ for } j = 1, \ldots, k \tag{7.3.25}$$

we get the reduced model:

$$E_R\left[\hat{\delta}_i(j)\right] = \delta_i(j),\ V_R\left[\hat{\delta}_i(j)\right] = \frac{\delta_i(j)(\sigma_{gj}^2 - \sigma_{hj}^2) + \sigma_{hj}^2}{(\mu_{gj} - \mu_{hj})^2} = \phi_{Bi}(j)$$
$$\text{and } C_R\left[\hat{\delta}_i(j), \left[\hat{\delta}_{i'}(j')\right]\right] = 0 \text{ for } (i, j) \neq (i', j') \tag{7.3.26}$$

Estimation of Proportion π_F

SRWR Sampling

Barabesi et al. (2012) considered the situation where the sample s is of size n is selected by the SRSWR sampling method. The proposed estimator of π_F is

$$\hat{\pi}_{WF} = \frac{1}{n}\sum_{i=1}^{n} \hat{T}_{Fi} \tag{7.3.27}$$

where $\hat{T}_{Fi} = \prod_{j \in F} \hat{\delta}_i(j)$

Equation (7.3.26) yields,

$$E_R(\hat{T}_{Fi}) = \prod_{j \in F} E_R\{\hat{\delta}_i(j)\} = \prod_{j \in F} \delta_i(j) = \delta_i(\mathrm{F}), \text{ say} \tag{7.3.28}$$

and

$$V_R(\hat{T}_{Fi}) = E_R(\hat{T}_{Fi})^2 - \left(\prod_{j \in F} \delta_i(j)\right)^2 = \prod_{j \in F}\left[\delta_i(j) + \phi_{Bi}(j)\right] - \left(\prod_{j \in F} \delta_i(j)\right)^2$$
$$= \phi_{Bi}(\mathrm{F}) \tag{7.3.29}$$

Furthermore, we note

$$\begin{aligned} E(\hat{\pi}_{WF}) &= E_p[E_R(\hat{\pi}_{WF})] \\ &= E_p\left[\frac{1}{n}\sum_{i=1}^{n}\prod_{j \in F} \delta_i(j)\right] \\ &= \frac{1}{N}\sum_{i \in U} \delta_i(\mathrm{F}) \end{aligned}$$

$$= \pi_{\mathrm{F}} \tag{7.3.30}$$

and

$$\begin{aligned}
\operatorname{Var}(\hat{\pi}_{\mathrm{WF}}) &= V_p[E_R(\hat{\pi}_{\mathrm{WF}})] + E_p[V_R(\hat{\pi}_{\mathrm{WF}})] \\
&= V_p\left(\frac{1}{n}\sum_{i=1}^{n}\delta_i(\mathrm{F})\right) + E_p\left(\frac{1}{n^2}\sum_{i=1}^{n}\phi_{\mathrm{B}i}(\mathrm{F})\right) \\
&= \frac{\pi_{\mathrm{F}}(1-\pi_{\mathrm{F}})}{n} + \frac{1}{n}\Phi_{\mathrm{BF}} \text{ with } \Phi_{\mathrm{BF}} = \sum_{i\in U}\phi_{\mathrm{B}i}(\mathrm{F})
\end{aligned} \tag{7.3.31}$$

Consider another category $\mathrm{G} = (q_1, q_2, \ldots, q_r) = A_{q_1} \cap A_{q_2} \cap \cdots \cap A_{q_r}$ with $(q_1, q_2, \ldots, q_r) \subset (1, \ldots, t)$. The proportion π_{G} can be estimated by

$$\hat{\pi}_{\mathrm{WG}} = \frac{1}{n}\sum_{i=1}^{n}\hat{T}_{\mathrm{G}i} \text{ with } \hat{T}_{\mathrm{G}i} = \prod_{j\in\mathrm{G}}\hat{\delta}_i(j) \tag{7.3.32}$$

The co-variance between $\hat{\pi}_{\mathrm{WF}}$ and $\hat{\pi}_{\mathrm{WF}}$ is given by

$$\begin{aligned}
\operatorname{Cov}(\hat{\pi}_{\mathrm{WF}}, \hat{\pi}_{\mathrm{WG}}) &= E_p[E_R(\hat{\pi}_{\mathrm{WF}}, \hat{\pi}_{\mathrm{WG}})] - \pi_{\mathrm{F}}\pi_{\mathrm{G}} \\
&= \frac{1}{n^2}E_p\left[\sum_{i=1}^{n}E_R(\hat{T}_{\mathrm{F}i}\hat{T}_{\mathrm{G}i}) + \sum_{i\neq}\sum_{k}T_{\mathrm{F}i}T_{\mathrm{G}k}\right] - \pi_{\mathrm{F}}\pi_{\mathrm{G}} \\
&= \frac{1}{n^2}E_p\left[\sum_{i=1}^{n}\left\{\left(\prod_{j\in F\cap G}\{\phi_{\mathrm{B}i}(j)+\delta_i(j)\}\right)\prod_{j\in(F-G)\cup(G-F)}\delta_i(j)\right\} + \sum_{i\neq}\sum_{k}T_{Fi}T_{\mathrm{G}k}\right] - \pi_{\mathrm{F}}\pi_{\mathrm{G}} \\
&= \frac{1}{n^2}E_p\left[\sum_{i=1}^{n}\psi_i + \sum_{i\neq}\sum_{k}T_{Fi}T_{\mathrm{G}k}\right] - \pi_{\mathrm{F}}\pi_{\mathrm{G}} \\
&= \frac{\pi_\psi - \pi_{\mathrm{F}}\pi_{\mathrm{G}}}{n}
\end{aligned} \tag{7.3.33}$$

where $\psi_i = \left(\prod_{j\in F\cap G}(\phi_{\mathrm{B}i}(j)+\delta_i(j))\right)\prod_{j\in(F-G)\cup(G-F)}\delta_i(j)$ and $\pi_\psi = \frac{1}{N}\sum_{i\in U}\psi_i$

Finally, noting that $z_1', \ldots, z_n'$ are iid random variables, the following results are obtained:

Theorem 7.3.4

(i) $\hat{\pi}_{\mathrm{WF}} = \frac{1}{n}\sum_{i=1}^{n}\hat{T}_{\mathrm{F}i}$ is an unbiased estimator of $\hat{\pi}_{\mathrm{F}}$
(ii) $\operatorname{Var}(\hat{\pi}_{\mathrm{WF}}) = \frac{\pi_{\mathrm{F}}(1-\pi_{\mathrm{F}})}{n} + \frac{1}{n}\phi_{\mathrm{BF}}$
(iii) $\operatorname{Cov}(\hat{\pi}_{\mathrm{WF}}, \hat{\pi}_{\mathrm{WG}}) = \frac{\pi_\psi - \pi_{\mathrm{F}}\pi_{\mathrm{G}}}{n}$
(iv) An unbiased estimator of $\operatorname{Var}(\hat{\pi}_{\mathrm{WF}})$ is $\widehat{\operatorname{Var}}(\hat{\pi}_{\mathrm{WF}}) = \frac{1}{n(n-1)}\sum_{i=1}^{n}\left(\hat{T}_{\mathrm{F}i} - \hat{\pi}_{\mathrm{WF}}\right)^2$

(v) An unbiased estimator of $\text{Cov}(\hat{\pi}_{\text{WF}}, \hat{\pi}_{\text{WG}})$ is $\widehat{\text{Cov}}(\hat{\pi}_{\text{WF}}, \hat{\pi}_{\text{WG}}) = \frac{1}{n(n-1)} \sum_{i=1}^{n} \left(\hat{T}_{\text{F}i} - \hat{\pi}_{\text{WF}}\right)\left(\hat{T}_{\text{G}i} - \hat{\pi}_{\text{WG}}\right)$

7.4 Comparison of the Models

The RR models proposed by Christofides (2005) and Lee et al. and's (2011, 2013) are applicable to only two categories, while Barabesi et al.'s (2012) method can be applied to any number of categories. All these methods, however, are limited to the SRSWR design only. The proposed Arnab's (2023) method can be applied to any number of categories, any sampling design, and any RR model. The derivations of variances of the estimators of the parameters proposed by Barabesi et al. (2012), Christofides (2005), and Lee et al. (2011, 2013) are not as simple as Arnab (2023). Since the proposed Barabesi et al. (2012), Christofides (2005), and Lee et al. and's (2011, 2013) estimators are based on the Christofides (2003), Franklin's (1989), and Warner's (1965) RR models, respectively, the performances of these proposed estimators depend on the respective RR models only. None of Christofides (2005), Warner's (1965), or Franklin's (1989) RR models are best in all situations. So, we cannot identify any of the proposed estimators as the best in all situations for estimating the population proportions and the conditional probabilities. However, we can always identify situations where one model performs better than others numerically, as was done by Barabesi et al. (2012) and Lee et al. (2011, 2013).

References

Abul-Ela, A. L. A., Greenberg, B. D., & Horvitz, D. G. (1967). Multi-proportions randomized response model. *Journal of the American Statistical Association, 62*, 990–1008.

Arnab, R. (2017). *Survey sampling theory and applications*. Academic Press.

Arnab, R. (2023). A note on randomized response technique for multiple sensitive attribute from complex surveys. *Communications in Statistics-Theory and Methods, 52*, 94–103.

Barabesi, L., Franceschi, S., & Marcheselli, M. (2012). A randomized response procedure for multiple sensitive attributes. *Statistical Papers, 53*, 703–718.

Burke, P. D. (1981). On the analysis of some multivariate randomized response designs for categorical data. *Journal of Statistical Planning and Inference, 5*, 165–170.

Burke, P. D. (1982). Randomized response multivariate designs for categorical data. *Communications in Statistics-Theory and Methods, 11*, 2889–2901.

Burke, P. D. (1990). Estimating a distribution for each category of sensitive variate. *Communications in Statistics-Theory and Methods, 19*, 3233–3241.

Chang, J. H., & Liang, D. H. (1996). A two stage unrelated randomized response procedure. *Australian Journal of Statistics, 33*, 43–51.

Christofides, T. (2003). A generalization of randomized response technique. *Metrika, 57*, 195–200.

Christofides, T. (2005). Randomized response technique for two sensitive characteristics at the same time. *Metrika, 62*, 53–63.

Cochran, W. G. (1977). *Sampling techniques* (3rd ed.). John Wiley and Sons.

Drane, W. (1976). On theory of randomized responses of two sensitive questions. *Communications in Statistics-Theory and Methods, 5*, 565–575.

Eriksson, S. A. (1973). A new model for randomized response. *International Statistical Review, 41*, 40–43.
Franklin, L. A. (1989). A comparison of estimators for randomized response sampling with continuous distribution from a dichotomous population. *Communications in Statistics-Theory and Methods, 18*, 489–505.
Greenberg, B. D., Abul-Ela, A. L. A., Simmons, W. R., & Horvitz, D. G. (1989). The unrelated question randomized response model: Theoretical framework. *Journal of the American Statistical Association, 64*, 520–539.
Hsieh, S.-H., Lee, S.-M., & Tu, S.-H. (2018). Randomized response technique for a multilevel attribute using a single sensitive question. *Statistical Papers, 59*, 291–306.
Kim, J. I., & Flueck, J. A. (1978). An additive randomized response model. In: *Proceedings of Social Statistics Section, American Statistical Association* (pp. 351–355).
Lee, C., Sedory, S. A., & Singh, S. (2011). A magical talk: Estimating at least seven measures of qualitative variables from a single sample using randomized response technique. In *International Statistical Institute: Proceedings of.58th World Statistical Congress, 2011, Dublin* (session STS001) (pp. 1948–1957).
Lee, C., Sedory, S. A., & Singh, S. (2013). Estimating at least seven measures of qualitative variables from a single sample using randomized response technique. *Statistical Probability Letters, 83*, 399–409.
Moshagen, M., & Musch, J. (2012). Surveying multiple sensitive attributes using extension of randomized response technique. *International Journal of Public Opinion Research, 24*, 508–523.
Mukerjee, R. (1981). Inference on conditional characters from survey data. *Calcutta Statistical Association Bulletin, 30*, 77–88.
Mukhopadhyaya, P. (1980). On the estimation of some confidential characters in survey. *Calcutta Statistical Association Bulletin, 29*, 77–88.
Thamane, A. C. (1981). Randomized response techniques for multiple attributes. *Journal of the American Statistical Association, 76*, 916–923.
Warner, S. L. (1965). Randomize response: A survey technique for eliminating evasive answer bias. *Journal of the American Statistical Association, 60*, 63–69.

Chapter 8
Randomized Response Techniques in Stages

Abstract A two-stage randomized response (RR) technique was proposed by Chang and Huang (2001) for estimating π, the proportion of individuals in a population that possess a certain sensitive attribute A and also estimating T, the sensitivity level of a question, which is defined as the proportion of individuals who possess the sensitive attribute A, answer "yes" to the direct sensitive question "Do you possess the attribute A?". Huang (2004), Marcheselli and Barabesi (2006) extended the method further by applying different RR methods in two stages and found that the two-stage RR procedure is more efficient than the single-stage RR procedure for estimating π. Quatember (2012) extended the two-stage procedure to the general $k(\geq 2)$ stages. Arnab (2023a) proposed two-stage RR procedures for complex survey designs and showed that the proposed procedure performed better than the corresponding single-stage RR procedures. He also showed that the results derived by Huang (2004), Barabesi and Marcheselli (2006) for the two-stage RR procedures can be obtained from his method as special cases. In this chapter, the works of Chang and Huang (2001), Huang (2004), Kim and Warde (2005), Marcheselli and Barabesi (2006), Quatember (2012), and Arnab (2023a) are reviewed.

Keywords Inclusion probabilities · Probability proportional to size · Randomized response · Sampling design · Two-stage randomized response · Unbiased estimators

8.1 Introduction

Consider a dichotomous population divided into a sensitive group A consisting of people with certain sensitive characteristics and a non-sensitive group $\overline{A}$ of people who don't have such sensitive features. The primary objective is to estimate π, the proportion of individuals that belong to the group A by applying a suitable randomized response (RR) procedure. However, an additional target might be to estimate T, the sensitivity level A which is defined as the probability of an individual belonging to A and providing a truthful answer about their membership in the group A under the

R. Arnab, *Indirect Methods of Data Collection and Analysis from Surveys*, Indian Statistical Institute Series, https://doi.org/10.1007/978-981-97-6005-3_8

direct question (DR) “Do you belong to the sensitive group A?”. Chang and Huang (2001), Huang (2004), and Marcheselli and Barabesi (2006) proposed two-stage RR methods for estimating both parameters π and T using the simple random sampling with replacement (SRSWR) sampling design, whereas Barabesi (2008) and Arnab (2023a) proposed methods for general sampling designs. The results of Chang and Huang (2001), Huang (2004), and Marcheselli and Barabesi (2006) can be derived as special cases of Arnab's (2023a) method. Details of the developments are given in the following sections. In the proposed two-stage RR procedure, respondents who answer “no” in direct questioning are required to respond again by using an RR device, which may cause embarrassment to respondents and hence may result in a reduction of cooperation from the respondents. Kim and Warde (2005) presented an alternative two-stage RR technique where all respondents, irrespective of whether they answered “yes” or “no” in stage one, are required to provide RR again by using suitable randomization devices. RR techniques with more than two stages were provided by Quatember (2012). Quatember's method yields various RR methods as special cases. The method preserves respondents' confidentiality greatly, but it is not applicable in practice because of its complexity.

8.2 Chang and Huang's Method

In Chang and Huang's (2001) method, two independent samples s_1 and s_2 of n_1 and $n_2(= n - n_1)$ units are selected from a population U of N units by the SRSWR method. In the first stage, each of the selected respondents (units) is asked a direct question: “Do you belong to the sensitive group A?”. If the answer is “yes”, the response is treated as a true response. If the answer is “no”, then the respondent is asked to go to the second stage. In the second stage, the respondents in the sample $s_j (j = 1, 2)$ are asked to provide an RR using Warner's (1965) randomization device with two statements: (i) I belong to the group A and (ii) I do not belong to the group A, with probabilities P_j and $1 - P_j$ respectively. The respondent completes the entire experiment without revealing the question answered. It is assumed that the respondents belonging to the non-sensitive group have no reason to tell a lie and, hence, provide a truthful answer under both direct questioning (DR) and Warner's RR method. Respondents from the sensitive group always provide a true response under the RR procedure, but only with a certain unknown probability $T(0 \leq T \leq 1)$ under the DR procedure. Under the proposed procedure, the probability of obtaining the answer “yes” from a respondent belonging to the sample s_j is

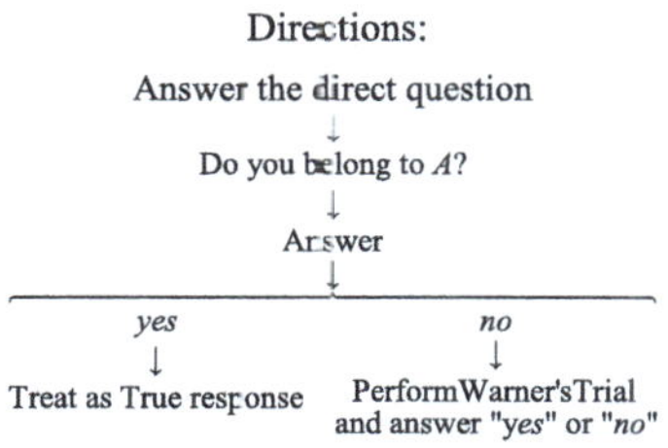

Fig. 8.1 Chang and Hung's (2001) method

$$\theta_j = \pi T + \pi(1-T)P_j + (1-\pi)(1-P_j);\ j = 1, 2 \tag{8.2.1}$$

where π is the proportion of persons in the population belonging to the sensitive group A (Fig. 8.1).

Denoting $\hat{\theta}_j$ as the proportion of "yes" answers in the sample s_j, Chang and Huang (2001) derived the estimators for π and T as

$$\hat{\pi}_{\text{CH}} = \frac{(1-P_2)\hat{\theta}_1 - (1-P_1)\hat{\theta}_2}{P_1 - P_2};\ P_1 \neq P_2$$

$$\text{and } \hat{T}_{\text{CH}} = \frac{(1-2P_2)\hat{\theta}_1 - (1-2P_1)\hat{\theta}_2 - (P_1 - P_2)}{(1-P_2)\hat{\theta}_1 - (1-P_1)\hat{\theta}_2} \tag{8.2.2}$$

The properties of the proposed estimators are given without derivation as follows.

Theorem 8.2.1

(i) $\hat{\pi}_{\text{CH}}$ is an unbiased estimator of π

(ii) $$\text{Var}(\hat{\pi}_{\text{CH}}) = \frac{1}{(P_1-P_2)^2}\left[\frac{(1-P_2)^2\theta_1(1-\theta_1)}{n_1} + \frac{(1-P_1)^2\theta_2(1-\theta_2)}{n_2}\right]$$

(iii) $\hat{T}_{\text{CH}}$ is a biased estimator of T with a bias up to order $O(n^{-1})$ is

$$\text{Bias}(\hat{T}_{\text{CH}}) = \frac{1}{(P_1-P_2)^2\pi^2}\left[\frac{(1-P_2)[T(1-P_2)+(2P_2-1)]\theta_1(1-\theta_1)}{n_1} + \frac{(1-P_1)[T(1-P_1)+(2P_1-1)]\theta_2(1-\theta_2)}{n_2}\right]$$

(iv) The mean-square error of $\hat{T}_{\text{CH}}$ up to order $O(n^{-1})$ is

$$\text{MSE}(\hat{T}_{\text{CH}}) = \frac{1}{(P_1-P_2)^2\pi^2}\left[\frac{[T(1-P_2)+(2P_2-1)]^2\theta_1(1-\theta_1)}{n_1} + \frac{[T(1-P_1)+(2P_1-1)]^2\theta_2(1-\theta_2)}{n_2}\right]$$

(v) The optimum value of n_j for a given n and the minimum variance with the optimum n_j s are

$$n_{j\text{opt}} = n\frac{\Delta_j}{\Delta_1 + \Delta_2}; j = 1, 2 \text{ and } V_{\min}(\hat{\pi}) = \frac{(\Delta_1 + \Delta_2)^2}{n(P_1 - P_2)^2}$$

where $\Delta_1 = (1 - P_2)\sqrt{\theta_1(1 - \theta_1)}$ and $\Delta_2 = (1 - P_1)\sqrt{\theta_2(1 - \theta_2)}$.

8.3 Huang's Method

In Huang's (2004) method, a single sample s of n respondents is selected by the SRSWR method. In stage-1, each of the respondents is asked the direct question, "Are you a member of the group A?" as in Chang–Huang's (2001) procedure. If the answer is "yes", the response is treated as a true response. But, if the answer given by a respondent is "no", then in stage-2, the respondent is required to provide a RR using Warner's method with a parameter P.

For this procedure, the probabilities of "yes" response under the DR and RR are

$$\theta_{\text{H}1} = \pi T \text{ and } \theta_{\text{H}2} = (2P - 1)\pi - P\pi T + (1 - P) \tag{8.3.1}$$

The proposed estimators of π and T are

$$\hat{\pi}_{\text{H}} = \frac{P\hat{\theta}_{\text{H}1} + \hat{\theta}_{\text{H}2} - (1 - P)}{2P - 1} \text{ and } \hat{T}_{\text{H}} = \frac{\hat{\theta}_{\text{H}1}}{\hat{\pi}_{\text{H}}} \tag{8.3.2}$$

where $\hat{\theta}_{\text{H}1}$ and $\hat{\theta}_{\text{H}2}$ are the proportion of "yes" answers under the DR and RR responses.

The properties of the estimators (8.3.1) are stated without derivation as follows.

Theorem 8.3.1

(i) $\hat{\pi}_{\text{H}}$ is unbiased for π with variance

$$\text{Var}(\hat{\pi}_{\text{H}}) = \frac{\pi(1 - \pi)}{n} + \frac{P(1 - P)(1 - \pi T)}{n(2P - 1)^2}$$

(ii) An unbiased estimator of $\text{Var}(\hat{\pi}_{\text{H}})$ is

$$\widehat{\text{Var}}(\hat{\pi}_{\text{H}}) = \frac{\hat{\pi}_{\text{H}}(1 - \hat{\pi}_{\text{H}})}{n - 1} + \frac{P(1 - P)}{(n - 1)(2P - 1)^2} - \frac{P(1 - P)\hat{\theta}_{\text{H}}}{n(2P - 1)^2}$$

(iii) $\hat{T}_{\text{H}}$ is a biased estimator of T with mean-square error up to the order $O(n^{-1})$ is

$$\text{MSE}(\hat{T}_{\text{H}}) = \frac{T(1-T)}{n\pi} + \frac{P(1-P)T^2(1-\pi T)}{n(2P-1)^2\pi^2}$$

Huang (2004) showed numerically that their proposed estimators for π and T fare better than those of Chang and Huang (2001)'s estimators.

8.4 Marcheselli and Barabesi's Method

Marcheselli and Barabesi (2006) generalized Huang's (2004) procedure by applying Franklin's (1989), Kuk's (1990), and Singh and Singh's (1993) RR procedures in the second stage. Marcheselli and Barabesi (2006) showed empirically that their proposed procedures provide more efficient estimators of π and T than those of Chang and Huang's (2001) method for some chosen values of the parameters. However, they did not provide any theoretical results for comparison. Marcheselli and Barabesi's (2006) method is described as follows.

In the first stage, a sample s of n units is selected using the SRSWR method. Each of the selected respondents in s is asked the direct question, "Do you belong to the sensitive group A?". If the answer is "yes", then it is treated as a true response, as in Chang and Huang's (2001), and Huang's (2004) methods. In case the answer is "no", the respondent is asked to respond by using Franklin's (1989) RR technique, where respondents are required to draw $k(\geq 1)$ random samples $g_{1j}(g_{2j})$, $i = 1, \ldots, k$ independently from the population $G_1(G_2)$, if the respondent bears the attribute $A(\overline{A})$. The mean μ_{tj} and variance σ_{tj}^2 of g_{tj} $(t = 1, 2; j = 1, \ldots, k)$ are assumed to be known. Let z_{ij} be the jth value of the random variable reported by the ith respondent. Denoting $\overline{Z}_i = \frac{1}{k}\sum_{j=1}^{k} z_{ij}$ and $\overline{Z} = \frac{1}{n}\sum_{i=1}^{n} \overline{Z}_i$, Marcheselli and Barabesi (2006), proposed the estimators of π and T as follows:

$$\hat{\pi}_{\text{MB}} = \frac{\overline{\mu}_1\hat{\theta}_1 + \overline{Z} - \overline{\mu}_2}{\overline{\mu}_1 - \overline{\mu}_2} \text{ and } \hat{T}_{\text{MB}} = \frac{\hat{\theta}_1}{\hat{\pi}_{\text{MB}}} \tag{8.4.1}$$

where $\hat{\theta}_1$ is the proportion of "yes" answers in direct questions and $\overline{\mu}_t = \frac{1}{k}\sum_{j=1}^{k} \mu_{tj}$; $t = 1, 2$. Marcheselli and Barabesi (2006) derived the following results.

Theorem 8.4.1

(i) $\hat{\pi}_{\text{MB}} = \frac{\mu_1\hat{\theta}_1 + \overline{Z} - \overline{\mu}_2}{\overline{\mu}_1 - \overline{\mu}_2}$, $\overline{\mu}_1 \neq \overline{\mu}_2$ is an unbiased estimator of π
(ii) The variance of $\hat{\pi}_{\text{MB}}$ is

$$\text{Var}(\hat{\pi}_{\text{MB}}) = \frac{\pi(1-\pi)}{n} + \frac{\overline{\sigma}_1^2(1-T)\pi + \overline{\sigma}_2^2(1-\pi)}{nk(\overline{\mu}_1 - \overline{\mu}_2)^2}$$

(iii) An unbiased estimator of $\text{Var}(\hat{\pi}_{\text{MB}})$ is

$$\widehat{\text{Var}}(\hat{\pi}_{\text{MB}}) = \frac{\hat{\pi}_{\text{MB}}(1-\hat{\pi}_{\text{MB}})}{n-1} + \frac{\overline{\sigma}_1^2(\hat{\pi}_{\text{MB}} - \overline{Z}) + \overline{\sigma}_2^2(1-\hat{\pi}_{\text{MB}})}{k(n-1)(\overline{\mu}_1 - \overline{\mu}_2)^2}$$

(iv) $\text{Var}(\hat{T}_{\text{MB}}) \cong \dfrac{T(1-T)}{n\pi} + \dfrac{T^2}{\pi^2}\dfrac{\overline{\sigma}_1^2(1-T)\pi + \overline{\sigma}_2^2(1-\pi)}{nk(\overline{\mu}_1 - \overline{\mu}_2)^2}$

(v) $\widehat{\text{Var}}(\hat{T}_{\text{MB}}) \cong \dfrac{\hat{T}_{\text{MB}}(1-\hat{T}_{\text{MB}})}{n\hat{\pi}_{\text{MB}}} + \dfrac{\hat{T}_{\text{MB}}^2}{\hat{\pi}_{\text{MB}}^2}\dfrac{\overline{\sigma}_1^2(\hat{\pi}_{\text{MB}} - \overline{Z}) + \overline{\sigma}_2^2(1-\hat{\pi}_{\text{MB}})}{nk(\overline{\mu}_1 - \overline{\mu}_2)^2}$

Marcheselli and Barabesi (2006) also generalized Huang's (2004) two-stage RR procedure further by using Kuk's (1990) and Singh and Sigh's (1993) RR procedures in stage two and observed in each case that the two-stage procedure performed better than the corresponding one-stage RR procedure.

8.5 Arnab's Method

Two-stage RR procedures proposed by Chang and Huang (2001), Huang (2004), and Marcheselli and Barabesi (2006) are limited to the SRSWR method and to certain specific randomization procedures such as Warner's (1965) and Franklin's (1989), Kuk's (1990), and Singh and Singh's (1993) methods. Arnab (2023a) proposed generalizations of the existing two-stage sampling procedures to general sampling designs, wider classes of estimators, and almost all the RR techniques that satisfy the general model (3.9.1) proposed by Arnab (1990).

8.5.1 General Sampling Design

A sample s of n units is selected from a finite population U of N units (respondents) with probability $p(s)$ by using a sampling design p so that the inclusion probabilities for the ith, and ith and jth$(j \neq i)$ units π_i and π_{ij} are positive.

Let us define

$$y_i = \begin{cases} 1 \text{ for } i \in A \\ 0 \text{ for } i \notin A \end{cases} \quad (8.5.1)$$

so that $\pi = \frac{1}{N}\sum\limits_{i \in U} y_i$.

8.5.2 Two-Stage RR Procedure

Each selected unit in s is asked the direct question, "Are you a member of A?". If a respondent answers "yes", the response is treated as a true response and retained. Otherwise, if a respondent answers "no", they are requested to supply a randomized response using a suitable randomization device. Here, we assume that the ith respondent provides a randomized response z_i that yields the revised randomized response, r_i (an estimate of y_i), which satisfies the following model proposed by Arnab (1990, 1994, 1996):

$$\text{Model } R : E_R(r_i|y_i) = y_i,\ V_R(r_i|y_i) = \phi_i \text{ and } C_R(r_i, r_j|y_i, y_j) = 0 \text{ for } i \neq j \quad (8.5.2)$$

where ϕ_i is a function of y_i; E_R, V_R and C_R denote respectively expectation, variance, and co-variance with respect to the model R.

Most of the randomization procedures available in the literature, such as Warner (1965), Horvitz et al. (1967), Franklin (1989), Mangat and Singh (1990), Kuk (1990), and Christofides (2003), satisfy the model (8.5.2). For more information, see Table 3.1.

Here, we attach a random variable x_i such that $x_i = 1$, if the ith respondent possesses the attribute A and gives a true response "yes" and $x_i = 0$ if the ith respondent belongs to A but provides a false response "no". The probabilities of $x_i = 1$ and $x_i = 0$ are T and $1 - T$ respectively. Here, we assume, as Huang (2004) and Marcheselli and Barabesi (2006) do, that the respondents belonging to the non-sensitive group $\bar{A}$ always provide true responses for the direct and randomized response procedures, while respondents belonging to the sensitive group A always provide a true answer for the RR procedure, but for the direct interview, they provide a true response with a probability T that is generally unknown.

In case the ith respondent responds "no" in the first stage of the direct interview, then in the second stage, he or she provides r_i, a revised randomized response which follows the model (8.5.2). Thus, the ith respondent yields $y_i x_i$ as the direct response in stage-1 and $(1 - y_i x_i) r_i$ as the RR in stage-2.

$$\tilde{r}_i = y_i x_i + (1 - y_i x_i) r_i \quad (8.5.3)$$

Denoting $E_x(E_R)$ as the expectation over $x_i(r_i)$ and $E_{xR}(V_{xR})$ as overall expectation and (variance) with respect to x_i and r_i, one finds

$$\begin{aligned} E_{xr}(\tilde{r}_i) &= y_i\{E_x(x_i)\} + \{1 - y_i E_x(x_i)\} E_R(r_i) \\ &= y_i T + (1 - y_i T) y_i \\ &= y_i (\text{noting } y_i^2 = y_i) \end{aligned} \quad (8.5.4)$$

$$\begin{aligned} V_{xr}(\tilde{r}_i|y_i) &= E_{xR}\{y_i x_i + (1 - y_i x_i) r_i\}^2 - y_i \\ &= y_i T + (1 - y_i T)(\phi_i + y_i) - y_i \end{aligned}$$

$$= (1 - y_i T)\phi_i$$
$$= \tilde{\phi}_i, \text{ say} \tag{8.5.5}$$

8.5.3 Estimation of Population Proportion π

The class of linear homogeneous unbiased estimators for the population proportion π in a direct response survey (responses obtained directly from the respondents) is

$$\hat{\pi}_{\text{D}} = \frac{1}{N}\sum_{i\in s} b_{si}y_i \tag{8.5.6}$$

where b_{si} s are constants free from y_i, r_i, $\tilde{r}_i$ and subject to $\sum_{s\supset i} b_{si}p(s) = 1$.

Replacing y_i by $\tilde{r}_i$ in (8.5.6), a homogeneous unbiased estimator of π for the two-stage RR model is obtained as

$$\hat{\pi}_{\text{TR}} = \frac{1}{N}\sum_{i\in s} b_{si}\tilde{r}_i \tag{8.5.7}$$

The variance of $\hat{\pi}_{\text{TR}}$ is given by

$$\begin{aligned}\text{Var}(\hat{\pi}_{\text{TR}}) &= V_p\{\text{E}_{xR}(\hat{\pi}_{\text{TR}})\} + E_p\{\text{V}_{xR}(\hat{\pi}_{\text{TR}})\}\\ &= \frac{1}{N^2}\left[\sum_{i\in U}(\alpha_{ii}-)y_i + \sum\sum_{i\neq j\in U}(\alpha_{ij}-1)y_iy_j\right] + \frac{1}{N^2}\left[\sum_{i\in U}\alpha_{ii}\tilde{\phi}_i\right]\end{aligned} \tag{8.5.8}$$

where $\alpha_{ii} = \sum_{s\supset i} b_{si}^2 p(s)$ and $\alpha_{ij} = \sum_{s\supset i,j} b_{si}b_{sj}p(s)$.

An unbiased estimator of $\text{Var}(\hat{\pi}_{\text{TR}})$ is obtained as

$$\widehat{\text{Var}}(\hat{\pi}_{\text{TR}}) = \frac{1}{N^2}\left[\sum_{i\in s} c_{si}\tilde{r}_i^2 + \sum\sum_{i\neq j\in s} c_{sij}\tilde{r}_i\tilde{r}_j\right] + \frac{1}{N^2}\sum_{i\in s} b_{si}\hat{\tilde{\phi}}_i \tag{8.5.9}$$

where c_{si}, c_{sij} are constants satisfy

$$\sum_{s\supset i} c_{si}p(s) = \alpha_{ii} - 1 \text{ and } \sum_{s\supset i} c_{sij}p(s) = \alpha_{ij} - 1 \tag{8.5.10}$$

and $\hat{\tilde{\phi}}_i$ is an unbiased estimator of $\tilde{\phi}_i$ satisfying $E_{xR}(\hat{\tilde{\phi}}_i) = \tilde{\phi}_i$.

The constants c_{si} and c_{sij} can be chosen in various ways to satisfying (8.5.10). Simple choices are

$$c_{si} = \frac{\alpha_{ii} - 1}{\pi_i} \text{ and } c_{sij} = \frac{\alpha_{ij} - 1}{\pi_{ij}} \tag{8.5.11}$$

The above findings are summarized as follows.

Theorem 8.5.1

(i) $\hat{\pi}_{\mathrm{TR}} = \frac{1}{N}\sum_{i\in s} b_{si}\tilde{r}_i$ is an unbiased estimator of π where $\sum_{s\supset i} b_{si}p(s) = 1$

(ii) The variance of $\hat{\pi}_{\mathrm{TR}}$ is

$$\mathrm{Var}(\hat{\pi}_{\mathrm{TR}}) = \frac{1}{N^2}\left[\sum_{i\in U}(\alpha_{ii} - 1)y_i + \sum\sum_{i\neq j\in U}(\alpha_{ij} - 1)y_i y_j\right] + \frac{1}{N^2}\left[\sum_{i\in U}\alpha_{ii}\tilde{\phi}_i\right]$$

(iii) An unbiased estimator of $\mathrm{Var}(\hat{\pi}_{\mathrm{TR}})$ is

$$\widehat{\mathrm{Var}}(\hat{\pi}_{\mathrm{TR}}) = \frac{1}{N^2}\left[\sum_{i\in s} c_{si}\tilde{r}_i^2 + \sum\sum_{i\neq j\in U} c_{sij}\tilde{r}_i\tilde{r}_j\right] + \frac{1}{N^2}\sum_{i\in s} b_{si}\hat{\tilde{\phi}}_i$$

where $\sum_{s\supset i} c_{si}p(s) = \alpha_{ii} - 1$, $\sum_{s\supset i} c_{sij}p(s) = \alpha_{ij} - 1$ and $E_{xR}(\hat{\tilde{\phi}}_i) = \tilde{\phi}_i$.

8.5.4 Comparison with Direct and Single Stage RR Method

Consider a single-stage RR survey where all the selected respondents in the sample s provide RR that satisfies the model (8.5.2). In this case, an unbiased estimator of π is given by

$$\hat{\pi}_{\mathrm{R}} = \frac{1}{N}\sum_{i\in s} b_{si}r_i \tag{8.5.12}$$

The variance of $\hat{\pi}_{\mathrm{R}}$ is

$$\mathrm{Var}(\hat{\pi}_{\mathrm{R}}) = \frac{1}{N^2}\left[\sum_{i\in U}(\alpha_{ii} - 1)y_i - \sum\sum_{i\neq j\in U}(\alpha_{ij} - 1)y_i y_j\right] + \frac{1}{N^2}\left(\sum_{i\in U}\alpha_{ii}\phi_i\right) \tag{8.5.13}$$

Now noting,

$$\text{Var}(\hat{\pi}_{\text{D}}) = \text{Var}\left(\sum_{i\in s} b_{si}y_i\right) = \frac{1}{N^2}\left[\sum_{i\in U}(\alpha_{ii}-1)y_i + \sum_{i\neq}\sum_{j\in U}(\alpha_{ij}-1)y_iy_j\right],$$

we derive the following results.

Theorem 8.5.2 The direct response technique is more efficient than the two-stage RR procedure, and the two-stage RR procedure is more efficient than the single-stage RR procedure in the sense that

$$\text{Var}(\hat{\pi}_{\text{D}}) \le \text{Var}(\hat{\pi}_{\text{TR}}) \le \text{Var}(\hat{\pi}_{\text{R}}).$$

Proof Consider the differences

(i) $$\text{Var}(\hat{\pi}_{\text{TR}}) - \text{Var}(\hat{\pi}_{\text{D}}) = \frac{1}{N^2}\sum_{i\in U}\alpha_{ii}\tilde{\phi}_i \ \ge 0$$

and

(ii) $$\text{Var}(\hat{\pi}_{\text{R}}) - \text{Var}(\hat{\pi}_{\text{TR}}) = \frac{1}{N^2}\sum_{i\in U}\alpha_{ii}(\phi_i - \tilde{\phi}_i)$$
$$= \frac{T}{N^2}\sum_{i\in U}\alpha_{ii}y_i\phi_i \ge 0 \text{ (nothing } \tilde{\phi}_i = (1-y_iT)\phi_i)$$

8.5.5 *Specific Two-Stage Strategies*

8.5.5.1 Horvitz-Thompson Estimator for an Arbitrary Sampling Design

The Horvitz-Thompson (1952) type estimator for π based on an arbitrary sampling design is obtained by replacing b_{si} with $1/\pi_i$ in $\hat{\pi}_{\text{TR}}$ and it is given by

$$\hat{\pi}_{\text{HT}} = \frac{1}{N}\sum_{i\in s}\frac{\tilde{r}_i}{\pi_i} \tag{8.5.14}$$

For a fixed effective size sampling design (where each sample contains exactly n distinct units), $\text{Var}(\hat{\pi}_{\text{HT}})$ can be expressed as

$$\text{Var}(\hat{\pi}_{\text{HT}}) = \frac{1}{N^2}\left[\sum_{i<}\sum_{j\in U}(\pi_i\pi_j - \pi_{ij})\left(\frac{y_i}{\pi_i} - \frac{y_j}{\pi_j}\right)^2 + \sum_{i\in U}\frac{\tilde{\phi}_i}{\pi_i}\right] \tag{8.5.15}$$

In this case, Yates–Grundy's (1953) type unbiased estimator of $\text{Var}(\hat{\pi}_{\text{HT}})$ comes out as

$$\hat{V}_{YG}(\hat{\pi}_{HT}) = \frac{1}{N^2}\left[\sum_{i<}\sum_{j\in s}\frac{(\pi_i\pi_j - \pi_{ij})}{\pi_{ij}}\left(\frac{\tilde{r}_i}{\pi_i} - \frac{\tilde{r}_j}{\pi_j}\right)^2 + \sum_{i\in s}\frac{\hat{\tilde{\phi}}_i}{\pi_i}\right] \tag{8.5.16}$$

In particular, the variances of Var($\hat{\pi}_{HT}$) with Warner's (1965) and Franklin's model (1989) in stage two reduces to

$$\mathrm{Var}(\hat{\pi}_{HT})_W = \frac{1}{N^2}\left[\sum_{i<}\sum_{j\in U}(\pi_i\pi_j - \pi_{ij})\left(\frac{y_i}{\pi_i} - \frac{y_j}{\pi_j}\right)^2 + \sum_{i\in U}\frac{P(1-P)(1-y_iT)}{\pi_i(2P-1)^2}\right] \tag{8.5.17}$$

and

$$\mathrm{Var}(\hat{\pi}_{HT})_F = \frac{1}{N^2}\left[\sum_{i<}\sum_{j\in U}(\pi_i\pi_j - \pi_{ij})\left(\frac{y_i}{\pi_i} - \frac{y_j}{\pi_j}\right)^2 + \sum_{i\in U}\frac{\tilde{\phi}_{Fi}}{\pi_i}\right] \tag{8.5.18}$$

where $\tilde{\phi}_{Fi} = \frac{\overline{\sigma}_1^2(1-T)y_i + \overline{\sigma}_2^2(1-y_i)}{k(\overline{\mu}_1 - \overline{\mu}_2)^2}$.

8.5.5.2 Simple Random Sampling Without Replacement (SRSWOR) Design

For a SRSWOR sampling, $\pi_i = n/N$ and $\pi_{ij} = n(n-1)/\{N(N-1)\}$. In this case $\hat{\pi}_{HT}$ reduces to

$$\hat{\pi}_{WOR} = \frac{1}{n}\sum_{i\in s}\tilde{r}_i \tag{8.5.19}$$

The expressions of the variance and unbiased estimator of the variance of $\hat{\pi}_{WOR}$ are obtained from (8.5.17) and (8.5.18) as follows:

$$\mathrm{Var}(\hat{\pi}_{WOR}) = \frac{N-n}{n(N-1)}\pi(1-\pi) + \frac{\overline{\tilde{\phi}}}{n} \text{ where } \overline{\tilde{\phi}} = \frac{1}{N}\sum_{i\in U}\tilde{\phi}_i \tag{8.5.20}$$

and

$$\widehat{\mathrm{Var}}(\hat{\pi}_{\mathrm{WOR}}) = \frac{N-n}{nN(n-1)}\sum_{i\in s}(\tilde{r}_i - \hat{\pi}_{\mathrm{WOR}})^2 + \frac{1}{nN}\sum_{i\in s}\hat{\hat{\phi}}_i \tag{8.5.21}$$

For Warner's and Franklin's two-stage models, the expressions (8.2.20) and (8.2.21) reduce to

$$\mathrm{Var}(\hat{\pi}_{\mathrm{WOR}})_{\mathrm{W}} = \frac{N-n}{n(N-1)}\pi(1-\pi) + \frac{P(1-P)(1-\pi T)}{n(2P-1)^2} \tag{8.5.22}$$

and

$$\mathrm{Var}(\hat{\pi}_{\mathrm{WOR}})_{\mathrm{F}} = \frac{N-n}{n(N-1)}\pi(1-\pi) + \frac{\overline{\sigma}_1^2(1-T)\pi + \overline{\sigma}_2^2(1-\pi)}{nk(\overline{\mu}_1 - \overline{\mu}_2)^2} \tag{8.5.23}$$

Finally, the above discussions lead to the following results.

Theorem 8.5.3 The two-stage RR model, with SRSWOR design in stage-1, and Warner's model in stage-2, provides a more efficient estimator than Huang's (2004) RR model, while SRSWOR design in stage-1 and Franklin's model in stage-2 provide a more efficient result than Barabesi et al.'s (2006) model.

Proof Theorems 8.3.1 and 8.4.1, Eqs. (8.5.22) and (8.5.23) yield

$$\mathrm{Var}(\hat{\pi}_{\mathrm{H}}) - \mathrm{Var}(\hat{\pi}_{\mathrm{WOR}})_{\mathrm{W}} \geq 0 \text{ and } \mathrm{Var}(\hat{\pi}_{\mathrm{MB}}) - \mathrm{Var}(\hat{\pi}_{\mathrm{WOR}})_{\mathrm{F}} \geq 0$$

8.5.5.3 With Replacement Sampling Designs

A respondent may be chosen more than once in a sample using a replacement sampling design such as probability proportional to size with replacement (PPSWR) or SRSWR. In this case, we assume that the selected respondent responds independently for each of the selections. For the PPSWR sampling, we will denote $\tilde{r}_{(j)}(p_{(j)})$ the revised randomized response (normed size measure) for the respondent selected at the jth draw ($j = 1, \ldots, n$). Thus, $\tilde{r}_{(j)} = \tilde{r}_i$ and $p_{(j)} = p_i$, if the ith respondent is selected at the jth draw with probability p_i. The Hansen–Hurwitz (1943) type estimator for the proportion π may be denoted by

$$\hat{\pi}_{\mathrm{HH}} = \frac{1}{Nn}\sum_{j=1}^{n}\frac{\tilde{r}_{(j)}}{p_{(j)}} \tag{8.5.24}$$

The estimator $\hat{\pi}_{\mathrm{HH}}$ is unbiased for π as

$$\begin{aligned} E(\hat{\pi}_{\mathrm{HH}}) &= \frac{1}{N} E_p \left[\frac{1}{n} \sum_{j=1}^{n} \frac{E_{xR}\{r_{(j)}\}}{p_{(j)}} \right] \\ &= \frac{1}{N} \sum_{j \in U} y_i \\ &= \pi \end{aligned} \tag{8.5.25}$$

The variance of $\hat{\pi}_{\mathrm{HH}}$ is

$$\begin{aligned} \mathrm{Var}(\hat{\pi}_{\mathrm{HH}}) &= V_p[E_{xR}(\hat{\pi}_{\mathrm{HH}})] + E_p[V_{xR}(\hat{\pi}_{\mathrm{HH}})] \\ &= \frac{1}{n}\left[\left(\frac{1}{N^2} \sum_{i \in U} \frac{y_i}{p_i} - \pi^2 \right) + \frac{1}{N^2} \sum_{j \in U} \frac{\tilde{\phi}_j}{p_j} \right] \end{aligned} \tag{8.5.26}$$

Noting, $\frac{\tilde{r}_{(j)}}{p_{(j)}}$ for $j = 1, \ldots, n$ are *iid* random variables, one finds an unbiased estimator of the variance $\mathrm{Var}(\hat{\pi}_{\mathrm{HH}})$ as

$$\widehat{\mathrm{Var}}(\hat{\pi}_{\mathrm{HH}}) = \frac{1}{N^2 n(n-1)} \sum_{j=1}^{n} \left(\frac{\tilde{r}_{(j)}}{p_{(j)}} - N\hat{\pi}_{\mathrm{HH}} \right)^2 \tag{8.5.27}$$

Finally, Eqs. (8.5.25), (8.5.26), and (8.5.27) lead.

Theorem 8.5.4 For a PPSWR sampling,

(i) $\hat{\pi}_{\mathrm{HH}} = \frac{1}{Nn} \sum\limits_{j=1}^{n} \frac{\tilde{r}_{(j)}}{p_{(j)}}$ is an unbiased estimator of π

(ii) The variance of $\hat{\pi}_{\mathrm{HH}}$ is

$$\mathrm{Var}(\hat{\pi}_{\mathrm{HH}}) = \frac{1}{n}\left[\left(\frac{1}{N^2} \sum_{i \in U} \frac{y_i}{p_i} - \pi^2 \right) + \frac{1}{N^2} \sum_{j \in U} \frac{\tilde{\varphi}_j}{p_j} \right]$$

(iii) An unbiased estimator of the variance $\mathrm{Var}(\hat{\pi}_{\mathrm{HH}})$ is

$$\widehat{\mathrm{Var}}(\hat{\pi}_{\mathrm{HH}}) = \frac{1}{N^2 n(n-1)} \sum_{j=1}^{n} \left(\frac{\tilde{r}_{(j)}}{p_{(j)}} - N\hat{\pi}_{\mathrm{HH}} \right)^2$$

8.5.5.4 SRSWR Sampling

PPSWR reduces to SRSWR if $p_i = 1/N \,\forall i \in U$. Substituting $p_i = 1/N$ in Theorem 8.5.4, one obtains.

Theorem 8.5.5 For a SRSWR sampling,

(i) $\hat{\pi}_{\text{SWR}} = \frac{1}{n}\sum_{j=1}^{n} \tilde{r}_{(j)}$ is an unbiased estimator of π

(ii) The variance of $\hat{\pi}_{\text{SWR}}$ is

$$\text{Var}(\hat{\pi}_{\text{SWR}}) = \frac{\pi(1-\pi)}{n} + \frac{\overline{\tilde{\phi}}}{n} \tag{8.5.28}$$

(where $\overline{\tilde{\phi}} = \frac{1}{N}\sum_{j\in U} \tilde{\phi}_j$)).

(iii) An unbiased estimator of the variance $\text{Var}(\hat{\pi}_{\text{SWR}})$ is

$$\widehat{\text{Var}}(\hat{\pi}_{\text{SWR}}) = \frac{1}{n(n-1)}\sum_{j=1}^{n}\left[\tilde{r}_{(j)} - \hat{\pi}_{\text{SWR}}\right]^2$$

8.5.5.5 SRSWR Sampling with Specific RR Models

Warner's (1965) Model

For Warner's (1965), $r_i = \frac{z_i-(1-P)}{2P-1}$ and $\phi_i = \frac{P(1-P)}{(2P-1)^2} = \phi_{\text{W}}$. In this case, Theorem 8.5.5 reproduces to Huang results, viz.

$$\hat{\pi}_{\text{SWR}} = \hat{\pi}_{\text{H}} = \frac{P\hat{\theta}_1 + \hat{\theta}_2 - (1-P)}{2P-1},$$

$$\text{Var}(\hat{\pi}_{\text{SWR}}) = \text{Var}(\hat{\pi}_{\text{H}}) = \frac{\pi(1-\pi)}{n} + \frac{P(1-P)(1-\pi T)}{n(2P-1)^2} \text{ and}$$

$$\widehat{\text{Var}}(\hat{\pi}_{\text{SWR}}) = \widehat{\text{Var}}(\hat{\pi}_{\text{SWR}}) = \frac{\hat{\pi}_{\text{H}}(1-\hat{\pi}_{\text{H}})}{n-1} + \frac{P(1-P)}{(n-1)(2P-1)^2} - \frac{P(1-P)\hat{\theta}_1}{(n-1)(2P-1)^2}$$

Franklin's (1989) Model

For Franklin's (1989), $r_i = \frac{\overline{z}_i-\overline{\mu}_2}{\overline{\mu}_1-\overline{\mu}_2}, \phi_i = \frac{y_i\overline{\sigma}_1^2+(1-y_i)\overline{\sigma}_2^2}{k(\overline{\mu}_1-\overline{\mu}_2)^2} = \phi_{\text{Fi}}$. In this case Theorem 8.5.5, yields

$$\hat{\pi}_{\text{SWR}} = \hat{\pi}_{\text{MB}} = \frac{\mu_1\hat{\theta}_1 + \overline{Z} - \overline{\mu}_2}{\overline{\mu}_1 - \overline{\mu}_2},$$

$$\text{Var}(\hat{\pi}_{\text{SWR}}) = \text{Var}(\hat{\pi}_{\text{MB}}) = \frac{\pi(1-\pi)}{n} + \frac{\overline{\sigma}_1^2(1-T)\pi + \overline{\sigma}_2^2(1-\pi)}{nk(\overline{\mu}_1-\overline{\mu}_2)^2} \text{ and}$$

$$\widehat{\text{Var}}(\hat{\pi}_{\text{SWR}}) = \widehat{\text{Var}}(\hat{\pi}_{\text{MB}}) = \frac{\hat{\pi}_{\text{MB}}(1-\hat{\pi}_{\text{MB}})}{n-1} + \frac{\overline{\sigma}_1^2(\hat{\pi}_{\text{MB}} - \overline{Z}) + \overline{\sigma}_2^2(1-\hat{\pi}_{\text{MB}})}{k(n-1)(\overline{\mu}_1-\overline{\mu}_2)^2}$$

Remark 8.5.1 It can be easily shown that Marcheselli and Barabesi's (2006) results concerning the second stage RR procedures with Kuk's (1990), and Singh and Singh's (1993) can be obtained as special cases from Theorem 8.5.1.

8.5.6 Estimation of the Level of Sensitivity T

Chang and Huang (2001) defined T, the probability that a respondent belonging to the sensitive group will tell the truth whenever questioned directly. The quantity T is a measurement instrument of sensitivity as it appraises the efficiency of different survey plans. The higher the value of T, the lower the sensitivity level of the question. Hence, an estimator of T may be used to decide whether a RR technique is more advantageous in efficiency than a DR scheme.

Since, $\hat{\pi}_{\text{DRT}} = \frac{1}{N}\sum_{i\in s} b_{si}y_i x_i$, is an unbiased estimator πT and $\hat{\pi}_{\text{TR}} = \frac{1}{N}\sum_{i\in s} b_{si}\tilde{r}_i$ is an unbiased estimator of π, the proposed ratio estimator of T is

$$\hat{T} = \frac{\hat{\pi}_{\text{DRT}}}{\hat{\pi}_{\text{TR}}} \tag{8.5.29}$$

The ratio estimator $\hat{T}$ is biased for T. Approximate expressions of the bias and mean-square error (MSE) can be obtained by using the formulae provided by Murthy (1967) and Arnab (2017) as follows:

$$\text{Bias}(\hat{T}) \cong \frac{T\ \text{Var}(\hat{\pi}_{\text{TR}}) - \text{Cov}(\hat{\pi}_{\text{DRT}}, \hat{\pi}_{\text{TR}})}{\pi^2} \tag{8.5.30}$$

and

$$\text{MSE}(\hat{T}) \cong \frac{\text{Var}(\hat{\pi}_{\text{DRT}}) - 2T\text{Cov}(\hat{\pi}_{\text{DRT}}, \hat{\pi}_{\text{TR}}) + T^2\text{Var}(\hat{\pi}_{\text{TR}})}{\pi^2} \tag{8.5.31}$$

The approximate unbiased estimators of $\text{Bias}(\hat{T})$ and $\text{MSE}(\hat{T})$ are given by

$$\widehat{\text{Bias}}(\hat{T}) \cong \frac{\hat{T}\ \widehat{\text{Var}}(\hat{\pi}_{\text{TR}}) - \widehat{\text{Cov}}(\hat{\pi}_{\text{DRT}}, \hat{\pi}_{\text{TR}})}{\hat{\pi}^2} \text{ and}$$

$$\widehat{\text{MSE}}(\hat{T}) \cong \frac{\widehat{\text{Var}}(\hat{\pi}_{\text{DRT}}) - 2\hat{T}\ \widehat{\text{Cov}}(\hat{\pi}_{\text{DRT}}, \hat{\pi}_{\text{TR}}) + \hat{T}^2\widehat{\text{Var}}(\hat{\pi}_{\text{TR}})}{\hat{\pi}^2}$$

where $\widehat{\text{Var}}(\hat{\pi}_{\text{DRT}})$, $\widehat{\text{Cov}}(\hat{\pi}_{\text{DRT}}, \hat{\pi}_{\text{TR}})$ and $\widehat{\text{Var}}(\hat{\pi}_{\text{TR}})$ are unbiased estimators of $\text{Var}(\hat{\pi}_{\text{DRT}})$, $\text{Cov}(\hat{\pi}_{\text{DRT}}, \hat{\pi}_{\text{TR}})$ and $\text{Var}(\hat{\pi}_{\text{TR}})$ respectively.

Expressions of the $\text{Bias}(\hat{T})$ and $\text{MSE}(\hat{T})$ are in general very complex. For the SRSWR, the expressions are obtained as follows.

Theorem 8.5.6 Under SRSWR,

(i) $\text{Bias}(\hat{T}) = \dfrac{T\overline{\tilde{\phi}}}{n\pi^2}$

(ii) $\text{MSE}(\hat{T}) = \dfrac{\pi(1-T)}{n\pi} + \dfrac{T^2\overline{\tilde{\phi}}}{n\pi^2}$

where $\overline{\tilde{\phi}} = \frac{1}{N}\sum\limits_{j \in U} \tilde{\phi}_j$.

For Warner's model, $\tilde{\phi}_j = (1 - y_i T)\phi_W = (1 - y_i T)\frac{P(1-P)}{(2P-1)^2}$. Substituting $\overline{\tilde{\phi}} = \frac{1}{N}\sum\limits_{i \in U}(1 - y_i T)\frac{P(1-P)}{(2P-1)^2} = (1 - \pi T)\frac{P(1-P)}{(2P-1)^2}$ in Theorem 8.5.6, one finds

$$\text{Bias}(\hat{T}) = \frac{T(1-\pi T)}{n\pi^2}\frac{P(1-P)}{(2P-1)^2} \text{ and } \text{MSE}(\hat{T}) = \frac{\pi(1-T)}{n\pi} + \frac{T^2(1-\pi T)}{n\pi^2}\frac{P(1-P)}{(2P-1)^2} \tag{8.5.32}$$

Remark 8.5.2 It is important to note that the expressions of Bias (8.5.31) and MSE (8.5.32) are the same as those obtained by Arnab (2017). Similarly, expressions for bias and mean-square errors for the two-stage RR model with Franklin's (1989), Kuk's (1990), and Singh and Singh's (1993) models can be obtained from Theorem 8.5.6 as special cases.

8.5.7 Comparisons of the Models

Theoretical comparisons between the different sampling strategies are, in general, very complex. However, from expressions (8.5.20) and (8.5.28), we note that irrespective of the RR techniques used in the second stage,

$$\text{Var}(\hat{\pi}_{\text{SWR}}) - \text{Var}(\hat{\pi}_{\text{WOR}}) \geq 0 \tag{8.5.33}$$

This implies that two-stage RR models with SRSWOR sampling designs are more efficient than two-stage RR models with SRSWR sampling designs.

For an IPPS (or πps) sampling design, $\pi_i = np_i (i \in U)$ where p_i is the normed size measure for the ith unit. In this case, one finds that for the several IPPS sampling designs (e.g., Hartley–Rao (1962) and Sampford (1967) sampling designs), the Horvitz-Thompson type estimator $\hat{\pi}_{\text{HT}}$ (8.5.14) becomes more efficient than the Hansen–Hurwitz (1943) estimator (8.5.24) in the sense that

$$\text{Var}(\hat{\pi}_{\text{HH}}) - \text{Var}(\hat{\pi}_{\text{HT}}) \geq 0 \tag{8.5.34}$$

Thus, the two-stage RR method with SRSWOR and IPPS sampling designs in the first stage yields a more efficient estimator than the corresponding SRSWR and PPSWR sampling designs used in stage one. Arnab (2023a) compared the performances of various sampling strategies based on real data from World Bank Enterprise Surveys. Details are omitted here.

8.6 Barabesi's Method

Barabesi (2008) proposed a two-stage RR procedure similar to the method proposed by Arnab (2023a) described above. The only difference in Barbasi (2008) is that it used a slightly different assumption than that made by Arnab (2004) n the optional RR technique (ORR). Under the assumptions, a section of respondents A_T belonging to the sensitive group A always provide a true direct response (DR), while the respondents belonging to $A_{\overline{T}}(\subset A)$ always provide a false direct response (DR). Under this model, the response obtained from the ith respondent in the sample is

$$\tilde{r}_{Bi} = y_i w_i + (1 - y_i w_i) r_i \tag{8.6.1}$$

where $w_i = 1$ for $i \in A_T$ and $w_i = 0$ for $i \in A_{\overline{T}}$, $T = \sum_{i \in A} w_i / N_A$, $N_A = \sum_{i \in A} =$ total number of persons belonging to the sensitive group A and r_i is the revised RR obtained from the ith respondent in the event he or she is provided with RR. The response $\tilde{r}_{Bi}$ follows the model analogous to (8.5.2), viz.

$$E_R(\tilde{r}_{Bi}|y_i) = y_i,\ V_R(\tilde{r}_{Bi}|y_i) = \tilde{\varphi}_{Bi} \text{ and } C_R(\tilde{r}_{Bi}, \tilde{r}_{Bj}|y_i, y_j) = 0 \text{ for } i \neq j.$$

Using Theorem (8.5.1), one finds the following results that were obtained by Barabesi (2008).

Theorem 8.6.1

(i) $\hat{\pi}_B = \frac{1}{N} \sum_{i \in s} b_{si} \tilde{r}_{Bi}$ is an unbiased estimator of π where $\sum_{s \supset i} b_{si} p(s) = 1$

(ii) The variance of $\hat{\pi}_B$ is

$$\text{Var}(\hat{\pi}_B) = \frac{1}{N^2}\left[\sum_{i \in U} (\alpha_{ii} - 1) y_i + \sum_{i \neq} \sum_{j \in U} (\alpha_{ij} - 1) y_i y_j\right] + \frac{1}{N^2}\left[\sum_{i \in U} \alpha_{ii} \tilde{\phi}_{Bi}\right]$$

(iii) An unbiased estimator of $\text{Var}(\hat{\pi}_B)$ is

$$\widehat{\text{Var}}(\hat{\pi}_B) = \frac{1}{N^2}\left[\sum_{i \in s} c_{si} \tilde{r}_{Bi}^2 + \sum_{i \neq} \sum_{j \in U} c_{sij} \tilde{r}_{Bi} \tilde{r}_{Bj}\right] + \frac{1}{N^2} \sum_{i \in s} b_{si} \hat{\tilde{\phi}}_{Bi}$$

where $\hat{\tilde{\phi}}_{Bi} = \tilde{r}_{Bi}(\tilde{r}_{Bi} - 1)$.

Barabesi (2008) also considered the estimators of π using Franklin's (1989), Kuk's (1990), and other RR procedures in stage two and showed that the proposed two-stage procedure fares better than the corresponding one-stage RR procedure that was used in stage two. Arnab and Olaomi (2017) showed that Barabesi's (2008) two-stage procedure is less efficient than the optional randomized response (ORT) procedure proposed by Arnab (2004).

8.7 Kim and Warde's Model

In Kim and Warde's (2005) model, a sample s of n respondents is selected by the SRSWR method. At first, each of the selected respondents is asked the direct question Q: "Are you a member of the innocuous trait group B?". If the answer to Q was "yes", the respondent was asked to perform the randomization device R_1. Otherwise, if the response was "no", the respondent was asked to use the randomization device R_2. The proportion of individuals in the population π_B that belong to the innocuous group B is assumed to be known. The device R_1 comprises the statements "I am a member of the sensitive group A" and "I am a member of the innocuous group B" with proportions P_1 and $(1-P_1)$, while the device R_2 comprises the statements "I am a member of the sensitive group A" and "I am not a member of the sensitive group A" with proportions P and $(1-P)$. Let n_1 and $n_2(=n-n_1)$ be the number of respondents in the sample s who answered "yes" and "no" for the direct question Q; m_1 and m_2 be the number of "yes" answers obtained from the RR devices R_1 and R_2. Denoting $\hat{\lambda}_j = \frac{m_j}{n_j}(j=1,2)$, one finds

$$E(\lambda_1) = \pi P_1 + (1-P_1) = \lambda_1 \text{ and } E(\hat{\lambda}_2) = \pi P + (1-\pi)(1-P) = \lambda_2 \quad (8.7.1)$$

where π is the proportion of persons in the population that belong to the sensitive group A.

Equation (8.7.1) yields unbiased estimators of π from the devices R_1 and R_2 are

$$\hat{\pi}_{\text{K1}} = \frac{\hat{\lambda}_1 - (1-P_1)}{P_1} \text{and } \hat{\pi}_{\text{K2}} = \frac{\hat{\lambda}_2 - (1-P)}{2P-1} \quad (8.7.2)$$

The variances of $\hat{\pi}_{\text{K1}}$ and $\hat{\pi}_{\text{K2}}$ are

$$\text{Var}(\hat{\pi}_{\text{K1}}) = \frac{\lambda_1(1-\lambda_1)}{n_1 P_1^2} = \frac{(1-\pi)[P_1\pi + (1-P_1)]}{n_1 P_1} \quad (8.7.3)$$

and

$$\text{Var}(\hat{\pi}_{\text{K2}}) = \frac{\lambda_2(1-\lambda_2)}{n_2 P^2} = \frac{\pi(1-\pi)}{n-n_1} + \frac{P(1-P)}{(n-n_1)(2P-1)^2} \quad (8.7.4)$$

Kim and Warde (2005) proposed a combined estimator of π as

$$\hat{\pi}_{\mathrm{K}} = \frac{n_1}{n}\hat{\pi}_{\mathrm{K}1} + \frac{n-n_1}{n}\hat{\pi}_{\mathrm{K}2} \tag{8.7.5}$$

Equations (8.7.2) to (8.7.5) yield.

Theorem 8.7.1

(i) $\hat{\pi}_{\mathrm{K}}$ is an unbiased estimator of π
(ii) The variance of $\hat{\pi}_{\mathrm{K}}$ is

$$\mathrm{Var}(\hat{\pi}_{\mathrm{K}}) = \frac{\lambda_2(1-\lambda_2)}{n_2P^2} = \frac{\pi(1-\pi)}{n-n_1} + \frac{P(1-P)}{(n-n_1)(2P-1)^2}$$

To maintain an equal level of privacy protection for the two RR devices R_1 and R_2, Kim and Warde (2005), recommended the value of P following Lanke (1976) as

$$P = \frac{1}{2} + \frac{P_1}{2P_1 + 4(1-P_1)} = \frac{1}{2-P_1} \tag{8.7.6}$$

Substituting $P = \frac{1}{2-P_1}$ in the expression of variance $\mathrm{Var}(\hat{\pi}_{\mathrm{K}})$, Kim and Warde (2005) obtained

$$\mathrm{Var}(\hat{\pi}_{\mathrm{K}}) = \frac{\pi(1-\pi)}{n} + \frac{(1-P_1)[wP(1-\pi)+(1-w)]}{(n-n_1)(2P-1)^2} \tag{8.7.7}$$

where $w = n_1/n$.

It was originally noted by Kim and Warde (2005) and later by Chaudhrui (2011) that the expression of the variance (8.7.7) is incomplete because of the involvement of the random variable w. Hence, the relative efficiency of Kim and Warde's model compared with Moors's (1971) model is inappropriate. None of the authors, Kim and Warde (2005) and Chaudhuri (2011), provided an unbiased estimator of the variance. Further, the proposed model is limited only to SRSWR sampling. To overcome such difficulties, the following modification is proposed by Arnab (2023b):

Exact Expression of Variance

Let $x_i = \begin{cases} 1, \text{ if } i\text{th individual} \in B \\ 0, \text{ if } i\text{th individual} \notin B \end{cases}$, $y_i = \begin{cases} 1, \text{ if } i\text{th individual} \in A \\ 0. \text{ if } i\text{th individual} \notin A \end{cases}$,

$$z_{i1} = \begin{cases} 1, \text{ if } i\text{th resondent of the trial } R_1 \text{ answers "yes"} \\ 0, \text{ if } i\text{th resondent of the trial } R_1 \text{ answers "no"} \end{cases} \text{ and}$$

$$z_{i2} = \begin{cases} 1, & \text{if } i\text{th resondent of the trial } R_2 \text{ answers "yes"} \\ 0, & \text{if } i\text{th resondent of the trial } R_2 \text{ answers "no"} \end{cases} \tag{8.7.8}$$

Then,

$$\begin{aligned} &E_R(z_{i1}) = P_1 y_i + (1 - P_1)x_i = E_R(z_{i1}^2), E_R(z_{i2}) = Py_i + (1 - P)(1 - y_i) = E_R(z_{i2}^2), \\ &V_R(z_{i1}) = \phi_i(1) = P_1 y_i + (1 - P_1)x_i - \{P_1 y_i + (1 - P_1)x_i\}^2 = P_1(1 - P_1)(x_i - y_i)^2 \text{ and} \\ &V_R(z_{i2}) = \phi_i(2) = Py_i + (1 - P)(1 - y_i) - \{Py_i + (1 - P)(1 - y_i)\}^2 = P(1 - P) \end{aligned} \tag{8.7.9}$$

Further, defining $r_{i1} = \frac{z_{i1} - (1 - P_1)x_i}{P_1}$ and $r_{i2} = \frac{z_{i2} - (1 - P)}{2P - 1}$, one finds

$$\begin{aligned} E_R(r_{i1}) = y_i,\ V_R(r_{i1}) &= \frac{(1 - P_1)(x_i - y_i)^2}{P_1} = \phi_i(1), \\ E_R(r_{i2}) = y_i \text{ and } V_R(r_{i2}) &= \frac{P(1 - P)}{(2P - 1)^2} = \phi_i(2) \end{aligned} \tag{8.7.10}$$

Now writing

$$r_i = x_i r_{i1} + (1 - x_i) r_{i2} \tag{8.7.11}$$

the following model is obtained:

$$\begin{aligned} E_R(r_i) &= x_i E_R(r_{i1}) + (1 - x_i) E_R(r_{i2}) = y_i \\ \text{and } V_R(r_i) &= x_i \phi_i(1) + (1 - x_i)\phi_i(2) = \phi_{\mathrm{K}i} \end{aligned} \tag{8.7.12}$$

SRSWR Sampling

For the SRSWR method, the proposed estimator of π is

$$\hat{\pi}_{\mathrm{K}o} = \frac{1}{n} \sum_{i \in s} r_i \tag{8.7.13}$$

where $\sum_{i \in s}$ denotes the sum over all the n units in s.

Theorem 8.7.2

(i) $\hat{\pi}_{Ko}$ is an unbiased estimator of π
(ii) The variance of $\hat{\pi}_{Ko}$ is

$$\mathrm{Var}(\hat{\pi}_{Ko}) = \frac{\pi(1-\pi)}{n} + \frac{\overline{\phi}_K}{n}$$

where $\overline{\phi}_K = \frac{P(1-P)}{(2P-1)^2} + \pi_B\left\{\frac{(1-P_1)(1-\pi)}{P_1} - \frac{P(1-P)}{(2P-1)^2}\right\}$
(iii) An unbiased estimator of $\mathrm{Var}(\hat{\pi}_{Ko})$ is

$$\widehat{\mathrm{Var}}(\hat{\pi}_{Ko}) = \frac{1}{n(n-1)}\sum\nolimits_{i\in s}(r_i - \hat{\pi}_{Ko})^2.$$

Proof

(i)
$$\begin{aligned}E(\hat{\pi}_{Ko}) &= E_p\left[\frac{1}{n}\sum\nolimits_{i\in s}E_R(r_i)\right]\\ &= E_p\left[\frac{1}{n}\sum\nolimits_{i\in s}y_i\right]\\ &= \frac{1}{N}\sum_{i\in U}y_i\\ &= \pi\end{aligned}$$

(ii) The variance of $\hat{\pi}_{Ko}$ is

$$\begin{aligned}\mathrm{Var}(\hat{\pi}_{Ko}) &= V_p[E_R(\hat{\pi}_{Ko})] + E_p[V_R(\hat{\pi}_{Ko})]\\ &= V_p\left(\frac{1}{n}\sum\nolimits_{i\in s}y_i\right) + E\left(\frac{1}{n}\sum\nolimits_{i\in s}\phi_{Ki}\right)\\ &= \frac{\pi(1-\pi)}{n} + \frac{1}{nN}\sum_{i\in U}\phi_{Ki}\end{aligned} \tag{8.7.14}$$

The second component of the above Eq. (8.7.14) is

$$\begin{aligned}\frac{1}{nN}\sum_{i\in U}\phi_{Ki} &= \frac{1}{nN}\sum_{i\in U}\left\{x_i\frac{(1-P_1)(x_i+y_i-2x_iy_i)}{P_1} + (1-x_i)\frac{P(1-P)}{(2P-1)^2}\right\}\\ &= \frac{1}{n}\left\{\frac{(1-P_1)(\pi_B-\pi_{AB})}{P_1} + \frac{P(1-P)}{(2P-1)^2}(1-\pi_B)\right\}\end{aligned}$$

(where $\pi_{AB} = \frac{1}{N}\sum\limits_{i\in U}x_iy_i$ = proportion of the persons belong to both the groups A and B).

Since the attributes A and B are independent, we write $\pi_{\mathrm{AB}} = \pi\pi_B$ and in this case

$$\overline{\phi}_{\mathrm{K}} = \frac{1}{N}\sum_{i\in U}\phi_{\mathrm{K}i} = \frac{P(1-P)}{(2P-1)^2} + \pi_{\mathrm{B}}\left\{\frac{(1-P_1)(1-\pi)}{P_1} - \frac{P(1-P)}{(2P-1)^2}\right\} \tag{8.7.15}$$

Finally, Eqs. (8.7.14) and (8.7.15) yield $\mathrm{Var}(\hat{\pi}_{\mathrm{K}o}) = \frac{\pi(1-\pi)}{n} + \frac{\overline{\phi}_{\mathrm{K}}}{n}$.

(iii) Since r_i s are independently and identically distributed random variables, an unbiased estimator of $\mathrm{Var}(\hat{\pi}_{\mathrm{K}o})$ is $\widehat{\mathrm{Var}}(\hat{\pi}_{\mathrm{K}o}) = \frac{1}{n(n-1)}\sum_{i\in s}(r_i - \hat{\pi}_{\mathrm{K}o})^2$.

General Sampling Design

Let a sample s of n distinct units be selected from a finite population U by using an arbitrary sampling design p with inclusion probabilities π_i and π_{ij} for the ith, and ith and $j(\neq i)$th unit. Then the Horvitz-Thompson estimator for the population proportion is given by

$$\hat{\pi}_{\mathrm{HT}} = \frac{1}{N}\sum_{i\in s}\frac{r_i}{\pi_i} \tag{8.7.16}$$

Properties of $\hat{\pi}_{\mathrm{HT}}$ are given as follows.

Theorem 8.7.3

(i) $\hat{\pi}_{\mathrm{HT}}$ is an unbiased estimator of π

(ii) The variance of $\hat{\pi}_{\mathrm{HT}}$ is

$$\mathrm{Var}(\hat{\pi}_{\mathrm{HT}}) = \frac{1}{N^2}\left[\sum_{i<}\sum_{j\in U}(\pi_i\pi_j - \pi_{ij})\left(\frac{y_i}{\pi_i} - \frac{y_j}{\pi_j}\right)^2 + \sum_{i\in U}\frac{\phi_{\mathrm{K}i}}{\pi_i}\right]$$

(iii) An unbiased estimator of $\mathrm{Var}(\hat{\pi}_{\mathrm{HT}})$ is

$$\widehat{\mathrm{Var}}(\hat{\pi}_{\mathrm{HT}}) = \frac{1}{N^2}\left[\sum_{i<}\sum_{j\in s}\frac{\pi_i\pi_j - \pi_{ij}}{\pi_{ij}}\left(\frac{r_i}{\pi_i} - \frac{r_j}{\pi_j}\right)^2 + \sum_{i\in s}\frac{\hat{\hat{\phi}}_{\mathrm{K}i}}{\pi_i}\right]$$

where $\hat{\hat{\phi}}_{\mathrm{K}i} = r_i(r_i - 1)$.

Proof

(i) $$E(\hat{\pi}_{\mathrm{HT}}) = \frac{1}{N}E_p\left(E_R\sum_{i\in s}\frac{r_i}{\pi_i}\right)$$
$$= \frac{1}{N}\sum_{i\in U} y_i$$
$$= \pi$$

(ii) $$\mathrm{Var}(\hat{\pi}_{\mathrm{HT}}) = V_p[E_R(\hat{\pi}_{\mathrm{HT}})] + E_p[V_R(\hat{\pi}_{\mathrm{HT}})]$$
$$= \frac{1}{N^2}\left[V_p\left(\sum_{i\in s}\frac{y_i}{\pi_i}\right) - E_p\sum_{i\in s}\frac{\phi_{\mathrm{K}i}}{\pi_i^2}\right]$$
$$= \frac{1}{N^2}\left[\sum\sum_{i<j\in U}(\pi_i\pi_j - \pi_{ij})\left(\frac{y_i}{\pi_i} - \frac{y_j}{\pi_j}\right)^2 + \sum_{i\in U}\frac{\phi_{\mathrm{K}i}}{\pi_i}\right]$$

(iii) $$\frac{1}{N^2}E\left[\sum\sum_{i<j\in s}\frac{\pi_i\pi_j - \pi_{ij}}{\pi_{ij}}\left(\frac{r_i}{\pi_i} - \frac{r_j}{\pi_j}\right)^2 + \sum_{i\in s}\frac{\hat{\phi}_{\mathrm{K}i}}{\pi_i}\right]$$
$$= \mathrm{Var}(\hat{\pi}_{\mathrm{HT}})$$

Remarks 8.7.1 Kim and Warde (2005) applied the mixed RR model for the stratified sampling and derived the stratum sizes under Neyman's allocation. They also compared the efficiency of the mixed model with respect to the Moors (1971) model numerically. The proposed comparisons, however, are not valid because the derived variance formula is incorrect because it involves random sample sizes. Following Kim and Warde (2005), Singh and Tarray (2013) proposed a few alternative mixed RR models and applied them to stratified sampling. The optimal sample sizes derived by them are also incorrect, as their variance formulas involve random variables. The optimal sample sizes derived by them are also incorrect because they involve random variables.

8.8 Quatember RR Techniques for More Than Two Stages

Quatember (2012) proposed a randomization procedure involving several stages $h(\geq 1)$. In stage-1, the selected respondent is asked to either answer "yes" or "no" to the question Q: "Are you a member of A?" with probability P_{11} or to go to stage-2 with probability $1 - P_{11}$. In stage-2, the respondent is directed either to answer the same question Q: "Are you a member of A?" with the probability P_{21} or go to stage-3 with probability $1 - P_{21}$. The procedure is repeated untill a predetermined number of stage say, h is reached. At the final stage h, the respondent is directed to respond by using

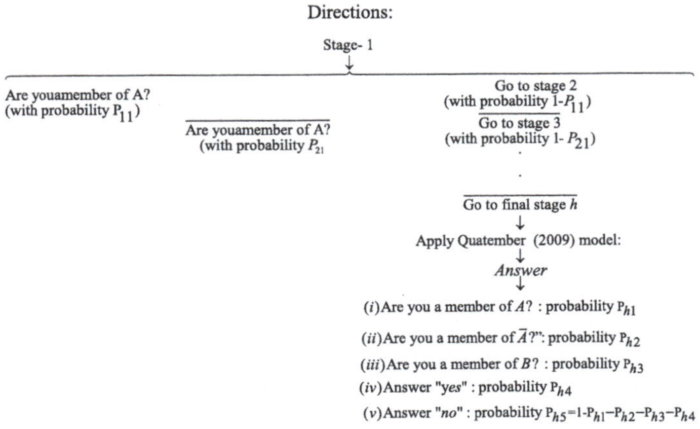

Fig. 8.2 Quatember (2012) method of stage h

Quatember (2009) RR method (see Sect. 5.6.5) with parameters $P_{h1}, P_{h2}, P_{h3}, P_{h4}$ and $P_{h5} = 1 - P_{h1} - P_{h2} - P_{h3} - P_{h4}$, meaning that a respondent should answer either Q: "Are you a member of A?" with probability P_{h1}; or $\overline{Q}$: "Are you a member of $\overline{A}$?" with probability P_{h2}; or Q_B: "Are you a member of innocuous group B?" with probability P_{h3}; or answer "yes" with probability P_{h4}; or finally answer "no" with probability P_{h5} (Fig. 8.2).

Let y_i, x_i and z_{hi} denote respectively value of the sensitive characteristic y, innocuous variable B and randomized response obtained at stage h from the ith unit of the population U. Define

$$y_i = \begin{cases} 1 \text{ for } i \in A \\ 0 \text{ for } i \notin A \end{cases}, \ x_i = \begin{cases} 1 \text{ for } i \in B \\ 0 \text{ for } i \notin B \end{cases} \text{ and } z_{hi} = \begin{cases} 1 \text{ for answer "yes"} \\ 0 \text{ for answer "no"} \end{cases} \tag{8.8.1}$$

so that $\overline{X} = \frac{1}{N} \sum\limits_{i \in U} x_i = \pi_{\text{B}}$ and $\overline{Y} = \frac{1}{N} \sum\limits_{i \in U} y_i = \pi_A$.

Then,

$$\begin{aligned}
\text{Prob}\ (z_{hi} = 1) &= p_{11}y_i + (1 - p_{11})p_{21}y_i + (1 - p_{11})(1 - p_{21})p_{31}y_i + \cdots \\
&\quad + (1 - p_{11})(1 - p_{21}) \ldots (1 - p_{h-11})p_{h-1,1}y_i \\
&\quad + (1 - p_{11})(1 - p_{21}) \ldots (1 - p_{h-11}) \\
&\quad [p_{h1}y_i + p_{h2}(1 - y_i) + p_{h3}x_i + p_{h4}] \\
&= [p_{11} + (1 - p_{11})p_{21} + (1 - p_{11})(1 - p_{21})p_{31} + \cdots \\
&\quad + (1 - p_{11})(1 - p_{21}) \ldots (1 - p_{h-11})(p_{h1} - p_{h2})]y \\
&\quad + (1 - p_{11})(1 - p_{21}) \ldots (1 - p_{h-11}) \\
&\quad \{p_{h2} + p_{h3}(x_i - \pi_B) + p_{h3}\pi_B + p_{h4}\} \\
&= u_h y_i + w_h \tilde{x}_i + v_h; \quad \tilde{x}_i = (x_i - \pi_B)
\end{aligned}$$

$$= E_R(z_{hi}) \tag{8.8.2}$$

where

$$\begin{aligned} u_h &= p_{11} + (1-p_{11})p_{21} + (1-p_{11})(1-p_{21})p_{31} + \ldots \\ &\quad + (1-p_{11})(1-p_{21})\ldots(1-p_{h-11})(p_{h1}-p_{h2}) \\ v_h &= (1-p_{11})(1-p_{21})\ldots(1-p_{h-11})\{p_{h2} + p_{h3}\pi_B + p_{h4}\} \\ w_h &= (1-p_{11})(1-p_{21})\ldots(1-p_{h-11})p_{h3} \end{aligned}$$

$$\begin{aligned} V_R(z_{hi}) &= E_R(z_{hi}^2) - (u_h y_i + w_h \tilde{x}_i + v_h)^2 \\ &= u_h y_i + w_h \tilde{x}_i + v_h - \left\{u_h^2 y_i + w_h^2 \tilde{x}_i^2 + v_h^2 + 2(u_h w_h y_i \tilde{x}_i + u_h v_h y_i + v_h w_h \tilde{x}_i)\right\} \\ &= u_h\{(1-u_h) - 2v_h\} y_i + w_h \tilde{x}_i (1 - w_h \tilde{x}_i) \\ &\quad + v_h(1-v_h) - 2(u_h w_h y_i \tilde{x}_i + v_h w_h \tilde{x}_i) \\ &= \phi_{zhi} \end{aligned} \tag{8.8.3}$$

Writing $r_{hi} = \frac{z_{hi} - v_h}{u_h}$, an unbiased estimator of y_i, one obtains the following RR model

$$E_R(r_{hi}) = y_i + \alpha_h \tilde{x}_i,\ V_R(r_{hi}) = \phi_{Qhi} \text{ and } C_R(r_{hi}, r_{hj}) = 0 \text{ for } i \neq j \tag{8.8.4}$$

where $\alpha_h = w_h/u_h$ and $\phi_{Qhi} = \phi_{zhi}/u_h^2$.

An obvious unbiased estimator of ϕ_{Qhi} is

$$\hat{\phi}_{Qai} = r_{hi}(r_{hi} - 1).$$

Estimation of π_A

General Sampling Scheme

Let a sample of fixed effective size n is selected from the population U by an arbitrary sampling design p with the inclusion probabilities π_i and π_{ij} for the ith, and ith and jth $(i \neq j)$ units respectively. Then the Horvitz-Thompson type estimator of π_A is

$$\hat{\pi}_{AHT} = \frac{1}{N} \sum_{i \in s} \frac{r_{hi}}{\pi_i} \tag{8.8.5}$$

Properties of the estimator $\hat{\pi}_{AHT}$ are as follows.

Theorem 8.8.1

(i) $\hat{\pi}_{\text{AHT}}$ is an unbiased estimator of π_{A}
(ii) The variance of $\hat{\pi}_{\text{AHT}}$ is

$$\text{Var}(\hat{\pi}_{\text{AHT}}) = \frac{1}{N^2}\left[V_{\text{HT}}(y) + \alpha_h^2 V_{\text{HT}}(\tilde{x}) + \sum_{i\in U}\frac{\phi_{\text{Q}hi}}{\pi_i}\right]$$

(iii) An unbiased estimator of $\text{Var}(\hat{\pi}_{\text{AHT}})$ is

$$\widehat{\text{Var}}(\hat{\pi}_{\text{AHT}}) = \frac{1}{N^2}\left[\sum_{i<}\sum_{j\in s}\frac{\pi_i\pi_j - \pi_{ij}}{\pi_{ij}}\left(\frac{r_{hi}}{\pi_i} - \frac{r_{hj}}{\pi_j}\right)^2 + \sum_{i\in s}\frac{\hat{\phi}_{\text{Q}hi}}{\pi_i}\right]$$

where $V_{\text{HT}}(d) = \sum_{i<}\sum_{j\in U}(\pi_i\pi_j - \pi_{ij})\left(\frac{d_i}{\pi_i} - \frac{d_j}{\pi_j}\right)^2$; $d = \tilde{x}, y$ and $\hat{\phi}_{\text{Q}hi}$ is as in (8.8.5).

Proof

(i)
$$\begin{aligned}E(\hat{\pi}_{\text{AHT}}) &= E_p\left[\frac{1}{N}\sum_{i\in s}\frac{E_R(r_{hi})}{\pi_i}\right]\\
&= E_p\left[\frac{1}{N}\sum_{i\in U}\frac{y_i + \alpha_h\tilde{x}_i}{\pi_i}\right]\\
&= \overline{Y}\ \left(\text{since } \frac{1}{N}\sum_{i\in U}\tilde{x}_i = 0\right)\\
&= \pi_{\text{A}}\end{aligned}$$

(ii)
$$\begin{aligned}\text{Var}(\hat{\pi}_{\text{AHT}}) &= V_p\left[\frac{1}{N}\sum_{i\in s}\frac{E_R(r_{hi})}{\pi_i}\right] + E_p\left[\frac{1}{N^2}\sum_{i\in s}\frac{V_R(r_{hi})}{\pi_i^2}\right]\\
&= V_p\left[\frac{1}{N}\sum_{i\in s}\frac{y_i + \alpha_h\tilde{x}_i}{\pi_i}\right] + E_p\left[\frac{1}{N^2}\sum_{i\in s}\frac{\phi_{\text{Q}hi}}{\pi_i^2}\right]\\
&= V_p\left[\frac{1}{N}\sum_{i\in s}\frac{y_i + \alpha_h\tilde{x}_i}{\pi_i}\right] + E_p\left[\frac{1}{N^2}\sum_{i\in s}\frac{\phi_{\text{Q}hi}}{\pi_i^2}\right]\\
&= \left[V_p\left(\frac{1}{N}\sum_{i\in U}\frac{y_i}{\pi_i}\right) + \alpha_h^2 V_p\left(\frac{1}{N}\sum_{i\in U}\frac{\tilde{x}_i}{\pi_i}\right)\right.\\
&\quad \left. + 2\alpha_h C_p\left(\frac{1}{N}\sum_{i\in U}\frac{y_i}{\pi_i}, \frac{1}{N}\sum_{i\in U}\frac{\tilde{x}_i}{\pi_i}\right)\right] + \frac{1}{N^2}\sum_{i\in U}\frac{\phi_{\text{Q}hi}}{\pi_i}\end{aligned}$$

Assuming $\tilde{x}_i$ and y_i are independent, one finds $C_p\left(\frac{1}{N}\sum_{i\in U}\frac{y_i}{\pi_i}, \frac{1}{N}\sum_{i\in U}\frac{\tilde{x}_i}{\pi_i}\right) = 0$ and

$$\begin{aligned}
\mathrm{Var}(\hat{\pi}_{\mathrm{AHT}}) &= \frac{1}{N^2}\Bigg[\sum_{i<}\sum_{j\in U}(\pi_i\pi_j - \pi_{ij}) \\
&\quad \left\{\left(\frac{y_i}{\pi_i}-\frac{y_i}{\pi_j}\right)^2 + \alpha_h^2\left(\frac{x_i}{\pi_i}-\frac{x_i}{\pi_j}\right)^2\right\} + \sum_{i\in U}\frac{\phi_{Qhi}}{\pi_i}\Bigg] \\
&= \frac{1}{N^2}\left[V_{HT}(y) + \alpha^2 V_{HT}(x) + \sum_{i\in U}\frac{\phi_{hQi}}{\pi_i}\right]
\end{aligned}$$

(iii) $E[\hat{\mathrm{Var}}(\hat{\pi}_{\mathrm{AHT}})] = E_p[E_R\{\hat{\mathrm{Var}}(\hat{\pi}_{\mathrm{AHT}})\}]$

$$\begin{aligned}
&= \frac{1}{N^2}E_p\left[\sum_{i<}\sum_{j\in s}\frac{\pi_i\pi_j - \pi_{ij}}{\pi_{ij}}E_R\left(\frac{r_{hi}}{\pi_i}-\frac{r_{hj}}{\pi_j}\right)^2 + \sum_{i\in s}\frac{E_R(\hat{\phi}_{Qhi})}{\pi_i}\right] \\
&= \frac{1}{N^2}E_p\Bigg[\sum_{i<}\sum_{j\in s}\frac{\pi_i\pi_j - \pi_{ij}}{\pi_{ij}}\Bigg\{\left(\frac{y_i+\alpha_h\tilde{x}_i}{\pi_i}-\frac{y_j-\alpha_h\tilde{x}_j}{\pi_j}\right)^2 \\
&\quad + \left(\frac{\phi_{Qhi}}{\pi_i^2}+\frac{\phi_{Qhj}}{\pi_j^2}\right)\Bigg\} + \sum_{i\in s}\frac{\phi_{Qhi}}{\pi_i}\Bigg] \\
&= \mathrm{Var}(\hat{\pi}_{\mathrm{AHT}}) \quad \left(\text{using } \sum_{j(\neq i)\in U}\pi_{ij} = (n-1)\pi_i\right)
\end{aligned}$$

SRSWR Sampling Scheme

For the SRSWR sampling scheme of size n, an unbiased estimator of π_A is

$$\hat{\pi}_{\mathrm{Aw}} = \frac{1}{n}\sum_{i=1}^{n} r_{hi} \tag{8.8.6}$$

In this case we have the following results.

Theorem 8.8.2

(i) $\hat{\pi}_{\mathrm{AW}}$ is an unbiased estimator of π_{A}
(ii) The variance of $\hat{\pi}_{\mathrm{AW}}$ is

$$\mathrm{Var}(\hat{\pi}_{\mathrm{Aw}}) = \frac{1}{n}\left[\pi_{\mathrm{A}}(1-\pi_{\mathrm{A}}) + \alpha_h^2\pi_{\mathrm{B}}(1-\pi_{\mathrm{B}}) + \overline{\phi}_{Qh}\right]$$

where $\overline{\phi}_{Qh} = \frac{1}{N}\sum_{i\in U}\phi_{Qhi}$

(iii) An unbiased estimator of $\text{Var}(\hat{\pi}_{AHT})$ is

$$\widehat{\text{Var}}(\hat{\pi}_{Aw}) = \frac{1}{n(n-1)}\sum_{i=1}^{n}(r_{hi} - \hat{\pi}_{Aw})^2$$

Proof

(i) $$\begin{aligned} E(\hat{\pi}_{AW}) &= E_p[E_R(\hat{\pi}_{AW})] \\ &= E_p\left[\frac{1}{n}\sum_{i=1}^{n}(y_i + \alpha_h\tilde{x}_i)\right] \\ &= \pi_A \end{aligned}$$

(ii) $$\begin{aligned} \text{Var}(\hat{\pi}_{AW}) &= V_p\left[\frac{1}{n}\sum_{i=1}^{n}(y_i + \alpha_h\tilde{x}_i)\right] + E_p\left(\frac{1}{n^2}\sum_{i=1}^{n}\phi_{hQi}\right) \\ &= \frac{1}{n}\left[\pi_A(1-\pi_A) + \alpha_h^2\pi_B(1-\pi_B) + \overline{\phi}_{hQ}\right] \end{aligned}$$

(assuming x and y are independent).

(iii) Results follow from the fact that r_{hi} s are iid random variables.

Remark 8.8.1 For $h = 1$, $u_1 = p_{11} - p_{12}$, $v_1 = p_{12} + p_{13}\pi_B + p_{14}$, $w_1 = p_{13}$,

$$\phi_{z1i} = u_1\{(u_1-1) - 2v_1\}y_i + w_1\tilde{x}_i(1 - w_1\tilde{x}_i) + v_1(1-v_1) - 2(u_1w_1y_i\tilde{x}_i + v_1w_1\tilde{x}_i)$$

$$\overline{\phi}_{Q1} = \frac{1}{u_1^2}\left[u_1\{(1-u_1) - 2v_1\}\pi_A - w_1^2\pi_B(1-\pi_B) + v_1(1-v_1)\right]$$

Under the SRSWR sampling, $\hat{\pi}_{Aw}$ reduces to $\hat{\pi}_{Aw} = \frac{1}{n}\sum_{i=1}^{n} r_{1i}$ where $r_{1i} = \frac{z_{1i}-v_1}{u_1}$. In this case, the expression of variance of $\hat{\pi}_{Aw}$ becomes

$$\text{Var}(\hat{\pi}_{Aw}) = \frac{1}{n}\left[\pi_A(1-\pi_A) + \overline{\phi}_{1Q}\right] \tag{8.8.7}$$

The expression (8.8.7) was obtained by Quatember (2009).

Remark 8.8.2 The expression of $E_R(z_{hi})$ that was obtained by Quatember (2012), Chaudhuri and Christofied (2013) is incorrect because it depends only on y_i but is independent of x_i. Hence, the expressions of variance and their unbiased estimators provided by Quatember for the general sampling design are also incorrect. But, for $h = 1$ and SRSWR sampling, the expression of variance $\text{Var}(\hat{\pi}_{AW})$ provided in Theorem 8.8.2 is the same as was obtained by Quatember (2009).

8.9 Discussions

Chang and Huang (2001) proposed the two-stage RR technique for estimating π, the proportion of individuals in a population belonging to a certain sensitive group A as well T, as the sensitivity level of the question Q, which is the proportion of persons belonging to A that provide true responses in direct questioning. The procedure was based on two independent samples selected by the SRSWR method Huang (2004) and Marcheselli and Barabesi (2006) extended Chang and Huang's procedure by introducing the two-stage RR procedure based on a single sample. In both procedures, at the first stage, a sample is selected by the SRSWR method, and each of the selected respondents is asked a direct question about whether they belong to the sensitive group. If the answer was "yes", the response was treated as true and retained. In case, a respondent answered "no", the respondent was asked to provide a RR using Warner's (1965) method. Huang (2004) showed that his proposed method is more efficient than the single-stage RR procedure. Marcheselli and Barabesi (2006) generalized Huang's (2004) method further by replacing Warner's (1965) model with Kuk's (1990), Franklin's (1989), and Singh and Singh's (1993) models for the selection of samples in the second stage. They showed that their proposed two-stage procedures are more efficient than the corresponding one-stage RR procedure. Marcheselli and Barabesi (2006) showed empirically that their proposed two-stage procedure performed better than Huang's (2004) model for some chosen values of parameters, but no theoretical result was obtained in support of the results. Arnab (2023a) generalized the existing methods by selecting the initial sample in the first stage using an arbitrary sampling design and applying a general RR technique that satisfies the general RR model in the second stage. Arnab (2023a) showed that the proposed two-stage model fares better than the corresponding single-stage RR procedure. Arnab's (2023a) procedure is applicable to a wider class of sampling designs and estimators. Hence, the results derived by Huang (2004) and Barbaresi et al. (2006) become special cases of the proposed method. It is also shown that the two-stage RR model with SRSWOR and IPPS sampling designs at the first stage provides more efficient estimators than the SRSWR and PPSWR sampling designs, irrespective of the RR model used in the second stage. However, the main drawback of the two-stage RR procedure is that a section of people who answer 'no" to a direct question are required to respond again by using the RR device, which may cause a reduction in cooperation from the respondents. Kim and Warde (2005) proposed an alternative two-stage procedure where each respondent is required to perform one of the two randomization devices depending on the answer to the direct question "Do you belong to sensitive group A?". However, the proposed formula for the variance was incorrect. The corrected version is presented in Sect. 8.7. Quatember (2012) proposed RR models for more than two stages that cover many RR models. Although the proposed RR procedure protects the confidentiality of the responses, it cannot be used in practice because of its complexity. The respondents may be prone to responding incorrectly because of a misunderstanding of the randomization procedure.

References

Arnab, R. (1990). On commutativity of design and model expectations in randomized response surveys. *Communications in Statistics—Thesory and Methods, 19*, 3751–3757.

Arnab, R. (1994). Non-negative variance estimation in randomized response surveys. *Communications in Statistics-Theory and Methods, 23*, 1743–1752.

Arnab, R. (1996). Randomized response trials: A unified approach for qualitative data. *Communications in Statistics-Theory and Methods, 25*, 1173–1183.

Arnab, R. (2004). Optional randomized response techniques for complex survey designs. *Biometrical Journal, 46*, 114–124.

Arnab, R. (2017). *Survey sampling theory and applications*. Academic Press.

Arnab, R. (2023a). Two stage randomized response technique. Unpublished manuscript.

Arnab, R. (2023b). Two stage stratified randomized response technique. Unpublished manuscript.

Arnab, R., & Olaomi, O. J. (2017). Optional randomized response technique in two-phase sampling. *Communications in Statistics-Theory and Methods, 46*, 8973–8986.

Barabesi, L. (2008). A design based randomized response procedure for the estimation of population proportion and sensitivity level. *Journal of Statistical Planning and Inference, 138*, 2398–2408.

Barabesi, L., & Marcheselli, M. (2006). A practical implementation and Bayesian estimation in Franklin's randomized response procedure. *Communications in Statistics-Simulation and Computations, 35*, 563–573.

Chang, H. J., & Huang, K. C. (2001). Estimation of proportion and sensitivity of a qualitative character. *Metrika, 53*, 269–280.

Chaudhuri, A. (2011). *Randomized response and indirect questioning techniques in surveys*. Chapman & Hall, CRC Press, Taylor & Francis Group.

Chaudhuri, A., & Christofides, T. (2013). *Indirect questioning in sample surveys*. Springer.

Christofides, T. C. (2003). A generalized randomized response technique. *Metrika, 57*, 195–200.

Franklin, L. A. (1989). A comparison of estimators for randomized response sampling with continuous distribution from dichotomous populations. *Communications in Statistics-Theory and Methods, 18*, 489–505.

Hansen, M. H., & Hurwitz, W. N. (1943). On the theory of sampling from finite populations. *Annals of the Mathematical Statistics, 14*, 333–362.

Hartley, H. O., & Rao, J. N. K. (1962). Sampling with unequal probabilities and without replacement. *Annals of the Mathematical Statistics, 33*, 350–374.

Horvitz, D. G., & Thompson, D. J. (1952). A generalization of sampling without replacement from a finite population. *Journal of the American Statistical Association, 47*, 663–685.

Horvitz, D. G., Shah, B. V., & Simmons, W. R. (1967). The unrelated question randomized response model. In *Proceedings of the Survey Research Methods Section of the American Statistical Association* (pp. 65–72).

Huang, K. C. (2004). A survey technique for estimating the proportion and sensitivity in a dichotomous finite population. *Statistica Neeralandica, 58*, 75–82.

Kim and Warde. (2005). A mixed randomized response model. *Journal of Statistical Planning and Inference, 133*, 211–221.

Kuk, A. Y. C. (1990). Asking sensitive question indirectly. *Biometrika, 77*, 436–438.

Lanke, J. (1976). On the degree of protection in randomized interviews. *International Statistical Review., 44*, 197–203.

Mangat, N. S., & Singh, R. (1990). An alternative randomized response procedure. *Biometrika, 77*, 349–442.

Marcheselli, M., & Barabesi, L. (2006). A generalization of Huang's randomized response procedure for the estimation of population proportion and sensitivity level. *Metron, LXIV*, 145–159.

Moors, J. A. (1971). Optimization of the unrelated question randomized response model. *Journal of the American Statistical Association, 66*, 627–629.

Murthy, M. N. (1967). *Sampling theory and methods*. Statistical Publishing Society.

Sampford, M. R. (1967). On sampling without replacement with unequal probability selection. *Biometrika, 67*, 639–650.
Singh, S., & Singh, R. (1993). Generalized Franklin's method for randomized response sampling. *Communications in Statistics-Theory and Methods, 22*, 741–755.
Singh, H.P., and Tarray, A. (2013). An alternative to Kim and Warde's mixed randomized response model. *Statistica,* anno LXXIII, 379–401.
Quatember, A. (2009). A standardization of randomized response strategies. *Survey Methodology, 35*, 143–212.
Quatember, A. (2012). An extension of the standardized response technique to a multi-stage setup. *Statistical Methods and Applications, 21*, 475–484.
Warner, S. L. (1965). Randomize response: A survey technique for eliminating evasive answer bias. *Journal of the American Statistical Association, 60*, 63–69.
Yates, F., & Grundy, P. M. (1953). Selection without replacement from within strata with probability proportional to size. *Journal of the Royal Statistical Society, Series B, 15*, 235–261.

Chapter 9
Optional Randomized Response Techniques

Abstract The optional randomized response technique (ORT) is used when most respondents feel that the subject of the inquiry is sensitive. Still, a minority feels that the inquiry is not sensitive enough and wishes to provide a direct response (DR). A randomized response technique that allows respondents to choose one of the two options: to use RR or DR for providing their responses. ORTs are classified into three categories: full, partial, and mixed. Estimators of population characteristics, their variances, and unbiased estimators of the variances are proposed for the various ORTs under different sampling designs and assumptions. Comparisons of the different types of ORTs are also provided.

Keywords Direct response · Level of sensitivity · Optional randomized response · Relative efficiency · Sampling designs

9.1 Introduction

RR techniques are used to collect information relating to highly sensitive and stigmatized issues. But the perception of sensitivity is subjective and differs from person to person. For example, while HIV/AIDS infection status is a sensitive issue for most people, some are nevertheless willing to spontaneously reveal their status to the interviewer. Here, we consider a situation where most respondents feel that the subject of the inquiry is sensitive, but a minority may think that it is not sensitive and are therefore willing to provide a direct response (DR) without any hesitation. A randomized response technique (RR) that provides the opportunity to give DR instead of making RR compulsory for all the respondents is known as an optional randomized response technique (ORT). Accordingly, in an ORT, respondents are given the option to either provide RR if they feel that the subject of the inquiry is sensitive or to provide DR if they feel that the inquiry is not sufficiently sensitive to require anonymity (Arnab, 2019). ORT was introduced by Chaudhuri and Mukherjee (1988) and was extended by several researchers, including Arnab (2004, 2019), Chaudhuri and Saha (2005), Chaudhuri and Dihidar (2009), Gupta et al. (2002), Huang (2008), Pal (2008), and

R. Arnab, *Indirect Methods of Data Collection and Analysis from Surveys*, Indian Statistical Institute Series, https://doi.org/10.1007/978-981-97-6005-3_9

Singh and Singh (1993). Comprehensive reviews are given by Arnab and Rueda (2016), Arnab (2017), Chaudhuri (2011), Chaudhuri and Christofides (2013), and Mukerjee (2016). A few authors classified two-stage RR procedures such as those described by Mangat and Singh (1994), Singh and Joarder (1997), and Singh and Mathur (2003) where respondents are asked to provide DR with a certain probability, as ORTs. But in the present discussion, their works are not classified as ORT as their direct responses are not spontaneous; they are made compulsory under the instructions of the RR procedures. Details are given in the two stage RR procedures described in Chap. 8.

ORT can be classified into three categories: full ORT (FORT), partial ORT (PORT), and mixed ORT (MORT). In a FORT, respondents are categorized into two groups: G and $\overline{G}$ (complement of G). Respondents belonging to the group G feel that the subject of inquiry is sensitive and hence provide RR. On the other hand, respondents belonging to $\overline{G}$ feel that the subject of investigation is not overly sensitive and hence always provide DR. FORT was considered by several researchers, e.g., Arnab (2004, 2019), Chaudhuri and Mukerjee (1988), and Chaudhuri and Saha (2005). In a PORT, a respondent provides either RR or DR depending on their assessment of the sensitivity of the inquiry subject at the specific time of responding (Arnab, 2019). PORT was considered by Chaudhuri and Dihidar (2009), Gupta (2001), Gupta et al. (2002, 2010), Huang (2010), Hussain et al. (2014), Kalucha et al. (2015), Pal (2008), and Singh and Singh (1993), among others. In the PORTs proposed by Gupta et al. (2002, 2010), Huang (2010), and Singh and Singh (1993), it was assumed that W, the probability of choosing the option of providing RR for the sensitive question is the same for all the respondents in the population. They termed W as a measure of the sensitivity level of the question. Accordingly, W is expected to be high for a more sensitive question and low for a less sensitive question. The value of W is unknown. Gupta et al. (2002) estimated W using a single sample. The proposed estimator of W is biased, and the expression of mean square error (MSE) is complex. Gupta et al. (2010) and Huang (2010) used two independent samples for the estimation of W and the mean $\overline{Y}$ of the sensitive characteristic y. In a MORT, a respondent is forced to provide RRs in some stages using a randomization device. However, they are given the option to choose RR or DR in some other stages of the randomization device. However, the importance of MORT has not been articulated by the researchers. MORT was considered by Chaudhuri (2011) and Gupta et al. (2010), among others.

Most of the ORTs are limited to SRSWR sampling only. A few of the ORTs, e.g., Arnab (2004, 2019), Chaudhuri and Saha (2005), Chaudhuri and Dihidar (2009), and Pal (2008), are developed for complex surveys. ORTs are more efficient than compulsory RRT (CRT) because the probability of obtaining true responses in an ORT is much higher than in a CRT (Arnab, 2004). Under the general sampling design, Arnab and Olaomi (2019) introduced a two-phase ORT. The proposed two-phase ORT performs better than the single-phase ORT.

9.2 Full ORT (FORT)

In a full ORT, respondents belonging to the group G always provide RR and respondents belonging to the complementary group $\overline{G}$ always provide DR. In both situations, respondents reveal their identities.

9.2.1 *Chaudhri and Mukerjee (1985)*

Chaudhuri and Mukherjee (1985) proposed a FORT for estimating π, the proportion of individuals in a population belonging to a particular sensitive group A. A sample s of n units was selected from a population U by the SRSWR method Each of the selected respondents was asked to provide a "yes" or "no" answer either directly or by using Warner's (1965) method with a parameter P about their membership in the sensitive group A. Respondents who opted for the direct or RR approach were identified.

Let y_i be the true value of the sensitive characteristic y associated with the ith unit of the population U and $y_i = 1(0)$ if $i \in A(\overline{A})$ where $\overline{A}$ complement of A. Then the ith unit selected in the sample s should provide the direct response y_i if he or she belongs to $\overline{G}$; otherwise, if he or she belongs to G, should provide a RR z_i using Warner's model.

Let the sub-sample $s_{\overline{G}}(\in \overline{G})$ of $n_{\overline{G}}$ respondents be provided with the DR, and the remaining n_G $(= n - n_{\overline{G}})$ respondents belonging to $s_G(= s - s_{\overline{G}} \subset G)$ provided with the RR using Warner's randomization device with parameter P. By denoting W_G $(W_{\overline{G}})$ the proportion of individuals in the population belonging to the group $G(\overline{G})$, the following results were obtained.

Theorem 9.2.1

(i) $\hat{\pi}_{\text{OW}} = \frac{1}{n}\left[\sum_{i\in s_G} r_i + \sum_{i\in s_{\overline{G}}} y_i\right]$ is an unbiased estimator of π

where $r_i = \frac{z_i-(1-P)}{2P-1}$ and $\Sigma_{i\in s_j}$ denotes sum over the units in $s_j, j = G, \overline{G}$ with repetition.

(ii) $\text{Var}(\hat{\pi}_{\text{OW}}) = \frac{1}{n}[\pi(1-\pi) + W_G\phi_{\text{W}}]$; $\phi_{\text{W}} = \frac{P(1-P)}{(2P-1)^2}$

(iii) $\hat{\text{Var}}(\hat{\pi}_{\text{OW}}) = \frac{1}{n(n-1)}\left[\sum_{i\in s_G} r_i^2 + \sum_{i\in s_{\overline{G}}} y_i - \hat{\pi}_{\text{OW}}^2\right]$ is an unbiased estimator of $\text{Var}(\hat{\pi}_{\text{OW}})$.

Proof

$$\text{(i)}\quad E(\hat{\pi}_{\text{OW}}) = \frac{1}{n}E\left[\sum\nolimits_{i\in s_G} E_R(r_i) + \sum\nolimits_{i\in s_{\overline{G}}} y_i\right]$$
$$= \frac{1}{n}E\left[\sum\nolimits_{i\in s_G} y_i + \sum\nolimits_{i\in s_{\overline{G}}} y_i\right]$$
$$= E\left(\frac{1}{n}\sum\nolimits_{i\in s} y_i\right)$$
$$= \pi$$

$$\text{(ii)}\quad \text{Var}(\hat{\pi}_{\text{OW}}) = V[E_R(\hat{\pi}_{\text{OW}})] + E[\text{V}_R(\hat{\pi}_{\text{OW}})]$$
$$= V\left(\frac{1}{n}\sum\nolimits_{i\in s} y_i\right) + \frac{\phi_W}{n^2}E(n_G)$$
$$= \frac{\pi(1-\pi)}{n} + \frac{W_G\phi_W}{n}$$

$$\text{(iii)}\quad \text{E}[\text{V}\hat{\text{a}}\text{r}(\hat{\pi}_{\text{OW}})] = E[E_R\{\text{V}\hat{\text{a}}\text{r}(\hat{\pi}_{\text{OW}})\}]$$
$$= \frac{1}{(n-1)}\begin{bmatrix} \frac{1}{n}E\left\{\sum\nolimits_{i\in s_G}(y_i+\phi_W) + \sum\nolimits_{i\in s_{\overline{G}}} y_i\right\} \\ -\{\text{Var}(\hat{\pi}_{\text{OW}}) + \pi^2\} \end{bmatrix}$$
$$= \frac{1}{(n-1)}\left[E\left\{\frac{1}{n}\sum\nolimits_{i\in s} y_i + \phi_W\frac{n_G}{n}\right\} - \{\text{Var}(\hat{\pi}_{\text{OW}}) + \pi^2\}\right]$$
$$= \frac{1}{(n-1)}[\pi(1-\pi) + \phi_W W_G - \text{Var}(\hat{\pi}_{\text{OW}})]$$
$$= \text{Var}(\hat{\pi}_{\text{OW}})$$

Remark 9.2.1 The variance of Warner's estimator $\hat{\pi}_{\text{w}}$(given in Theorem 1.2.1), based on the CRR (complete randomized response), where each of the n individuals in the selected sample provided with RR is $\text{Var}(\hat{\pi}_{\text{w}}) = \frac{\pi(1-\pi)}{n} + \frac{\phi_W}{n}$. Since, $\text{Var}(\hat{\pi}_{\text{w}}) \geq \text{Var}(\hat{\pi}_{\text{OW}})$ as $0 \leq W_G \leq 1$, we conclude that the ORR is more efficient than the CRR based on Warner's model. Chaudhuri and Mukherjee (1985, 1988) and Chaudhuri and Christofides (2013), however, did not supply any unbiased estimators of $\text{Var}(\hat{\pi}_{\text{OW}})$.

9.2.2 Arnab's (2004) ORT

Arnab (2004) proposed a general method of estimating population characteristics for the FORT. The proposed method is applicable to any survey design and to both qualitative and quantitative characteristics. The method is described as follows.

Consider a finite population $U = (1, .., i, .., N)$ of N identifiable units from which a sample s of n units is selected with probability $p(s)$ by using the sampling design p. The inclusion probabilities for the ith, and i and $j(\neq i)$th units of the sampling design are denoted by π_i and $\pi_{ij}(> 0)$. If the $i(\in U)$ th unit is selected in the sample s, the respondent is asked to choose one of two options: (i) provide the true response (DR) y_i or (ii) provide a RR z_i using a suitable randomization device. Let $s_G(= s \cap G)$ of size n_G be the set of respondents in s who provided the RR, and $s_{\overline{G}}(= s \cap \overline{G})$ of size $n_{\overline{G}}$ be the set of those respondents who provided the DR. From the RR z_i, an estimate of y_i as a function of z_i is obtained, and it will be denoted by r_i. We will call r_i as the revised or reduced response obtained from the RR z_i. We will assume r_i follows the model

$$E_R(r_i) = y_i,\ V_R(r_i) = \phi_i \text{ and } C_R(r_i, r_j) = 0 \text{ for } i \neq j \tag{9.3.9.1}$$

Let us define

$$\tilde{r}_i = \delta_i y_i + (1 - \delta_i) r_i \tag{9.2.1}$$

where,

$$\delta_i = 1 \ \text{ if } \ i \in \overline{\text{G}} \ \text{ and } \ \delta_i = 0 \ \text{ if } \ i \in \text{G}. \tag{9.2.2}$$

Then, $\tilde{r}_i$ follows the model

$$E_R(\tilde{r}_i) = y_i, \ \ V_R(\tilde{r}_i) = (1 - \delta_i)\phi_i \ \text{ and } \ C_R(\tilde{r}_i, \tilde{r}_j) = 0 \ \text{ for } \ i \neq j \tag{9.2.3}$$

The model (9.2.3) leads to a linear homogeneous unbiased estimator for the population total $Y = \sum_{i\in U} y_i$ for the ORT as

$$t_{os} = \sum_{i\in s} b_{si}\tilde{r}_i \tag{9.2.4}$$

where b_{si} s are constants that satisfy the unbiasedness condition

$$\sum_{s\supset i} b_{si} p(s) = 1\, \forall\, i \in U \tag{9.2.5}$$

Arnab (2004) derived the following results:

Theorem 9.2.2

(i) $t_{os} = \sum_{i\in s} b_{si}\tilde{r}_i$ is an unbiased estimator of Y

(ii) $\text{Var}(t_{os}) = \sum_{i\in U} (\alpha_i - 1)y_i^2 + \sum\sum_{i\neq j\in U} (\alpha_{ij} - 1)y_i y_j + \sum_{i\in G} \alpha_i\phi_i$

where $\alpha_i = \sum_{s \supset i} b_{si}^2 p(s)$ and $\alpha_{ij} = \sum_{s \supset i,j} b_{si} b_{sj} p(s)$.

(iii) $\hat{\text{Var}}(t_{os}) = \sum_{i \in s} c_{si} \tilde{r}_i^2 + \sum_{i \neq} \sum_{j \in s} c_{sij} \tilde{r}_i \tilde{r}_j + \sum_{i \in s_G} b_{si} \hat{\phi}_i$ is an unbiased estimator of $\text{Var}(t_{os})$

where $\hat{\phi}_i$ is an unbiased estimator of ϕ_i,the constants c_{si} and c_{sij} are subjected to

$$\sum_{s \supset i} c_{si} p(s) = \alpha_i - 1 \text{ and } \sum_{s \supset i,j} c_{sij} p(s) = \alpha_{ij} - 1 \tag{9.2.6}$$

Proof

(i)
$$\begin{aligned} E(t_{os}) &= E\left[\sum_{i \in s} b_{si} E_R(\tilde{r}_i)\right] \\ &= E\left(\sum_{i \in s} b_{si} y_i\right) \\ &= \sum_{i \in U} y_i \sum_{s \supset i} b_{si} p(s) \\ &= Y \text{ (using } \sum_{s \supset i} b_{si} p(s) = 1) \end{aligned}$$

(ii)
$$\begin{aligned} \text{Var}(t_{os}) &= V\left[\sum_{i \in s} b_{si} E_R(\tilde{r}_i)\right] + E\left[\sum_{i \in s} b_{si} V_R(\tilde{r}_i)\right] \\ &= V\left(\sum_{i \in s} b_{si} y_i\right) + E\left(\sum_{i \in s} b_{si}^2 (1 - \delta_i) \phi_i\right) \\ &= \sum_{i \in U} y_i^2 \left\{\sum_{s \supset i} b_{si}^2 p(s) - 1\right\} + \sum_{i \neq} \sum_{j \in U} \left\{\sum_{s \supset i,j} b_{sj} b_{sj} p(s) - 1\right\} y_i y_j \\ &\quad + E\left\{\sum_{i \in U} (1 - \delta_i) \phi_i \sum_{s \supset i} b_{si}^2 p(s)\right\} \\ &= \sum_{i \in U} (\alpha_i - 1) y_i^2 + \sum_{i \neq} \sum_{j \in U} (\alpha_{ij} - 1) y_i y_j + \sum_{i \in U} \alpha_i (1 - \delta_i) \phi_i \\ &= \sum_{i \in U} (\alpha_i - 1) y_i^2 + \sum_{i \neq} \sum_{j \in U} (\alpha_{ij} - 1) y_i y_j + \sum_{i \in G} \alpha_i \phi_i \end{aligned}$$

(iii) $E[\hat{\text{Var}}(t_{os})] = E\left[\sum_{i\in s} c_{si}E_R(\tilde{r}_i^2) + \sum\sum_{i\neq j\in s} c_{sij}E_R(\tilde{r}_i\tilde{r}_j) + \sum_{i\in s_G} b_{si}E_R(\hat{\phi}_i)\right]$

$$= E\left[\sum_{i\in s} c_{si}E_R(\tilde{r}_i^2) + \sum\sum_{i\neq j\in s} c_{sij}E_R(\tilde{r}_i\tilde{r}_j) + \sum_{i\in s} b_{si}(1-\delta_i)E_R(\hat{\phi}_i)\right];$$

$\delta_i = 1$ for i $\in s_{\overline{G}}$

$$= E\left[\sum_{i\in U}\{y_i^2 + \phi_i(1-\delta_i)\}\sum_{s\ni i} c_{si}p(s) + \sum\sum_{i\neq j\in U} y_iy_j\sum_{s\supset i,j} c_{sij}p(s) + \sum_{i\in U}(1-\delta_i)\phi_i\sum_{s\supset i} b_{si}p(s)\right]$$

$$= \text{Var}(t_{os})$$

Remark 9.2.1

(i) The constants c_{si} and c_{sij} can be chosen in various ways to satisfy (9.2.6). The obvious choices are

$$c_{si} = \frac{\alpha_i - 1}{\pi_i} \text{ and } c_{sij} = \frac{\alpha_{ij} - 1}{\pi_{ij}}.$$

Theorem 9.2.3 The optional RR technique (ORT) is more efficient the compulsory RRT.

Proof Linear homogeneous unbiased estimators for the population total Y for the ORT and compulsory RRT are $t_{os} = \sum_{i\in s} b_{si}\tilde{r}_i$ and $t_s = \sum_{i\in s} b_{si}r_i$, respectively. The difference of their variances is

$$\text{Var}\left(\sum_{i\in s} b_{si}r_i\right) - \text{Var}\left(\sum_{i\in s} b_{si}\tilde{r}_i\right)$$

$$= \left[\sum_{i\in U}(\alpha_i - 1)y_i^2 + \sum\sum_{i\neq j\in U}(\alpha_{ij} - 1)y_iy_j + \sum_{i\in U}\alpha_i\phi_i\right]$$

$$- \left[\sum_{i\in U}(\alpha_i - 1)y_i^2 + \sum\sum_{i\neq j\in U}(\alpha_{ij} - 1)y_iy_j + \sum_{i\in U}\alpha_i(1-\delta_i)\phi_i\right]$$

$$= \sum_{i\in G}\alpha_i\phi_i$$

$$\geq 0 \text{ as } \alpha_i > 0, \phi_i \geq 0$$

Arnab (2004) provided the expressions of t_{os},Var(t_{os}) and $\hat{\text{Var}}(t_{os})$ for various sampling strategies. A few of them are given below.

9.2.2.1 Horvitz-Thomson Estimator (HTE) for a General Sampling Design

Substituting $b_{si} = \frac{1}{\pi_i}$ in t_{os}, the expression of the HTE of the mean $\overline{Y} = Y/N$ is obtained as

$$\hat{\overline{Y}}_{\text{oHT}} = \frac{1}{N}\sum_{i\in s}\frac{\tilde{r}_i}{\pi_i} \tag{9.2.7}$$

The expressions of $\text{Var}(\hat{\overline{Y}}_{\text{oHT}})$ and $\hat{\text{Var}}(\hat{\overline{Y}}_{\text{oHT}})$ are obtained from Theorem 9.2.2 as

$$\text{Var}(\hat{\overline{Y}}_{\text{oHT}}) = \frac{1}{N^2}\left[\sum_{i<}\sum_{j\in U}(\pi_i\pi_j - \pi_{ij})\left(\frac{y_i}{\pi_i} - \frac{y_j}{\pi_j}\right)^2 + \sum_{i\in G}\frac{\phi_i}{\pi_i}\right] \tag{9.2.8}$$

$$\hat{\text{Var}}(\hat{\overline{Y}}_{\text{oHT}}) = \frac{1}{N^2}\left[\sum_{i<}\sum_{j\in s}\frac{\pi_i\pi_j - \pi_{ij}}{\pi_{ij}}\left(\frac{\tilde{r}_i}{\pi_i} - \frac{\tilde{r}_j}{\pi_j}\right)^2 + \sum_{i\in s_G}\frac{\hat{\phi}_i}{\pi_i}\right] \tag{9.2.9}$$

9.2.2.2 SRSWOR

Substituting $\pi_i = \frac{n}{N}$ in $\hat{\overline{Y}}_{\text{oHT}}$, the estimator of $\overline{Y}$ for the SRSWOR sampling is obtained as

$$\overline{\tilde{r}}_{\text{o}} = \frac{1}{n}\sum_{i\in s}\tilde{r}_i \tag{9.2.10}$$

Expressions of the $\text{Var}(\overline{\tilde{r}}_{\text{o}})$ and its unbiased estimator $\hat{\text{Var}}(\overline{\tilde{r}}_{\text{o}})$ are obtained as

$$\text{Var}(\overline{\tilde{r}}_{\text{o}}) = (1-f)\frac{S_y^2}{n} + \frac{1}{nN}\sum_{i\in G}\phi_i \text{ and } \hat{\text{Var}}(\overline{\tilde{r}}_{\text{o}}) = (1-f)\frac{s_{\tilde{r}}^2}{n} + \frac{1}{nN}\sum_{i\in s_G}\hat{\phi}_i \tag{9.2.11}$$

where $S_y^2 = \frac{1}{N-1}\sum_{i\in U}(y_i - \overline{Y})^2$, $s_{\tilde{r}}^2 = \frac{1}{n-1}\sum_{i\in s}(\tilde{r}_i - \overline{\tilde{r}})^2$ and $f = \frac{n}{N}$.

9.2.2.3 SRSWR

For the SRSWR sampling design, an unbiased estimator of the population mean is given by

$$\bar{\tilde{r}}_{0w} = \frac{1}{n}\sum\nolimits_{i\in s}\tilde{r}_i \quad (9.2.12)$$

where $\Sigma_{i\in s}$ is the sum over units in s with repetition.

Expressions of the variance $\mathrm{Var}(\bar{\tilde{r}}_{0w})$ and its unbiased estimators $\widehat{\mathrm{Var}}(\bar{\tilde{r}}_{0w})$ are obtained as

$$\mathrm{Var}(\bar{\tilde{r}}_{0w}) = \frac{1}{n}\left[\sigma_y^2 + \frac{1}{N}\sum_{i\in G}\phi_i\right] \text{ and } \widehat{\mathrm{Var}}(\bar{\tilde{r}}_{0w}) = \frac{1}{n(n-1)}\sum\nolimits_{i\in s}(\tilde{r}_i - \bar{\tilde{r}}_{0w})^2$$

where $\sigma_y^2 = \frac{1}{N}\sum\limits_{i\in U}(y_i - \overline{Y})^2$.

Arnab (2004) has also given expressions for the estimators of the mean, variance, and unbiased estimators of the variances for various sampling designs, including PPSWR, Rao-Hartley-Cochran (1962), and Midzuno (1952) and Sen (1953). Chaudhuri and Saha (2005) considered the situations for the qualitative characteristics only. Hence, their results may be derived from Arnab's (2004) results as special cases.

9.2.3 *Huang's Methods*

Huang (2008) proposed two methods for estimating the mean $\overline{Y}$ and proportion W_G of a population belonging to the sensitive group G. For both methods, two independent samples s_1 and s_2 of sizes n_1 and n_2 are selected by the SRSWR sampling designs. The methods are as follows:

Method 1: The ith($\in U$) respondent selected in the sample $s_j (j = 1, 2)$ is given the option to provide either (i) the true response (DR)y_i or (ii) a RR $z_j = y_i u_{ij}$ where u_{ij} is a random sample from a population with a known mean $\mu_u = 1$ and known variance σ_{uj}^2. Assuming that the respondents belonging to the group G provide RR with probability one and the respondents belonging to the group $\overline{G}$ provide DR with probability one, the response obtained from the ith respondent can be expressed as

$$z_{oij} = \delta_i y_i + (1 - \delta_i) y_i u_{ij} \quad (9.2.13)$$

where $\delta_i = 1$ for $i \in \overline{\mathrm{G}}$ and $\delta_i = 0$ for $i \in \mathrm{G}$.

Equation (9.2.13) yields

$$E_R(z_{oij}) = y_i \text{ and } V_R(z_{oij}) = (1 - \delta_i)\sigma_{uj}^2 y_i^2 \quad (9.2.14)$$

Denoting $\overline{z}_{oj} = \frac{1}{n_j}\sum_{j\in s_j} z_{oij}$, one obtains

$$E(\overline{z}_{oj}) = E_p\left[\frac{1}{n_j}\sum_{j\in s_j} E_R(z_{oij})\right] = \overline{Y} \tag{9.2.15}$$

and

$$\begin{aligned}\mathrm{Var}(\overline{z}_{oj}) &= V_p\left[\frac{1}{n_j}\sum_{j\in s_j} E_R(z_{oij})\right] + E_p\left[\frac{1}{n_j^2}\sum_{j\in s_j} V_R(z_{oij})\right]\\ &= \frac{\sigma_y^2 + W_G\sigma_{uj}^2(\sigma_{yG}^2 + \overline{Y}_G^2)}{n_j}\end{aligned} \tag{9.2.16}$$

where $\overline{Y}_{\mathrm{G}} = \sum\limits_{i\in G} y_i/N_G$, $\sigma_{y\mathrm{G}}^2 = \sum\limits_{i\in G}(y_i - \overline{Y}_{\mathrm{G}})^2/N_{\mathrm{G}}$ and N_{G} is the total number of individuals in group G.

Equation (9.2.15) and (9.2.16) yield.

Theorem 9.2.4

(i) $\hat{\overline{Y}}_{H1} = \alpha\overline{z}_{o1} + (1-\alpha)\overline{z}_{o2}$ with $0 \leq \alpha \leq 1$ is an unbiased estimator of $\overline{Y}$

(ii) $\mathrm{Var}(\hat{\overline{Y}}_{H1}) = \alpha^2\frac{\sigma_{\mathrm{H1}}^2(1)}{n_1} + (1-\alpha)^2\frac{\sigma_{\mathrm{H1}}^2(2)}{n_2}$ and

(iii) $\hat{\mathrm{Var}}(\hat{\overline{Y}}_{H1}) = \alpha^2\frac{s_1^2}{n_1} + (1-\alpha)^2\frac{s_2^2}{n_2}$ where $\sigma_{\mathrm{H1}}^2(j) = \sigma_y^2 + W_G\sigma_{uj}^2(\sigma_y^2 + \overline{Y}^2)$ and $s_j^2 = \frac{1}{n_j-1}\sum_{i\in s_j}(z_{oij}-\overline{z}_{oi})^2$; $j = 1, 2$.

Method 2: It is similar to method 1, but with a little difference: the scramble response u_{ij} has a known mean $\mu_{uj}(\neq 1)$. In this case,

$$E_R(z_{oij}) = \delta_i y_i + (1-\delta_i)y_i\mu_{uj} \text{ and } V_R(z_{oij}) = (1-\delta_i)\sigma_{uj}^2 y_i^2$$

and

$$\begin{aligned}E(\overline{z}_{oj}) &= E_p\left[\frac{1}{n_j}\sum_{j\in s_j} E_R(z_{oij})\right]\\ &= \overline{Y} + (\mu_{uj} - 1)W_G\overline{Y}_G\end{aligned} \tag{9.2.17}$$

The variance of $\overline{z}_{oj}$ is

$$\begin{aligned}\mathrm{Var}(\overline{z}_{oj}) &= V_p\left[\frac{1}{n_j}\sum_{j\in s_j}\left\{\delta_i y_i + (1-\delta_i)y_i\mu_{uj}\right\}\right] + \sigma_{uj}^2 E_p\left[\frac{1}{n_j^2}\sum_{j\in s_j}(1-\delta_i)y_i^2\right]\\ &= \frac{1}{n_j}\left[\begin{array}{l}\sigma_y^2 + \left\{(\mu_{uj}^2 - 1) + \sigma_{uj}^2\right\}W_G\left(\sigma_{yG}^2 + \overline{Y}_G^2\right) - (\mu_{uj} - 1)\\ W_G\overline{Y}_G\left\{(\mu_{uj} - 1)W_G\overline{Y}_G + 2\overline{Y}\right\}\end{array}\right]\end{aligned}$$

$$= \sigma^2_{\mathrm{H2}}(j) \tag{9.2.18}$$

In this case, an unbiased estimator of $\overline{Y}$ is

$$\hat{\overline{Y}}_{\mathrm{H2}} = \frac{(1-\mu_{u2})\overline{z}_{o1} - (1-\mu_{u1})\overline{z}_{o2}}{\mu_{u1}-\mu_{u2}};\ \mu_{u1} \neq \mu_{u2} \tag{9.2.19}$$

The variance of $\hat{\overline{Y}}_{\mathrm{H2}}$ and its unbiased estimator come out as.
$\mathrm{Var}(\hat{\overline{Y}}_{H2}) = \frac{1}{(\mu_{u1}-\mu_{u2})^2}\left[(1-\mu_{u2})^2\frac{\sigma^2_{\mathrm{H2}}(1)}{n_1} + (1-\mu_{u1})^2\frac{\sigma^2_{\mathrm{H2}}(2)}{n_2}\right]$ and

$$\hat{\mathrm{Var}}(\hat{\overline{Y}}_{H2}) = \frac{1}{(\mu_{u1}-\mu_{u2})^2}\left[(1-\mu_{u2})^2\frac{s_1^2}{n_1} + (1-\mu_{u1})^2\frac{s_2^2}{n_2}\right]$$

Remark 9.2.3.1 The proposed estimators of W_G are biased, and their mean square errors are very complex. Hence, these are omitted in the present discussion.

9.3 Partial ORT (PORT)

In a PORT, respondents are given options to report either the true value of the sensitive characteristic y or a RR using a suitable randomization device without divulging the option exercised to the investigator. The choice between DR and RR is governed entirely by nature in the sense that a particular respondent may choose DR on some occasions and RR on other occasions depending on his or her judgment at the time of responding. Most researchers, including Gupta (2001), Gupta et al. (2002, 2006, 2010, 2013), and Huang (2004, 2008, 2010), assumed that the probability of providing the RR response $W(0 \leq W \leq 1)$ is unknown and is the same for all the respondents, while Chaudhuri and Dihidar (2009), Pal (2008), and others assumed that the likelihood of providing the RR varies from respondent to respondent. The unknown quantity W was referred to by Gupta et al. (2002) as the sensitivity level of the question. The value of W is expected to be higher for more sensitive questions and lower for less sensitive ones.

9.3.1 Fixed W

Here, we will assume that W, the probability of providing the DR and RR for all the respondents is the same. Let the ith respondent provide the response

$$z_{oi} = \begin{cases} y_i, \text{ with probability } 1-W \\ z_i, \text{ with probability } W \end{cases} \tag{9.3.1}$$

where y_i is the value of sensitive characteristic y and z_i is the RR obtained from the ith respondent. The range of z_i is expected to be the same as y_i so that one cannot identify whether the respondent is providing with the true or a scrambled response.

To develop a general theory, let us assume that z_i follows the model:

$$E_R(z_i) = \alpha y_i + \beta,\ V_R(z_i) = \phi_i \text{ and } C_R(z_i, z_j) = 0 \text{ for } i \neq j \tag{9.3.2}$$

where α and β are known constants.

Equation (9.3.1) can be written as

$$z_{oi} = x_i z_i + (1 - x_i) y_i \tag{9.3.3}$$

where x_i is a Bernoulli variable with $P(x_i = 1) = W$ and $P(x_i = 0) = 1 - W$.

Denoting $E_{Rx}(V_{Rx})$ as overall expectation (variance) with respect to x and R, one finds

$$\begin{aligned} E(z_{oi}) &= E_X\{E_R(z_{oi})\} \\ &= \{(\alpha - 1)W + 1\} y_i + \beta W \\ \text{and} & \\ V(z_{oi}) &= E_X\{V_R(z_{oi})\} + V_X\{E_R(z_{oi})\} \\ &= E_X\{x_i\phi_i\} + V_X[x_i\{(\alpha - 1)y_i + \beta\} + y_i] \\ &= W\phi_i + W(1 - W)[(\alpha - 1)y_i + \beta]^2 \end{aligned} \tag{9.3.4}$$

The revised RR

$$r_{oi} = \frac{z_{oi} - \beta W}{c}$$

follows the model

$$\begin{aligned} & E_{Rx}(r_{oi}) = y_i,\ V_{Rx}(r_{oi}) = \psi_i \\ & = \frac{W(1 - W)\{(\alpha - 1)y_i + \beta\}^2 + W\phi_i}{c^2} \text{ and } C_{Rx}(r_{oi}, r_{oj}) = 0 \text{ for } i \neq j \end{aligned} \tag{9.3.5}$$

The model (9.3.5) yields an unbiased estimator for the population mean $\overline{Y}$ for various sampling strategies. In particular, expressions for the HTE for the general sampling design and the sample mean estimator for the SRSWR become

$$\hat{\overline{Y}}_{\text{cHT}} = \frac{1}{N}\sum_{i\in s}\frac{r_{\text{o}i}}{\pi_i} \tag{9.3.6}$$

and

$$\hat{\overline{Y}}_{\text{oWR}} = \frac{1}{n}\sum\nolimits_{i\in s} r_{\text{o}i} \tag{9.3.7}$$

9.3.1.1 Specific PORT Models

Gupta et al. (2002)

Consider the situation where a sample s of n units is selected by the SRSWR method. The ith respondent reports true value y_i with probability $1 - W$ and a scrambled response $z_i = y_i u_i/\mu_u$ that follows the multiplicative model proposed by Eichorn and Hayre (1983) with probability W. Here u_i is a random sample from a population with a known mean μ_u and known variance σ_u^2. In this case

$$z_{\text{o}i} = \begin{cases} y_i \text{ with probability } 1 - W \\ y_i u_i/\mu_u \text{ with probability } W \end{cases}$$
$$= x_i z_i + (1 - x_i) y_i \tag{9.3.3.8}$$

follows the model (9.3.5) with $\alpha = 1, \beta = 0, c = 1$, $\phi_i = y_i^2\sigma_u^2/\mu_u^2$ and $\psi_i = W y_i^2\sigma_u^2/\mu_u^2 = \psi_{Gi}$.

Since $z_{\text{o}i}$ s are independently and identically distributed (iid) random variables, one finds the following results:

Theorem 9.3.1

(i) $\overline{z}_{\text{oG}} = \frac{1}{n}\sum_{i\in s} z_{\text{o}i}$ is an unbiased estimator of $\overline{Y}$.

(ii) $\text{Var}(\overline{z}_{\text{oG}}) = \frac{1}{n}\left[\sigma_y^2 + W\frac{\sigma_u^2}{\mu_u^2}(\sigma_y^2 + \overline{Y}^2)\right]$

(iii) An unbiased estimator $\text{Var}(\overline{z}_{\text{oG}})$ is

$$\hat{\text{Var}}(\overline{z}_{\text{oG}}) = \frac{1}{n(n-1)}\sum\nolimits_{i\in s}(z_{\text{o}i} - \overline{z}_{\text{oG}})^2$$

Estimation of *W*

Since, W is unknown, Gupta et al. (2002) provided an unbiased estimator of W by writing

$$z_{\text{o}i} = z_i^{x_i} y_i \tag{9.3.10}$$

where x_i is a Bernoulli variable with parameter W. Equation (9.3.10) yields,

$$E_x(\log z_{oi}) = W\log(u_i/\mu_u) + \log y_i \tag{9.3.11}$$

Using (9.3.11), Gupta et al. (2002) proposed an estimator of W as

$$\hat{W}_{\mathrm{G}} = \frac{\frac{1}{n}\sum_{i\in s}\log z_{oi} - \log \bar{z}_{o\mathrm{G}}}{E\left(\log\frac{u_i}{\mu_u}\right)} \tag{9.3.12}$$

The proposed estimator $\hat{W}_{\mathrm{G}}$ is biased and its mean square error is difficult to calculate (Huang, 2010). Moreover, $E\left(\log\frac{u_i}{\mu_u}\right)$ is not in closed form for most of the distributions, e.g., chi-square, Poisson.

Gupta et al. (2014)

Gupta et al. (2014) also considered the situation where the sample s is selected by the SRSWOR method and each of the selected respondents in the sample provided with the true response y with probability $1 - W$ and a scrambled response using the additive model $z = y + t$ with probability W. The random variable t has a mean zero and a known variance σ_t^2. In this case, model (9.3.5) becomes $z_{oi} = r_{oi}$, $\alpha = 1$, $\beta = 0$ and $\psi_i = W\sigma_t^2$.

Remark 9.3.1.1 Singh and Singh (1993) proposed an ORT for estimating π, the proportion of persons belonging to certain sensitive group A. In this method, each respondent was free to provide an answer "yes" or "no" either by using (i) Warner's (1965) method or (ii) directly, without using any RR method.

9.3.1.2 Estimation of $\overline{Y}$ and W

To derive both the unbiased estimators of $\overline{Y}$ and W from the same experiment, several methods have been proposed for the SRSWR designs, which include Gupta et al. (2006, 2010, 2013) and Huang (2004, 2010). The reduced responses from methods can be expressed in the form (9.3.5) as follows:

Two independent samples s_1 and s_2 of sizes n_1 and n_2 are selected by the SRSWR method. If the ith respondent is selected in the sample $s_j (j = 1, 2)$, he or she provides the response, z_{oij} which is either the DR y_i with probability $1 - W$ or a RR z_{ij} using a randomization device R_j with probability W.

$$z_{0ij} = \begin{cases} y_i \text{ with probability } 1 - W \\ z_{ij} \text{ with probability } W \end{cases}$$

The response z_{ij} is assumed to follow the model

$$E_R(z_{ij}) = \alpha_j y_i + \beta_j,\ V_R(z_{ij}) = \phi_{ij} \text{ and } C_R(z_{ik}, z_{jl}) = 0 \text{ for } i, k \neq j, l \tag{9.3.13}$$

In this case,

$$z_{oij} = x_i z_{ij} + (1 - x_i) y_i \tag{9.3.14}$$

where x_i is a Bernoulli variable with $P(x_i = 1) = W$ and $P(x_i = 0) = 1 - W$.

Equation (9.3.14) yields

$$E_{Rx}(z_{oij}) = \theta_j y_i + \beta_j W,\ V_{Rx}(z_{oij}) = \gamma_{ij} \text{ and } C_{Rx}(r_{oij}, r_{okl}) = 0 \text{ for } (i, j) \neq (k, l) \tag{9.3.15}$$

where $\theta_j = \{1 - (1 - \alpha_j)W\}$ and $\gamma_{ij} = W(1 - W)[(\alpha_j - 1)y_i + \beta_j]^2 + W\phi_{ij}$

Denoting, $\bar{z}_{oj} = \frac{1}{n_j}\sum_{i\in s_j} z_{oij}$, one can easily verify the following Theorems.

Theorem 9.3.2

(i) $\bar{z}_o = \frac{\beta_2 \bar{z}_{o1} - \beta_1 \bar{z}_{o2}}{\theta_1\beta_2 - \theta_2\beta_1}$ is an unbiased estimator of $\overline{Y}; \theta_1\beta_2 \neq \theta_2\beta_1$.

(ii) $\mathrm{Var}(\bar{z}_o) = \frac{1}{(\theta_1\beta_2 - \theta_2\beta_1)^2}\left[\frac{\beta_2^2\sigma_{o1}^2}{n_1} + \frac{\beta_1^2\sigma_{o2}^2}{n_2}\right]$

(iii) $\widehat{\mathrm{Var}}(\bar{z}_o) = \frac{1}{(\theta_1\beta_2 - \theta_2\beta_1)^2}\left[\frac{\beta_2^2 s_{o1}^2}{n_1} + \frac{\beta_1^2 s_{o2}^2}{n_2}\right]$

where $\sigma_{oj}^2 = \theta_j^2\sigma_y^2 + \overline{\gamma}_j, \overline{\gamma}_j = \frac{1}{N}\sum_{i\in U}\gamma_{ij}$ and $s_{oj}^2 = \frac{1}{n_j - 1}\sum_{i\in s_j}(z_{oij} - \bar{z}_{oj})^2; j = 1, 2$

Theorem 9.3.3

(i) $\hat{W} = \frac{\theta_2\bar{z}_{o1} - \theta_1\bar{z}_{o2}}{\beta_1\theta_2 - \beta_2\theta_1}$ is an unbiased estimator of $W; \beta_1\theta_2 \neq \beta_2\theta_1$

(ii) $\mathrm{Var}(\hat{W}) = \frac{1}{(\theta_2\beta_1 - \theta_1\beta_2)^2}\left[\frac{\theta_2^2\sigma_{o1}^2}{n_1} + \frac{\theta_1^2\sigma_{o2}^2}{n_2}\right]$

(iii) $\widehat{\mathrm{Var}}(\hat{W}) = \frac{1}{(\theta_2\beta_1 - \theta_1\beta_2)^2}\left(\frac{\theta_2^2 s_{o1}^2}{n_1} + \frac{\theta_1^2 s_{o1}^2}{n_2}\right)$

9.3.1.3 Examples

Huang (2010) considered the special case of the model (9.3.13) where $z_{ij} = u_{ij}y_i + t_{ij}$ with $E_R(u_{ij}) = 1$, $V_R(u_{ij}) = \sigma_{uj}^2$, $E_R(t_{ij}) = \mu_{tj}$, and $V_R(t_{ij}) = \sigma_{tj}^2$. In this case, $\alpha_j = 1, \beta_j = \mu_{tj}$,

$$\phi_{ij} = y_i^2\sigma_{uj}^2 + \sigma_{tj}^2,\ \theta_j = 1,\ \gamma_{ij} = W(1 - W)\mu_{tj}^2 \text{ and}$$
$$\overline{\gamma}_{Hj} = W(1 - W)\mu_{tj}^2 + W\left\{\sigma_{uj}^2(\sigma_y^2 + \overline{Y}^2) + \sigma_{tj}^2\right\}.$$

Using Theorems 9.3.2 and 9.3.3, one obtains the following:

Theorem 9.3.4

(i) $\overline{z}_{\mathrm{oH}} = \frac{\mu_{t2}\overline{z}_{o1} - \mu_{t1}\overline{z}_{o2}}{\mu_{t2} - \mu_{t1}}$ is an unbiased estimator of $\overline{Y}$

(ii) $\mathrm{Var}(\overline{z}_{\mathrm{oH}}) = \frac{1}{(\mu_{t2} - \mu_{t1})^2}\left[\frac{\mu_{t2}^2\sigma_{\mathrm{oH1}}^2}{n_1} + \frac{\mu_{t1}^2\sigma_{\mathrm{oH2}}^2}{n_2}\right]$, (iii)$\hat{\mathrm{Var}}(\overline{z}_{\mathrm{oH}}) = \frac{1}{(\mu_{t2} - \mu_{t1})^2}\left[\frac{\mu_{t2}^2}{n_1}s_{0z1}^2 + \frac{\mu_{t1}^2}{n_2}s_{0z2}^2\right]$

(iii) $\hat{W}_{\mathrm{H}} = \frac{\overline{z}_{o1} - \overline{z}_{o2}}{\mu_{t1} - \mu_{t2}}$ is an unbiased estimator of W,

(iv) $\mathrm{Var}(\hat{W}_{\mathrm{H}}) = \frac{1}{(\mu_{t2} - \mu_{t1})^2}\left[\frac{\sigma_{\mathrm{oH1}}^2}{n_1} + \frac{\sigma_{\mathrm{oH2}}^2}{n_2}\right]$ and $\hat{\mathrm{Var}}(\hat{W}_{\mathrm{H}}) = \frac{1}{(\mu_{t2} - \mu_{t1})^2}\left[\frac{s_{0z1}^2}{n_1} + \frac{s_{0z2}^2}{n_2}\right]$.

where $\sigma_{\mathrm{oH}j}^2 = \sigma_y^2 + \overline{\gamma}_{\mathrm{H}j}$ and $\overline{\gamma}_{\mathrm{H}j} = W(1-W)\mu_{tj}^2 + W\left\{\sigma_{uj}^2(\sigma_y^2 + \overline{Y}^2) + \sigma_{tj}^2\right\}$.

Gupta and Shabbir (2004) considered Huang's (2010) model with $t_{ij} = 0$. Ryu et al. (2006) proposed a two-stage ORT using a multiplicative scrambling response, while Shim and Gupta (2015) considered a three-stage PORT for qualitative characteristics.

9.3.2 *Varying W's*

Chaudhuri and Dihidar (2009), and Pal (2008) extended PORT, where the probabilities of providing DR and RR were allowed to vary from respondent to respondent and were applicable to the general sampling design. In the proposed methods, ith respondent provides a DR with probability $1 - W_i$ and a RR with probability W_i. The probabilities W_i s are unknown and are governed by nature.

$$z_{0ij} = \begin{cases} y_i, \text{ with probability } 1 - W_i \\ z_{ij}, \text{ with probability } W_i \end{cases}$$

9.3.2.1 Pal (2008)

In Pal's PORT (2008), the ith respondent provides two independent responses z_{oi1} and z_{oi2}. The first response z_{oi1} is the true response, y_i with probability $1 - W_i$ and the scrambled response, $z_{i1} = \frac{y_i h_{i1}}{\theta_1} + u_{i1}$ with probability W_i.The second response z_{oi2} is either y_i with probability $1 - W_i$ or the scrambled response z_{i2} with probability W_i. The scrambled response z_{i2} is z_{oi1} with probability p and $\frac{y_i h_{i2}}{\theta_2} + u_{i2}$ with probability $1 - p$. The random samples h_{ij} and u_{ij} have known means μ_{hj} and μ_{uj}, and known variances σ_{hj}^2 and σ_{uj}^2 respectively, $j = 1, 2$. In this case,

$$E_R[(z_{0i}(1)] = \alpha_{0i}(1)y_i + \beta_{0i}(1) \text{ and } E_R[(z_{0i}(2)] = \alpha_{0i}(2)y_i + \beta_{0i}(2) \quad (9.3.16)$$

where $\alpha_{0i}(1) = (1 - W_i) + W_i\frac{\mu_{h1}}{\theta_1}$, $\beta_{0i}(1) = W_i\mu_{u1}$, $\alpha_{02} = (1 - W_i)(1 + pW_i) + \frac{\mu_{h1}}{\theta_1}W_i^2p + \frac{\mu_{h2}}{\theta_2}W_i(1-p)$,, and $\beta_{02} = \mu_{u1}W_i^2p + \mu_{u2}W_i(1-p)$.

The expression (9.3.13) yields an unbiased estimator of y_i as

$$r_{0i} = \frac{z_{0i}(2)\beta_{0i}(1) - z_{0i}(1)\beta_{0i}(2)}{\alpha_{0i}(2)\beta_{0i}(1) - \alpha_{0i}(1)\beta_{0i}(2)};\ \alpha_{0i}(2)\beta_{0i}(1) \neq \alpha_{0i}(1)\beta_{0i}(2) \tag{9.3.17}$$

Since the revised response r_{0i} satisfies the model (9.3.15), one can find an unbiased estimator of the total, variance, and an unbiased estimator of the variance by using Theorem 9.3.2.

9.3.2.2 Chaudhuri and Dihidar (2009)

Chaudhuri and Dihidar proposed a few PORTs that are applicable to both qualitative and quantitative characteristics as well as to general sampling designs. All the proposed models are based on similar techniques. For the estimation of the variance, each of the selected respondents in the sample was asked to report two responses independently. For the proposed techniques, *i* th respondent selected in the sample is required to supply the responses

$$z_{0i}(j) = \begin{cases} y_i, \text{ with probability } 1 - W_i \\ z_{ij}, \text{ with probability } W_i \end{cases}, j = 1, 2$$

where y_i is the true response and z_{ij} is a scrambled response obtained from the RR technique $R_j (j = 1, 2)$ that satisfies the model (9.3.15). The estimators of the mean, variance, and unbiased estimator of the variance can be obtained as usual from Theorem 9.3.2. The main disadvantage of this model is that the anonymity of the option chosen by the respondents is not protected, as $z_{0i}(1) \neq z_{0i}(2)$ indicates that the ith respondent opted at least once for the RR model.

9.3.3 Mixed ORT

In a mixed ORT (MORT), respondents are not given the complete option to provide the DR. However, he or she is given the option to choose DR at a certain stage of the randomization procedure. The MORT was proposed by Chaudhuri (2011) and Gupta et al. (2010), among others. However, the motivation behind the proposed models was not articulated clearly.

9.3.3.1 Gupta et al. (2010)

In this method, two independent samples s_1 and s_2 of sizes n_1 and n_2 are selected by the SRSWR method. If the $i(\in U)th$ respondent is selected in $s_j(j = 1, 2)$, then he or she is asked to provide z_{oij} which is the true response y_i with probability p and a scrambled response z_{ij}^* with probability $1 - p$ using an ORT. Response z_{ij}^* is either the true response y_i with unknown probability (determined by nature) $1 - W$ or a RR $\tilde{z}_{ij} = y_i + u_{ij}$ with probability W where u_{ij} is a random sample from a population with the known mean μ_{uj} and variance σ_{uj}^2. Thus,

$$z_{oij} = \begin{cases} y_i, \text{ with probability } p \\ z_{ij}^*, \text{ with probability } 1 - p \end{cases} \text{ and } z_{ij}^* = \begin{cases} y_i, \text{ with probability } 1 - W \\ y_i + u_{ij}, \text{ with probability } W \end{cases} \quad (9.3.18)$$

Equation (9.3.18) can be expressed as

$$z_{oij} = \begin{cases} y_i \text{ with probability } 1 - \tilde{W} \\ z_{ij} \text{ with probability } \tilde{W} \end{cases} \quad (9.3.19)$$

where $z_{ij} = y_i + u_{ij}$ and $\tilde{W} = (1 - p)W$.

Denoting $z_{oij} = x_i z_{ij} + (1 - x_i)y_i$ with x_i is a Bernoulli variable with parameter $\tilde{W}$, one finds that z_{oij} follows the model (9.3.15) with.

$$\theta_j = 1,\ \beta_j = \mu_{uj} \text{ and } \gamma_{ij} = \tilde{W}\sigma_{uj}^2 + \tilde{W}(1 - \tilde{W})\mu_{uj}^2$$
$$= p(1 - W)(\sigma_{uj}^2 + \mu_{uj}^2) - \{p(1 - W)\}^2\mu_{uj}^2 = \overline{\gamma}_{Gj}$$

i.e.,

$$E_R(z_{oij}) = y_i + \tilde{W}\mu_{uj},\ V_R(z_{oij}) = \overline{\gamma}_{Gj} \text{ and } Cov(z_{oij}, z_{okl}) = 0 \text{ for } i, j \neq k, l \quad (9.3.20)$$

Equation (9.3.18), Theorems 9.3.2 and 9.3.3 yield.

Theorem 9.3.4

(i) $\overline{z}_{oG} = \frac{\mu_{u2}\overline{z}_{o1} - \mu_{u1}\overline{z}_{o2}}{\mu_{u2} - \mu_{u1}}$ is an unbiased estimator of $\overline{Y}$,

(ii) $\hat{W} = \frac{\overline{z}_{o1} - \overline{z}_{o2}}{(1-p)(\mu_{u1} - \mu_{u2})}$ is an unbiased estimator of W

(iii) $\text{Var}(\overline{z}_{OG}) = \frac{1}{(\mu_{u2} - \mu_{u1})^2}\left[\frac{\mu_{u2}^2\sigma_{oG1}^2}{n_1} + \frac{\mu_{u1}^2\sigma_{oG2}^2}{n_2}\right]$,

(iv) $\text{Var}(\hat{W}) = \frac{1}{(1-p)^2(\mu_{u1} - \mu_{u2})^2}\left[\frac{\sigma_{oG1}^2}{n_1} + \frac{\sigma_{oG2}^2}{n_2}\right]$

(v) $\hat{\text{Var}}(\overline{z}_{OG}) = \frac{1}{(\mu_{u2} - \mu_{u1})^2}\left[\frac{\mu_{u2}^2 s_{Gz1}^2}{n_1} + \frac{\mu_{u1}^2 s_{Gz2}^2}{n_2}\right]$ and

(vi) $\hat{\text{Var}}(\hat{W}) = \frac{1}{(1-p)^2(\mu_{u1}-\mu_{u2})^2}[s^2_{\text{Gz}1} + s^2_{\text{Gz}2}]$

where $\sigma^2_{\text{oG}j} = \sigma^2_y + \overline{\gamma}_{\text{G}j}$ and $s^2_{\text{Gz}j} = \frac{1}{n_j-1}\sum_{i\in s_j}(z_{\text{o}ij} - \overline{z}_{\text{o}j})^2; j = 1, 2$

9.4 Comparison Between Full and Partial ORT

In an ORT, respondents are given the choice between one of two options: providing a DR or providing a RR using a randomization device. FORT assumes that the entire population is divided into two mutually exclusive and exhaustive groups G and $\overline{G}$. Respondents belonging to the group G always provide RR, while respondents belonging to $\overline{G}$ always provide DR. The PORT assumes that a respondent can provide RR on some occasions and DR on other occasions with a certain unknown probability, which is governed by nature. It is natural to investigate whether the two assumptions cause any difference in making inferences for estimating the parameters of interest. Arnab (2019) showed that for the SRSWR sampling design, the unbiased estimators of $\overline{Y}$ and W are the same, but their variances differ significantly. In this section, Arnab's (2019) results concerning the performances of the estimators of population mean for the Gupta et al. (2002) and Huang et al. (2010) ORT under the assumptions of FORT and PORT are described.

9.4.1 Gupta et al. (2002) Model

Let a sample s of size n be selected by the SRSWR method. If the $i(\in U)$th respondent is selected in s, then he or she is requested to provide either the true response (DR) y_i or a RR $z_i = y_i\frac{u_i}{\mu_u}$, where u_i is a random sample from a population with a known mean μ_u and known variance σ^2_u. Let the response obtained from the ith respondent be denoted by z_{oi}.

Under the assumption of FORT,

$$z_{oi} = \begin{cases} y_i \text{ for } i \in \overline{G} \\ y_i\frac{u_i}{\mu_u} \text{ for } i \in G \end{cases} \tag{9.4.1}$$

The expression (9.4.1) can be expressed as

$$z_{oi} = y_i\delta_i + (1-\delta_i)\left(y_i\frac{u_i}{\mu_u}\right) \text{ with } \delta_i = 1 \text{ for } i \in \overline{G} \text{ and } \delta_i = 0 \text{ for } i \in G \tag{9.4.2}$$

The following Theorem follows from Eq. (9.4.2).

Theorem 9.4.1

(ii) $\overline{z}_{\mathrm{oGF}} = \frac{1}{n}\sum_{i\in s} z_{oi}$ is an unbiased estimator of $\overline{Y}$

(iii) $\mathrm{Var}(\overline{z}_{\mathrm{oGF}}) = \frac{\sigma^2_{\mathrm{oz}}}{n}$

(iv) An unbiased estimator of $\mathrm{Var}(\overline{z}_{\mathrm{oGF}})$ is

$$\hat{\mathrm{Var}}(\overline{z}_{\mathrm{oGF}}) = \frac{1}{n(n-1)}\sum_{i\in s}(z_{oi} - \overline{z}_{\mathrm{oGF}})^2$$

where $\sigma^2_{\mathrm{oz}} = \sigma^2_y + C^2_u W_G \overline{Y}^2_G(1+C^2_{yG})$, $C_u = \sigma_u/\mu_u$ is the CV (coefficient of variation) of u for the entire population, C_{yG} is the CV of y for the group G, $\overline{Y}_G$ is mean of y for the group G, and W_G proportion of persons belong to the group G.

Proof

$$\begin{aligned}
\text{(i) } E(\overline{z}_{\mathrm{oGF}}) &= E_p\left[\frac{1}{n}\sum_{i\in s} E_R(z_{oi})\right] \\
&= E_p\left(\frac{1}{n}\sum_{i\in s} y_i\right) \\
&= \overline{Y}
\end{aligned}$$

$$\begin{aligned}
\text{(ii) } V(\overline{z}_{\mathrm{oGF}}) &= V_p\left[\frac{1}{n}\sum_{i\in s} E_R(z_{oi})\right] + E_p\left[\frac{1}{n^2}\sum_{i\in s} V_R(z_{oi})\right] \\
&= V_p\left[\frac{1}{n}\sum_{i\in s} y_i\right] + E_p\left[\frac{C^2_u}{n^2}\sum_{i\in s}\delta_i y_i^2\right] \\
&= \frac{1}{n}\left[\sigma^2_y + \frac{C^2_u}{N}\sum_{i\in U}\delta_i y_i^2\right] \\
&= \frac{1}{n}\left[\sigma^2_y + C^2_u W_G\left(\sigma^2_{yG} + \overline{Y}^2_G\right)\right] \\
&= \frac{1}{n}\left[\sigma^2_y + C^2_u W_G\left(\sigma^2_{yG} + \overline{Y}^2_G\right)\right]
\end{aligned}$$

$$
\begin{aligned}
\text{(iii) } E[\hat{Var}(\overline{z}_{oGF})] &= \frac{1}{n(n-1)} E\Big[\sum\nolimits_{i\in s} z_{oi}^2 - n\overline{z}_{oGF}^2\Big] \\
&= \frac{1}{n(n-1)} E\Big[\sum\nolimits_{i\in s} z_{oi}^2 - n\overline{z}_{oGF}^2\Big] \\
&= \frac{1}{n(n-1)} \Big[E_p \sum\nolimits_{i\in s} \{V_R(z_{oi}) + y_i^2\} - n\Big\{V(\overline{z}_{oGF}) + \overline{Y}^2\Big\}\Big] \\
&= \frac{1}{n(n-1)} \left[\frac{n}{N} \sum_{i\in U} \{\delta_i y_i^2 C_u^2 + y_i^2\} - n\Big\{V(\overline{z}_{oGF}) + \overline{Y}^2\Big\}\right] \\
&= \frac{1}{n(n-1)} \left[\frac{n}{N} \left(C_u^2 \sum_{i\in G} y_i^2 + \sum_{i\in U} y_i^2\right) - n\Big\{V(\overline{z}_{oGF}) + \overline{Y}^2\Big\}\right] \\
&= \frac{1}{(n-1)} \Big[C_u^2 W_G\Big(\sigma_{yG}^2 + \overline{Y}_G^2\Big) + \sigma_y^2 - V(\overline{z}_{oGF})\Big] \\
&= Var(\overline{z}_{oGF})
\end{aligned}
$$

From Theorem 9.3.1, we note that the unbiased of $\overline{Y}$ for Gupta et al.'s (2002) model under the assumption of PORT and FORT are the same, i.e., $\overline{z}_{oG} = \overline{z}_{oGF} = \frac{1}{n}\sum_{i\in s} z_{oi} = \overline{z}_o$. But the expression of the variance of $\overline{z}_{oG}$ under the assumption of PORT is significantly different from the variance of $\overline{z}_{oGF}$ under the assumption of FORT. From Theorems 9.3.1 and 9.4.1, one finds their difference is.

$$
Var(\overline{z}_{oGF}) - Var(\overline{z}_{oG}) = \frac{C_u^2}{n}\Big[W_G\overline{Y}_G^2(1 + C_{yG}^2) - W\overline{Y}^2(1 + C_y^2)\Big] \tag{9.4.3}
$$

i.e.,$Var(\overline{z}_{oGF}) >=$ or $< Var(\overline{z}_{oG})$ if $W_G\overline{Y}_G^2(1 + C_{yG}^2) >=$ or $< W\overline{Y}^2(1 + C_y^2)$

In particular, if $W = W_G$ and $C_y = C_{yG}$, the variance of $\overline{z}_o$ under the assumption of FORT will be larger than the PORT if $\overline{Y}_G \geq \overline{Y}$. The condition $\overline{Y}_G \geq \overline{Y}$ holds if the mean of the sensitive characteristics y for the group of the people who opted for RR technique is larger than the entire population, such as questions pertaining to personal income, the incidence of domestic violence, the number of sexual partners, etc. (see Arnab, 2019).

9.4.2 Huang (2010) Model

In Sect. 9.3.1.3., the expressions of estimators of $\overline{Y}$ and W, their variances, and unbiased estimators of the variances have been obtained for Huang's (2010) RR model under the assumptions of the PORT and SRSWR sampling. In this section, we will consider such derivations under the assumption of FORD. Let z_{oij} denote the response obtained from the ith respondent from the sample $s_j (j = 1, 2)$. Then, under the assumption of FORD,

$$z_{oij} = \begin{cases} y_i \text{ if the respondent } \in \overline{\mathrm{G}} \\ u_{ij}y_i + t_{ij} \text{ if the respondent } \in \mathrm{G} \end{cases} \tag{9.4.4}$$

where $E_R(u_{ij}) = 1$, $V_R(u_{ij}) = \sigma_{uj}^2$, $E_R(t_{ij}) = \mu_{tj}$ and $V_R(t_{ij}) = \sigma_{tj}^2$. Equation (9.4.4) can be expressed as

$$z_{oij} = y_i\delta_i + (u_{ij}y_i + t_{ij})(1 - \delta_i) \tag{9.4.5}$$

with $\delta_i = 1$ for $i \in \overline{G}$ and $\delta_i = 0$ for $i \in G$.

Writing $\overline{z}_{oj} = \frac{1}{n_j}\sum_{i \in s_j} z_{oij}$, and using Arnab's (2019) results, we obtain.

Theorem 9.4.2 Under the SRSWR sampling,

(i) $\overline{z}_{\text{oHF}} = \frac{\mu_{t2}\overline{z}_{o1} - \mu_{t1}\overline{z}_{o2}}{\mu_{t2} - \mu_{t1}}$ is an unbiased estimator of $\overline{Y}$

(ii) $\text{Var}(\overline{z}_{\text{oHF}}) = \frac{1}{(\mu_{t2} - \mu_{t1})^2}\left(\mu_{t2}^2 \frac{\sigma_{\text{HF1}}^2}{n_1} + \mu_{t1}^2 \frac{\sigma_{\text{HF2}}^2}{n_2}\right)$,

(iii) $\hat{\text{Var}}(\overline{z}_{\text{oHF}}) = \frac{1}{(\mu_{t2} - \mu_{t1})^2}\left[\frac{\mu_{t2}^2}{n_1} s_{\text{oHF1}}^2 + \frac{\mu_{t1}^2}{n_2} s_{\text{oHF2}}^2\right]$

(iv) $\hat{W}_{\text{HF}} = \frac{\overline{z}_{o1} - \overline{z}_{o2}}{\mu_{t1} - \mu_{t2}}$ is an unbiased estimator of W,

(v) $\text{Var}(\hat{W}_{\text{HF}}) = \frac{1}{(\mu_{t2} - \mu_{t1})^2}\left(\frac{\sigma_{\text{HF1}}^2}{n_1} + \frac{\sigma_{\text{HF2}}^2}{n_2}\right)$,

(vi) and$\hat{\text{Var}}(\hat{W}_{\text{HF}}) = \frac{1}{(\mu_{t2} - \mu_{t1})^2}\left(\frac{s_{\text{oHF1}}^2}{n_1} + \frac{s_{\text{oHF2}}^2}{n_2}\right)$.

where $\sigma_{\text{HF}j}^2 = \sigma_y^2 + W_G(1 - W_G)\mu_{tj}^2 + 2\mu_{tj}W_G(\overline{Y}_G - \overline{Y}) + W_G\{C_{uj}^2(\sigma_{yG}^2 + \overline{Y}_G^2) + \sigma_{tj}^2\}$.
and $s_{\text{oHF}j}^2 = \frac{1}{n_j - 1}\sum_{i \in s_j}(z_{oij} - \overline{z}_{oj})^2; j = 1, 2$

The estimators $\overline{z}_{\text{oHF}}$ and $\hat{W}_{\text{HF}}$ for FORT are the same as $\overline{z}_{\text{oH}}$ and $\hat{W}_{\text{H}}$ for PORT. But their variances are significantly different. The sufficient conditions for superiority of $\overline{z}_{\text{oH}}$ and $\hat{W}_{\text{H}}$ under PORT, over $\overline{z}_{\text{oHF}}$ and $\hat{W}_{\text{HF}}$ under FORD, are the same, e.g., $\sigma_{\text{HF}j}^2 \geq \sigma_{\text{oH}j}^2$. However, such a condition is generally difficult to verify. In particular, if $W = W_{\text{G}}$ and C_y(cv of y for the entire population) $\geq C_{y\text{G}}$(cv of y for the group G), the variances of $\overline{z}_{\text{oHF}}$ and $\hat{W}_{\text{HF}}$ under FORT provide higher, equal, and lower variances than those of $\overline{z}_{\text{oHF}}$ and $\hat{W}_{\text{HF}}$ under PORT, respectively. In case $W = W_{\text{G}}$ and $C_y = C_{y\text{G}}$, variance estimators under the assumption of FORD and PORT are the same. Details comparisons under the general sampling design are provided by Arnab (2019).

9.5 ORR in Two-Phase Sampling

Arnab and Olaomi (2019) proposed two-phase sampling, where an initial sample $\tilde{s}$ of size $\tilde{n}$ was selected from a population $U = (U_1, ..., U_N)$ of N units by using an arbitrary sampling design $\tilde{p}$ with inclusion probabilities $\tilde{\pi}_i$ and $\tilde{\pi}_{ij}(> 0)$ for the i th, and i and $j(\neq i)$ th units. In phase 1, each of the selected units in $\tilde{s}$ was asked about his or her options for providing the RR or DR method of interview for the sensitive question. Let $\tilde{s}_g$ and $\tilde{s}_{\overline{g}}(= \tilde{s} - \tilde{s}_g)$ be the sets of respondents of sizes $\tilde{n}_g$ and $\tilde{n}_{\overline{g}}(= \tilde{n} - \tilde{n}_g)$ were opted for the RR and DR methods of interviewing. In phase 2, sub-samples s_g and $s_{\overline{g}}$ of sizes $n_g = \gamma_g \tilde{n}_g$ and $n_{\overline{g}} = \gamma_{\overline{g}} \tilde{n}_{\overline{g}}$ are selected from s_g and $s_{\overline{g}}$ by using sampling designs $p_{g|\tilde{s}}$ and $p_{\overline{g}|\tilde{s}}$ respectively. Here, γ_g.and $\gamma_{\overline{g}}$ are suitably chosen fractions, and n_g and $n_{\overline{g}}$ are assumed to be integers (Rao, 1973). The inclusion probabilities for the ith, and ith and $j(\neq i)$ th units of the sampling designs $p_{g|\tilde{s}}(p_{\overline{g}|\tilde{s}})$ are denoted by $\pi_{gi|\tilde{s}}(\pi_{\overline{g}i|\tilde{s}})$ and $\pi_{gij|\tilde{s}}(\pi_{\overline{g}i|\tilde{s}})$. If the ith respondent is selected in s_g, a RR z_i is obtained from him or her by using a suitable randomization device. On the other hand, if the respondent is selected in $s_{\overline{g}}$, the DR y_i is obtained from him or her. Here we assume that the revised randomized response r_i (an estimate of y_i) obtained from z_i satisfies the model proposed by Arnab (2004) as follows:

$$E_R(r_i) = y_i,\ V_R(r_i) = \phi_{Ai} \text{ and } C_R(r_i, r_j) = 0 \text{ for } i \neq j \tag{9.5.1}$$

Estimation of the Mean and Variance

An estimator for the population mean $\overline{Y} = \sum_{i \in U} y_i/N$ is

$$\hat{\hat{Y}}_{\text{OT}} = \frac{1}{N}\left(\hat{Y}_g + \hat{Y}_{\overline{g}}\right) \tag{9.5.2}$$

where $\hat{Y}_g = \sum_{i \in s_g} \frac{r_i}{\pi^*_{gi|\tilde{s}}}$, $\hat{Y}_{\overline{g}} = \sum_{i \in s_{\overline{g}}} \frac{y_i}{\pi^*_{\overline{g}i|\tilde{s}}}$, $\pi^*_{gi|\tilde{s}} = \tilde{\pi}_i \pi_{gi|\tilde{s}}$ and $\pi^*_{\overline{g}i|\tilde{s}} = \tilde{\pi}_i \pi_{\overline{g}i|\tilde{s}}$.

Theorem 9.5.1

(i) $\hat{\hat{Y}}_{\text{OT}}$ is an unbiased estimator of $\overline{Y}$

(ii) $\text{Var}(\hat{\hat{Y}}_{\text{OT}}) = \frac{1}{N^2}\left[V_{\text{YG}} + E\left(V_{\text{YG}|\tilde{s}_g} + V_{\text{YG}|\tilde{s}_{\overline{g}}} + \sum_{i \in \tilde{s}_g} \frac{\phi_{Ai}}{\pi_i^2 \pi_{gi|\tilde{s}}}\right)\right]$

(iii) An unbiased estimator of $\text{Var}(\hat{\hat{Y}}_{\text{OT}})$ is

$$\hat{\text{Var}}(\hat{\hat{Y}}_{\text{OT}}) = \frac{1}{N^2}\left[\hat{V}_{\text{YG}}(r, y) + \hat{R}\right]$$

where

$$V_{\mathrm{GY}} = \sum_{i<} \sum_{j \in U} \Delta_{ij} \left(\frac{y_i}{\pi_i} - \frac{y_j}{\pi_j} \right)^2; \; V_{\mathrm{GY}|\tilde{s}_g} = \sum_{i<j} \sum_{\in \tilde{s}_g} \Delta_{gij} \left(\frac{y_i}{\pi^*_{gi|\tilde{s}}} - \frac{y_j}{\pi^*_{gj|\tilde{s}}} \right)^2,$$

$$V_{\mathrm{GY}|\tilde{s}_{\overline{g}}} = \sum_{i<j} \sum_{\in \tilde{s}_{\overline{g}}} \Delta_{\overline{g}ij} \left(\frac{y_i}{\pi^*_{\overline{g}i|\tilde{s}}} - \frac{y_j}{\pi^*_{\overline{g}j|\tilde{s}}} \right)^2;$$

$$\Delta_{ij} = \pi_i \pi_j - \pi_{ij}; \; \Delta_{gij} = \pi^*_{gi|\tilde{s}} \pi^*_{gj|\tilde{s}} - \pi^*_{gij|\tilde{s}}$$

$$\text{and} \Delta_{\overline{g}ij} = \pi^*_{\overline{g}i|\tilde{s}} \pi^*_{\overline{g}j|\tilde{s}} - \pi^*_{\overline{g}ij|\tilde{s}}$$

$$\hat{V}_{\mathrm{YG}}(r, y) = \sum_{i<} \sum_{j \in s_g} \frac{\Delta_{ij} Q_{ij}(r)}{\pi_{ij} \pi_{gij|\tilde{s}}} + \sum_{i<} \sum_{j \in s_{\overline{g}}} \frac{\Delta_{ij} Q_{ij}(y)}{\pi_{ij} \pi_{\overline{g}ij|\tilde{s}}}$$

$$+ 2 \sum_{i \in s_g} \sum_{j s_{\overline{g}}} \frac{\Delta_{ij} Q_{ij}(r, y)}{\pi_{ij} \pi_{gi|\tilde{s}} \pi_{\overline{g}j|\tilde{s}}};$$

$$Q_{ij}(r) = \left(\frac{r_i}{\pi_i} - \frac{r_i}{\pi_i} \right)^2 - \left(\frac{\hat{\phi}_{Ai}}{\pi_i^2} - \frac{\hat{\phi}_{Aj}}{\pi_j^2} \right);$$

$$Q_{ij}(y) = \left(\frac{y_i}{\pi_i} - \frac{y_j}{\pi_i} \right)^2; \; Q_{ij}(r, y) = \left(\frac{r_i^2 - \hat{\phi}_{Ai}}{\pi_i^2} + \frac{y_j^2}{\pi_j^2} - 2 \frac{r_i}{\pi_i} \frac{y_j}{\pi_j} \right);$$

$\hat{\phi}_{Ai}$is an unbiased estimator of ϕ_{Ai} and

$$\hat{R} = \sum_{i<} \sum_{j \in s_g} \frac{\Delta_{gij}}{\pi_{gij}} \left\{ \left(\frac{r_i}{\pi^*_{gi|\tilde{s}}} - \frac{r_j}{\pi^*_{gj|\tilde{s}}} \right)^2 - \left(\frac{\hat{\phi}_{Ai}}{\pi^{*2}_{gi|\tilde{s}}} + \frac{\hat{\phi}_{Aj}}{\pi^{*2}_{gj|\tilde{s}}} \right) \right\}$$

$$+ \sum_{i<} \sum_{j \in s_{\overline{g}}} \frac{\Delta_{\overline{g}ij}}{\pi_{\overline{g}ij}} \left\{ \left(\frac{y_i}{\pi^*_{\overline{g}i|\tilde{s}}} - \frac{y_j}{\pi^*_{\overline{g}j|\tilde{s}}} \right)^2 + \sum_{i \in s_g} \frac{\hat{\phi}_{Ai}}{\pi_i^2 \pi_{gi|\tilde{s}}} \right\}$$

Proof

(i)
$$E\left(\hat{\overline{Y}}_{\mathrm{OT}}\right) = \frac{1}{N}\left(\hat{Y}_g + \hat{Y}_{\overline{g}}\right)$$
$$= \frac{1}{N} E\left[E\left\{ \sum_{i \in s_g} \frac{E_R(r_i)}{\pi^*_{gi|\tilde{s}}} + \sum_{i \in s_{\overline{g}}} \frac{y_i}{\pi^*_{\overline{g}i|\tilde{s}}} |\tilde{s} \right\} \right]$$
$$= \frac{1}{N} E\left(\sum_{i \in \tilde{s}} \frac{y_i}{\pi_i} \right)$$
$$= \overline{Y}$$

(ii) $V\left(\hat{\overline{Y}}_{\text{OT}}\right) = V\left[E_R\left(\hat{\overline{Y}}_{\text{OT}}\right)\right] + E\left[V_R\left(\hat{\overline{Y}}_{\text{OT}}\right)\right]$

$$= \frac{1}{N^2}\left[V\left\{\sum_{i\in s_g}\frac{y_i}{\pi^*_{gi|\tilde{s}}} + \sum_{i\in s_{\overline{g}}}\frac{y_i}{\pi^*_{\overline{g}i|\tilde{s}}}\right\} + E\left(\sum_{i\in s_g}\frac{\phi_{Ai}}{\pi^{*2}_{gi|\tilde{s}}}\right)\right]$$

$$= \frac{1}{N^2}\left[\begin{array}{l} V\left\{E\left(\sum_{i\in s_g}\frac{y_i}{\pi^*_{gi|\tilde{s}}} + \sum_{i\in s_{\overline{g}}}\frac{y_i}{\pi^*_{\overline{g}i|\tilde{s}}}|\tilde{s}\right)\right\} + \\ E\left\{V\left(\sum_{i\in s_g}\frac{y_i}{\pi^*_{gi|\tilde{s}}} + \sum_{i\in s_{\overline{g}}}\frac{y_i}{\pi^*_{\overline{g}i|\tilde{s}}}|\tilde{s}\right)\right\} + E\left(\sum_{i\in \tilde{s}_g}\frac{\phi_{Ai}}{\pi_i^2\pi_{gi|\tilde{s}}}\right)\end{array}\right]$$

$$= \frac{1}{N^2}\left[\begin{array}{l} V\left(\sum_{i\in s}\frac{y_i}{\pi_i}\right) + E\left\{V\left(\sum_{i\in s_g}\frac{y_i}{\pi^*_{gi|\tilde{s}}} + \sum_{i\in s_{\overline{g}}}\frac{y_i}{\pi^*_{\overline{g}i|\tilde{s}}}|\tilde{s}\right)\right\} \\ +E\left(\sum_{i\in \tilde{s}_g}\frac{\phi_{Ai}}{\pi_i^2\pi_{gi|\tilde{s}}}\right)\end{array}\right]$$

$$= \frac{1}{N^2}\left[V_{\text{GY}} + E\left\{V_{\text{GY}|\tilde{s}_g} + V_{\text{GY}|\tilde{s}_{\overline{g}}} + \sum_{i\in \tilde{s}_g}\frac{\phi_{Ai}}{\pi_i^2\pi_{gi|\tilde{s}}}\right\}\right]$$

(iii) Noting $\hat{V}_{\text{YG}}(r, y)$ and $\hat{R}$ are unbiased estimators of V_{GY} and $E\left\{V_{\text{GY}|\tilde{s}_g} + V_{\text{GY}|\tilde{s}_{\overline{g}}} + \sum_{i\in \tilde{s}_g}\frac{\phi_{Ai}}{\pi_i^2\pi_{gi|\tilde{s}}}\right\}$, one can verify the result.

SRSWOR Sampling

Consider the case when the samples $\tilde{s}$, s_g, and $s_{\overline{g}}$ are selected by the simple random sampling without replacement (SRSWOR) method. In this case,

$$\pi_i = \frac{\tilde{n}}{N}, \pi_{ij} = \frac{\tilde{n}(\tilde{n}-1)}{N(N-1)}, \pi_{gi|\tilde{s}} = \frac{n_g}{\tilde{n}_g}, \pi_{gij|\tilde{s}} = \frac{n_g(n_g-1)}{\tilde{n}(\tilde{n}-1)}, \pi_{\overline{g}i|\tilde{s}} = \frac{n_{\overline{g}}}{\tilde{n}_{\overline{g}}},$$
$$\pi_{gij|\tilde{s}} = \frac{n_{\overline{g}}(n_{\overline{g}}-1)}{\tilde{n}_{\overline{g}}(\tilde{n}_{\overline{g}}-1)} \quad (9.5.3)$$

Further writing,

$$w_g = \frac{\tilde{n}_g}{\tilde{n}} = 1 - w_{\overline{g}}, \overline{r}_g = \frac{1}{n_g}\sum_{i\in s_g} r_i, \overline{y}_{\overline{g}} = \frac{1}{n_{\overline{g}}}\sum_{i\in s_{\overline{g}}} y_i$$

$$\overline{Y}_g = \frac{1}{N_G}\sum_{i\in G} y_i, \overline{Y}_{\overline{g}} = \frac{1}{N_{\overline{G}}}\sum_{i\in \overline{G}} y_i,$$

$$S_y^2 = \frac{1}{N-1}\sum_{i\in U}(y_i-\overline{Y})^2,\ S_{gy}^2 = \frac{1}{N_g-1}\sum_{i\in G}(y_i-\overline{Y}_g)^2,$$
$$S_{\overline{g}y}^2 = \frac{1}{N_{\overline{g}}-1}\sum_{i\in\overline{G}}(y_i-\overline{Y}_{\overline{g}})^2, \tag{9.5.4}$$

$\overline{\phi}_g = \frac{1}{N_G}\sum_{i\in G}\phi_{Ai}$, N_g = total number of respondents in the stratum G who opted for RR and $N_{\overline{g}} = N - N_g$ = number of respondents opted for direct response,

$$W_g = \frac{N_g}{N} = 1 - W_{\overline{g}},$$

one finds,

$$\hat{\overline{Y}} = w_g\overline{r}_g + w_{\overline{g}}\overline{y}_{\overline{g}}, \tag{9.5.5}$$

$$V_{YG} = N^2\left(\frac{1}{\tilde{n}}-\frac{1}{N}\right)S_y^2,\ E_p(V_{YG|\tilde{s}_g}) = \frac{N^2}{\tilde{n}}\left(\frac{1}{\gamma_g}-1\right)W_gS_{gy}^2,$$
$$E_p(V_{YG|\tilde{s}_{\overline{g}}}) = \frac{N^2}{\tilde{n}}\left(\frac{1}{\gamma_{\overline{g}}}-1\right)W_{\overline{g}}S_{\overline{g}y}^2 \tag{9.5.6}$$

and

$$E_p\left(\sum_{i\in s_{\overline{g}}}\frac{\phi_{Ai}}{\pi_i^2\pi_{gi}}\right) = \frac{N^2}{\tilde{n}\gamma_g}W_g\overline{\phi}_g \tag{9.5.7}$$

Using (9.5.5), (9.5.6), (9.5.7), and Theorem 9.5.1, Arnab and Olaomi (2019) derived the following results:

Theorem 9.5.2 Under the SRSWOR sampling,

(i) $\hat{\overline{Y}} = w_g\overline{r}_g + w_{\overline{g}}\overline{y}_{\overline{g}}$ is an unbiased estimator of $\overline{Y}$.

(ii) The variance of $\hat{\overline{Y}}$ is

$$\mathrm{Var}(\hat{\overline{Y}}) = \left(\frac{1}{\tilde{n}}-\frac{1}{N}\right)S_y^2 + \frac{1}{\tilde{n}}\left\{\begin{array}{l}\left(\frac{1}{\gamma_g}-1\right)W_gS_{gy}^2+\\ \left(\frac{1}{\gamma_{\overline{g}}}-1\right)W_{\overline{g}}S_{\overline{g}y}^2\end{array}\right\} + \frac{1}{\tilde{n}\gamma_g}W_g\overline{\phi}_g$$

(iii) An unbiased estimator of $\mathrm{Var}(\hat{\overline{Y}})$ is

$$\hat{\text{Var}}(\hat{\overline{Y}}) = \frac{1}{N}\left[\begin{array}{l}(N-1)\left\{\begin{array}{l}\left(\dfrac{\tilde{n}_g-1}{\tilde{n}-1}-\dfrac{n_g-1}{N-1}\right)\dfrac{w_g\left(s_{gr}^2-\overline{\hat{\phi}}_g\right)}{n_g}\\+\left(\dfrac{\tilde{n}_{\overline{g}}-1}{\tilde{n}-1}-\dfrac{n_{\overline{g}}-1}{N-1}\right)\dfrac{w_{\overline{g}}s_{\overline{g}y}^2}{n_{\overline{g}}}\end{array}\right\}\\+\dfrac{N-\tilde{n}}{\tilde{n}-1}w_g w_{\overline{g}}\left\{\left(\overline{r}_g-\overline{y}_{\overline{g}}\right)^2-\dfrac{\overline{\hat{\phi}}_g}{n_g}\right\}\end{array}\right]+\frac{\overline{\hat{\phi}}_g}{\tilde{n}\gamma_g}$$

where $s_{gr}^2 = \frac{1}{n_g-1}\sum_{i\in s_g}(r_i-\overline{r}_g)^2$, $s_{gy}^2 = \frac{1}{n_g-1}\sum_{i\in s_g}(y_i-\overline{y}_g)^2$, $s_{\overline{g}y}^2 = \frac{1}{n_{\overline{g}}-1}\sum_{i\in s_{\overline{g}}}(y_i-\overline{y}_{\overline{g}})^2$ and $\overline{\hat{\phi}}_g = \frac{1}{n_g}\sum_{i\in s_g}\hat{\phi}_{Ai}$

Arnab and Olaomi (2019) delivered the optimum values of $\tilde{n}$, γ_g and $\gamma_{\overline{g}}$ under a suitable cost function. Under the proportional allocation, with $n_g + n_{\overline{g}} = n$ is fixed, we have $\gamma_g = \gamma_{\overline{g}} = n/\tilde{n}$. In this case, $\text{Var}(\hat{\overline{Y}})$ reduces to

$$\text{V}_{\text{prop}}(\hat{\overline{Y}}) = \left(\frac{1}{\tilde{n}}-\frac{1}{N}\right)S_y^2 + \left(\frac{1}{n}-\frac{1}{\tilde{n}}\right)\left(W_g S_{gy}^2 + W_{\overline{g}}S_{\overline{g}y}^2\right) + \frac{1}{n}\frac{1}{N}\sum_{i\in G}\phi_{Ai} \quad (9.5.8)$$

Comparison with Arnab's (2004) Estimators

For the SRSWOR sampling, Arnab (2004)'s estimator (9.2.10) for $\overline{Y}$ is $\overline{r}_o = \frac{1}{n}\sum_{i\in s}\tilde{r}_i$. The variance of $\overline{r}_o$ with $\phi_{Ai} = \phi$ is obtained from Eq. (9.2.11) as

$$\text{V}_A = \text{Var}(\overline{r}_o) = (1-f)\frac{S_y^2}{n} + W_g\frac{\phi}{n} \quad (9.5.9)$$

For large N,

$$S_y^2 \geq W_g S_{gy}^2 + W_{\overline{g}}S_{\overline{g}y}^2 + W_g(\overline{Y}_g-\overline{Y})^2 + W_{\overline{g}}(\overline{Y}_{\overline{g}}-\overline{Y})^2 \geq W_g S_{gy}^2 - W_{\overline{g}}S_{\overline{g}y}^2 \quad (9.5.10)$$

Equations (9.5.8) (9.5.9), and (9.5.10) yield

$$\text{V}_A - V_{prop} = \left(\frac{1}{n}-\frac{1}{\tilde{n}}\right)\left\{S_y^2 - (W_g S_{gy}^2 + W_{\overline{g}}S_{\overline{g}y}^2)\right\} \geq 0 \quad (9.5.11)$$

From the above results, we obtain the following theorem:

Theorem 9.5.3 If the cost of the first-phase sample is ignored and N is large, the proposed two-phase ORT is more efficient than Arnab's (2004) ORR under SRSWOR sampling with proportional allocation, i.e.,

$$\mathrm{V}_A - V_{prop} \geq 0$$

9.6 Discussions

Randomized response techniques are used to collect data relating to sensitive items where direct questioning is not possible because of the sensitivity of the questions. However, the perception of sensitivity is subjective in the sense that the question may be considered sensitive for most of the respondents. Still, it may not be treated as so sensitive by others. ORT allows respondents to choose one of the two options, viz., either to provide DR or RR. The ORT is classified into three categories: full ORT (FORT), partial ORT (PORT), and mixed ORT (MORT). In a FORT, respondents are divided into two mutually exclusive and exhaustive groups G and $\overline{G}$. Respondents belonging to the group G always provide RR but never provide DR, while respondents belonging to $\overline{G}$ provide DR but never provide RR. In a PORT, a respondent sometimes provides DR and other times RR, depending on the particular state of their mind while answering the question. In a MORT, respondents are asked compulsory to provide RR and are given the option to choose DR at certain stages of the randomization procedure. The motivations behind the MORT are not clearly articulated in the literature. Most ORTs are limited to the SRSWR designs. A few ORTs, such as Arnab (2004), Chaudhuri and Saha (2005), and Chaudhuri and Dihidar (2009), are applicable to complex survey designs. In a PORT, most researchers assumed that the probability of a respondent providing RR was fixed for all the respondents. The assumption of constant probability is restrictive. However, Arnab (2004), Chaudhuri Pal (2008), and Chaudhuri and Dihidar (2009) relax such constraints. Gupta et al. (2002) termed W as the sensitivity level of the question answered. The value of W is unknown, and it is higher for a more sensitive question and lower for a less sensitive question. Several ORTs are proposed for estimating the population mean $\overline{Y}$ and W from a survey. Regarding efficiency, FORT is more efficient than the corresponding RRT (Arnab, 2004). However, such a conclusion is not valid for PORTs; it depends on the parameters of the RR models. Arnab (2019) compared the performances of estimators of $\overline{Y}$ and W under the assumptions of PORT and FORT and concluded that the variance of the estimators depends highly on the assumptions used. Arnab and Olaomi (2019) introduced two-phase ORT under the general sampling scheme and found that the two-phase ORT is more efficient than the single-phase ORT proposed by Arnab (2004).

References

Arnab, R. (2004). Optional randomized response techniques for complex survey designs. *Biometrical Journal, 46*(1), 114–124.

Arnab, R. (2017). *Survey sampling theory and Applications*. Elsevier.

Arnab, R. (2019). Optional randomized response techniques for quantitative characteristics. *Communications in statistics—theory and methods, 48*(16), 4154–4170.

Arnab, R., & Olaomi, J. O. (2019). Optional randomized response technique for quantitative characteristics. *Communications in Statistics-Theory and Methods, 48*, 4154–4170.

Arnab, R., & Rueda, M. (2016). Optional randomized response: A critical review. In A. Chaudhuri, T. C. Christofieds, & C. R. Rao (Eds.), *Hand book of statistics* (Vol. 34, pp. 253–271). Elsevier.

Chaudhuri, A. (2011). *Randomized response and Indirect questioning techniques in surveys*. Chapman & Hall/CRC.

Chaudhuri, A., & Dihidar, K. (2009). Estimating means of stigmatizing qualitative and quantitative variables from discretionary responses randomized or direct. In *Sankhyā* (Vol 71-B, pp. 123–136).

Chaudhuri, A., & Mukherjee, R. (1985). Optionally randomized response techniques. *Calcutta Statistical Association Bulletin, 34*, 225–229.

Chaudhuri, A., & Mukherjee, R. (1988). *Randomized response: Theory and techniques*. Marcel Dekker.

Chaudhuri, A., & Saha, A. (2005). Optional versus compulsory randomized response techniques in complex surveys. *Journal of the Statistical Planning and Inference, 135*, 516–527.

Chaudhuri, A., & Christofides, T. C. (2013). *Indirect questioning in sample surveys*. Springer.

Eichorn, B. H., & Hayre, L. S. (1983). Scrambled randomized response method for obtaining sensitive quantitative data. *Journal of Statistical Planning and Inference, 7*, 307–316.

Gupta, S. (2001). Qualifying the sensitivity level of binary response personal interview survey questions. *Journal of Combinatorics, Information & System Sciences, 26*(1–4), 101–109.

Gupta, S., & Shabbir, J. (2004). Sensitivity estimation for personal interview survey questions. *Statistica, 64*(3), 643–653.

Gupta, S., Gupta, B., & Singh, S. (2002). Estimation of sensitivity level of personal interview survey question. *Journal of Statistical Planning and Inference, 100*, 239–247.

Gupta, S., Kalucha, G., Shabbir, J., & Dass, B. K. (2014). Estimation of finite population mean using optional RRT models in the presence of nonsensitive uxiliary information. *American Journal of Mathematical and Management Sciences, 33*(2), 147–159.

Gupta, S., Mehta, S., Shabbir, J., & Dass, B. K. (2013). Generalized scrambling in quantitative optional randomized response models. *Communications in Statistics—Theory and Methods*, 42, 4034–4042.

Gupta, S., Shabbir, J., & Sehra, S. (2010). A comparison of multiplicative and additive optional RRT models. *Journal of Statistical Planning and Inference, 140*(10), 2870–2874.

Gupta, S., Thornton, B., Shabbir, J., & Singhal, S. (2006). A comparison of multiplicative and additive optional RRT models. *Journal of Statistical Theory and Applications, 64*, 226–239.

Huang, K. C. (2004). A survey technique for estimating the proportion and sensitivity in a dichotomous finite population. *Statistica Neeralandica, 58*, 75–82.

Huang, K. C. (2010). Unbiased estimators of mean, variance and sensitivity level for quantitative characteristics in finite population sampling. *Metrika, 71*, 341–352.

Huang, K. C. (2008). Estimation of sensitive characteristics using optional randomized technique. *Quality & Quantity, 42*, 679–686.

Hussain, Z., Al-Sobhi, M. M., & Al-Zahrani, B. (2014). Additive and subtractive scrambling in optional randomized response modeling. *PLoS One*, 9(1), e83557.

Kalucha, G., Gupta, S., & Dass, B. K. (2015). Ratio estimation of finite population mean using optional randomized response models. *Journal of Statistical Theory and Practice, 9*, 633–645.

Mangat, N. S., & Singh, S. (1994). Optional randomized response model. *Journal of Indian Statistical Association, 32*, 71–75.

Midzuno, H. (1952). On the sampling system with probability proportional to sum of sizes. *Annals of the Institute of Mathematical Statistics, 54*, 594–612.

Mukherjee, R. (2016). Optional randomized response revisited. In A. Chaudhuri, T. C. Christofieds, & C. R. Rao (Eds.), Volume 34 of *handbook of statistics* (pp. 331–339). Elsevier.

Pal, S. (2008). Unbiasedly estimating the total of a stigmatizing variables from a complex survey on permitting options for direct or randomized responses. *Statistical Papers, 49*, 157–164.

Rao, J. N. K., Hartley, H. O., & Cochran, W. G. (1962). On the simple procedure of unequal probability sampling without replacement. *Journal of the Royal Statistical Society, Series B, 24*, 482–491.

Ryu, J. B., Kim, J. M., Heo, T. Y., & Park, C. G. (2006). On stratified randomized response sampling. *Model Assisted Sampling and Applications, 1*, 31–36.

Sen, A. R. (1953). On estimate of variance in sampling with varying probabilities. *Journal of the Indian Society of Agricultural Statistics, 5*, 119–127.

Singh, S., & Joarder, A. H. (1997). Unknown repeated trials in randomized response model. *Journal of the Indian Society of Agricultural Statistics, 38*, 201–211.

Singh, H. P., & Mathur, N. (2003). Modified optional randomized response sampling techniques. *Journal of the Indian Society of Agricultural Statistics, 56*, 199–206.

Warner, S. L. (1965). Randomize response: A survey technique for eliminating evasive answer bias. *Journal of the American Statistical Association, 60*, 63–69.

Chapter 10
Item Count and Item Sum Techniques

Abstract The RR methods are used to collect data relating to highly stigmatizing characteristics that are not generally available through direct surveys. However, such methods are not totally satisfactory. They are subjected to certain limitations, such as a lack of reproducibility, a high cost, and being time-consuming. Respondents very often suspect randomization devices and feel there might be some hidden trick in the randomization process through which their secrecy might be somehow revealed. Several alternative methods have been proposed in the literature to overcome such difficulties. The methods include the item count technique (ICT), the item sum technique (IST), the non-randomized response (NRR) method, the parallel method, the hidden sensitivity model, and the three-card method. Details of the proposed techniques, along with their theoretical results, have been presented in this chapter. Bayesian methods for analyzing sensitive questions for the NRR models have also been proposed.

Keywords Bayesian method · Crosswise model · Dirichlet distribution · EM algorithm · Hidden sensitivity model · Item count technique (ICT) · Item sum technique (IST) · Maximum likelihood estimator (MLE) · Non-randomized response model (NRR) · Parallel model · Posterior mean · Posterior mode · Posterior variance · Restricted MLE · Triangular model

10.1 Introduction

In previous chapters, we have discussed the role of randomized response (RR) techniques in sensitive question surveys for obtaining reliable data by reducing socially desirable response bias, protecting respondents' confidentiality, and avoiding unacceptable rates of non-response. However, the approach is not totally acceptable to survey statisticians and practitioners. Some of the limitations are identified by Geurts (1980), Dalton and Metzger (1992), Hubbard et al. (1989), Chaudhuri and Christifides (2007), Tian et al. (2009a, 2009b), Hussain et al. (2012), Rueda et al. (2018),

R. Arnab, *Indirect Methods of Data Collection and Analysis from Surveys*, Indian Statistical Institute Series, https://doi.org/10.1007/978-981-97-6005-3_10

and Arnab et al. (2019), among others, as follows: (i) In RR techniques, investigators are required to explain randomization devices to the respondents, such as drawings of cards, the use of random numbers, etc., which makes the survey more expensive, time-consuming, and complex compared to the direct questioning (DQ) method; (ii) Lack of reproducibility: If asked to repeat the experiment, the same respondent may provide different responses; (iii) Some of the questions are alternatives to sensitive questions (e.g., Warner's (1965) model); (iv) Unsuitable for mail questionnaire; (v) Estimates are subject to lower efficiency than the DQ estimates; and (vi) Respondents do not understand the mathematical logic behind the technique and, hence, become suspicious that there is some trick behind the technique that may reveal their true status. To overcome the limitations of the RR technique, several alternative indirect questioning methods have been proposed for collecting sensitive information while preserving the confidentiality of the respondents. Such methods include the "Block total response method" proposed by Raghavarao and Federer (1979), "The unmatched count technique" by Dalton et al. (1994), the "Item count technique (ICT)" proposed by Droitcour et al. (1991), which is also known as the "List of experiments", "the nominative technique" proposed by Miller (1984, 1985), "the three-card method" proposed by Droitcour and Larson (2002), and the "Non-randomized response technique (NRR)" proposed by Tian and Tang (2014).

Applications of ICT in real surveys were cited by several researchers, including "drug use" by Droitcour et al. (1991), "racial hatred against black people" by Sinderman et al. (1997), "self-reports of racial prejudice" by Kulinski et al. (1997), "risky sexual behavior" by LaBrie and Earleywine (2000), "shoplifting" by Tsuchiya et al. (2007), and "attitudes toward immigration" by Janus (2010).

The validity of ICT has not yet been rigorously investigated (Imai, 2011). However, some of the researchers, e.g., Tsuchiyae et al. (2007), Holbrook and Krosnick (2010), and Coutts and Jann (2011), reported that the ICT elicits truthful answers to sensitive questions and outperforms the RRT by reducing social desirability bias in surveys relating to sensitive attributes. Kuha and Jackson (2014) reported ICT advantages over direct questioning (DQ) and RR techniques. Wimbush and Dalton (1997), Tsuchiya et al. (2007), and Walter and Lair (2014) reported that ICT provides higher estimates of sensitive items than DQ. However, some of the researchers, e.g., Droitcour et al. (1991), Biemer and Wright (2004), and Biemer and Brown (2005), found that DQ provides higher estimates compared to ICT for studying prevalence rate estimates of intravenous drug and cocaine use, respectively. However, Coutts and Jan (2011) and Trappmann et al. (2014) pointed out that the significant disadvantage of ICT over RRT is that it provides larger standard errors compared to the RRT estimates of the same sample size.

The application of ICT for quantitative sensitive characteristics was first introduced by Chaudhuri and Christofides (2013). Trappmann et al. (2014) used ICT in a nationwide computer-assisted telephone interview (CATI) survey on undeclared work in Germany and renamed it the item sum technique (IST). They reported that IST impressively outperformed DQ. Shaw (2015) extended the IST for varying probability sampling designs, while Hussain et al. (2017) used auxiliary information through ratio, product, and regression methods of estimation. Rueda et al. (2018)

applied a calibration estimation procedure for estimating the population mean for the IST in a complex survey design.

10.2 Item Count Techniques for Qualitative Variables

10.2.1 Pioneering Method

Droitcour et al. (1991) first proposed the item count technique (ICT) to estimate π_A, the proportion of individuals belonging to a population U that possess a certain sensitive attribute A. In this technique, two independent samples s_1 (the treatment group) and s_2 (the control group) of sizes n_1 and n_2 are selected by the simple random sampling with replacement (SRSWR) method. Respondents belonging to the treatment group s_1 are provided with a questionnaire containing a long list (LL) of $g + 1$ dichotomous questions, of which the first g of the questions are nonkey questions, which are not of direct interest and are not meant to be sensitive. The remaining $g + 1$-st question is the sensitive (key) question relating to the sensitive attribute A which is the focus of interest. The key question should be unrelated to the nonkey questions. The respondents belonging to the control group (sample s_2) are provided with a short list (SL) containing the same set of g innocuous questions that are present in the long list (LL). All the questions should have either a "yes" (if applicable to the respondent) or a "no" (if not applicable to the respondent) answer. Respondents are requested to report only the total score, which is the number of "yes" answers, but not the answer to any of the individual questions. Respondents' privacy is protected in the sense that the answer to the sensitive question of the LL (treatment group) remains undisclosed to the investigator unless the respondents provide a total score zero ("no" answer to all the questions) or $g + 1$ ("yes" answer to all the questions) to the LL containing the sensitive question.

Example 10.2.1 Control group (SL):

Instruction: State the total number of "yes" answers to the following three questions:

(i) Do you have a brother?
(ii) Are you a vegetarian?
(iii) Have you traveled outside of the country this year?

Treatment group (LL):

Instruction: State the total number of "yes" answers to the following four questions:

(i) Do you have a brother?
(ii) Are you a vegetarian?
(iii) Have you traveled outside of the country this year?
(iv) Have you ever bought a stolen item?

10.2.1.1 Estimation of the Proportion

Let the control group comprises g non-sensitive or innocuous attributes (nonkey variables) $B_1, \ldots, B_g$. Denote

$$x_{ij} = \begin{cases} 0, & \text{if the } i\text{th respondent does not possess the attribute } B_j \\ 1, & \text{if the } i\text{th respondent possess the attribute } B_j \end{cases} ; j = 1, \ldots, g \tag{10.2.1}$$

and

$$y_i = \begin{cases} 0, & \text{if the } i\text{th respondent does not possess the sensitive attribute } A \\ 1, & \text{if the } i\text{th respondent possess the sensitive attribute } A \end{cases} \tag{10.2.2}$$

Then, the total score of the ith individual belonging to the control group is

$$x_i = \sum_{j=1}^{g} x_{ij} \tag{10.2.3}$$

Response from the ith respondent can be expressed as

$$\begin{aligned} z_{1i} &= x_i + y_i, && \text{if the ith unit belongs to the treatment group } s_1 \\ \text{and } z_{2i} &= x_i, && \text{if the ith unit belongs to the control group } s_2 \end{aligned} \tag{10.2.4}$$

From Eq. (10.2.4), Droitcour et al. (1991) proposed an unbiased estimate of the proportion π_A as

$$\hat{\pi}_A = \bar{z}_1 - \bar{z}_2 \tag{10.2.5}$$

where $\bar{z}_1 = \frac{1}{n_1}\sum_{i=1}^{n_1} z_{1i} = \frac{1}{n_1}\sum_{i=1}^{n_1}(x_i + y_i)$, $\bar{z}_2 = \frac{1}{n_2}\sum_{i=1}^{n_2} z_{2i} = \frac{1}{n_2}\sum_{i=1}^{n_2} x_i$, and $\sum_{i=1}^{n_i}$ is the sum of all n_i units with repetition in the sample s_i, $i = 1, 2$.

The properties of $\hat{\pi}_A$ are stated as follows.

Theorem 10.2.1

(i) $\hat{\pi}_A$ is an unbiased estimator of π_A.
(ii) The variance of $\hat{\pi}_A$ is

$$\mathrm{Var}(\hat{\pi}_A) = \frac{\pi_A(1-\pi_A)}{n_1} + \frac{(n_1+n_2)}{n_1 n_2}\sum_{j=1}^{g}\theta_j(1-\theta_j)$$

where $\theta_j = \frac{1}{N}\sum_{i \in U} x_{ij}$ is the proportion of persons belonging to the innocuous group B_j.

(iii) An unbiased estimator of $\text{Var}(\hat{\pi}_A)$ is

$$\hat{\text{Var}}(\hat{\pi}_A) = \frac{\sum_{i=1}^{n_1} (z_{1i} - \bar{z}_1)^2}{n_1(n_1 - 1)} + \frac{\sum_{i=1}^{n_2} (z_{2i} - \bar{z}_2)^2}{n_2(n_2 - 1)}.$$

Proof

(i) $E(\hat{\pi}_A) = E(\bar{z}_1) - E(\bar{z}_2)$

$$= \frac{1}{N}\sum_{i \in U} (y_i + x_i) - \frac{1}{N}\sum_{i \in U} x_i$$

$$= \pi_A$$

(ii) $\text{Var}(\hat{\pi}_A) = \text{Var}(\bar{z}_1) + \text{Var}(\bar{z}_2)$

$$= \left[\text{Var}\left(\frac{1}{n_1}\sum_{i=1}^{n_1} y_i\right) + \text{Var}\left(\frac{1}{n_1}\sum_{i=1}^{n_1} x_i\right) + 2\text{Cov}\left(\frac{1}{n_1}\sum_{i=1}^{n_1} y_i, \frac{1}{n_1}\sum_{i=1}^{n_1} x_i\right)\right] + \text{Var}\left(\frac{1}{n_2}\sum_{i=1}^{n_2} x_i\right)$$

Now noting y_i and $x'_{ij}s$ are independent random variables, we find

$$\text{Var}(\hat{\pi}_A) = \frac{1}{n_1}\left[\pi_A(1 - \pi_A) + \sum_{j=1}^{g} \theta_j(1 - \theta_j)\right] + \frac{1}{n_2}\sum_{j=1}^{g} \theta_j(1 - \theta_j)$$

$$= \frac{\pi_A(1 - \pi_A)}{n_1} + \frac{(n_1 + n_2)}{n_1 n_2}\sum_{j=1}^{g} \theta_j(1 - \theta_j)$$

(iii) This follows from the fact that the samples s_1 and s_2 are selected independently from the population U by the SRSWR method.

10.2.1.2 Efficiency of the Estimator $\hat{\pi}_A$

The estimator $\hat{\pi}_A$ is a moment estimator. It can take inadmissible values outside of the parameter range [0, 1]. Further, the estimator of $\hat{\pi}_A$ is less efficient than the direct estimator $\hat{\pi}_d = \frac{1}{n}\sum_{i=1}^{n} y_i$ based on a sample of the same size n as the difference $\text{Var}(\hat{\pi}_A) - \text{Var}(\hat{\pi}_d)$ is positive.

10.2.1.3 Choice of g

If g is large, a respondent feels safe reporting their status for the key item because the actual status of the key item cannot be easily discovered. But a large g increases the variance $\text{Var}(\hat{\pi}_A)$, as indicated in Theorem 10.2.1, and the non-sampling error associated with counting the number of items also increases. For a small value of g, the respondent may feel that their response to the key item may not be kept undisclosed. Normally, the value of g is chosen between 2 and 6.

10.2.1.4 Selection of Nonkey Items

Most researchers recommended that the nonkey variables be mutually unrelated to one another and the key variables. Chaudhuri and Christofides (2013) suggested that the nonkey variables should not be totally unrelated to one another as well as to the key variable to give respondents the impression that each of the questions provides important information relating to the key variable. However, such recommendations were not supported by any real data.

10.2.2 Two Lists with General Sampling Design

Chaudhuri and Chistofides (2007) proposed an ICT with two sets, say, set1 and set2, of questionnaires, each comprised of $g+1$ questions. The first g questions of both sets are identical and innocuous in nature, e.g., the jth question is related to the innocuous attribute $B_j; j=1,\ldots,g$ while the last $g+1$-st question for set1 and set2 is different and is a mixture of the sensitive item A and an innocuous item F (which is different from the first g questions). The $g+1$-st question of set1 was framed as "Are you a member of the sensitive group A or the non-sensitive group F or both $A\&F$ (i.e., Do you $\in A\cup F$?)?" while the $g+1$-st question of set2 was framed as "Do you not belong to A or not belong to F (i.e., do you $\in A^c\cup F^c$)?". The prevalence θ_F of the non-sensitive group F in the population is assumed to be known.

Chaudhuri and Chistofides' (2007) method is based on two independent samples s_1 and s_2 of the same average size ν that were selected from a population U using the same sampling design p with inclusion probabilities π_i for the ith unit and $\pi_{ij}(>0)$ for the ith and jth $(i\neq j)$ units. Respondents belong to s_1 and s_2 are administered with set1 and set2 of questionnaires, respectively. Respondents were requested to report the total number of "yes" answers for the questionnaire as their responses, which can take values $0,1,\ldots,g+1$. The confidentiality of the responses is ensured for both the sensitive and non-sensitive items, as the interviewer will not know whether a respondent belongs to the sensitive group A or its complement A^c.

Denoting α_i and β_j as the responses from the ith unit of the sample s_1 and the jth unit of the sample s_2, $T_1 = \frac{1}{N}\sum_{i\in s_1}\frac{\alpha_i}{\pi_i}$ and $T_2 = \frac{1}{N}\sum_{i\in s_2}\frac{\beta_j}{\pi_j}$, Chaudhuri and Christofides (2007) derived the following theorem.

Theorem 10.2.2

(i) $\hat{\pi}_{\text{AHT}} = T_1 - T_2 + 1 - \pi_F$ is an unbiased estimator of π_A where π_F is the proportion of persons belonging the innocuous group F.
(ii) The variance of $\hat{\pi}_{\text{AHT}}$ is

$$\text{Var}(\hat{\pi}_{\text{AHT}}) = \frac{1}{N^2}\left[\sum_{i<}\sum_{j\in U}(\pi_i\pi_j - \pi_{ij})\left\{\left(\frac{\alpha_i}{\pi_i} - \frac{\alpha_j}{\pi_j}\right)^2 + \left(\frac{\beta_i}{\pi_i} - \frac{\beta_j}{\pi_j}\right)^2\right\}\right.$$
$$\left. + \sum_{i\in U}\frac{\alpha_i^2 + \beta_i^2}{\pi_i}\delta_i\right]$$

(iii) An unbiased estimator of $\text{Var}(\hat{\pi}_{\text{AHT}})$ is

$$\hat{\text{Var}}(\hat{\pi}_{\text{AHT}}) = \frac{1}{N^2}\left[\sum_{i<}\sum_{j\in s_1}\left(\frac{\pi_i\pi_j - \pi_{ij}}{\pi_{ij}}\right)\left(\frac{\alpha_i}{\pi_i} - \frac{\alpha_j}{\pi_j}\right)^2\right.$$
$$\left. + \sum_{i<}\sum_{j\in s_2}\left(\frac{\pi_i\pi_j - \pi_{ij}}{\pi_{ij}}\right)\left(\frac{\beta_i}{\pi_i} - \frac{\beta_j}{\pi_j}\right)^2 + \sum_{i\in s_1}\frac{\alpha_i^2}{\pi_i^2}\delta_i + \sum_{j\in s_2}\frac{\beta_j^2}{\pi_j^2}\delta_j\right]$$

where $\delta_k = 1 + \frac{1}{\pi_k}\sum_{j(\neq k)}\pi_{ij} - \nu$.

10.2.3 An Alternative Model

The alternative model is similar to Chaudhuri and Chistofides's (2007) ICT model in every respect except the last $g + 1$-st question of set 2. The last question of the proposed set 2 is "Do you not belong to A or belong to F (i.e., do you $\in A^c \cup F$)?".

Let $\tilde{\alpha}_i = \alpha_i$ and $\tilde{\beta}_j$ be the responses obtained from the ith unit of s_1 and the jth unit of s_2, and

$$f_i = \begin{cases} 1, & \text{if the } i\text{th unit belongs to innocous group } F \\ 0, & \text{if the } i\text{th unit does not belong to the innocous group } F \end{cases}$$

Then,

$$\tilde{\alpha}_i = \alpha_i = x_i + y_i + f_i - y_i f_i \text{ and}$$
$$\tilde{\beta}_i = x_i + (1 - y_i) + f_i - (1 - y_i)f_i = x_i - y_i + y_i f_i + 1 \qquad (10.2.6)$$

where y_i and x_i are given in Eqs. (10.2.2) and (10.2.3) respectively.

Let

$$T_1 = \frac{1}{N}\sum_{i\in s_1}\frac{\alpha_i}{\pi_i} \text{ and } T_3 = \frac{1}{N}\sum_{i\in s_2}\frac{\tilde{\beta}_j}{\pi_j} \tag{10.2.7}$$

Then,

$$E(T_1) = \frac{1}{N}\sum_{i\in U}\tilde{\alpha}_i = \frac{1}{N}\sum_{i\in U}\left[y_i + f_i + \left(\sum_{j=1}^{g} x_{ij}\right) - y_i f_i\right] = \pi_A + \pi_F + \sum_{j=1}^{g}\theta_j - \pi_{AF} \tag{10.2.8}$$

$$E(T_3) = \frac{1}{N}\sum_{i\in U}\tilde{\beta}_j = \frac{1}{N}\sum_{i\in U}\left[\left(\sum_{j=1}^{g} x_{ij}\right) - y_i + y_i f_i + 1\right] = \sum_{j=1}^{g}\theta_j - \pi_A + \pi_{AF} + 1 \tag{10.2.9}$$

where $\pi_{AF} = \frac{1}{N}\sum_{i\in U} y_i f_i$ = proportion of persons belonging to both the sensitive group A and innocuous group F.

Noting that $\pi_{AF} = \pi_A\pi_F$ because the sensitive group A is independent of the innocuous group F, an unbiased estimator of π_A is obtained as

$$\hat{\pi}_{\text{AHT}}(\text{M}) = \frac{T_1 - T_3 + (1 - \pi_F)}{2(1 - \pi_F)} \tag{10.2.10}$$

The variance of $\hat{\pi}_{\text{AHT}}(\text{M})$ and its unbiased estimator come out as

$$\begin{aligned}\text{Var}[\hat{\pi}_{\text{AHT}}(\text{M})] &= \frac{1}{4(1-\pi_F)^2}[\text{Var}(T_1) + \text{Var}(T_3)] \\ &= \frac{1}{4(1-\pi_F)^2}\frac{1}{N^2}\left[\sum_{i\in U}\left(\frac{1}{\pi_i} - 1\right)(\tilde{\alpha}_i^2 + \tilde{\beta}_i^2)\right. \\ &\quad \left. + \sum_{i\neq}\sum_{j\in U}\left(\frac{\pi_{ij}}{\pi_i\pi_j} - 1\right)(\tilde{\alpha}_i\tilde{\alpha}_j + \tilde{\beta}_i\tilde{\beta}_j)\right]\end{aligned} \tag{10.2.11}$$

and

$$\begin{aligned}&\hat{\text{Var}}[\hat{\pi}_{\text{AHT}}(\text{M})] \\ &= \frac{1}{4(1-\pi_F)^2}\frac{1}{N^2}\left[\left\{\sum_{i\in s_1}\left(\frac{1}{\pi_i} - 1\right)\frac{\tilde{\alpha}_i^2}{\pi_i} + \sum_{i\neq}\sum_{j\in s_1}\left(\frac{\pi_{ij}}{\pi_i\pi_j} - 1\right)\frac{\alpha_i\alpha_j}{\pi_{ij}}\right\}\right. \\ &\quad \left. + \left\{\sum_{i\in s_2}\left(\frac{1}{\pi_i} - 1\right)\frac{\tilde{\beta}_i^2}{\pi_i} + \sum_{i\neq}\sum_{j\in s_2}\left(\frac{\pi_{ij}}{\pi_i\pi_j} - 1\right)\frac{\tilde{\beta}_i\tilde{\beta}_j}{\pi_{ij}}\right\}\right]\end{aligned} \tag{10.2.12}$$

10.2.4 Comparison with Chaudhuri and Christofides Model

The proposed model will be more efficient than the Chaudhuri and Christofides (2007) model if

$$\mathrm{Var}(\hat{\pi}_{\mathrm{AHT}}(\mathrm{M})) \leq \mathrm{Var}(\hat{\pi}_{\mathrm{AHT}}) \tag{10.2.13}$$

i.e.,

$$\mathrm{V}(T_3) \leq \mathrm{V}(\mathrm{T}_1)\left\{4(1-\pi_F)^2 - 1\right\} + 4(1-\pi_F)^2\mathrm{V}(\mathrm{T}_2) \tag{10.2.14}$$

It is not easy to check the situations where the condition (10.2.14) holds for a general sampling design.

10.2.4.1 SRSWR Sampling

For the SRSWR, the estimators T_1, T_2 and T_3 become

$$T_1 = \frac{1}{n}\sum\nolimits_{s_1} \alpha_i,\ T_2 = \frac{1}{n}\sum\nolimits_{s_2} \beta_i \text{ and } T_3 = \frac{1}{n}\sum\nolimits_{s_2} \tilde{\beta}_i \tag{10.2.15}$$

where $\sum_{s_j}$ denotes the sum over all the n units of $s_j (j = 1, 2)$ with repetition.

Now writing,

$$\begin{aligned}
\alpha_i &= x_i + y_i + f_i - y_i f_i,\\
\beta_i &= x_i + (1-y_i) + (1-f_i) - (1-y_i)(1-f_i) = 1 + x_i - f y_i,\\
\tilde{\beta}_i &= x_i - y_i + y_i f_i + 1
\end{aligned}$$

and assuming that the sensitive variable y is unrelated to each of the innocuous variables and the innocuous variables are mutually independent, we find

$$\begin{aligned}
V(T_1) &= V\left[\frac{1}{n}\sum\nolimits_{s_1}(x_i + y_i + f_i - y_i f_i)\right]\\
&= V\left(\frac{1}{n}\sum\nolimits_{s_1} y_i\right) + V\left(\frac{1}{n}\sum\nolimits_{s_1} x_i\right) + V\left(\frac{1}{n}\sum\nolimits_{s_1} f_i\right) + V\left(\frac{1}{n}\sum\nolimits_{s_1} y_i f_i\right)\\
&\quad - 2Cov\left(\frac{1}{n}\sum\nolimits_{s_1} y_i, \frac{1}{n}\sum\nolimits_{s_1} y_i f_i\right) - 2Cov\left(\frac{1}{n}\sum\nolimits_{s_1} f_i, \frac{1}{n}\sum\nolimits_{s_1} y_i f_i\right)\\
&= \frac{1}{n}\left[\pi_A(1-\pi_A) + \sum_{j=1}^{g}\theta_j(1-\theta_j) + \pi_F(1-\pi_F)\right.\\
&\qquad\left. + \pi_{AF}(1-\pi_{AF}) - 2\pi_{AF}(1-\pi_A) - 2\pi_{AF}(1-\pi_F)\right]
\end{aligned}$$

$$= \frac{1}{n}\left[\pi_A(1-\pi_A)+\sum_{j=1}^{g}\theta_j(1-\theta_j)+\pi_F(1-\pi_F)\right.$$
$$\left.+\pi_{AF}(1-\pi_{AF})-2\pi_{AF}(2-\pi_F-\pi_A)\right] \tag{10.2.16}$$

$$\begin{aligned}V(T_2) &= V\left[\frac{1}{n}\sum\nolimits_{s_1}(1+x_i-f_iy_i)\right]\\ &= V\left(\frac{1}{n}\sum\nolimits_{s_1}x_i\right)+V\left(\frac{1}{n}\sum\nolimits_{s_1}f_iy_i\right)\\ &= \frac{1}{n}\left[\sum_{j=1}^{g}\theta_j(1-\theta_j)+\pi_{AF}(1-\pi_{AF})\right]\end{aligned} \tag{10.2.17}$$

and

$$\begin{aligned}V(T_3) &= V\left[\frac{1}{n}\sum\nolimits_{s_2}(x_i-y_i+y_if_i+1)\right]\\ &= V\left(\frac{1}{n}\sum\nolimits_{s_2}x_i\right)+V\left(\frac{1}{n}\sum\nolimits_{s_2}y_i\right)+V\left(\frac{1}{n}\sum\nolimits_{s_2}f_iy_i\right)\\ &\quad -2Cov\left(\frac{1}{n}\sum\nolimits_{s_2}y_i,\frac{1}{n}\sum\nolimits_{s_2}f_iy_i\right)\\ &= \frac{1}{n}\left[\pi_A(1-\pi_A)+\sum_{j=1}^{g}\theta_j(1-\theta_j)+\pi_{FA}(1-\pi_{FA})-2\pi_{FA}(1-\pi_A)\right]\end{aligned}$$

Now writing $\pi_{AF}=\pi_A\pi_F$ in (10.2.17), one obtains

$$\begin{aligned}V(T_3) &= \frac{1}{n}\left[\pi_A(1-\pi_A)(1-2\pi_F)+\sum_{j=1}^{g}\theta_j(1-\theta_j)+\pi_{AF}(1-\pi_{AF})\right]\\ &= V(T_2)+\frac{\delta}{n}\end{aligned} \tag{10.2.18}$$

where

$$\delta=\pi_A(1-\pi_A)(1-2\pi_F) \tag{10.2.19}$$

The estimators of π_A for the Chaudhuri and Christofides (2007) and the proposed model become

$$\hat{\pi}_{\text{ACH}}=\frac{1}{n}\left[\sum\nolimits_{s_1}\alpha_i-\sum\nolimits_{s_2}\beta_i+1-\pi_F\right] \text{ and}$$

$$\hat{\pi}_{\mathrm{AP}} = \frac{1}{2(1-\pi_F)}\frac{1}{n}\left[\sum\nolimits_{s_1}\alpha_i - \sum\nolimits_{s_2}\tilde{\beta}_i + 1 - \pi_F\right] \tag{10.2.20}$$

The variances of $\hat{\pi}_{\mathrm{ACH}}$ and $\hat{\pi}_{\mathrm{AP}}$ are

$$\mathrm{Var}(\hat{\pi}_{\mathrm{ACH}}) = \frac{1}{n}\begin{bmatrix} \pi_A(1-\pi_A) + 2\sum_{j=1}^{g}\theta_j(1-\theta_j) + \pi_F(1-\pi_F) \\ +2\pi_{AF}(1-\pi_{AF}) - 2\pi_{AF}(2-\pi_A-\pi_F) \end{bmatrix} \tag{10.2.21}$$

and

$$\mathrm{Var}(\hat{\pi}_{\mathrm{AP}}) = \frac{1}{4(1-\pi_F)^2}\frac{1}{n}\begin{bmatrix} \pi_A(1-\pi_A) + 2\sum_{j=1}^{g}\theta_j(1-\theta_j) + \pi_F(1-\pi_F) \\ +2\pi_{AF}(1-\pi_{AF}) - 2\pi_{AF}(2-\pi_A-\pi_F) + \delta \end{bmatrix} \tag{10.2.22}$$

Remark 10.2.1 From Theorem 10.2.2, Chaudhuri and Christofides (2007, 2013) concluded that the number of nonkey questions g has no role in making inferences on π_A which is incorrect, as Eq. (10.2.21) shows that $\mathrm{Var}(\hat{\pi}_{\mathrm{ACH}})$ increases with the total number of innocuous variables g.

Remark 10.2.2 Chaudhuri and Christofides (2013) did not provide an explicit expression of variance of $\mathrm{Var}(\hat{\pi}_{\mathrm{ACH}})$ and remarked that $\mathrm{Var}(\hat{\pi}_{\mathrm{ACH}})$ is unknown.

10.2.5 Three List ICT

In this method, three lists of questionnaires are prepared. The first list L_1 consists of g innocuous questions similar to the method described in Sect. 10.2.1. The second list L_2 consists of $g+1$ question. The first g questions are the same as the first list L_1 and the last $g+1$-st question is "Do you possess the sensitive characteristic A or innocuous characteristic F?". The third list L_3 consists of $g+1$ questions of which the first g questions are the same as the list L_1 and the last $g+1$-st question is "Do you possess the sensitive characteristic A or F^c?".

Three independent samples s_1, s_2 and s_3 of sizes n_1, n_2 and n_3 are selected by the SRSWR method. The respondents belonging to s_j are served the list L_j ($j = 1, 2, 3$) and requested to provide the total scores only. Let $z_i(j)$ denotes the scores obtained from the ith respondent if he or she receives the list L_j. Then,

$$z_i(1) = x_i = \sum_{j=1}^{g} x_{ij},\ z_i(2) = x_i + y_i + f_i - f_i y_i \text{ and}$$
$$z_i(3) = x_i + y_i + 1 - f_i - (1-f_i)y_i = x_i + 1 - f_i + f_i y_i \tag{10.2.23}$$

where x_{ij} is given in (10.2.1).

Denoting, $\overline{z}(1) = \frac{1}{n_1}\sum_{i=1}^{n_1} z_i(1)$, $\overline{z}(2) = \frac{1}{n_2}\sum_{i=1}^{n_2} z_i(2)$ and $\overline{z}(3) = \frac{1}{n_3}\sum_{i=1}^{n_3} z_i(3)$, we find

$$E(\overline{z}_1) = \sum_{j=1}^{g}\theta_j,\ E(\overline{z}_2) = \sum_{j=1}^{g}\theta_j + \pi_A + \pi_F - \pi_{FA} = \sum_{j=1}^{g}\theta_j + \pi_A(1-\pi_F) + \pi_F,$$

$$E(\overline{z}_3) = \sum_{j=1}^{g}\theta_j + 1 - \pi_F + \pi_{FA} = \sum_{j=1}^{g}\theta_j - \pi_F(1-\pi_A) + 1, \tag{10.2.24}$$

$$V(\overline{z}_1) = \frac{\gamma_1}{n_1},\ V(\overline{z}_2) = \frac{1}{n_2}\Big[\gamma_1 + \pi_A(1-\pi_A) + \pi_F(1-\pi_F)$$
$$+ \pi_A\pi_F\{1 - \pi_A\pi_F - 2(2-\pi_A-\pi_F)\}\Big] = \frac{\gamma_2}{n_2} \text{ and}$$
$$V(\overline{z}_3) = \frac{1}{n_3}\Big[\gamma + \pi_F(1-\pi_F) - 2\pi_A\pi_F(1-\pi_F)\Big] = \frac{\gamma_3}{n_3}$$
$$\text{where } \gamma_1 = \sum_{j=1}^{g}\theta_j(1-\theta_j) \tag{10.2.25}$$

Equations (10.2.24) and (10.2.25) lead to the following theorem.

Theorem 10.2.3

(i) $\tilde{\pi}_A = \overline{z}_2 + \overline{z}_3 - 2\overline{z}_1 - 1$ is an unbiased estimator of π_A
(ii) $\text{Var}(\tilde{\pi}_A) = \frac{\gamma_1}{n_1} + \frac{\gamma_2}{n_2} + \frac{\gamma_3}{n_3}$
(iii) An unbiased estimator of $\text{Var}(\tilde{\pi}_A)$ is

$$\hat{\text{Var}}(\tilde{\pi}_A) = \frac{s_1^2}{n_1} + \frac{s_2^2}{n_2} + \frac{s_3^2}{n_3} \text{ where } s_j^2 = \frac{1}{n_j-1}\sum_{i=1}^{n_j}\left(z_i(j) - \overline{z}_j\right)^2;\ j = 1, 2, 3.$$

10.2.6 Single List ICT

Hussain et al. (2011, 2012) proposed an ICT sampling scheme where each respondent selected in the sample was asked to fill in a list containing g questions. The $j(=1,\ldots,g)$th question is a mixture of a sensitive characteristic (A) and an innocuous item (F_j) such as "Do you possess at least one of the attributes A or F_j?". The respondent is directed to count 1 if the answer is "yes". Otherwise, they should count 0 as their response to the jth question. The respondent should report only the total score for the entire set of g questions. For this method, it is assumed that the proportion of persons belonging to each of the innocuous g items F_j is known, and that the innocuous attributes are unrelated among themselves and to the sensitive attribute A. Writing,

$$x_{ij} = \begin{cases} 1 \text{ if } i \in F_j \\ 0 \text{ if } i \notin F_j \end{cases}, x_i = \sum_{j=1}^{g} x_{ij} \text{ and } y_i = \begin{cases} 1 \text{ if } i \in A \\ 0 \text{ if } i \notin A \end{cases}, \tag{10.2.26}$$

one finds the total score from the ith respondent as

$$z_i = \sum_{j=1}^{g} \left(y_i + x_{ij} - x_{ij}y_i \right) = gy_i + x_i(1 - y_i). \tag{10.2.27}$$

Let a sample s of size n be selected from a population by the SRSWR method and $\overline{z} = \frac{1}{n} \sum_{i=1}^{n} z_i$ be the sample mean. Then,

$$\begin{aligned} E(\overline{z}) &= \left[g \frac{1}{N} \sum_{i \in U} y_i - \sum_{j=1}^{g} \frac{1}{N} \sum_{i \in U} x_{ij} - \sum_{j=1}^{g} \frac{1}{N} \sum_{i \in U} x_{ij} y_i \right] \\ &= g\pi_A + \sum_{j=1}^{g} \theta_j - \sum_{j=1}^{g} \pi_{Aj} \end{aligned}$$

where π_{Aj} = proportion of persons belonging to both the group A and F_j.

Noting, $\pi_{Aj} = \pi_A \theta_j$ as A and F_j are independent, one finds

$$\begin{aligned} E(\overline{z}) &= g\pi_A + \sum_{j=1}^{g} \theta_j (1 - \pi_A) \\ &= \left(g - \sum_{j=1}^{g} \theta_j \right) \pi_A - \sum_{j=1}^{g} \theta_j \end{aligned} \tag{10.2.28}$$

The variance of $\overline{z}$ is

$$\begin{aligned} V(\overline{z}) &= g^2 V\left(\frac{1}{n} \sum_{i=1}^{n} y_i \right) + V\left(\frac{1}{n} \sum_{i=1}^{n} x_i \right) + V\left(\frac{1}{n} \sum_{i=1}^{n} y_i x_i \right) \\ &\quad - 2gCov\left(\frac{1}{n} \sum_{i=1}^{n} y_i, \frac{1}{n} \sum_{i=1}^{n} y_i x_i \right) \\ &= \frac{1}{n} \left[g^2 \pi_A (1 - \pi_A) + \sum_{j=1}^{g} \theta_j (1 - \theta_j) + \sum_{j=1}^{g} \pi_{AF_j} \left(1 - \pi_{AF_j} \right) \right. \\ &\quad \left. + \sum_{j \neq} \sum_{k} \left(\pi_{AF_j F_k} - \pi_{AF_j} \pi_{AF_k} \right) - 2g \sum_{j=1}^{g} \left(\pi_{AF_j} - \pi_A \pi_{AF_j} \right) \right] \end{aligned}$$

$$= \frac{1}{n}\left[g^2\pi_A(1-\pi_A) + \sum_{j=1}^{g}\theta_j(1-\theta_j) + \sum_{j=1}^{g}\pi_A\theta_j(1-\pi_A\theta_j)\right.$$

$$\left. + \pi_A(1-\pi_A)\sum_{j\neq}^{g}\sum_{k=1}^{g}\theta_j\theta_k - 2g\sum_{j=1}^{g}\pi_A\theta_j(1-\pi_A)\right]$$

$$= \frac{1}{n}\left[\left\{g - \left(\sum_j \theta_j\right)\right\}^2 \pi_A(1-\pi_A) + (1-\pi_A)\sum_{j=1}^{g}\theta_j(1-\theta_j)\right] \quad (10.2.29)$$

Equations (10.2.5) and (10.2.6) lead the following theorem.

Theorem 10.2.4

(i) $\hat{\pi}_{AH} = \frac{\bar{z}-\sum_{j=1}^{g}\theta_j}{g-\sum_{j=1}^{g}\theta_j}$ is an unbiased estimator of π_A

(ii) The variance of $\hat{\pi}_{AH}$ is

$$\mathrm{Var}(\hat{\pi}_{AH}) = \frac{1}{n}\left[\pi_A(1-\pi_A) + \frac{(1-\pi_A)\sum_{j=1}^{g}\theta_j(1-\theta_j)}{\left(g-\sum_{j=1}^{g}\theta_j\right)^2}\right]$$

(iii) An unbiased estimator of $\mathrm{Var}(\hat{\pi}_{AH})$ is

$$\hat{\mathrm{Var}}(\hat{\pi}_{AH}) = \frac{1}{n(n-1)}\sum_{i=1}^{n}(z_i-\bar{z})^2 \Big/ \left(g-\sum_{j=1}^{g}\theta_j\right)^2.$$

Remark 10.2.3 Apart from the usual drawback of the possibility of an inadmissible value of the estimate $\hat{\pi}_{AH}$, the method provides limited privacy protection to the respondents as they are required to answer all the g questions, each of which is related to the sensitive item.

10.2.7 Christofides's ICT Model

In Christofides's (2015) ICT model, two independent samples s_1 and s_2 of sizes n_1 and n_2 are selected from the population by the SRSWR method. Each person in the sample s_1 is provided with an LL of $g+1$ items or statements, of which the first g of them are innocuous and the $g+1$-st item is sensitive. The respondent is requested to count the total number of items applicable to him or her and is directed to report the exact number if it is less than $g+1$. If all the $g+1$ items are applicable to them, they should report zero as a response. Each person in the sample s_2 is presented with a SL containing the first g innocuous items of LL only and asked to report the number of items applicable to him or her. In either case, the respondent should not reveal any information relating to a particular item, whether the item is applicable or not.

Let.

q_k = Probability of obtaining a response k from a respondent belonging to s_1, $k = 0, \ldots, g$

p_k = Probability that exactly k of the innocuous items is applicable to an individual, $k = 0, \ldots, g$ $n_k^{(1)}$ = Number of individuals in s_1 reporting the number k as response, $k = 0, \ldots, g$

$n_k^{(2)}$ = Number of individuals in s_2 reporting the number k as response, $k = 0, \ldots, g$

$$C_1 = \frac{1}{n_1}\sum_{k=1}^{g} kn_k^{(1)} \text{ and } C_2 = \frac{1}{n_1}\sum_{k=1}^{g} kn_k^{(2)}.$$

Then,

$$\begin{aligned} q_0 &= p_0(1-\pi_A) + p_g\pi_A \\ q_k &= p_k(1-\pi_A) + p_{k-1}\pi_A;\ k = 1, \ldots, g \end{aligned} \tag{10.2.30}$$

Further, noting that the vectors $\left(n_0^{(1)}, \ldots, n_g^{(1)}\right)$ and $\left(n_0^{(2)}, \ldots, n_g^{(2)}\right)$ follow multinomial distributions with parameters $n_1, q_0, \ldots, q_g$ and $n_2, p_0, \ldots, p_g$, Christofides suggested the following estimator of $\hat{\pi}_A$ as

$$\hat{\pi}_{AC} = \frac{C_1 - C_2}{1-(g+1)\hat{p}_g} \tag{10.2.31}$$

where $(g+1)\hat{p}_g \neq 1$ and $\hat{p}_g = n_g^{(2)}/n_2$.

Christofides obtained the following theorem relating to the properties of the estimator $\hat{\pi}_{AC}$: The proof of the theorem is omitted here as it is highly complex.

Theorem 10.2.5

(i) $\hat{\pi}_{AC}$ is a bias estimator of π_A
(ii) The amount of bias of $\hat{\pi}_{AC}$ is

$$B(\hat{\pi}_{AC}) \cong \frac{(g+1)p_g}{n_2\left[1-(g+1)p_g\right]^2}\left[\pi_A(g+1)(1-p_g) + H - g\right]$$

where $H = E(C_2) = \sum_{k=1}^{g} kp_k$

(iii) For large n_2,

$$\operatorname{Var}(\hat{\pi}_{AC}) \cong \frac{1}{\left[1-(g+1)p_g\right]^2}\left(\frac{\sigma_1^2}{n_1}+\frac{\sigma_2^2}{n_2}\right)$$

$$\text{where } \sigma_1^2 = \left[\sum_{k=0}^{g} k^2 q_k - \left(\sum_{k=0}^{g} k q_k\right)^2\right] \text{ and } \sigma_2^2 = \left[\sum_{k=0}^{g} k^2 p_k - \left(\sum_{k=0}^{g} k p_k\right)^2\right].$$

(iv) An approximate unbiased estimator of $\operatorname{Var}(\hat{\pi}_{AC})$ is

$$\widehat{\operatorname{Var}}(\hat{\pi}_{AC}) \cong \frac{1}{\left[1-(g+1)p_g\right]^2}\left(\frac{S_1^2}{n_1}+\frac{S_2^2}{n_2}\right)$$

where S_1^2 and S_2^2 are the sample variances of the numbers reported by the first and second sample respectively.

Remark 10.2.4 Christofides's (2015) method provides privacy for the respondents, as the respondents do not reveal whether the participants belong to the sensitive category or not. It also deters people from reporting the number zero. In the original version, some respondents deliberately reported zero to be on the safe side, regardless of what the actual number is.

10.2.8 Tian et al. (2017) ICT Model

In the original version of ICT provided by Droitcour et al. (1991), a respondent with an LL could provide a response $g+1$ in the event that all the g innocuous items as well as the sensitive item were applicable to him or her. In this situation, respondents' privacy is not protected. To overcome such difficulties, Tian et al. (2017) have proposed two ICTS using Poisson and negative binomial distributions.

Let y be the answer to the sensitive question (e.g., are you HIV+?) which is either "yes" or "no". Let $y = 1$, if the respondent possesses the sensitive characteristic A and $y = 0$, otherwise. Let X be an answer to the innocuous question (e.g., how many times did you go to a church last month?) that is unrelated to y.

Two independent samples s_1 and s_2 of sizes n_1 and n_2 are selected from a population by the SRSWR method. Respondents belong to s_1 are provided with a short list (SL) of the non-sensitive questions.

SL:

Q1: How many times did you go to church last week?

Respondents belonging to the sample s_2 were provided with a long list (LL) consisting of two questions; the first one, Q1, is the same as in s_1 but the second one, Q2, is a sensitive question.

LL:

Q1: How many times did you go to the last week?

Q2: Do you have HIV?

The respondents are required to provide the sum of the answers ($z = X + y$) of the two questions to protect the privacy of the answer to question Q2.

Let $\{x_i\}_{i=1}^{n_1}$ and $\{z_i\}_{i=1}^{n_2}$ be the observed responses from SL and LL respectively. Tian et al. (2017) considered the following two models:

10.2.8.1 Poisson Model

Let X follow Poisson distribution with a parameter λ and y follows Bernoulli distribution with a parameter π_A. Then,

$$E\left(\frac{1}{n_1}\sum_{i=1}^{n_1} x_i\right) = \lambda \tag{10.2.32}$$

and

$$E\left(\frac{1}{n_2}\sum_{i=1}^{n_2} z_i\right) = \lambda + \pi_A \tag{10.2.33}$$

From (10.2.29) and (10.2.30), Tian et al. (2017) derived the following theorem: The proof of the theorem is omitted as it is simple.

Theorem 10.2.6

(i) $\hat{\pi}_{\mathrm{AT}} = \frac{1}{n_2}\sum\limits_{i=1}^{n_2} z_i - \frac{1}{n_1}\sum\limits_{i=1}^{n_1} x_i$ is an unbiased estimator of π_A

(ii) The variance of $\hat{\pi}_{\mathrm{T}}$ is

$$\mathrm{Var}(\hat{\pi}_{\mathrm{AT}}) = \frac{\lambda + \pi_A(1-\pi_A)}{n_2} + \frac{\lambda}{n_1}$$

(iii) An unbiased estimator of $\mathrm{Var}(\hat{\pi}_{\mathrm{T}})$ is

$$\hat{\mathrm{Var}}(\hat{\pi}_{\mathrm{AT}}) = \frac{\hat{\lambda} + \hat{\pi}_{\mathrm{AT}}(1-\hat{\pi}_{\mathrm{AT}})}{n_2} + \frac{\hat{\lambda}}{n_1}$$

where $\hat{\lambda} = \overline{x} = \frac{1}{n_1}\sum\limits_{i=1}^{n_1} x_i$.

Remark 10.2.5 The estimator $\hat{\pi}_{\mathrm{AT}}$ can take an inadmissible value outside the interval [0, 1]. In this case, one can use the EM algorithm to find the MLE of π_A which always lies within [0, 1].

Remark 10.2.6 Wald $100(1-\alpha)\%$ confidence interval (CI) of π_{A} is given by $\hat{\pi}_{\text{AT}} \pm z_{1-\alpha/2}\sqrt{\hat{\text{V}}\text{ar}(\hat{\pi}_{\text{AT}})}$. For the small sample size, the interval may include inadmissible values of π_{A}. In this case, one can use the bootstrap method to get admissible CI.

Remark 10.2.7 For a given total sample size n, the optimum value of $n_2(= n - n_1)$ that minimizes the $\text{Var}(\hat{\pi}_{\text{AT}})$ is

$$n_2 = n\frac{\sqrt{\pi(1-\pi)+\lambda}}{\sqrt{\pi(1-\pi)+\lambda}+\sqrt{\lambda}} \tag{10.2.34}$$

In order to use the optimum n_2, one needs to know the values of π and λ which are generally unknown. In this case, one may use the values of π and λ from past surveys.

10.2.8.2 Negative Binomial Model

Here, X is assumed to follow a negative binomial distribution with parameters r and p. The probability mass function of X is

$$\Pr(X = x) = \frac{\Gamma(x+r)}{x!\Gamma(r)}(1-p)^r p^x; x = 0, 1, 2, \ldots. \tag{10.2.35}$$

The value of $r(> 0)$ was assumed to be unknown, and it was estimated by using Bliss and Fisher (1953) formula as

$$\hat{r} = \frac{\overline{x}}{S_x^2 - \overline{x}} \tag{10.2.36}$$

where $\overline{x} = \frac{1}{n_1}\sum_{i=1}^{n_1} x_i$ and $S_x^2 = \sum_{i=1}^{n_1} (x_i - \overline{x})^2/(n_1 - 1)$.

Noting, $E(X) = \frac{rp}{1-p}$ and $E(z) = \pi_A + \frac{rp}{1-p}$, a moment estimate of π_A was proposed as

$$\hat{\pi}_{\text{AT}}(1) = \overline{z} - \frac{r\hat{p}}{1-\hat{p}} = \frac{1}{n_2}\sum_{i=1}^{n_2} z_i - \frac{1}{n_1}\sum_{i=1}^{n_1} x_i = \hat{\pi}_{\text{AT}} \tag{10.2.37}$$

Remark 10.2.8 For the negative binomial model, Tian et al. (2017) proposed modified admissible point and interval estimators of π_A by using the EM algorithm. The optimum allocation of the sample size was also obtained, assuming the Poisson model.

10.2.9 ICT for Multiple Sensitive Items

Blair and Imai (2012) proposed ICT for multiple sensitive items. Suppose that the individuals in a population may possess any of the $k(\geq 2)$ sensitive attributes $A_1, \ldots, A_k$. The attributes A_j's may not be disjoint, i.e., an individual may possess more than one attribute. In this case, $k+1$ independent samples $s_0, s_1, \ldots, s_k$ of sizes $n_0, n_1, \ldots, n_k$ are selected from the population by the SRSWR method, and $k+1$ lists $L_0, L_1, \ldots, L_k$ of questionnaires are prepared. The list L_0 is the control list, consisting of g innocuous questions, while the list $L_i (i = 1, \ldots, k)$ consists of $g+1$ questions, of which the first g questions is the same as the list L_0 and the $g+1$-st question is sensitive, which relates to the sensitive attribute A_i such as "Do you possess the attribute A_i?". Respondents in the samples s_j $(j = 0, \ldots, k+1)$ are requested to report the total number of items applicable to them. Let z_{ij} be the response from the ith respondent if he or she belongs to the sample s_j and $y_i(r) = 1(0)$ if the respondent possesses (does not possess) the sensitive attribute $A_r (r = 1, \ldots, k)$. In this case, an unbiased estimator of π_r, the proportion of individuals in the population possessing the attribute A_r is

$$\hat{\pi}_r = \bar{z}_r - \bar{z}_0, \; r = 1, \ldots, k \tag{10.2.38}$$

where $\bar{z}_r = \frac{1}{n_r} \sum_{i \in s_r} z_{ir}$ and $\bar{z}_0 = \frac{1}{n_0} \sum_{i \in s_0} z_{i0}$.

$$\text{Var}(\hat{\pi}_r) = \frac{\pi_r(1-\pi_r)}{n_r} + \left(\frac{1}{n_0} + \frac{1}{n_r}\right) \sum_{j=1}^{g} \pi_g(1-\pi_g) \tag{10.2.39}$$

$$\hat{\text{Var}}(\hat{\pi}_r) = \frac{\hat{S}_r^2}{n_r} + \frac{\hat{S}_0^2}{n_0} \tag{10.2.40}$$

where $\hat{S}_0^2 = \frac{1}{n_0 - 1} \sum_{j \in s_0} (z_{i0} - \bar{z}_0)$ and $\hat{S}_r^2 = \frac{1}{n_r - 1} \sum_{j \in s_r} (z_{ir} - \bar{z}_r)$.

10.3 The Item Sum Technique (IST)

Like ICT for qualitative characteristics, Chaudhuri and Christofides (2013), and Trappmann et al. (2014) proposed IST for quantitative characteristics. In this method, two independent samples s_1 and s_2 of sizes n_1 and n_2 are selected from a population U by the SRSWR method. Respondents in the sample s_1 are confronted with a short list (SL) of g questions relating to innocuous quantitative variables $x_1, \ldots, x_g$ only. Respondents in the sample s_2 are provided with a long list (LL) of $g+1$ questions, where the first g questions are the same as the SL and $g+1$-st question is a sensitive question, related to the quantitative variable y of interest. The respondents are asked to report only the sum of the answers of the list, say, z as their response. The respondent should not report the values of the individual items. For example, to estimate the consumption of marijuana, the IST questionnaire may be framed as follows:

Short-list (SL)	Long-list (LL)
Q1: How much did you spend on food last week?	Q1: How much did you spend on food last week?
Q2: How much did you spend on transportation?	Q2: How much did you spend on transportation?
	Q3: How much did you spend on marijuana?
Total score	Total score

Let us denote z_i the response from ith respondent, x_{ij} be the value of the innocuous variable x_j $(j = 1, \ldots, g)$, and y_i denote the value of the sensitive variable y for the ith respondent. For analyzing IST data, the following assumptions were made by Imai (2011), and Blair and Imai (2012).

(i) There is no design effect: adding a sensitive item does not affect the aggregate response for control items.
(ii) No liars: respondents give truthful answers regarding sensitive items when giving aggregate responses to the treatment list.

Then, an unbiased estimator of the population mean μ_y of y is

$$\hat{\mu}_y = \overline{z}_1 - \overline{z}_2 \tag{10.3.1}$$

where $\overline{z}_1 = \frac{1}{n_1}\sum_{i\in s_1} z_i$ and $\overline{z}_2 = \frac{1}{n_2}\sum_{i\in s_2} z_i$

The variance of $\hat{\mu}_y$ is

$$\begin{aligned}\mathrm{Var}(\hat{\mu}_y) &= \frac{1}{n_1}\left(\sigma_y^2 + \sum_{j=1}^{g}\sigma_{xj}^2\right) + \frac{1}{n_2}\left(\sum_{j=1}^{g}\sigma_{xj}^2\right)\\ &= \frac{1}{n_1}\sigma_y^2 + \left(\frac{1}{n_1} + \frac{1}{n_2}\right)\left(\sum_{j=1}^{g}\sigma_{xj}^2\right)\end{aligned} \tag{10.3.2}$$

where σ_y^2 is the variance of the sensitive variable y and σ_{xj}^2 is the variance of the innocuous variable $x_j; j = 1, \ldots, g$.

An unbiased estimator of $\mathrm{Var}(\hat{\mu}_y)$ is

$$\widehat{\mathrm{Var}}(\hat{\mu}_y) = \frac{S_{z1}^2}{n_1} + \frac{S_{z2}^2}{n_2} = \frac{S_{z1}^2}{n_1} + \frac{S_{z2}^2}{n_2} \tag{10.3.3}$$

where $S_{zj}^2 = \sum_{i\in s_j}\left(z_i - \overline{z}_j\right)^2/\left(n_j - 1\right); j = 1, 2.$

Remark 10.3.1 Equation (10.3.1) indicates that $\hat{\mu}_y$ may take a negative value. Although there is no restriction on the number of innocuous variables g, Eq. (10.3.2) indicates that $\text{Var}(\hat{\mu}_y)$ increases with the increase of the number of innocuous variables. In addition, the summing of individual answers for quantitative variables requires more cognitive demand than the binary items. Trappmann et al. (2014) recommend that the lists be as short as possible. Trappmann et al. (2014) used only one innocuous variable in their CATI survey on undeclared work in Germany. However, fewer items may reduce the privacy of the respondents. Further, both the sensitive and non-sensitive questions should be quantitative, and they should be measured on the same scale.

10.3.1 IST for Complex Surveys

Shaw (2015) and Rueda et al. (2018) extended IST for general sampling designs. In this method, two independent samples s_1 and s_2 each of a fixed effective sample size (number of distinct units) n are selected from the population by using the same sampling design so that the inclusion probabilities for the ith unit, and the ith and $j(\neq i)$ units are π_i and $\pi_{ij}(>0)$ respectively.

Respondents in the sample s_1 are given a LL of a questionnaire consisting of $g+1$ questions, the first g of which is innocuous and the last of which is the key question referring to the sensitive quantitative variable that is the subject of our study. Respondents belonging to the second sample s_2 are provided with a SL containing the first g innocuous questions of the LL. All the $g+1$ questions are quantitative in nature. Respondents are requested to provide the total of all the items without revealing the individual score of any item. Let y_i and x_{ij} be the value of the sensitive variable y and innocuous variable $x_j (j = 1, \ldots, g)$ for the $i(\in U)$ th unit; $\alpha_i = y_i + \sum_{j=1}^{g} x_{ij}$ be the response from the ith unit if it belongs to s_1 and $\beta_i = \sum_{j=1}^{g} x_{ij}$ be the response from the ith respondent belonging to the sample s_2. Then, we have the following theorem relating to the estimator of the population mean $\overline{Y} = \frac{1}{N}\sum_{i\in U} y_i$. The proof of the theorem is omitted as it is very simple.

Theorem 10.3.1

(i) $\hat{\overline{Y}}_{\text{HT}} = \frac{1}{N}\left(\sum_{i\in s_1} \frac{\alpha_i}{\pi_i} - \sum_{i\in s_2} \frac{\beta_i}{\pi_i}\right)$ is an unbiased estimator of $\overline{Y}$.

(ii) The variance of $\hat{\overline{Y}}_{\text{HT}}$ is

$$\text{Var}(\hat{\overline{Y}}_{\text{HT}}) = \frac{1}{N^2}\sum_{i<}\sum_{j\in U}(\pi_i\pi_j - \pi_{ij})\left\{\left(\frac{\alpha_i}{\pi_i} - \frac{\alpha_j}{\pi_j}\right)^2 + \left(\frac{\beta_i}{\pi_i} - \frac{\beta_j}{\pi_j}\right)^2\right\}$$

(iii) An unbiased estimator of $\mathrm{Var}(\hat{\overline{Y}}_{\mathrm{HT}})$ is

$$\hat{\mathrm{Var}}(\hat{\overline{Y}}_{\mathrm{HT}}) = \frac{1}{N^2}\left[\sum_{i<}\sum_{j\in s_1}\frac{\pi_i\pi_j - \pi_{ij}}{\pi_{ij}}\left(\frac{\alpha_i}{\pi_i} - \frac{\alpha_j}{\pi_j}\right)^2 + \sum_{i<}\sum_{j\in s_2}\frac{\pi_i\pi_j - \pi_{ij}}{\pi_{ij}}\left(\frac{\beta_i}{\pi_i} - \frac{\beta_j}{\pi_j}\right)^2\right]$$

10.3.2 IST from a Single Sample

Hussain et al. (2012) and Shaw (2015) provided IST using a single sample. In Hussain's IST, a sample s of n units is selected by the SRSWR method. Each of the respondents selected in the sample was given a list of g items. The $j(=1,\ldots,g)$th item is the addition of the queries for the sensitive item y and the innocuous item x_j. Respondents are required to report the sum of the scores of all the items without disclosing the individual scores of each item. Let y_i and x_{ij} be the values of y and x_j for the ith unit of the population. Then the response from the ith respondent, if he or she is included in the sample s is

$$z_i = gy_i + \sum_{j=1}^{g} x_{ij} \tag{10.3.4}$$

Assuming μ_{xj}, the population mean of the jth innocuous variable is known, Hussain et al. (2012) provided an unbiased estimator of μ_y, the mean of the sensitive variable y as

$$\hat{\mu}_{y\mathrm{H}} = \frac{\bar{z} - \sum_{j=1}^{g}\mu_{xj}}{g} \tag{10.3.5}$$

The variance of $\hat{\mu}_{y\mathrm{H}}$ and its unbiased estimators are respectively given by

$$\mathrm{Var}(\hat{\mu}_{y\mathrm{H}}) = \frac{1}{n}\left(\sigma_y^2 + \frac{1}{g^2}\sum_{j=1}^{g}\sigma_{xj}^2\right) \tag{10.3.6}$$

and

$$\hat{\mathrm{Var}}(\hat{\mu}_{y\mathrm{H}}) = \frac{\hat{S}_z^2}{n} \tag{10.3.7}$$

where σ_y^2 and σ_{xj}^2 are the population variance of y and x_j while $\hat{S}_z^2 = \sum_{i \in s} (z_i - \bar{z})^2/(n-1)$.

Remark 10.3.2 The main demerits of Hussain's (2012) IST are that (i) the population means of all the innocuous variables should be known; (ii) cooperation from the respondents is questionable since all the questions are mixed with the sensitive variable; and (iii) respondents are required to make several totals, which might be a significant source of response error.

10.3.3 IST for Multiple Sensitive Quantitative Variables

Perri et al. (2018) proposed an IST for $q(\geq 2)$ quantitative variables. In this method, $q+1$ independent samples $s_0, s_1, \ldots, s_q$ of sizes $n_0, n_1, \ldots, n_q$ are selected from a population U by using a suitable varying probability sampling design. The inclusion probabilities for the ith, and the pair $i, j(\neq i)$th unit for the selection of sample s_k are $\pi_{i|k}$ and $\pi_{ij|k}$; $k = 0, \ldots, q$ respectively. Respondents belonging to the sample s_0 were provided with a short list (SL), say, L_0 of g innocuous quantitative items, and they were asked to provide the total $z_0 = x_0$ of the g items as their response. Respondents belonging to s_1 are provided with a long list (LL) of $g+1$ items. The first g items are the same as in the short list L_0 and the $g+1$st item is related to the sensitive variable $y_{(1)}$ of interest, and they were asked to report $z_1 = x_0 + y_{(1)}$ as their response. In general, respondents belonging to the sample s_k are provided with the list L_0 plus $k(=1, \ldots, q)$ sensitive items $y_{(1)}, y_{(2)}, \ldots, y_{(k)}$ of interest and asked to provide the sum $z_k = x_0 + y_{(1)} + \cdots + y_{(k)}$ as their responses. Let z_{ik} be the response from the $i(\in U)$th respondent if he or she belongs to the sample s_k and $y_{i(k)}$ be the associated value of the sensitive variable $y_{(k)}(k = 1, \ldots, q)$. Then, the Horvitz-Thompson type estimator for the population mean $\overline{Y}_{(k)} = \frac{1}{N}\sum_{i \in U} y_{i(k)}$ is

$$\hat{\overline{Y}}_{HT(k)} = \hat{\overline{Y}}_{(k)} - \hat{\overline{Y}}_{(k-1)} \tag{10.3.8}$$

where
$\hat{\overline{Y}}_{HT(k)} = \frac{1}{N}\sum_{i \in s_k} \frac{z_{ik}}{\pi_{i|k}}$, $\hat{\overline{Y}}_{HT(0)} = \frac{1}{N}\sum_{i \in s_0} \frac{z_{i0}}{\pi_{i|0}} = \frac{1}{N}\sum_{i \in s_0} \frac{x_{i0}}{\pi_{i|0}}$ and x_{0i} = value of x_0 for the ith unit.

Theorem 10.3.2

(i) $\hat{\overline{Y}}_{HT(k)}$ is an unbiased estimator of $\overline{Y}_{(k)}$
(ii) The variance of $\hat{\overline{Y}}_{HT(k)}$ is

$$\mathrm{Var}(\hat{\overline{Y}}_{HT(k)}) = \frac{1}{N^2}\left[\sum_{i<}\sum_{j \in U}\left\{(\pi_{i|k}\pi_{j|k} - \pi_{ij|k})\left(\frac{z_{ik}}{\pi_{i|k}} - \frac{z_{jk}}{\pi_{j|k}}\right)^2\right.\right.$$

$$+\left(\pi_{i|k-1}\pi_{j|k-1}-\pi_{ij|k-1}\right)\left(\frac{z_{ik-1}}{\pi_{i|k-1}}-\frac{z_{jk-1}}{\pi_{j|k-1}}\right)^{2}\Bigg\}\Bigg]$$

(iii) An unbiased estimator of $\mathrm{Var}(\hat{\overline{Y}}_{\mathrm{HT}(k)})$ is

$$\hat{\mathrm{Var}}\left(\hat{\overline{Y}}_{\mathrm{HT}(k)}\right)=\frac{1}{N^2}\left[\sum_{i<}\sum_{j\in s_k}\left(\frac{\pi_{i|k}\pi_{j|k}-\pi_{ij|k}}{\pi_{ij|k}}\right)\left(\frac{z_{ik}}{\pi_{i|k}}-\frac{z_{jk}}{\pi_{j|k}}\right)^2\right.$$
$$\left.+\sum_{i<}\sum_{j\in s_{k-1}}\left(\frac{\pi_{i|k-1}\pi_{j|k-1}-\pi_{ij|k-1}}{\pi_{ij|k-1}}\right)\left(\frac{z_{ik-1}}{\pi_{i|k-1}}-\frac{z_{jk-1}}{\pi_{j|k-1}}\right)^2\right]$$

Remark 10.3.3 Perri et al. (2018) provided the optimum allocation of sample sizes for a given cost function and extended their IST technique for stratified sampling. The main demerit of this method is that the respondents belonging to the sample s_k have to provide the sum of the k sensitive questions, which may cause a reduction in cooperation from the respondents. Further, the variance of $\hat{\overline{Y}}_{\mathrm{HT}(k)}$ is likely to increase with the number of variables k. Perri et al.'s (2018) IST provides an unbiased estimator mean of the first t sensitive items $\overline{Y}(t)=\frac{1}{t}\sum_{k=1}^{t}\overline{Y}_{(k)}$ as.

$$\hat{\overline{Y}}_{\mathrm{HT}}(t)=\frac{1}{t}\left(\frac{1}{N}\sum_{i\in s_t}\frac{z_{it}}{\pi_{i|t}}-\frac{1}{N}\sum_{i\in s_0}\frac{z_{0i}}{\pi_{i|0}}\right)\tag{10.3.9}$$

10.4 Modeling ICT and IST

Blair and Imai (2012) pointed out that the standard difference-in-estimators in Eq. (10.3.1) do not use multivariate relationships between preferences over the sensitive items and respondents' characteristics. Imai (2011) proposed the following additive nonlinear regression model, which includes the difference in means estimator and the linear least square estimator as special cases:

$$Y_i=f(\mathbf{X}_i,\boldsymbol{\gamma})+k_ih(\mathbf{X}_i,\boldsymbol{\delta})+\in_i\tag{10.4.1}$$

where Y_i is the response from the ith respondent, k_i is the treatment category, viz., $k_i=1(0)$ indicates that the ith respondent received the LL (SL) of $g+1(g)$ items, $E(\in_i|X_i,k_i)=0$ and $(\boldsymbol{\gamma},\boldsymbol{\delta})$ is a vector of unknown parameters and $\mathbf{X}_i$ is the vector of covariates associated with the ith$(\in U)$ respondent.

For example, for the logistic regression model

$$f(\mathbf{X}_i, \gamma) = g \text{ logit}^{-1}(\mathbf{X}_i^{\mathrm{T}}\gamma) \text{ and } h(\mathbf{X}_i, \delta) = \text{logit}^{-1}(\mathbf{X}_i^{\mathrm{T}}\delta) \quad (10.4.2)$$

For the linear regression model

$$Y_i = \mathbf{X}_i^{\mathrm{T}}\gamma + k_i\mathbf{X}_i^{\mathrm{T}}\delta + \in_i \quad (10.4.3)$$

For estimating the parameters of the model (10.4.3), Imai (2011) used the maximum likelihood estimator using an expectation–maximization (EM) algorithm, treating the values of the sensitive items of the SL as missing data. Kuha and Jackson (2014) modeled item counts as categorical data and expanded the results of Imai (2011). They provided explicit variance estimates for the parameters. Further details are available in Holland (1986), Corstange (2009), Trappmann et al. (2014), and Aronow et al. (2015), among others.

Rueda et al. (2018) provided an improvement of the Horvitz-Thompson type difference-in-estimators by applying a calibrating method using auxiliary information for complex survey designs, while Hussain et al. (2017) used ratio, product, and regression methods for the improvement of the difference-in-estimators under SRSWR sampling.

10.5 The Three-Card Method

The three-card method was developed by the U.S. General Accounting Office (GAO) to collect sensitive data on immigration status based on large-scale personal-interview surveys (vide Droitcour and Larson (2002)). In this method, the population is divided into six mutually exclusive and exhaustive categories A, F_1, F_2, F_3, F_4 and O. The category A is the sensitive category and interest of study. The remaining categories F_1, F_2, F_3, F_4 and O are related but innocuous. Three independent samples s_1, s_2 and s_3 of sizes n_1, n_2 and n_3 are selected from the population by the SRSWR method. The respondents belong to the samples s_1, s_2 and s_3 are provided with the three boxes of statements as follows:

Sample s_1:

Box A: I belong to the category F_1(i.e. I $\in F_1$);
Box B: I belong to the category A or F_2 or F_3 or F_4(i.e. I $\in A \cup F_2 \cup F_3 \cup F_4$);
Box C: I belong to a category other than *Box A* or *Box B* (i.e., $I \in O$).

Sample s_2:

Box A: I belong to the category F_2(i.e. I $\in F_2$);
Box B: I belong to the category A or F_1 or F_3 or F_4(i.e. I $\in A \cup F_1 \cup F_3 \cup F_4$); and
Box C: I belong to the category other than *Box A or Box B* (i.e., I $\in O$).

Sample s_3:

Box A: I belong to the category F_2 or F_4(i.e. I $\in F_2 \cup F_4$);
Box B: I belong to the category A or F_1 or F_2 (i.e. I $\in A \cup \mathrm{F}_1 \cup F_2$);
Box C: I belong to a category other than Box A or Box B (i.e., $I \in O$).

Respondents are requested to report the box whose statement is applicable to them. Let n_{i1}, n_{i2} and n_{i3} be number of respondents belonging to the sample $s_i (i = 1, 2, 3)$ declared their membership to *Box A*, *Box B* and *Box C* respectively. Then an unbiased estimator of the proportion of individuals belonging to the sensitive group A in the population is given by

$$\hat{\pi}_{\mathrm{DL}} = \frac{1}{3n}[n_{12} + n_{22} + n_{32} - 2(n_{11} + n_{21} + n_{31})] \tag{10.5.1}$$

The variance of $\hat{\pi}_{\mathrm{DL}}$ and its unbiased estimator can be easily obtained by using the properties of the multinomial distribution.

10.6 Non-Randomized Response Model

Unlike the RR models, non-randomized response (NRR) models, also known as parallel models, replace randomization devices with non-sensitive or innocuous questions. NRR was introduced by Swensson (1974), and Takahasi and Sakasegawa (1977). Thus, the NRR models bear reproducibility properties, receive greater cooperation from respondents, are cheaper to administer, protect respondents' confidentiality, and are feasible for face-to-face or mail-inquiry. This section will consider a few important NRR techniques and their properties.

10.6.1 Triangular Model

The Triangular model was proposed by Tian et al. (2007a, 2007b, 2009a, 2009b) and Yu et al. (2008). Let $y_i = 1$ if the ith member of the population U possesses the sensitive characteristic A (e.g., drug-taking) and $y_i = 0$ if the ith member belongs to the complementary category $\overline{A}$ (e.g., non-drug-taking). Let x be a dichotomous variable such that $x_i = 1$ if the ith member of the population belongs to some non-sensitive or innocuous group B (e.g., born between January and June) and $x_i = 0$ if the ith member belongs to the complementary group $\overline{B}$ (e.g., born between July and December). It is assumed that the variables x and y are independent, and π_x, the proportion of individuals in the population belonging to the class B is either known or can easily be estimated. Our goal is to estimate $\pi_y = \sum_{i \in U} y_i / N$, the proportion of the population that belongs to the sensitive group A.

A sample s of n respondents is selected by the SRSWR method. Each of the selected respondents in the sample is requested to mark with a tick ($\surd$) whether

Table 10.1 The triangular model and cell probabilities

Categories	$x = 0$	$x = 1$	Categories	$x = 0$	$x = 1$	Total
$y = 0$	O	•	$y = 0$	$(1 - \pi_y)(1 - \pi_x)$	$(1 - \pi_y\ \pi_x$	$1 - \pi_y$
$y = 1$	•	•	$y = 1$	$\pi_y(1 - \pi_x)$	$\pi_y\pi_x$	π_y
			Total	$1 - \pi_x$	π_x	1

he or she belongs to the category $(y = 0, x = 0)$ or its complementary category $(y = 0, x = 1) \cup (y = 1)$. The sub-class $(y = 0, x = 0)$ is non-sensitive as it comprises respondents who are not drug users and were born in the first half of the year (i.e., from January to June). The sub-class $(y = 0, x = 1) \cup (y = 1)$ is a mixture of non-sensitive and sensitive categories as it contains respondents who are either drug users or born in the second half of the year (i.e., July to December). Hence, the response "yes" or "no" will not reveal the respondent's membership in the category A. Such a design is expected to encourage the respondents' participation in the survey and extract truthful answers from them. In a face-to-face interview, the investigator may use the format given on the left-hand side of Table 10.1 and either put a tick (✓) inside the circle or in the triangle formed by the three solid dots.

Instruction: Please truthfully put a ✓ either in the circle or in the triangle formed by the three solid dots.

10.6.1.1 Estimation of Proportion π_y

Let out of the n respondents, n_1 of them marked a tick in the circle and $n - n_1$ marked a tick in the triangle. Then, n_1 follows a binomial distribution with parameters n and $\theta = (1 - \pi_y)(1 - \pi_x)$. Hence,

$$E(n_1) = n\theta = n(1 - \pi_y)(1 - \pi_x) \tag{10.6.1}$$

Equation (10.6.1) yields a moment estimator of π_y as

$$\hat{\pi}_{\mathrm{T}y} = 1 - \frac{\hat{\theta}}{1 - \pi_x} \tag{10.6.2}$$

where $\hat{\theta}$ is the proportion of respondents who put tick marks in the circle.

The variance of $\hat{\pi}_{\mathrm{T}y}$ is

$$\begin{aligned} \mathrm{Var}(\hat{\pi}_{\mathrm{T}y}) &= \frac{V(\hat{\theta})}{(1 - \pi_x)^2} \\ &= \frac{1}{n}\frac{\theta(1 - \theta)}{(1 - \pi_x)^2} \end{aligned}$$

$$= \frac{\pi_y(1-\pi_y)}{n} + \frac{\pi_x(1-\pi_y)}{n(1-\pi_x)} \tag{10.6.3}$$

An unbiased estimator of $\text{Var}(\hat{\pi}_{\text{Ty}})$ is

$$\hat{\text{Var}}(\hat{\pi}_{\text{Ty}}) = \frac{1}{(n-1)} \frac{\hat{\theta}(1-\hat{\theta})}{(1-\pi_x)^2} \tag{10.6.4}$$

10.6.2 Crosswise Model

The crosswise model was also proposed by Tian et al. (2007a, 2007b, 2009a, 2009b) and Yu et al. (2008). In this model, the respondents are asked to place a tick (√) in either the diagonal of the two circles or off-diagonal with the solid dots according to their truthful status. It should be noted that the subclasses $(y = 0, x = 0) \cup (y = 1, x = 1)$ and $(y = 0, x = 1) \cup (y = 1, x = 0)$ are non-sensitive in the sense that the response will not reveal whether the respondent possesses the sensitive attribute A or not.

Instruction: Please truthfully place a √ either in the main diagonal or off-diagonal.

10.6.2.1 Estimation of Proportion π_y

Let $\tilde{n}$ and $n - \tilde{n}$ be the total number of ticks observed in the main and off-diagonals. Then, $\tilde{n}$ follows a binomial distribution with parameters n and $\tilde{\theta} = (1-\pi_y)(1-\pi_x) + \pi_x\pi_y$. Now denoting, $\hat{\tilde{\theta}} = \tilde{n}/n$, the proportion of ticks in the main diagonal, one finds

$$\hat{\pi}_{\text{Cy}} = \frac{\hat{\tilde{\theta}} - (1-\pi_x)}{2\pi_x - 1} \tag{10.6.5}$$

as an unbiased estimator of π_y.

The variance of $\hat{\pi}_{\text{cy}}$ is

$$\text{Var}(\hat{\pi}_{\text{Cy}}) = \frac{\tilde{\theta}(1-\tilde{\theta})}{n(2\pi_x-1)^2} = \frac{\pi_y(1-\pi_y)}{n} + \frac{\pi_x(1-\pi_x)}{n(2\pi_x-1)^2} \tag{10.6.6}$$

An unbiased estimator of $\text{Var}(\hat{\pi}_{\text{Cy}})$ is

$$\hat{\text{Var}}(\hat{\pi}_{\text{Cy}}) = \frac{\hat{\tilde{\theta}}(1-\hat{\tilde{\theta}})}{(n-1)(2\pi_x-1)^2} \tag{10.6.7}$$

Remark 10.6.1 Both the estimators $\hat{\pi}_{\mathrm{T}y}$ and $\hat{\pi}_{\mathrm{C}y}$ are moment estimators that can take values outside of the parameter space (0, 1). The maximum likelihood estimators (MLE) of π_y for both the triangular and crosswise models are provided by Yu et al. (2008) and are given as follows:

MLE of π_y for the Triangular model is

$$\hat{\pi}_{\mathrm{T}y}^{\mathrm{MLE}} = \max\left(0, \hat{\pi}_{\mathrm{T}y}\right) = \begin{cases} \hat{\pi}_{\mathrm{T}y}, & \text{if } 0 \le \hat{\theta} \le 1 - \pi_x \\ 0, & \text{if } 1 - \pi_x < \hat{\theta} \le 1 \end{cases} \tag{10.6.8}$$

MLE of π_y for the crosswise model is

$$\hat{\pi}_{\mathrm{C}y}^{\mathrm{MLE}} = \min\left\{1, \max\left(0, \hat{\pi}_{\mathrm{C}y}\right)\right\} = \begin{cases} 0, & \text{if } 0 \le \hat{\theta} < \min(1-\pi_x, \pi_x) \\ \hat{\pi}_{\mathrm{C}y}, & \text{if } \min(1-\pi_x, \pi_x) \le \hat{\theta} \le \max(1-\pi_x, \pi_x) \\ 1, & \text{if } \max(1-\pi_x, \pi_x) < \hat{\theta} \le 1. \end{cases} \tag{10.6.9}$$

10.6.3 *Comparison of Efficiencies of Triangular and Crosswise Designs*

Equations (10.6.3) and (10.6.6) yield

$$\mathrm{Var}\left(\hat{\pi}_{\mathrm{C}y}\right) - \mathrm{Var}\left(\hat{\pi}_{\mathrm{T}y}\right) = \frac{\pi_x}{n(1-\pi_x)(2\pi_x - 1)^2} g(\pi_x) \tag{10.6.10}$$

where $g(\pi_x) = \left(4\pi_y - 3\right)\pi_x^2 + 2\left(1 - 2\pi_y\right)\pi_x + \pi_y$.

The following comparisons are provided by Yu et al. (2008). Details proofs are omitted here.

Theorem 10.6.1

(i) If $\pi_y = 3/4$, then

$$\mathrm{Var}\left(\hat{\pi}_{\mathrm{C}y}\right) \ge \mathrm{Var}\left(\hat{\pi}_{\mathrm{T}y}\right) \text{ if } 0 \le \pi_x \le 3/4 (\pi_x \ne 1/2)$$
$$\mathrm{Var}\left(\hat{\pi}_{\mathrm{C}y}\right) < \mathrm{Var}\left(\hat{\pi}_{\mathrm{T}y}\right) \text{ if } 3/4 \le \pi_x \le 1$$

(ii) If $\pi_y \ne 3/4$, then

$$\mathrm{Var}\left(\hat{\pi}_{\mathrm{C}y}\right) \ge \mathrm{Var}\left(\hat{\pi}_{\mathrm{T}y}\right) \text{ if } 0 \le \pi_x \le \pi_{x0}, (\pi_x \ne 1/2)$$
$$\mathrm{Var}\left(\hat{\pi}_{\mathrm{C}y}\right) < \mathrm{Var}\left(\hat{\pi}_{\mathrm{T}y}\right) \text{ if } \pi_{x0} \le \pi_x \le 1$$

where $\pi_{x0} = 2\pi_y - 1 - \sqrt{1 - \pi_y}/\left(4\pi_y - 3\right)$ is an increasing function of π_y.

Remark 10.6.2 Tian (2014) pointed out that both the triangular and crosswise designs are applicable to one sensitive category only, viz., category $A(y = 1)$ is sensitive while its complementary category $\overline{A}$ $(y = 0)$ is non-sensitive. So, the designs are not applicable where both categories A and $\overline{A}$ are sensitive, such as the number of sex partners ≤ 2 or >2. The efficiency of the designs is not always high. Further, the triangular model provides lower privacy protection. To remove such limitations, Tian (2014) proposed the following alternative parallel model, which is applicable to two sensitive characteristics.

10.6.4 The Parallel Model

The parallel model, which is the non-randomized version of the unrelated question model, was proposed by Tian (2014). Let A be a group of people who possess certain sensitive characteristics y (e.g., members of a certain political party A, $y = 1$) and $\overline{A}(y = 0)$ denote complementary group, that is also sensitive. Our objective is to estimate π_A, the proportion of people belonging to the group A. Let x and w be two non-sensitive dichotomous variables such that x, y and w are mutually independent of one another. It is assumed that the prevalence π_x and π_w of x and w are known. For example, $x_i = 1$, if the ith member of the population is born between 1 and 15 days of a month and $x_i = 0$, otherwise (i.e., born after the 15th day of a month). Similarly, $w_i = 0$, if the ith member is born in the first half of a year (i.e., between January and June) and $w_i = 1$, otherwise (i.e., between July and December). Here the variables x and w are independent of y; $\pi_x = \text{Prob}(x = 1) \cong 1/2$ and $\pi_w = \text{Prob}(w = 1) \cong 1/2$.

In Tian's (2014) method, a sample of size n is selected from the population by the SRSWR method. Each of the selected respondents was asked to answer "yes" or "no" to the non-sensitive unrelated question Q1 if he or she was born in the first half of the year (i.e., $w = 0$). The respondent should answer "yes" or "no" to the sensitive question Q2 if he or she was born in the second half of the year (i.e., $w = 1$). The questions are as follows:

Q1: Is your birthday between the first and the fifteenth of the month?

Q2: Do you belong to the group A?

The privacy of the respondent is maintained as the respondent reveals only the answer "yes" or "no" without disclosing which of the questions Q1 or Q2 he or she has answered.

For this model, the probability of obtaining a "yes" answer from a respondent is

$$\begin{aligned}\lambda_T &= \text{Prob}(w = 0 \cap x = 1) + \text{Prob}(w = 1 \cap y = 1)\\ &= (1 - \pi_w)\pi_x + \pi_w \pi_A \end{aligned} \tag{10.6.11}$$

Let $\hat{\lambda}_T$ be the proportion of "yes" answers. Then, noting that $\hat{\lambda}_T$ is a binomial variable with parameters n and λ_T, the following theorem is derived (proof is omitted because it is simple).

Theorem 10.6.2

(i) $\hat{\pi}_T = \frac{\hat{\lambda}_T - (1-\pi_w)\pi_x}{\pi_w}$ is an unbiased estimator of π_A.
(ii) The variance of $\hat{\pi}_T$ is

$$\mathrm{Var}(\hat{\pi}_T) = \frac{\lambda_T(1-\lambda_T)}{n\pi_w^2} = \frac{\pi_A(1-\pi_A)}{n} + \frac{1-\pi_w}{n\pi_w^2}g \tag{10.6.12}$$

where $g = \pi_A\pi_w + (1-2\pi_w\pi_A)\pi_x - (1-\pi_w)\pi_x^2$.

(iii) An unbiased estimator of $\mathrm{Var}(\hat{\pi}_T)$ is

$$\hat{\mathrm{Var}}(\hat{\pi}_T) = \frac{\hat{\lambda}_A(1-\hat{\lambda}_A)}{(n-1)\pi_w^2}$$

Remark 10.6.3 Theorems 1.3.1 and 10.6.2 indicate the expressions $\hat{\pi}_G$ and $\hat{\pi}_T$ are the same. Hence, the parallel model can be regarded as a non-randomized version of the unrelated question model (U-model).

The estimator $\hat{\pi}_T$ is a moment estimator and can take values outside of the parametric range [0, 1]. Hence, the MLE of π_A is

$$\hat{\pi}_T^{\text{mle}} = \begin{cases} 0, & \text{if } \hat{\pi}_T < 0 \\ \hat{\pi}_T & \text{if } 0 \le \hat{\pi}_T < 0 \\ 1, & \text{if } \hat{\pi}_T > 1 \end{cases}$$

Tian (2014) provided a comparison of the parallel model with the crosswise and triangular models. Since the comparisons are not simple, they are omitted. Interested readers are referred to Tian (2014). Tian (2014) also provided expressions for the degree of privacy protection of the parallel model compared with the crosswise and triangular models.

10.6.4.1 Parallel Model for Complex Survey Designs

Arnab et al. (2019) proposed a methodology for estimating population proportions from the parallel model under complex survey designs. For this model, a sample s of size n is selected from a population U with probability $p(s)$ using an arbitrary sampling design p. Let $\pi_i = \sum_{s \supset i} p(s)$ and $\pi_{ij} = \sum_{s \supset i,j} p(s)$ be the inclusion probabilities for the ith, and ith and $j(\neq i)$th unit of the population U. From each of the selected units of the population, data were collected using the parallel model described in Sect. 10.6.4. Let $B(\overline{B})$ be the group of people who are born in the first half of a month, i.e., the 01 and 15 days of a month (after the 15th of the month), $W(\overline{W})$ be a group of people who are born in the second half of a year, i.e., between July and December (1st half of a year, i.e., between January and June). Define,

$$x_i = \begin{cases} 0, & \text{if } i \in \overline{B} \\ 1, & \text{if } i \in B \end{cases}, \quad w_i = \begin{cases} 0, & \text{if } i \in \overline{W} \\ 1, & \text{if } i \in W \end{cases}, \quad y_i = \begin{cases} 0, & \text{if } i \in \overline{A} \\ 1, & \text{if } i \in A \end{cases}$$

$$\text{and } z_i = \begin{cases} 0, & \text{if the } i\text{th unit answers "no"} \\ 1, & \text{if the } i\text{th unit answers "yes"} \end{cases} \tag{10.6.13}$$

For the parallel model, respondents belonging to $\overline{W}$ answer question Q1. In this case, a respondent answers "yes" if he or she was born in the first half of a month. Otherwise, he or she answered "no",

$$\text{i.e., } z_i = x_i \text{ if } i \in \overline{W} \tag{10.6.14}$$

Similarly, if a respondent belongs to the group W, then he or she answers the sensitive question Q2. In this case, the respondent answers "yes" if the respondent possesses the sensitive characteristic A. Otherwise, if the respondent belongs to the complementary group $\overline{A}$, the respondent answers "no". Hence, in this case

$$z_i = y_i \text{ if } i \in W \tag{10.6.15}$$

From Eqs. (10.6.14) and (10.6.15), one can express the response from the $i(\in U)$ th unit as

$$z_i = w_i y_i + (1 - w_i) x_i \tag{10.6.16}$$

Equation (10.6.16) yields

$$\begin{aligned} \overline{Z} &= \frac{1}{N} \sum_{i \in U} z_i = \frac{1}{N} \sum_{i \in U} y_i w_i + \frac{1}{N} \sum_{i \in U} (1 - w_i) x_i \\ &= \pi_{WA} + \pi_{\overline{W}B} \end{aligned} \tag{10.6.17}$$

where π_{WA} and $\pi_{\overline{W}B}$ denote respectively proportions of persons that belong to groups $W \cap A$ and $\overline{W} \cap B$.

Further, assuming the groups A, B and W are mutually independent, we find

$$\overline{Z} = \pi_w \pi_A + (1 - \pi_w) \pi_x \tag{10.6.18}$$

where π_w, π_A and π_x denote respectively proportion of individuals in the population that belong to the group W, A and B respectively.

From Eq. (10.6.16), we propose a linear homogeneous unbiased estimator of $\overline{Z}$ as.

$$\hat{\overline{Z}}_{\text{LHU}} = \frac{1}{N} \sum_{i \in s} b_{si} z_i \tag{10.6.19}$$

where b_{si}'s are known constants satisfying the unbiasedness condition

$$\sum_{s\supset i} b_{si} = 1 \tag{10.6.20}$$

The variance of $\hat{\hat{Z}}_{\text{LHU}}$ is

$$\begin{aligned}
\operatorname{Var}\left(\hat{\hat{Z}}_{\text{LHU}}\right) &= \frac{1}{N^2}\left[E\left(\sum_{i\in s} b_{si}z_i\right)^2 - \left(\sum_{i\in U} z_i\right)^2\right] \\
&= \frac{1}{N^2}\left[\sum_{i\in U} (\alpha_{ii} - 1)z_i^2 + \sum_{i\in U} (\alpha_{ij} - 1)z_i z_j\right]
\end{aligned} \tag{10.6.21}$$

where

$$\alpha_{ii} = \sum_{s\supset i} b_{si}^2 p(s) \text{ and } \alpha_{ij} = \sum_{s\supset i,j} b_{si} b_{sj} p(s) \tag{10.6.22}$$

From expression (10.6.21) an unbiased estimator of $\operatorname{Var}\left(\hat{\hat{Z}}_{\text{LHU}}\right)$ is obtained as

$$\widehat{\operatorname{Var}}\left(\hat{\hat{Z}}_{\text{LHU}}\right) = \frac{1}{N^2}\left[\sum_{i\in s} c_{si}z_i^2 + \sum_{i\neq}\sum_{j\in s} c_{sij}z_i z_j\right] \tag{10.6.23}$$

where c_{si} and c_{sij} are suitably chosen constant that satisfy

$$\sum_{s\supset i} c_{si}p(s) = \alpha_{ii} - 1 \text{ and } \sum_{s\supset i,j} c_{sij}p(s) = \alpha_{ij} - 1 \tag{10.6.24}$$

The constants c_{si} and c_{sij} can be chosen in various ways. One obvious choice is

$$c_{si} = (\alpha_{ii} - 1)/\pi_i \text{ and } c_{sij} = (\alpha_{ij} - 1)/\pi_{ij} \tag{10.6.25}$$

Now writing $z_i = w_i y_i + (1 - w_i)y_i$ in (10.6.21), we find

$$\begin{aligned}
\operatorname{Var}\left(\hat{\hat{Z}}_{LHU}\right) = \frac{1}{N^2}\Bigg[&\sum_{i\in U} (\alpha_{ii} - 1)\{w_i y_i + (1 - w_i)x_i\}^2 \\
&+ \sum_{i\in U} (\alpha_{ij} - 1)\{w_i y_i + (1 - w_i)x_i\}\{w_j y_j + (1 - w_j)x_j\}\Bigg]
\end{aligned}$$

Now noting x_i, y_i and w_i are indicator variables and writing

$$\overline{\alpha}_{ii} = (\alpha_{ii} - 1)/N^2 \text{ and } \overline{\alpha}_{ij} = (\alpha_{ij} - 1)/N^2 \tag{10.6.26}$$

we have

$$\begin{aligned}
\mathrm{Var}\left(\hat{\hat{Z}}_{\mathrm{LHU}}\right) &= \left(\sum_{i\in W} \overline{\alpha}_{ii} y_i + \sum_{i\neq}\sum_{j\in W} \overline{\alpha}_{ij} y_i y_j\right) + \left(\sum_{i\in \overline{W}} \overline{\alpha}_{ii} x_i + \sum_{i\neq}\sum_{j\in \overline{W}} \overline{\alpha}_{ij} x_i x_j\right) \\
&\quad + \left(\sum_{i\in W} y_i \sum_{j(\neq i)\in \overline{W}} \overline{\alpha}_{ij} x_j + \sum_{j\in W} x_i \sum_{i(\neq i)\in \overline{W}} \overline{\alpha}_{ij} y_j\right) \\
&= \left(\sum_{i\in W\cap A} \overline{\alpha}_{ii} + \sum_{i\neq}\sum_{j\in W\cap A} \overline{\alpha}_{ij}\right) + \left(\sum_{i\in \overline{W}\cap B} \overline{\alpha}_{ii} + \sum_{i\neq}\sum_{j\in \overline{W} B} \overline{\alpha}_{ij} x_i x_j\right) \\
&\quad + \left(\sum_{i\in W\cap A}\ \sum_{j(\neq i)\in \overline{W}\cap B} \overline{\alpha}_{ij} + \sum_{j\in W\cap B}\ \sum_{i(\neq j)\in \overline{W}\cap A} \overline{\alpha}_{ij}\right)
\end{aligned} \tag{10.6.27}$$

The above results lead to the following theorem:

Theorem 10.6.3

(i) $$\hat{\pi}_{\mathrm{LHU}} = \frac{\hat{\hat{Z}}_{\mathrm{LHU}} - (1 - \pi_w)\pi_x}{\pi_w} \tag{10.6.28}$$

is an unbiased estimator of π_A when π_x and π_w are known.

(ii) The variance of $\hat{\pi}_{\mathrm{LHU}}$ is

$$\mathrm{Var}(\hat{\pi}_{\mathrm{LHU}}) = \frac{1}{\pi_w^2}\left[\begin{aligned}&\left(\sum_{i\in W\cap A} \overline{\alpha}_{ii} + \sum_{i\neq}\sum_{j\in W\cap A} \overline{\alpha}_{ij}\right) + \left(\sum_{i\in \overline{W}\cap B} \overline{\alpha}_{ii} + \sum_{i\neq}\sum_{j\in \overline{W}\cap B} \overline{\alpha}_{ij}\right) \\ &+\left(\sum_{i\in W\cap A}\ \sum_{j(\neq i)\in \overline{W}\cap B} \overline{\alpha}_{ij} + \sum_{j\in W\cap B}\ \sum_{i(\neq j)\in \overline{W}\cap A} \overline{\alpha}_{ij}\right)\end{aligned}\right]$$

(iii) An unbiased estimator of $\mathrm{Var}(\hat{\pi}_{\mathrm{LHU}})$ is

$$\widehat{\mathrm{Var}}(\hat{\pi}_{\mathrm{LHU}}) = \frac{1}{\pi_w^2}\left[\sum_{i\in s} c_{si} z_i^2 + \sum_{i\neq}\sum_{j\in s} c_{sij} z_i z_j\right]$$

where c_{si} and c_{sij} are subject to (10.6.25).

10.6.4.2 Expressions for Various Sampling Strategies

In this section we will present expression of $\hat{\pi}_{\text{LHU}}$ for various sampling strategies.

Horvitz-Thompson Type Estimator Based on Arbitrary Sampling Design

Substituting $b_{si} = 1/\pi_i$ in (10.6.28), we have the following results:

Theorem 10.6.4

(i) The Horvitz-Thompson type estimator $\pi_A^{\text{HT}} = \frac{1}{\pi_w}\left[\frac{1}{N}\sum_{i\in s}\frac{z_i}{\pi_i} - (1-\pi_w)\pi_x\right]$ is an unbiased estimator of π_A.

(ii) $$\text{Var}\left(\pi_A^{\text{HT}}\right) = \frac{1}{N^2\pi_w^2}\left[\begin{array}{l}\left(\sum_{i\in W\cap A}\gamma_i + \sum_{i\neq}\sum_{j\in W\cap A}\gamma_{ij}\right) + \left(\sum_{i\in \overline{W}\cap B}\gamma_i + \sum_{i\neq}\sum_{j\in \overline{W}B}\gamma_{ij}\right)\\ +\left(\sum_{i\in W\cap A}\sum_{j(\neq i)\in \overline{W}\cap B}\gamma_{ij} + \sum_{j\in W\cap B}\sum_{i(\neq j)\in \overline{W}\cap A}\gamma_{ij}\right)\end{array}\right]$$

where $\gamma_i = \frac{1}{\pi_i} - 1$ and $\gamma_{ij} = \frac{\pi_i\pi_j}{\pi_{ij}} - 1$

(iii) $$\hat{\text{Var}}\left(\pi_A^{\text{HT}}\right) = \frac{1}{N^2\pi_w^2}\left[\sum_{i\in s}\gamma_i(\gamma_i - 1)z_i + \sum_{i,j\in s}\frac{\gamma_{ij}}{\pi_{ij}}z_iz_j\right]$$

Simple Random Sampling Without Replacement Sampling Design

For SRSWOR, $\pi_i = n/N$, $\pi_{ij} = n(n-1)/\{N(N-1)\}$, $\gamma_i = (N-n)/N$ and $\gamma_{ij} = -(N-n)/\{n(N-1)\}$. In this case, we have the following results:

Theorem 10.6.5 For SRSWOR sampling,

(i) $\pi_A^{\text{WOR}} = \frac{\overline{z}_s - (1-\pi_w)\pi_x}{\pi_w}$ is an unbiased estimator of π_A

(ii) $$\text{Var}(\hat{\pi}_A^{\text{WOR}}) = \frac{N(1-f)}{(N-1)}\left[\frac{\pi_A(1-\pi_A)}{n} + \frac{(1-\pi_w)}{n\pi_w^2}\{\pi_w\pi_A + (1-2\pi_w\pi_A)\pi_x\} - (1-\pi_w)\pi_x^2\right]$$

where $f = n/N$, $\gamma_i = \frac{1}{\pi_i} - 1$ and $\gamma_{ij} = \frac{\pi_i\pi_j}{\pi_{ij}} - 1$.

(iii) An unbiased estimator of $\text{Var}(\hat{\pi}_A^{\text{WOR}})$ is

$$\hat{Var}(\hat{\pi}_A^{WOR}) = \frac{1}{\pi_w^2}\frac{1-f}{n-1}\sum_{i\in s}(z_i-\bar{z}_s)^2 = \frac{1}{\pi_w^2}\frac{1-f}{n-1}\lambda_s(1-\lambda_s)$$

where λ_s is the proportion of "yes" answers in the sample s.

Probability Proportional to Size with Replacement

Let a sample s of size n be selected from a population U by the probability proportional to size with replacement (PPSWR) method using the normed size measure p_i attached to the i th unit. Let $z(r)$ and $p(r)$ be the response and normed size measure of the unit that was selected at the rth$(r = 1, \ldots, n)$ draw.

Then $z(r) = z_j$ and $p(r) = p_j$ if the rth draw selects jth unit. The Hansen-Hurwitz-type estimator for the population proportion π_A is given by

$$\hat{\pi}_A^{HH} = \frac{1}{\pi_w}\left[\frac{1}{Nn}\sum_{r=1}^{n}\frac{z(r)}{p(r)} - (1-\pi_w)\pi_x\right] \tag{10.6.29}$$

Now noting $\frac{z(r)}{p(r)}$ are iid random variables with mean $\sum_{i\in U} z_i = N\{\pi_w\pi_A + (1-\pi_w)\pi_x\}$ and variance $\sum_{i\in U}\frac{z_i}{p_i} - \left(\sum_{i\in U} z_i\right)^2$, the following results are obtained.

Theorem 10.6.6 For PPSWR sampling,

(i) $\hat{\pi}_A^{HH}$ is an unbiased estimator of π_A

(ii) $\text{Var}(\hat{\pi}_A^{HH}) = \frac{1}{N^2\pi_w^2 n}\left[\sum_{i\in U}\frac{w_i y_i + (1-w_i)x_i}{p_i} - N^2\{\pi_w\pi_A + (1-\pi_w)\pi_x\}^2\right]$

(iii) An unbiased estimator of $\text{Var}(\hat{\pi}_A^{HH})$ is

$$\hat{Var}(\hat{\pi}_A^{HH}) = \frac{1}{N^2\pi_w^2 n(n-1)}\sum_{r=1}^{n}\left\{\frac{z(r)}{p(r)} - \left(\sum_{r=1}^{n}\frac{z(r)}{p(r)}\right)\right\}^2$$

SRSWR Sampling

The PPSWR reduces to SRSWR when $p_i = 1/N$ for every $i \in U$. Substituting $p_i = 1/N$ in Theorem 10.6.6, we obtain.

Theorem 10.6.7 For a SRSWR sampling,

(i) $\hat{\pi}_A^{WR} = \dfrac{\hat{\lambda}_{WR} - (1-\pi_w)\pi_x}{\pi_w}$ (10.6.30) is an unbiased estimator of π_A where $\hat{\lambda}_{\text{WR}}$ is the proportion of "yes" answers in the sample.

$$\text{(ii) Var}\left(\hat{\pi}_A^{\text{WR}}\right) = \frac{\pi_A(1-\pi_A)}{n} + \frac{1-\pi_w}{n\pi_w^2}\left[\pi_A\pi_w + (1-2\pi_w\pi_A)\pi_x - (1-\pi_w)\pi_x^2\right] \tag{10.6.31}$$

(iii) An unbiased estimator of $\text{Var}\left(\hat{\pi}_A^{\text{HH}}\right)$ is

$$\hat{\text{Var}}\left(\hat{\pi}_A^{\text{HH}}\right) = \frac{\hat{\lambda}_{\text{WR}}\left(1-\hat{\lambda}_{\text{WR}}\right)}{n(n-1)} \tag{10.6.32}$$

Expressions (10.6.30), (10.6.31) and (10.6.32) are the same as was obtained in Theorem 10.6.2.

10.6.5 Hidden Sensitivity Model for Two Sensitive Attributes

Tian et al. (2007a, 2007b, 2009a, 2009b) proposed the hidden sensitive model for estimating two sensitive attributes x (taking drugs) and y(alcoholic). Let A and B be the groups of people who possess the attributes x and y respectively while $\overline{A}$ and $\overline{B}$ denote the complementary of the groups A and B.

Let us define $x_i = 1$ if the ith unit of the population U possesses the attribute x(i.e. $i \in A$) and $x_i = 0$ otherwise (i.e., $i \in \overline{A}$). Similarly, define $y_i = 1$ if the ith unit has the attribute y(i.e. $i \in B$) and $y_i = 0$ otherwise (i.e., $i \in \overline{B}$).

Further, define,

Category-1: $\overline{AB}$ = group of people who do not possesses either x or y

Category-2: $A\overline{B}$ = group of people who possesses the attribute x but not the attribute y

Category-3: $\overline{A}B$ = group of people does not possess the attribute x but possess the attribute y

Category-4: AB = group of people who possesses both the attributes x and y.

and

$$\theta_{00} = \frac{1}{N}\sum_{i\in U}(1-x_i)(1-y_i),\ \theta_{x0} = \frac{1}{N}\sum_{i\in U}x_i(1-y_i),$$
$$\theta_{0y} = \frac{1}{N}\sum_{i\in U}(1-x_i)y_i \text{ and } \theta_{xy} = \frac{1}{N}\sum_{i\in U}x_iy_i \tag{10.6.33}$$

be the proportion of persons belonging to the groups $\overline{AB}$, $A\overline{B}$, $\overline{A}B$ and AB respectively. Our objective is to make statistical inferences regarding

$$\theta_x = \theta_{x0} + \theta_{xy}, \theta_y = \theta_{0y} + \theta_{xy} \text{ and } \boldsymbol{\theta} = (\theta_{00}, \theta_{x0}, \theta_{0y}, \theta_{xy}) \tag{10.6.34}$$

To obtain reliable data from the selected respondents, Tian et al. (2007a, 2007b, 2009a, 2009b) introduced a non-sensitive variable w that can take four values: 1, 2, 3, and 4 with known probabilities p_1,p_2, p_3 and p_4. The variable w is assumed to be independent of both the characteristics x and y of interest. For example, w may be the date of birth of the respondent, such that $w = k$ if the respondent is born in the kth$(= 1, 2, 3, 4)$ quarter of a year where quarter-1: January, February, March; quarter-2: April, May, June; quarter-3: July, August, September; and quarter-4: October, November, December. In this case, one can assume $p_k \cong 1/4$ for $k = 1, 2, 3, 4$. Respondents belonging to the non-sensitive Category-1 ($\overline{AB}$) are directed to put a tick (√) in the Block $i(= 1, 2, 3, 4)$ if he or she is born in Quarter-$i(= 1, 2, 3, 4)$. It is expected that the respondents belonging to the non-sensitive Category-1 should report truthfully, as they have no reason to hide their actual status. Respondents belonging to sensitive categories 2, 3, and 4 may not feel comfortable revealing their status. Hence, respondents who belong to categories 2, 3, and 4 are respectively directed and forced to put a tick (√) in Block 2, Block 3, and Block-4 to somehow protect their privacy.

Question: In which quarter were you born?

Instructions: Category-1: Please put tick (√) against your Block-1; Category-2: Please put tick (√) against Block -2; Category-3: Please put tick (√) against Block -3, Category-4: Please put tick (√) against Block-4.

Let a sample s of size n be selected from the population U by the SRSWR method and n_i be the observed frequency of respondents putting a tick in Block i. Now, noting a tick in Block 1 is obtained from those respondents who belong to the non-sensitive Category-1 and were born in the 1st quarter of a year, while a tick mark in the Block $i(= 2, 3, 4)$ is obtained from those respondents who belong to the Category-1 and were born in the ith quarter of the year and from those who belong to sensitive category $i(= 2, 3, 4)$.

Table 10.2 gives the cell probabilities $\{\theta_i\}$, the observed frequencies $\{n_i\}$, and the unobservable frequencies $\{m_i\}$. The observed frequency n_i is the sum of the frequencies of the respondents belonging to the Block i and the Category $i(= 2, 3, 4)$.

The observed-data likelihood function for $\boldsymbol{\theta}$ is given by

Table 10.2 Questionnaire for the hidden sensitivity (HS) model

Categories	$w = 1$	$w = 2$	$w = 3$	$w = 4$	Total
1: $\overline{AB}$	$p_1\theta_1$	$p_2\theta_1$	$p_3\theta_1$	$p_4\theta_1$	$\theta_1(m_1)$
2: $\bar{A}B$					$\theta_2(m_2)$
3: $A\bar{B}$					$\theta_3(m_3)$
4: AB					$\theta_4(m_4)$
Total	$p_1(n_1)$	$p_2(n_2)$	$p_3(n_3)$	$p_4(n_4)$	$1(n)$

$$L(\boldsymbol{\theta}|Y_{\text{obs}}) = \frac{n!}{n_1!n_2!n_3!n_4}(p_1\theta_1)^{n_1}\prod_{i=2}^{4}(p_i\theta_1+\theta_i)^{n_i};\ \theta_i \geq 0, \sum_{i=1}^{4}\theta_i = 1,\ i = 1, 2, 3, 4. \tag{10.6.35}$$

where $Y_{\text{obs}} = (n; n_1, n_2, n_3, n_4)$.

The log-likelihood function of the observed-data is

$$\begin{aligned} L(\boldsymbol{\theta}|Y_{\text{obs}}) = \text{constant } &+ n_1\log(p_1\theta_1) + n_2\log(p_2\theta_2+\theta_2) \\ &+ n_3\log(p_3\theta_3+\theta_3) + n_4\log(p_4\theta_3+\theta_4) \end{aligned} \tag{10.6.36}$$

where $\theta_4 = 1 - \theta_1 - \theta_2 - \theta_3$.

The LME of θ_j's are as follows:

$$\hat{\theta}_1 = \frac{n_1}{np_1} \text{ and } \hat{\theta}_j = \frac{n_j}{n} - \hat{\theta}_1 p_j;\ j = 2, 3, 4 \tag{10.6.37}$$

Clearly, $\hat{\theta}_j$ can take values outside of the parameter range [0, 1] In case there exists a subset $\Gamma_1 \subset \{2, 3, 4\}$ such that $\hat{\theta}_j < 0$ for any $i \in \Gamma_1$, we obtain true MLE following Tang et al. (2009) as follows:

$$\begin{aligned} &\hat{\theta}_i = 0,\ i \in \Gamma_1,\ \hat{\theta}_1 = \frac{n_1 + \sum_{i\in\Gamma} n_i}{n\left(p_1 + \sum_{i\in\Gamma} p_i\right)}, \text{ and} \\ &\hat{\theta}_j = \frac{n_j - \left(n_1 + \sum_{i\in\Gamma} n_i\right)p_j}{n\left(p_1 + \sum_{i\in\Gamma} p_i\right)}, j \in \overline{\Gamma}_1 = \{2, 3, 4\} - \Gamma_1 \end{aligned} \tag{10.6.38}$$

Further, if there exists a subset $J \subset \overline{\Gamma}_1$ such that $\hat{\theta}_j < 0$ for $j \in J$, we have

$$\hat{\theta}_i = 0,\ i \in \{\Gamma_1, J\},\ \hat{\theta}_1 = \frac{n_1 + \sum_{i\in(\Gamma_1,J)} n_i}{n\left(p_1 + \sum_{i\in(\Gamma_1,J)} p_i\right)}, \text{ and}$$

$$\hat{\theta}_j = \frac{n_j - \left(n_1 + \sum_{i\in(\Gamma_1,J)} n_i\right)p_j}{n\left(p_1 + \sum_{i\in(\Gamma_1,J)} p_i\right)}, j \in \overline{\Gamma}_1 - J$$

If there still a subset $J^* \subset \overline{\Gamma}_1 - J$ such that $\theta_j < 0$ for $j \in J^*$, we can repeat the above process until the estimates $\hat{\theta}_j \geq 0$.

For other values of n_i s such that (i) $n_1 = 0$ and (ii) $n_i = 0$ for $i \in \Gamma_1$, similar corrections are provided by Tang et al. (2009).

10.6.5.1 MLE by EM Algorithm

Tian et al., (2007a, 2007b) obtained MLE by using an EM algorithm treating frequencies m_i's of the respondents' belonging Category i are missing. By denoting the missing data by $Y_{\text{mis}} = \{m_2, m_3, m_4\}$ and the complete-data by $Y_{\text{com}} = \{Y_{\text{obs}}, Y_{\text{mis}}\}$, the complete likelihood function is obtained as

$$L(\theta, Y_{\text{com}}) \propto \prod_{i=1}^{4} \theta_i^{m_i}, \; m_1 = n - m_2 - m_3 - m_4 \tag{10.6.39}$$

Equation (10.6.39) provides complete MLEs of $\boldsymbol{\theta}$ as

$$\hat{\theta}_i = m_i/n, \; i = 2, 3, 4 \tag{10.6.40}$$

Given $(Y_{\text{obs}}, \boldsymbol{\theta})$, m_i follows a binomial distribution with parameters n_i and $\theta_i/(p_i\theta_1 + \theta_i)$, $i = 2, 3, 4$, the EM algorithm's E-step computes the condition expectation

$$\begin{aligned} &E(m_i|Y_{\text{obs}}, \boldsymbol{\theta}) = n_i\theta_i/(p_i\theta_1 + \theta_i), \; i = 2, 3, 4 \text{ and} \\ &E(m_1|Y_{\text{obs}}, \boldsymbol{\theta}) = n - \sum_{i=2}^{4} E(m_i|Y_{\text{obs}}, \boldsymbol{\theta}) \end{aligned} \tag{10.6.41}$$

At the M-step updates (10.6.40) by replacing $\{m_i\}$ with their conditional expectations.

Remark 10.6.3 Tang et al. (2009) extended the Hidden sensitivity model proposed by Tian et al. (2007a, 2007b, 2009a, 2009b) for $k(\geq 4)$ categories and obtained an MLE estimator of the parameter $\boldsymbol{\theta} = (\theta_1, \theta_2, \ldots, \theta_k)$.

10.6.6 Bayesian Methods for NRR

In this section, we will consider the Bayesian inference for the NRR models proposed by Tian et al. (2009a, 2009b). The exact posterior distribution of the proportion and the expressions of the posterior moments and posterior mode are obtained by using the EM algorithm.

10.6.6.1 Triangular Model

Following Tian et al. (2009a, 2009b), let us define z_i^T as the response obtained from a respondent selected at the $i(=1, \ldots, n)$ th draw. Then,

$$z_i^T = \begin{cases} 1, & \text{with probability } \pi + (1-\pi)p, \\ & \text{if a tick is put in the triangle} \\ 0, & \text{with probability } (1-\pi)(1-p), \\ & \text{if a tick is put in the circle} \end{cases}$$

In this case, we obtain the observed-data and the likelihood function of π are $Z_{obs} = \left(z_1^T, \ldots, z_n^T\right)$ and

$$\begin{aligned} L_T(\pi|Z_{obs}) &= \prod_{i=1}^{n} \{\pi + (1-\pi)p\}^{z_i T}\{(1-\pi)(1-p)\}^{1-z_i T} \\ &= \{\pi + (1-\pi)p\}^t\{(1-\pi)(1-p)\}^{n-t}; 0 \le \pi \le 1 \end{aligned} \quad (10.6.41)$$

where $t = \sum_{i=1}^{n} z_i^T$ = total number of ticks that were placed in a triangle.

Assuming the prior distribution of π is a Beta(α, β) distribution, the posterior distribution of π comes out as

$$f(\pi|Z_{obs}) = \frac{1}{C_T(\alpha, \beta; t, n-t)}\pi^{\alpha-1}(1-\pi)^{\beta+n-t-1}\{\pi + (1-\pi)p\}^t \quad (10.6.42)$$

The explicit expressions for posterior mean and q th order moment for π come out as follows

$$E(\pi|Z_{obs}) = \frac{C_T(\alpha+1, \beta; t, n-t)}{C_T(\alpha, \beta; t, n-t)} \text{ and } E\left(\pi^q|Z_{obs}\right) = \frac{C_T(\alpha+q, \beta; t, n-t)}{C_T(\alpha, \beta; t, n-t)}; \; q \ge 1 \quad (10.6.43)$$

where $C_T(\alpha, \beta; t, n-t) = \sum_{j=0}^{t} \binom{t}{j} p^{t-j} B(\alpha + j, \beta + n - j)$.

EM Algorithm for Posterior Mode

For moderate symmetrical distributions of $f(\pi|Z_{obs})$, the posterior mean and variance adequately describe the location and dispersion. For highly skewed posterior distributions, the posterior mode is generally used for location.

Let $\tilde{Z}$ denotes the unobservable variable representing the number of respondents with sensitive attributes. In this case, the complete-data is $Z_{com} = \left(Z_{obs}, \tilde{Z}\right)$. The complete-data posterior distribution and the conditional predictive distributions are given by Tian et al. (2009a, 2009b) as follows:

$$f(\pi|Z_{obs}, \tilde{Z}) = \text{Beta}(\pi|a + \tilde{Z}, b + n - \tilde{Z}) \tag{10.6.44}$$

and

$$f(\tilde{Z}|Z_{obs}, \pi) = \text{Binomial}\left[\tilde{Z}|t, \pi/\{\pi + (1-\pi)p\}\right] \tag{10.6.45}$$

Therefore, the M-step of the EM algorithm computes the complete-data posterior mode as

$$\pi_{T,M} = \frac{a + \tilde{Z} - 1}{a + b + n - 2} \tag{10.6.46}$$

and the E-step replaces $\tilde{Z}$ with its conditional expectation

$$E(\tilde{Z}|Y_{obs}, \pi) = \frac{t\pi}{\pi + (1-\pi)p} \tag{10.6.47}$$

10.6.6.2 Crosswise Model

Let z_i^C be the response obtained from a respondent that is selected at the $i(= 1, \ldots, n)$ th draw. Then,

$$z_i^C = \begin{cases} 1, & \text{with probability } \pi p + (1-\pi)(1-p), \\ & \text{if a tick is put in the main diagonal with two circles} \\ 0, & \text{with probability } \pi(1-p) + \pi(1-p), \\ & \text{if a tick is put in the off - diagonal with the two dots} \end{cases} \tag{10.6.48}$$

The likelihood function of π is

$$\begin{aligned} L_c(\pi|Z_{obs}) &= \prod_{i=1}^{n} \{\pi p + (1-\pi)(1-p)\}^{z_i^c} \{\pi(1-p) + (1-\pi)p\}^{1-z_i^c} \\ &= \{\pi p + (1-\pi)(1-p)\}^s \{\pi(1-p) + (1-\pi)p\}^{n-s};\ 0 \le \pi \le 1 \end{aligned} \tag{10.6.49}$$

where $s = \sum_{i=1}^{n} z_i^c$.

Assuming, the prior distribution of π as Beta(α, β), the posterior distribution of π is obtained as

$$\begin{aligned} f(\pi|Z_{obs}) = &\frac{1}{C_C(\alpha, \beta; s, n-s)} \pi^{\alpha-1}(1-\pi)^{\beta-1} \\ &\{\pi p + (1-\pi)(1-p)\}^s \{\pi(1-p) + (1-\pi)p\}^{n-s} \end{aligned} \tag{10.6.50}$$

Table 10.3 The crosswise model and cell probabilities

Categories	$x = 0$	$x = 1$	Categories	$x = 0$	$x = 1$	Total
$y = 0$	O	•	$y = 0$	$(1 - \pi_y)(1 - \pi_x)$	$(1 - \pi_y)\pi_x$	$1 - \pi_y$
$y = 1$	•	O	$y = 1$	$\pi_y(1 - \pi_x)$	$\pi_y\pi_x$	π_y
			Total	$1 - \pi_x$	π_x	1

where $C_C(\alpha, \beta; s, n - s) = \frac{p^{n-s}(1-p)^s}{\Gamma(\alpha+\beta+n)} \sum_{j=0}^{s} \sum_{k=0}^{n-s} \binom{s}{j} \binom{n-s}{k} \left(\frac{p}{1-p}\right)^{j-k} \Gamma(\alpha + j + k)\Gamma(\beta + n - j - k)$.

The q th posterior moment of π is obtained as

$$E\left(\pi^q | Z_{\text{obs}}\right) = \frac{C_C(\alpha + q, \beta; s, n - s)}{C_C(\alpha, \beta; s, n - s)} \frac{\Gamma(\alpha + \beta + n)}{\Gamma(\alpha + \beta + n + q)}, \; q \geq 1 \tag{10.6.51}$$

EM Algorithm for Posterior Mode

Denoting the unobservable variables $\tilde{Z} = \left(\tilde{Z}_{11}, \tilde{Z}_{10}\right)$ with $\tilde{Z}_{11}$ and $\tilde{Z}_{10}$ as the counts of the cells (1, 1) and (1, 0) of Table 10.3, Tian et al. (2009a, 2009b) obtained the complete-data posterior and conditional predictive distributions as

$$f(\pi | Z_{\text{obs}}, \tilde{Z}) = \text{Beta}(\pi | \alpha + Z_{11} + Z_{10}, \beta + n - Z_{10} - Z_{11}) \tag{10.6.52}$$

and

$$\begin{aligned} f(\tilde{Z} | Z_{\text{obs}}, \pi) &= f(Z_{11} | Z_{\text{obs}}, \pi) \times f(Z_{10} | Z_{\text{obs}}, \pi) \\ &= \text{Binomial}\left(Z_{11} | s, \frac{\pi p}{\pi p + (1 - \pi)(1 - p)}\right) \\ &\quad \times \text{Binomial}\left(Z_{10} | n - s, \frac{\pi(1 - p)}{\pi(1 - p) + (1 - \pi)p}\right) \end{aligned} \tag{10.6.53}$$

The EM algorithm at the M-step yields the complete-data posterior mode

$$\pi_{\text{TC}} = \frac{a + \tilde{Z}_1 + \tilde{Z}_2 - 1}{\alpha + \beta + n - 2} \tag{10.6.54}$$

while at the E-step to replace $\tilde{Z}_1 + \tilde{Z}_2$ by its conditional expectation

$$E\left(\tilde{Z}_1 + \tilde{Z}_2 | Y_{\text{obs}}, \pi\right) = \frac{s\pi p}{\pi p + (1 - \pi)(1 - p)} + \frac{(n - s)\pi(1 - p)}{\pi(1 - p) + (1 - \pi)p}; \; p \neq 1/2 \tag{10.6.55}$$

In case, $p = 1/2$, the mode π_{TC} of the prior distribution Beta(α, β) reduces to $\frac{\alpha-1}{\alpha+\beta-2}$.

10.6.6.3 Hidden Sensitivity Model

Suppose that out of the n respondents selected at random, n_i of them put a tick against the Block i, $i = 1, \ldots, 4$. In this case, the observed frequency is $\mathbf{n}_{\text{obs}} = (n_1, n_2, n_3, n_4)$, and the observed-data likelihood is

$$L_{\mathrm{H}}(\boldsymbol{\theta}|\mathbf{n}_{\text{obs}}) = \theta_1^{n_1} \prod_{i=2}^{4} (p_i\theta_1 + \theta_i)^{n_i} \tag{10.6.56}$$

where $\boldsymbol{\theta} = (\theta_1, \theta_2, \theta_3, \theta_4)$.

Assuming the prior distribution of $\boldsymbol{\theta}$ follows Dirichlet distribution with parameter $\mathbf{a} = (a_1, a_2, a_3, a_4)$, Tian et al. (2009a, 2009b) derived the posterior distribution of $\boldsymbol{\theta}$ as

$$f(\boldsymbol{\theta}|\mathbf{n}_{\text{obs}}^{\mathrm{H}}) = c_{\mathrm{H}}^{-1}(\mathbf{a}, \mathbf{n}) \left(\prod_{i=1}^{4} \theta_k^{a_i - 1} \right) L_{\mathrm{H}}(\boldsymbol{\theta}|\mathbf{n}_{\text{obs}}) \tag{10.6.57}$$

where

$\mathbf{a} = (a_1, \ldots, a_4)$, $c_{\mathrm{H}}(\mathbf{a}, \mathbf{n}) = c_{\mathrm{H}}^*(\mathbf{a}, \mathbf{n}) / \Gamma\left(\sum_{i=1}^{4} a_i + n \right)$ and $c_{\mathrm{H}}^*(\mathbf{a},\mathbf{n}) = \sum_{j_2=0}^{n_2} \sum_{j_3=0}^{n_3} \sum_{j_4=0}^{n_4} \Gamma(a_1 + n_1 + j_2 + j_3 + j_4) \prod_{\ell=2}^{4} \binom{n_\ell}{j_\ell} \Gamma(a_\ell + n_\ell - j_\ell) p_\ell^{j_\ell}$.

The posterior moment of $\boldsymbol{\theta}$ was obtained as

$$E\left(\theta_1^{t_1}\theta_2^{t_2}\theta_3^{t_3}\theta_4^{t_4}|\mathbf{n}_{\text{obs}}\right) = \frac{c_{\mathrm{H}}^*(\mathbf{a}+\mathbf{t},\mathbf{n})}{c_{\mathrm{H}}^*(\mathbf{a},\mathbf{n})} \frac{\Gamma\left(\sum_{i=1}^{4} a_i + n\right)}{\Gamma\left(\sum_{i=1}^{4} (a_i + t_i) + n\right)}; \mathbf{t} = (t_1, \ldots, t_4) \tag{10.6.58}$$

EM Algorithm for Posterior Mode

To derive the posterior mode of $\boldsymbol{\theta}$, Tian et al. (2009a, 2009b) treated the observed frequencies n_2, n_3 and n_4 as incomplete data, m_2, m_3 and m_4 as missing data and $m_1 = n - m_1 - m_2 - m_3$. In this case, the complete-data posterior and the conditional predictive distributions with $\mathbf{m} = (m_2, m_3, m_4)$, were obtained respectively as

$$f(\boldsymbol{\theta}|\mathbf{n}, \mathbf{m}) = \text{Dirichlet}(\boldsymbol{\theta}|a_1 + m_1, \ldots, a_4 + m_4) \tag{10.6.59}$$

and

$$f(\mathbf{m}|\mathbf{n}, \boldsymbol{\theta}) = \prod_{i=2}^{4} \text{Binomial}(m_i/n_i, \theta_i/\{p_i\theta_1 + \theta_i\}) \tag{10.6.60}$$

For the EM algorithm the M-step computes the complete-data posterior mode as

$$\tilde{\theta}_i^{\text{H}} = \frac{a_i + m_i - 1}{\sum_{i=1}^{4} a_i + n - 4},\ i = 2, 3, 4;\ \tilde{\theta}_1^{\text{H}} = 1 - \tilde{\theta}_2^{\text{H}} - \tilde{\theta}_3^{\text{H}} - \tilde{\theta}_4^{\text{H}} \tag{10.6.61}$$

and at the E-step, m_i is replace with its conditional expectations

$$E(m_i|\mathbf{n}, \boldsymbol{\theta}) = n_i \frac{\theta_i}{p_i\theta_i + \theta_i},\ i = 2, 3, 4 \tag{10.6.62}$$

10.6.7 Case Studies

Tian (2014) reported real surveys on premarital sexual behavior and plagiarism at the University of Hong Kong, while Tang et al. (2014) reported surveys on homosexual experience among college students. Tian used the same crosswise method for both surveys, while Tang et al. used a Triangular model for their survey. The summarized results of surveys on sexual behavior and homosexual experiences among students are presented as follows:

10.6.7.1 Premarital Sexual Behavior Survey at Wuham

Information relating to premarital sexual activity among students in China is considered a highly sensitive topic. Tian (2014) surveyed 500 college students about their sexual behavior at Central China University by using the NRR method (Crosswise model) with SRSWR sampling. The questionnaire for the survey was framed as follows:

1. If your birthday is in the first half of the year, please answer "yes" or "no" to the question:

Q1: Is your cell phone number's last digit even?

2. If your birthday is in the second half of the year, please answer "yes" or "no" to the question:

Q2: Have you ever engaged in premarital sexual behavior?

Among 500 respondents, 112 answered "yes" (112 lines connecting two circles), 248 answered "no" (248 lines connecting two solid dots), and 140 did not respond.

Assuming, $p =$ Prob (a respondent's birthday is in the second half of the year) $=$ $q =$ Prob (last digit of a respondent's cell phone number is even) $= 0.5$, the MLE (moment estimate) of π and its estimated variance were obtained as $\hat{\pi}_p = 0.1222$ and $\hat{\text{Var}}(\hat{\pi}_p) = 0.0024$. Assuming uniform prior ($\alpha = \beta = 1$), the posterior mean of π given Z_{obs} was obtained as $E(\pi|Z_{\text{obs}}) = 0.12487$. Further, using $\pi^{(0)} = 0.5$ an initial value, the EM algorithm yielded the posterior mode as $\tilde{\pi}_p = 0.1222$ which is exactly equal to the moment estimate.

10.6.7.2 Homosexual Experience Among College Students

Tang et al. (2014) conducted a survey using a triangular model to find out whether the proportion of college students having homosexual experiences is identical for men and women. In this survey, 50 males and 48 females were selected at random. The sensitive variable y represents the homosexual experience. The variable $y = 1$ if a respondent had homosexual experience before and $y = 0$ otherwise. The non-sensitive variable W was taken as the last digit of the telephone number. $W= 1$, if the last digit was even, and $W = 0$ if the last digit was odd. The interviewees were asked to tick ($\surd$) against the answer A if they belonged to the group $y = 0,\ W = 0$. Otherwise, they were asked to tick against B. From the survey, 26 male and 26 female students marked tick against A. Assuming that the proportion of the last digit of a phone number being even is identical for males and females, the estimated homosexual experience among males and females was obtained as $\hat{\pi}_1 = 0.04$, and $\hat{\pi}_2 = 0.0833$ respectively. For testing the equality of homosexuality among males and females Wald score and likelihood ratio statistics values were obtained as 0.2149, 0.2148, and 0.2131. The p-values of all three test statistics exceed 0.05, and hence all test statistics indicate that the difference in the experience of homosexuality is insignificant at a 5% level of significance.

10.6.8 Christofides Models

Christofides (2009) proposed two simple methods that can be applied when both categories A and $\overline{A}$ are sensitive. Both methods are based on SRSWR samplings of size n.

10.6.8.1 Method 1

At first, each person was asked a direct question:

Q1: Are you a member of the sensitive group A?

If the answer to Q1 is "yes", without answering, the respondent should provide an answer of "yes" or "no" to the following question.

Q2: Are you a member of the non-sensitive group B_1?

Otherwise (i.e., if the answer to Q1 is "no"), the respondent should answer.

Q3: Are you a member of the non-sensitive group B_2?

The group B_1 could be people born in the first half of a year, January to June and the group B_2 are those whose last digit of their telephone number is 5, 7, 8, or 9. It is assumed that the proportion of people in the population who belong to the groups B_1 and B_2 are known as P_1 and $P_2(\neq P_1)$.

Let $\hat{\lambda}_C$ be the proportion of respondent who answers "yes" in the sample. Then,

$$E(\hat{\lambda}_C) = \pi_A P_1 + (1 - \pi_A)P_2 = \lambda_C \text{ and } V(\hat{\lambda}_{C1}) = \lambda_C(1 - \lambda_C) \tag{10.6.63}$$

Equation (10.6.11) yields the following theorem, which is easy to verify.

Theorem 10.6.8

(i) $\hat{\pi}_C = \frac{\hat{\lambda}_C - P_2}{P_1 - P_2}$ is an unbiased estimator of π_A.
(ii) The variance of $\hat{\pi}_C$ is

$$\begin{aligned}\text{Var}(\hat{\pi}_C) &= \frac{\lambda_C(1 - \lambda_C)}{n(P_1 - P_2)^2} \\ &= \frac{1}{n}\left[\pi_A(1 - \pi_A) + \frac{P_2(1 - P_2) + (1 - P_1 - P_2)(P_1 - P_2)\pi}{(P_1 - P_2)^2}\right]\end{aligned}$$

(iii) An unbiased estimator of $\text{Var}(\hat{\pi}_C)$ is

$$\widehat{\text{Var}}(\hat{\pi}_C) = \frac{\hat{\lambda}_C(1 - \hat{\lambda}_C)}{(n - 1)(P_1 - P_2)^2}.$$

Remark 10.6.3

(i) The first component of $\text{Var}(\hat{\pi}_C)$ is the variance due to direct questioning, while the second component is the price, we pay by asking indirect questions.
(ii) The $\text{Var}(\hat{\pi}_C)$ can be controlled by choosing the values of P_1 and P_2 appropriately.
(iii) In particular, if Q2 is the complement of Q1, then, $P_2 = 1 - P$ .and method 1 reduces to Warner's (1965) model.
(iv) $\hat{\pi}_C$ is a moment estimator that can take value outside of parametric space [0, 1]. The MLE of π_C is

$$\hat{\pi}_C^{\text{mle}} = \begin{cases} 0, & \text{if } \hat{\lambda} \leq P_2 < P_1 \text{ or } P_1 < P_2 \leq \hat{\lambda} \\ \hat{\pi}_C, & \text{if } P_2 < \hat{\lambda} < P_1 \text{ or } P_1 < \hat{\lambda} < P_2 \\ 1, & \text{if } P_2 < P_1 < \hat{\lambda} \text{ or } \hat{\lambda} \leq P_1 < P_2 \end{cases} \tag{10.6.64}$$

10.6.8.2 Method 2

Method 2 is used in situations where the prevalence of one of the two innocuous characteristics, say, B_1 is unknown. In this method, two independent samples of sizes n_1 and n_2 are selected from the population by the SRSWR method. Each person in the first sample was asked to respond as in Method 1. While each person in the second sample follows the same procedure, except for the answer to question Q1 is "no", he or she is asked to provide an answer to the following question Q4 instead of Q3 of Method 1.

Q4: Are you a member of the group $\overline{B}_2$ (complement of B_2)?

Let $\hat{\lambda}_{C1}$ and $\hat{\lambda}_{C2}$ be the proportions of "yes" answers from the samples 1 and 2. Then,

$$E(\hat{\lambda}_{C1}) = \pi_A P_1 + (1-\pi_A)P_2 = \lambda_{C1} = \lambda_C,\ V(\hat{\lambda}_{C1}) = \lambda_{C1}(1-\lambda_{C1}),$$
$$E(\hat{\lambda}_{C2}) = \pi_A P_1 + (1-\pi_A)(1-P_2) = \lambda_{C2} \text{ and } V(\hat{\lambda}_{C2}) = \lambda_{C2}(1-\lambda_{C2}) \qquad (10.6.65)$$

Equation (10.6.65) yields.

Theorem 10.6.9

(i) $\hat{\pi}_{C1} = 1 - \frac{\hat{\lambda}_{C1}-\hat{\lambda}_{C2}}{2P_2-1};\ P_2 \neq 1/2$ is an unbiased estimator of π_A.
(ii) The variance of $\hat{\pi}_{C1}$ is

$$\text{Var}(\hat{\pi}_{C1}) = \frac{1}{(2P_2-1)^2}\left[\frac{\lambda_{C1}(1-\lambda_{C1})}{n_1} + \frac{\lambda_{C2}(1-\lambda_{C2})}{n_2}\right]$$

(iii) An unbiased estimator of $\text{Var}(\hat{\pi}_{C1})$ is

$$\widehat{\text{Var}}(\hat{\pi}_{C1}) = \frac{1}{(2P_2-1)^2}\left[\frac{\hat{\lambda}_{C1}(1-\hat{\lambda}_{C1})}{n_1-1} + \frac{\hat{\lambda}_{C2}(1-\hat{\lambda}_{C2})}{n_2-1}\right].$$

Remark 10.6.4 From Eq. (10.6.65), one can estimate the prevalence P_1.

10.7 Discussions

Warner (1965) introduced the RR method to collect data relating to sensitive items that are not available through direct surveys. Direct data collection from respondents on sensitive items frequently involves significant non-sampling errors and a high non-response rate because respondents frequently provide false or socially acceptable responses due to social stigma and fear. Various improvements to RR techniques are proposed in the literature to improve the efficiency of the estimates and preserve the privacy of the respondents. However, some limitations of the RR surveys are identified, such as their lack of reproducibility, high cost, and time-consuming. Sometimes, respondents feel that there might be a hidden trick, as the randomization process is generally complex, which may not be understandable to the respondents. Several alternative methods for indirect data collection have been proposed to overcome such limitations to some extent. Alternative methods include the item count technique (ICT), the item sum technique (IST), the non-Randomized method (NRR), the parallel Method, the three-card method, and the nominative method.

The ICT was proposed by Droitcour et al. (1991). In this technique, two SRSWR samples s_1 (the treatment group) and s_2 (the control group) are selected independently from the population. Respondents belonging to the treatment group s_1 are provided with a long list (LL) of $g + 1$ dichotomous questions, of which the first g of the questions is not sensitive. The remaining $g + 1$-st question is the sensitive (key) question relating to the sensitive attribute A of interest. Respondents belonging to the control group s_2 are provided with a short list (SL) containing the same set of g innocuous questions present in the LL. Respondents are requested to report only the total score, the total number of "yes" answers. They should not disclose the answer to any of the individual questions. An estimate of the proportion of sensitive items was obtained from the difference between the sample means of the LL and SL of the samples.

The IST is similar to the ICT, the only difference is that the IST deals with quantitative sensitive variables only. Extensions of ICT in various directions were proposed by Hussain et al. (2011, 2012), Chaudhuri and Christofied (2007, 2013), and Tian et al. (2017), among others. IST for multiple sensitive attributes was proposed by Blair and Imai (2012). Perri et al. (2018) proposed IST for $q(\geq 2)$ quantitative variables, while Shaw (2015) and Rueda (2018) extended IST to complex survey designs. Several researchers, including Holland (1986), Corstange (2009), Imai (2011), Blair and Imai (2012), Trappmann et al. (2014), and Aronow et al. (2015), proposed multivariate relationships between preferences over the sensitive items and respondents' characteristics through appropriate modeling to improve estimates of the parameters of interest.

Non-randomized response (NRR) was introduced by Swensson (1974) and Takahasi and Sakasegawa (1977). The NRR models replace randomization devices with non-sensitive or innocuous questions and bear the reproducibility property. The NRR models are cheaper to administer, protect respondents' confidentiality, increase efficiency, and are feasible for face-to-face or mail inquiries. Several NRR methods are proposed in the literature. The triangular and crosswise models were proposed by Tian et al. (2007a, 2007b, 2009a, 2009b) and Yu et al. (2008) for a single sensitive question, while the parallel model was proposed by Tian (2014). Tian et al. (2007a, 2007b, 2009a, 2009b) proposed the hidden sensitive model for estimating two sensitive attributes for the same survey design, while Christofides (2009) provided two alternative models for two sensitive characteristics. As far as efficiency is concerned, the parallel model is more efficient than the crosswise and triangular models for most situations (Tian, 2014). Bayesian methods for analyzing sensitive questions for the triangular, crosswise, and hidden sensitivity models have been proposed. EM algorithms were used to obtain the exact posterior distribution, posterior moments, and posterior modes.

References

Arnab, R., Shangodoyin, D. K., & Arcos, A. (2019). Nonrandomized response model for complex survey designs. *Statistics in Transitions, 20*, 67–86.

Aronow, P. M., Coppock, A., Crawford, F. W., & Green, D. P. (2015). Combining list experiment and direct question estimates of sensitive behavior prevalence. *Journal of Survey Statistics and Methodology, 3*, 43–66.

Biemer, P. P., & Wright, D. (2004). *Estimating cocaine use using the item count methodology, in 'Preliminary results from the national survey on drug use and health.'* Annual meeting of the American Association for Public Opinion Research.

Biemer, P. and Brown, G. (2005). Model based estimation of drug prevalence item count data. *JOurnal of Official Statistics*, 21, 287–308.

Blair, G., & Imai, K. (2012). Statistical analysis of list of experiments. *Political Analysis, 20*, 47–77.

Bliss, C. I., & Fisher, R. A. (1953). Fitting the negative binomial distribution to biological data. *Biometrics, 1953*(9), 176–200.

Chaudhuri, A., & Christofides, T. C. (2007). Item count technique in estimating the proportion of people with sensitive feature. *Journal of the Statistical Planning and Inference, 137*, 589–593.

Chaudhuri, A., and Christofides, T.C. (2013). *Indirect Questioning in Sample Surveys*. Berlin, Heidelberg: Springer. Berlin, Heidelberg.

Christofides, T. C. (2009). Randomized response without a randomization device. *Advances and Applications in Statistics, 11*, 15–28.

Christofides, T. C. (2015). A new version of the item count technique. *Model Assisted Statistics and Applications, 10*, 289–297.

Corstange, D. (2009). Sensitive questions, Truthful answers? Modeling the list experiment with LISTIT. *Political Analysis, 17*, 45–63.

Coutts, E., & Jann, B. (2011). Sensitive questions in online surveys: Experimental results for the randomized response technique (RRT) and the unmatched count technique (UCT). *Sociological Methods and Research, 40*, 169–193.

Dalton, D. R., & Metzger, M. (1992). Integrity testing for personal selections: An unsparing perspective. *Journal of Business Ethics, 12*, 147–156.

Dalton, D. R., Wimbush, J. C., & Daily, C. M. (1994). Using the unmatched count technique (UCT) to estimate base rates for sensitive behavior. *Personnel Psychology, 47*, 817–828.

Droitcour, J. A., Caspar, R. A., Hubbard, M. L., Parsley, T. L.,Visscher, W., & Ezzati, T. M. (1991). The item count technique as a method of indirect questioning: A review of its development and a case study application. *Measurement Errors in Surveys*, 185–210.

Droitcour, J. A., & Larson, E. M. (2002). An innovative technique for asking sensitive questions: The three-card method. *Bulletin De Méthodologie Sociologique, 75*, 5–23.

Geurts, M. D. (1980). Using a randomized response research design to determine non-response. *Journal of the Academy of Marketing Science, 8*, 83–90.

Holbrook, A. L., & Krosnick, J. A. (2010). Social desirability bias in voter turnout reports. *Public Opinion Quarterly, 74*, 37–67.

Holland, P. W. (1986). Statistics and casual inference. *Journal of the American Statistical Association, 81*, 945–960.

Hubbard, M. L., Casper, R. A., & Lessler, J. T. (1989). Respondents OCO reactions to item count lists and randomized response. In *Proceedings of Survey Research Section, American Statistical Association*, Washington D.C. 544OCo448.

Hussain, Z., & Shabbir, J. (2011). On item count technique in survey sampling. *Journal of the Informatics and Mathematical Sciences, 2*, 161–169.

Hussain, Z., Shah, E. A., & Shabbir, J. (2012). An alternative item count technique in sensitive surveys. *Revista Colombiana De Estadistica, 35*, 39–54.

Hussain, Z., Shah, E. A., & Shabbir, J. (2017). An alternative item count technique for improved estimators of population means in sensitive surveys. *Hacettepe Journal of Mathematics and Statistics, 46*, 907–934.

Imai, K. (2011). Multivariate regression analysis for the item count technique. *Journal of the American Statistical Association, 106*, 407–415.

Janus, A. L. (2010). The influence of social desirability pressures on expressed immigration attitudes. *Social Science Quarterly, 91*(4), 928–946.

Kuha, J., & Jackson, J. (2014). The item count method for sensitive survey questions: Modeling criminal behaviour. *Journal of the Royal Statistical Society: Series C, 63*, 321–341.

Kulinski, J. H., Cobb, M. D., & Gilens, M. (1997). Racial attitudes in the "New South." *The Journal of Politics, 59*, 323–349.

LaBrie, J. W., & Earleywine, M. (2000). Sexual risk behaviors and alcohol: Higher base rate revealed using the unmatched-count technique. *Journal of Sex Research, 37*, 321–326.

Miller, J.D. (1984). *A new survey technique for studying deviant behavior*. Ph.D. Thesis, The George Washington University.

Miller, J. D. (1985). The nominative technique: a new method of estimating heroin prevalence. *NIDA Research Monograph*, No. 57, 104–124.

Perri, P. F., Rueda, M., & Cobo, B. (2018). Multiple sensitive estimation and optimal sample size allocation in the item sum technique. *Biometrical Journal, 60*, 155–173.

Raghavarao, D., & Federer, W. T. (1979). Block total response as an alternate to randomized response method in surveys. *Journal of the Royal Statistical Society, Series B, 41*, 40–45.

Rueda, M., Perri, P. F., & Cobo, B. (2018). Advances in estimating by the item sum technique using auxiliary information in complex surveys. *AStA Advances in Statistical Analysis, 102*, 455–478.

Shaw, P. (2015). Estimating a finite population mean of sensitive quantitative variable from a single probability sample by the item count technique. *Model Assisted Statistics and Applications, 10*, 411–419.

Sinderman, P. M., & Carmines, E. G. (1997). *Reaching beyond race*. Harvard University Press.

Swensson, B. (1974). Combined questions: a new survey technique for eliminating evasive answer bias (I). basic theory. *Report No. 70 of errors in survey research project, Institute of Statistics, University of Stockholm.*

Takahasi, K., & Sakasegawa, K. (1977). A randomized response technique without making use of randomizing device. *Annals of the Institute of Statistical Mathematics, 29*, 1–8.

Tang, M. L., Tian, G. L., Tang, N. S., & Liu, N. T. (2009). A new non-randomized multi-category response model with a single sensitive question: Design and analysis. *Journal of Korean Statistical Society, 38*, 339–349.

Tang, M. L., Wu, Q., Tian, G. L., & Guo, J. H. (2014). Two-sample non randomized response techniques for sensitive questions. *Communications in Statistics-Theory and Methods, 43*, 408–425.

Tian, G. L. (2014). A new non-randomized response model: The parallel model. *Statistica Neerlandica, 68*, 294–323.

Tian, G. L., & Tang, M. L. (2014). *Incomplete categorical data design: Non-randomized response techniques for sensitive questions in surveys*. CRC Press, Taylor & Francis Group.

Tian, G. L., Tang, M. L., Wu, Q., & Liu, Y. (2017). Poisson and Negative binomial item count technique with sensitive question. *Statistical Methods in Medical Research, 26*, 931–947.

Tian, G. L., Tan, M., & NG, K. W. (2007a). An exact non-iterative sampling procedure for discrete missing data problems. *Statistica Neerlandica, 61*, 232–242

Tian, G. L., Yu, J. W., Tang, M. L., & Geng, Z. (2007b). A new nonrandomized model for analyzing sensitive questions with binary outcomes. *Statistics in Medicine, 26*, 4238–4252.

Tian, G. L., Yuen, K. C., Tang, M. L., & Tan, M. T. (2009a). Bayesian non-randomized response models for surveys with sensitive questions. *Statistics and Inference, 2*, 13–25.
Tian, M. T., Tian, G. L., & Tang, M. L. (2009b). Sample surveys with sensitive questions: A non-randomized response approach. *American Statistician, 63*, 9–16.
Trappmann, M., Krumpal, I., Kirchner, A., & Jann, B. (2014). Item sum: A new technique for asking quantitative sensitive questions. *Journal of Survey Statistics and Methodology, 2*, 58–77.
Tsuchiya, T., Hirai, Y., & Ono, S. (2007). A study of the properties of the item count technique. *Public Opinion Quarterly, 71*, 253–272.
Walter, F., & Lair, B. (2014). The effectiveness of the item count technique in eliciting valid answers to sensitive questions. An evaluation in the context of self-reported delinquency. *Survey Research Methods, 8*, 153–168.
Warner, S. L. (1965). Randomized response: A survey technique for eliminating evasive answer bias. *Journal of the American Statistical Association, 60*, 63–69.
Wimbush, J. C., & Dalton, D. R. (1997). Base rate for employee theft: Convergence of multiple methods. *Journal of Applied Psychology, 82*, 756–763.
Yu, J.-W., Tian, G.-L., & Tang, M.-L. (2008). Two new models for survey sampling with sensitive characteristics: Design and Analysis. *Metrika, 67*, 251–263.

Chapter 11
Measures of Privacy Protection

Abstract Randomized response (RR) techniques are designed to collect reliable data relating to sensitive and stigmatizing subjects indirectly to reduce nonresponse and obtain randomized responses from the respondents. Several methods of measuring privacy protection ($\mathcal{PP}$) for the RR techniques and comparisons of their efficiencies are proposed. Most methods of $\mathcal{PP}$ are limited to the simple random sampling with replacement (SRSWR) method only. In this chapter, various forms of $\mathcal{PP}$ and their efficiencies are reviewed. In addition, a new way of measuring $\mathcal{PP}$ for complex survey designs has been proposed.

Keywords Complex surveys · Efficiency · Measures of jeopardy · Privacy protection · Randomized response

11.1 Introduction

The primary purpose of the RR technique is to deliberately mask the true response of a respondent so that his or her true response remains undisclosed to the interviewer or third parties. Hence, the RR technique retains privacy, prevents social embarrassment, and protects the self-image of a respondent while answering sensitive and threatening questions. As a result, the RR technique garners more cooperation from the respondents and reduces untruthful answers and nonresponses considerably. However, RR techniques are blamed for being less efficient than the direct questioning (DR) method, as the variances of the estimates based on RR techniques are higher than those of the DR method. However, such a comparison is not totally valid as the DR method estimates are based on fewer observations (responses) and more untruthful answers. The degrees of privacy protection ($\mathcal{PP}$) for the respondents and the efficiencies of estimates vary from one RR technique to the other. An efficient RR provides good estimates of the population parameters of interest and, at the same time, retains a sufficient degree of privacy for the interviewees. However, the two main objectives, $\mathcal{PP}$ and efficiency, are fundamentally conflicting because a large $\mathcal{PP}$ yields a lower efficiency of the estimate and vice-versa. Hence the efficiencies

R. Arnab, *Indirect Methods of Data Collection and Analysis from Surveys*, Indian Statistical Institute Series, https://doi.org/10.1007/978-981-97-6005-3_11

of different RR techniques should be compared while keeping the degree of confidentiality fixed at a certain level. Various measures of $\mathcal{PP}$ are available in the literature. However, most methods are restricted to the SRSWR method only. The methods include Lanke (1975, 1976), Leysieffer and Warner (1976), and Anderson (1975a, 1975b, 1977), among others. A good review is given by Chaudhuri and Mukerjee (1988), Ljungvist (1993), and Arnab (2017). There is no satisfactory method of $\mathcal{PP}$ for the general sampling designs available in the literature. Some of the important measures are described in this section. In addition, a new method of $\mathcal{PP}$ has been introduced.

11.2 SRSWR Sampling for Qualitative Characteristics

Consider a finite population consisting of N units (individuals) that are classified into two categories A and $\overline{A}$ (complementary of A). The category A comprises those individuals who possess certain sensitive attributes. The proportions of individuals in the population that belong to the categories A and $\overline{A}$ are $\pi(>0)$ and $1-\pi$, respectively. A sample s of n individuals is selected from the population by the SRSWR sampling method. Each of the selected individuals in the sample was requested to perform a RR device R and provide a truthful response (r) "yes = Y" or "no = N" as the outcome of the experiment R. The conditional probabilities of obtaining an answer $r = Y$ from a respondent belonging to the sensitive categories A and $\overline{A}$ under the randomization device R were set apriori to $P(Y|A) = 1 - P(N|A)$ and $P(Y|\overline{A}) = 1 - P(N|\overline{A})$ respectively. The probabilities $P(Y|A)$ and $P(Y|\overline{A})$ depend on the parameter θ of the chosen RR model and can be controlled by the investigator by setting θ appropriately. The parameter θ is called a design parameter.

The posterior probabilities of classifying an individual in the group A and $\overline{A}$ after obtaining a response $r(= Y$ or $N)$ from a respondent are given by

$$P(A|r) = \frac{\pi P(r|A)}{\pi P(r|A) + (1-\pi)P(r|\overline{A})} \text{ and}$$

$$P(\overline{A}|r) = 1 - P(A|r) = \frac{(1-\pi)P(r|\overline{A})}{\pi P(r|A) + (1-\pi)P(r|\overline{A})} \tag{11.2.1}$$

The probabilities $P(A|r)$ and $P(\overline{A}|r)$ are termed revealing probabilities. The revealing probabilities are unknown because of the involvement of the unknown parameter π.

11.2.1 Leysieffer and Warner's Measure

Leysieffer and Warner (1976) argued that a response r jeopardizes with respect to the classification if the posterior probability of the classification increases. Thus, the response r is jeopardized with respect to A if the posterior probability of a classification

$$P(A|r) > \pi \tag{11.2.2}$$

Similarly, the response r is jeopardized with respect to $\overline{A}$ if the posterior probability

$$P(\overline{A}|r) > 1 - \pi \tag{11.2.3}$$

In case $P(A|r) = \pi$[i.e. $P(\overline{A}|r) = 1 - \pi$] the response jeopardies neither.

From the above expressions (11.2.2) and (11.2.3), one notes that a response r jeopardizing with respect to A implies that the response r is not jeopardizing with respect to $\overline{A}$ and vice-versa as $P(A|r) > \pi$ implies $P(\overline{A}|r) < 1 - \pi$. Leysieffer and Warner proposed measures (LW-measure) of jeopardy with respect to A and $\overline{A}$ as

$$g(r|A) = \frac{P(r|A)}{P(r|\overline{A})} \text{ and } g(r|\overline{A}) = \frac{P(r|\overline{A})}{P(r|A)} = \frac{1}{g(r|A)} \tag{11.2.4}$$

The function $g(r|A)$ depends on the design parameter and response r.

Equation (11.2.1) yields

$$g(r|A) = \frac{P(A|r)}{P(\overline{A}|r)}\frac{1-\pi}{\pi} = \frac{1}{g(r|\overline{A})} \tag{11.2.5}$$

The expression (11.2.5) indicates that the response r is jeopardized with respect to A if $g(r|A) > 1$ (in this case $P(A|r) > \pi$). The response r is jeopardized with respect to $\overline{A}$ if $g(r|\overline{A}) > 1$ (in this case $P(\overline{A}|r) > 1 - \pi$) and for $g(r|A) = 1$, r is neither jeopardized with respect to A nor $\overline{A}$.

Leysieffer and Warner (1976) provided the criteria for an efficient RR model. The model R^* is the most efficient if it provides the smallest sampling variance for a given sampling design at the same time that it provides the maximum allowable level of $g(r|A)$ and $g(r|\overline{A})$ which is consistent with considerations of ethics and practical cooperation.

11.2.1.1 Optimum Dichotomous-Response Model for Dichotomous Population

Suppose that each individual in the population is either a member of the sensitive group A or its complement $\overline{A}$ and that the proportion of individuals in the population that belong to the group A is π. A SRSWR sample of size n is selected from the population, and each of the selected individuals in the sample was asked to perform a RR trial, which produced a response $r = Y$ (yes) or $r = N$ (no). The probability of obtaining an answer Y is

$$\lambda = P(Y) = P(Y|A)\pi + P(Y|\overline{A})(1-\pi);\ P(Y|A) \neq P(Y|\overline{A}); \tag{11.2.6}$$

Let $\hat{\lambda}$, be the proportion of "Y" responses in the sample. Then, an unbiased estimate of π and its variance are

$$\hat{\pi} = \frac{\hat{\lambda} - P(Y|\overline{A})}{P(Y|A) - P(Y|\overline{A})} \tag{11.2.7}$$

and

$$\text{Var}(\hat{\pi}) = \frac{\pi(1-\pi)}{n} + \frac{\pi P(Y|A)[1-P(Y|A)] + (1-\pi)P(Y|\overline{A})\big[1-P(Y|\overline{A})\big]}{n\big[P(Y|A) - P(Y|\overline{A})\big]^2} \tag{11.2.8}$$

Without loss of generality, let us assume that $P(Y|A) > P(Y|\overline{A})$ so that answer "Y" is jeopardizing with respect to A i.e., "N" answer is jeopardizing with respect to $\overline{A}$. In this case

$$g(Y|A) = \frac{P(Y|A)}{P(Y|\overline{A})} > 1 \text{ and } g(r|\overline{A}) = \frac{P(r|\overline{A})}{P(r|A)} = \frac{1}{g(r|A)} \tag{11.2.9}$$

Equations (11.2.8) and (11.2.9) lead

$$\text{Var}(\hat{\pi}) = \frac{\pi(1-\pi)}{n} + \frac{\pi g(Y|A) + (1-\pi)g(N|\overline{A})}{n\{g(Y|A)-1\}\big\{g(N|\overline{A})-1\big\}} \tag{11.2.10}$$

Since $\frac{\partial \text{Var}(\hat{\pi})}{\partial g(Y|A)} < 0$ and $\frac{\partial \text{Var}(\hat{\pi})}{\partial g(N|\overline{A})} < 0$, the variance of $\text{Var}(\hat{\pi})$ decreases with increase of $g(Y|A)$ and $g(N|\overline{A})$. So, to minimize $\text{Var}(\hat{\pi})$ one should choose $g(Y|A)$ and $g(N|\overline{A})$ as large as possible. But such an increase is not practical for the protection of privacy since the increase in $g(Y|A)$, jeopardizing "Y" answer with respect to A increase of $g(N|\overline{A})$ jeopardizing "N" answer with respect to $\overline{A}$. Leysieffer and Warner (1976) suggested that one should choose an RR model that provides the maximum acceptable level of jeopardy in terms of $g(Y|A) = k_1$ and $g(N|\overline{A}) = k_2$ which yet allows cooperation.

For the choice of $k_1 = k_2 = k$, the design parameter of the optimum RR model R^* was obtained by Leysieffer and Warner (1976) as

$$P(Y|A) = k/(k+1) \text{ and } P(Y|\overline{A}) = 1/(k+1) \tag{11.2.11}$$

In particular, k_1 is finite but $k_2 = \infty$, reflecting no jeopardy in "N" answer, the optimum design parameter (11.2.11) reduces to

$$P(Y|A) = \underset{k_2\to\infty}{Lim} \frac{k_1k_2 - k_1}{k_1k_2 - 1} = 1 \text{ and } P(N|\overline{A}) = \underset{k_2\to\infty}{Lim} \frac{k_2 - 1}{k_1k_2 - 1} = \frac{1}{k_1} \tag{11.2.12}$$

For more details readers are referred to Leysieffer and Warner (1976).

11.2.2 Lanke's Measure

Lanke (1976) argued that respondents in a sensitive group A (such as HIV +ve) may feel uncomfortable disclosing their membership in the group A, whereas respondents in a non-sensitive group $\overline{A}$ (HIV −ve) have no reason to conceal their membership in the group $\overline{A}$. Lanke (1976) concluded that "embarrassment" must mean "suspicion of belonging to A". A reasonable conclusion would be that the greater the conditional probability of belonging to A, given a specific answer, the greater the embarrassment caused by providing the answer. Let $P(A|Y)$ and $P(A|N)$ be the posterior probabilities of a respondent belonging to the group A given that he or she provided the "Y" and "N" answers. Lanke's (1976) measure of privacy protection is

$$\mathcal{L} = \text{Max}[P(A|Y), P(A|N)] \tag{11.2.13}$$

The smaller value of $\mathcal{L}$, the larger is the privacy protection.

Now noting,

$$P(A|Y) = \frac{\pi P(Y|A)}{P(Y)} \text{ and } P(A|N) = \frac{\pi P(N|A)}{P(N)}$$

one finds

$$\begin{aligned} P(A|Y) - P(A|N) &= \pi\left[\frac{P(Y|A)}{P(Y)} - \frac{P(N|A)}{1-P(Y)}\right] (\text{as } P(N) = 1 - P(Y)) \\ &= \pi\left[\frac{P(Y|A) - P(Y|A)P(Y) - P(N|A)P(Y)}{P(Y)\{1-P(Y)\}}\right] \end{aligned}$$

$$= \pi \left[\frac{P(Y|A) - P(A)P(Y|A) - P(\overline{A})P(Y|\overline{A}}{P(Y)\{1 - P(Y)\}} \right]$$

$$= \pi(1 - \pi) \left[\frac{P(Y|A) - P(Y|\overline{A}}{P(Y)\{1 - P(Y)\}} \right] \tag{11.2.14}$$

Thus,

$$L = \begin{cases} P(A|Y) & \text{if } P(Y|A) > P(Y|\overline{A}) \\ P(A|Y) = P(A|N) & \text{if } P(Y|A) = P(Y|\overline{A}) \\ P(A|N) & \text{if } P(Y|A) < P(Y|\overline{A}) \end{cases} \tag{11.2.15}$$

11.2.3 Anderson's Measure

Anderson (1975a) suggested that privacy should be protected for all people, irrespective of their membership to A or $\overline{A}$. Hence, revealing probabilities $P(A|R)$ and $P(\overline{A}|R)$, risk of "suspicions" for a given response $R(= Y \text{ or } N)$ should be restricted to a certain level, i.e.,

$$P(A|R) \leq \xi_1 \text{ and } P(\overline{A}|R) \leq \xi_2 \tag{11.2.16}$$

where $0 \leq \xi_1, \xi_2 \leq 1$.

Since, $P(A|R) = 1 - P(\overline{A}|R)$, Eq. (11.2.16) reduces to

$$1 - \xi_2 \leq P(A|R) \leq \xi_1 \tag{11.2.17}$$

Further, (11.2.17) implies

$$g(Y|A) \leq \frac{1-\pi}{\pi} \frac{\xi_1}{1-\xi_1} \text{ and } g(N|\overline{A}) \leq \frac{\pi}{1-\pi} \frac{\xi_2}{1-\xi_2} \tag{11.2.18}$$

Thus, Anderson's (1975a) measure of privacy restricts the upper bound of $g(Y|A)$ and $g(N|\overline{A})$ to a certain level.

11.2.4 Efficiency Comparisons

Definition 11.2.1 For a given sampling design p and a given level of $\mathcal{PP}$, the RR model R_1 with a parameter $\theta_1 (\in \Omega_1)$ and an estimator $\hat{\pi}_1$ of the population proportion π is said to be more efficient than the model R_2 with a parameter $\theta_2(\in \Omega_2)$ and a given estimator $\hat{\pi}_2$ of π if the variance of $\hat{\pi}_1$ is smaller than that of $\hat{\pi}_2$ $\forall \theta_1 \in \Omega_1$ and $\theta_2 \in \Omega_2$.

$$\text{i.e., } \mathrm{Var}_p(\hat{\pi}_1|R_1, \theta_1) \leq \mathrm{Var}_p(\hat{\pi}_2|R_2, \theta_2)\ \forall \theta_1 \in \Omega_1, \theta_2 \in \Omega_2, \quad (11.2.19)$$

where Ω_1 and Ω_2 are the parameter spaces of θ_1 and θ_2, and Var_p denotes the variance with respect to the sampling design p.

The models R_1 and R_2 are equivalent if

$$\mathrm{Var}_p(\hat{\pi}_1|R_1, \theta_1) = \mathrm{Var}_p(\hat{\pi}_2|R_2, \theta_2)\ \forall \theta_1 \in \Omega_1, \theta_2 \in \Omega_2 \quad (11.2.20)$$

11.2.4.1 Warner's Versus Mangat–Singh Model

LW-Measure of $\mathcal{PP}$

For Warner's (1965) model (briefly, W-model), described in Sect. 1.2, a respondent answers "yes" or "no" to one of the questions "Are you a member of the sensitive group A?" and "Are you a member of the non-sensitive group $\overline{A}$?" with probability P_W and $1 - P_W$ respectively. In this case, the parameters of the model are $\theta = P_W, P(Y|A) = P_W, P(Y|\overline{A}) = 1 - P_W, P(N|A) = 1 - P_W$ and $P(N|\overline{A}) = P_W$. The measure of $\mathcal{PP}$ is

$$g(Y|A) = g_W(Y|A) = \frac{P_W}{1 - P_W} \text{ and } g(N|\overline{A}) = g_W(N|\overline{A}) = \frac{P_W}{1 - P_W}$$

The choice of $g_W(Y|A) = g_W(N|\overline{A}) = k$ yields

$$P_W = \frac{k}{1 + k} \quad (11.2.21)$$

Warner's estimator for π from a SRSWR sample of size n is

$$\hat{\pi}_W = \frac{\hat{\lambda}_W - (1 - P_W)}{2P_W - 1}$$

where $\hat{\lambda}_W$ is the proportion of "yes" answers obtained from Warner's randomization device. The variance of $\hat{\pi}_W$ with $P_W = \frac{k}{1+k}$ is given by

$$\mathrm{Var}_p(\hat{\pi}_W) = \frac{1}{n}\left[\pi(1 - \pi) + \frac{k}{(k - 1)^2}\right] \quad (11.2.22)$$

For Mangat and Singh's (1990) model (briefly, MS-model), described in Sect. 1.6, each of the respondents is provided with two randomization devices R_1 and R_2. The randomization R_1 bears two statements (i) "I belong to the sensitive group A" and (ii) "Go to randomization device R_2" represented with probabilities T and $1 - T$ respectively. The second randomization R_2 is the same as Warner's device with probability P_{MS}. For the MS-model, $\theta = \theta_{\text{MS}} = (T, P_{\text{MS}})$,

$$P(Y|A) = P^*_{\text{MS}}, P(Y|\overline{A}) = 1 - P^*_{\text{MS}}, P(N|A) = 1 - P^*_{\text{MS}}, P(N|\overline{A}) = P^*_{\text{MS}},$$
$$g(Y|A) = g_{\text{MS}}(Y|A) = \frac{P^*_{\text{MS}}}{1 - P^*_{\text{MS}}} \text{ and } g(N|\overline{A}) = g_{\text{MS}}(N|\overline{A}) = \frac{P^*_{\text{MS}}}{1 - P^*_{\text{MS}}} \quad (11.2.23)$$

where $P^*_{\text{MS}} = T + (1 - T)P_{\text{MS}}$.

The expression of the estimator of π for the MS-model based on a sample of size n selected by the SRSWR method is

$$\hat{\pi}_{\text{MS}} = \frac{\hat{\lambda}_{\text{MS}} - (1 - P^*_{\text{MS}})}{2P^*_{\text{MS}} - 1}$$

where $\hat{\lambda}_{\text{MS}}$ is the proportion number of "yes" answers obtained from the MS-model.

Keeping $g_{\text{MS}}(Y|A) = g_{\text{MS}}(N|\overline{A}) = k$, the expression variance of $\hat{\pi}_{\text{MS}}$ for the MS-model becomes

$$\text{Var}_p(\hat{\pi}_{\text{MS}}) = \frac{1}{n}\left[\pi(1 - \pi) + \frac{k}{(k - 1)^2}\right] \quad (11.2.24)$$

Equations (11.2.22) and (11.2.24) yield the following result:

Theorem 11.2.1 Mangat and Singh's model (1990) is equally efficient as Warner's model under the LW-measure of $\mathcal{PP}$.

L-measure of $\mathcal{PP}$

The posterior probabilities $P(A|Y)$ and $P(A|N)$ for the W-model are given by

$$P_{\text{W}}(A|Y) = \frac{\pi P_{\text{W}}}{\pi P_{\text{W}} + (1 - \pi)(1 - P_{\text{W}})} \text{ and } P_{\text{W}}(A|N) = \frac{\pi(1 - P_{\text{W}})}{\pi(1 - P_{\text{W}}) + (1 - \pi)P_{\text{W}}} \quad (11.2.25)$$

In this case, $\mathcal{L}$-measure is

$$\mathcal{L}_{\text{W}} = \text{Max}[P_{\text{W}}(A|Y), P_{\text{W}}(A|N)]$$
$$= \begin{cases} P_{\text{W}}(A|Y) = \frac{\pi P_{\text{W}}}{\pi P_{\text{W}} + (1-\pi)(1-P_{\text{W}})} & \text{for } P_{\text{W}} > 1/2 \\ P_{\text{W}}(A|N) = \frac{\pi(1-P_{\text{W}})}{\pi(1-P_{\text{W}}) + (1-\pi)P_{\text{W}}} & \text{for } P_{\text{W}} < 1/2 \end{cases} \quad (11.2.26)$$

Similarly, $\mathcal{L}$-measure for the MS-model is

$$\mathcal{L}_{\text{MS}} = \begin{cases} P_{\text{MS}}(A|Y) = \frac{\pi P^*_{\text{MS}}}{\pi P^*_{\text{MS}}+(1-\pi)(1-P^*_{\text{MS}})} \text{ if } P^*_{\text{MS}} > 1/2 \\ P_{\text{MS}}(A|N) = \frac{\pi(1-P^*_{\text{MS}})}{\pi(1-P^*_{\text{MS}})+(1-\pi)P^*_{\text{MS}}} \text{ if } P^*_{\text{MS}} < 1/2 \end{cases} \tag{11.2.27}$$

For situation (i) $P_{\text{W}} > 1/2$, $P^*_{\text{MS}} > 1/2$, condition $\mathcal{L}_{\text{W}} = \mathcal{L}_{\text{MS}}$ gives $P_{\text{W}} = P^*_{\text{MS}}$ which in turn yields $\text{Var}_p(\hat{\pi}_{\text{W}}) = \text{Var}_p(\hat{\pi}_{\text{MS}})$. For each of the other three scenarios (ii) $P_{\text{W}} > 1/2, P^*_{\text{MS}} < 1/2$, (iii) $P_{\text{W}} < 1/2, P^*_{\text{MS}} > 1/2$ and (iv) $P_{\text{W}} < 1/2, P^*_{\text{MS}} < 1/2$, the condition $\mathcal{L}_{\text{W}} = \mathcal{L}_{\text{MS}}$, yields $\text{Var}_p(\hat{\pi}_{\text{W}}) = \text{Var}_p(\hat{\pi}_{\text{MS}})$. Hence we have the following result.

Theorem 11.2.2 Mangat and Singh's model (1990) is equally efficient as Warner's (1965) model under the $\mathcal{L}$-measure of $\mathcal{PP}$.

11.2.4.2 Warner's Versus Unrelated Question Model

For Greenberg et al.'s (1969) unrelated question model, also known as Simmons' model (briefly, U-model) described in Sect. 1.3, respondents answer the questions "Do you belong to the sensitive group A?" with probability P_{G1} and "Do you belong to the group B?" with probability $P_{\text{G2}} = 1 - P_{\text{G1}}$ where B is totally non-sensitive and unrelated to A so that a truthful answer can never be embarrassing. For this U-model, the design parameter is $\theta = \theta_{\text{U}} = (P_{\text{G1}}, \pi_{\text{B}})$ where π_{B} is the proportion of individuals in the population that belong to the innocuous group B. The design probabilities are

$$P(Y|A) = P_{\text{G1}} + \pi_{\text{B}}(1 - P_{\text{G1}}) \text{ and } P^*_{\text{MS}} < 1/2, \tag{11.2.26}$$

LW-Measure of $\mathcal{PP}$

$$g(Y|A) = g_{\text{U}}(Y|A) = \frac{P_{\text{G1}} + (1 - P_{\text{G1}})\pi_{\text{B}}}{(1 - P_{\text{G1}})\pi_{\text{B}}} \text{ and}$$
$$g(N, \overline{A}) = g_{\text{U}}(N|\overline{A}) = \frac{1 - (1 - P_{\text{G1}})\pi_{\text{B}}}{(1 - P_{\text{G1}})(1 - \pi_{\text{B}})} \tag{11.2.27}$$

Equation (11.2.27) shows that Y and N jeopardizing for A and $\overline{A}$, respectively if $P_{\text{G1}}, \pi_{\text{B}} > 0.P_{\text{G1}}, \pi_{\text{B}} > 0$.

Setting maximum allowable limits of $g_{\text{U}}(Y|A) = k_{\text{U1}}$ and $g_{\text{U}}(N|\overline{A}) = k_{\text{U2}}$, one finds optimum values of P_{G1} and π_{B} as

$$P_{\text{G1}} = \frac{(k_{\text{U1}} - 1)(k_{\text{U2}} - 1)}{k_{\text{U1}}k_{\text{U2}} - 1} \text{ and } \pi_{\text{B}} = \frac{k_{\text{U2}} - 1}{(k_{\text{U1}} - 1) + (k_{\text{U2}} - 1)}$$

Leysieffer and Warner (1976) provided comparisons between the W-model and U-models for various values of the design parameters. In particular, where the magnitude of $g_{\mathrm{U}}(N|\overline{A})$ does not matter, Leysieffer and Warner (1976) have shown that the U-model is more efficient than the W-model for the same level of privacy $g_{\mathrm{W}}(Y|A) = g_{\mathrm{U}}(Y|A)$ when $\pi_{\mathrm{B}} > 1/2$.

$\mathcal{L}$-measure of $\mathcal{PP}$

The posterior probabilities $P(A|Y)$ and $P(A|N)$ for the U-model are given by

$$P_{\mathrm{U}}(A|Y) = \frac{\pi\{P_{\mathrm{G1}} + \pi_{\mathrm{B}}(1 - P_{\mathrm{G1}})\}}{\pi P_{\mathrm{G1}} + (1 - P_{\mathrm{G1}})\pi_{\mathrm{B}}} \text{ and}$$
$$P_{\mathrm{U}}(A|N) = \frac{\pi(1 - P_{\mathrm{G1}})(1 - \pi_{\mathrm{B}})}{P_{\mathrm{G1}}(1 - \pi) + (1 - P_{\mathrm{G1}})(1 - \pi_{\mathrm{B}})} \quad (11.2.28)$$

In this case, $\mathcal{L}$-measure is

$$\begin{aligned}\mathcal{L}_{\mathrm{W}} &= \mathrm{Max}[P_{\mathrm{U}}(A|Y), P_{\mathrm{U}}(A|N)] \\ &= \mathrm{Max}\left[\frac{\pi\{P_{\mathrm{G1}} + \pi_{\mathrm{B}}(1 - P_{\mathrm{G1}})\}}{\pi P_{\mathrm{G1}} + (1 - P_{\mathrm{G1}})\pi_{\mathrm{B}}}, \frac{\pi(1 - P_{\mathrm{G1}})(1 - \pi_{\mathrm{B}})}{P_{\mathrm{G1}}(1 - \pi) + (1 - P_{\mathrm{G1}})(1 - \pi_{\mathrm{B}})}\right] \\ &= \frac{\pi\{P_{\mathrm{G1}} + \pi_{\mathrm{B}}(1 - P_{\mathrm{G1}})\}}{\pi P_{\mathrm{G1}} + (1 - P_{\mathrm{G1}})\pi_{\mathrm{B}}}\end{aligned}$$

For $\mathcal{L}$-measure, W-model and U-model are equally protective (Lanke, 1976) if

$$P_{\mathrm{W}} = \frac{1}{2} + \frac{P_{\mathrm{G}}}{2P_{\mathrm{G}} + 4(1 - P_{\mathrm{G}})\pi} \quad (11.2.29)$$

11.2.4.3 Mangat Model

In Mangat's (1994) model (in brief, M-Model) described in Sect. 1.7, respondents should report "yes = Y" if he or she possesses the attribute A otherwise provide "Y" or "N" answers using the W-model with a parameter P_{M}. Hence, the M-Model design parameters are: $P(Y|A) = 1$ and $P(Y|\overline{A}) = 1 - P_{\mathrm{M}}$.

LW measures

$$g(Y|A) = g_{\mathrm{M}}(Y|A) = \frac{1}{1 - P_{\mathrm{M}}} \text{ and } g(N|\overline{A}) = \infty, \text{ as } P(N|A) = 0 \quad (11.2.30)$$

Equation (11.2.30) indicates that there is no jeopardy in the "N" answer. Bhargava and Singh (2002) claimed that the M-model is more efficient than the W-model when compared at the same level of $\mathcal{PP}$. However, such a claim is not valid since the "N" answer ensures the individual belongs to the group $\overline{A}$.

L-measure

$$\mathcal{L}_{\rm M} = Max\left[\frac{\pi}{\lambda_{\rm M}}, \frac{0}{1-\lambda_{\rm M}}\right] = \frac{\pi}{\lambda_{\rm M}} \tag{11.2.31}$$

where $\lambda_{\rm M} = \pi + (1-\pi)(1-P_{\rm M})$.

11.2.4.4 Mangat et al.'s Model B (1992)/Chang–Liang's Model (1996)

The model parameter of Mangat et al.'s model B (1992) or Chang–Liang's model (1996) described in Sect. 1.8 is $\theta = (T, P, \pi_B)$. Here, the design parameters are:

$$\begin{aligned} &P(Y|A) = T + (1-T)\{P + (1-P)\pi_{\rm B}\},\ P(N|A) = (1-T)(1-P)(1-\pi_{\rm B}), \\ &P(Y|\overline{A}) = (1-T)(1-P)\pi_{\rm B} \text{ and } P(N|A) = (1-T)(1-P)(1-\pi_{\rm B}) \end{aligned} \tag{11.2.32}$$

The LW and $\mathcal{L}$-measures are:

LW-Measure:

$$g(Y|A) = \frac{T + (1-T)\{P + (1-P)\pi_{\rm B}\}}{(1-T)(1-P)\pi_{\rm B}} \text{ and } g(N|\overline{A}) = \frac{1-(1-T)(1-P)\pi_{\rm B}}{(1-T)(1-P)(1-\pi_{\rm B})} \tag{11.2.33}$$

$\mathcal{L}$-measure:

$$Max[P(A|Y), P(A|N)] = \pi\frac{T + (1-T)\{P + (1-P)\pi_{\rm B}\}}{\lambda_{\rm CL}} \tag{11.2.34}$$

where $\lambda_{\rm CL} = \pi[T + (1-T)\{P\}] + (1-T)(1-P)\pi_{\rm B}$.

11.2.4.5 Chang et al. Model

In the Chang et al. (2004) model (C-model, in brief) described in Sect. 1.11, respondents are asked to answer "yes = Y" or "No = N" truthfully to the statements "I belong to the sensitive group A" and "I belong to the non-sensitive group $\overline{A}$" with probabilities $P_{\rm C1}$ and $P_{\rm C2}$ respectively. It also forced respondents to answer "yes = Y" and "no = N" with probabilities $P_{\rm C3}$ and $P_{\rm C4}(P_{\rm C1} + P_{\rm C2} + P_{\rm C3} + P_{\rm C4} = 1)$. For this model, $\theta = \theta_{\rm C} = (\theta_{\rm C1}, \theta_{\rm C2}, \theta_{\rm C3})$, $P(Y|A) = P_{\rm C1} + P_{\rm C3}$ and $P(Y|\overline{A}) = P_{\rm C2} + P_{\rm C3}$.

LW-Measure

$$g(Y|A) = g_{\rm C}(Y|A) = \frac{P_{\rm C1} + P_{\rm C3}}{P_{\rm C2} + P_{\rm C3}} \text{ and } g(N|\overline{A}) = g_{\rm C}(N|\overline{A}) = \frac{1 - P_{\rm C2} - P_{\rm C3}}{1 - P_{\rm C1} - P_{\rm C3}} \tag{11.2.35}$$

L-measure

$$L_C = Max[P(A|Y), P(A|N)] = Max\left[\frac{\pi(P_{C1} + P_{C3})}{\lambda_C}, \frac{\pi(1 - P_{C1} - P_{C3})}{1 - \lambda_C}\right] \tag{11.2.36}$$

where $\lambda_C = \pi P_{C1} + (1 - \pi)P_{C2} + P_{C3}$.

Guerriero and Sandri (2007) showed that

$$L_C = \begin{cases} \pi(P_{C1} + P_{C3})/\lambda_C & \text{for } \theta_{CH} \in \Omega_{CH} \\ \pi(1 - P_{C1} - P_{C3})/(1 - \lambda_C) & \text{for } \theta_{CH} \notin \Omega_{CH} \end{cases} \tag{11.2.37}$$

where $\Omega_{CH} = \left\{P_{C1}, P_{C2}, P_{C3} | P_{C2} < \frac{1}{2} - \left|\frac{1}{2} - P_{C1}\right|, P_{C3} < 1 - P_{C1} - P_{C2}\right\}$.

11.2.4.6 Kuk's Model Versus U-model

In Kuk's (1990) RR model (K-model), defined in Sect. 6.4.1, respondents belonging to the sensitive and non-sensitive groups A and $\overline{A}$ are requested to perform Bernoulli trials $c(\geq 1)$ times independently with success probabilities p_{K1} and $p_{K2}(\neq p_{K1})$ respectively. The respondent should report the total number of successes z as their RR. In this K-model, there is no straight-forward "yes" or "no" response.

$\mathcal{PP}$ **for** $c = 1$

Guerriero and Sandri (2007) considered $\mathcal{PP}$ measure for $c = 1$ while treating the outcomes "success" as "yes = Y" and failure as "no = N" response respectively. For this model design parameter $\theta = \theta_K = (p_{K1}, p_{K2})$, $P(Y|A) = p_{K1}$, $P(N|A) = 1 - p_{K1}$, $P(Y|\overline{A}) = p_{K2}$ and $P(N|\overline{A}) = 1 - p_{K2}$.

$\mathcal{LW}$**-measure**

$$g(Y|A) = g_K(Y|A) = \frac{p_{K1}}{p_{K2}} \text{ and}$$
$$g(N|\overline{A}) = g_K(N|\overline{A}) = \frac{1 - p_{K2}}{1 - p_{K1}} \tag{11.2.38}$$

$\mathcal{L}$**-measure**

$$\mathcal{L}_K = \text{Max}[P(A|Y), P(A|N)]$$
$$= \text{Max}\left[\frac{\pi P_{K1}}{\lambda_K}, \frac{\pi(1 - P_{K1})}{1 - \lambda_K}\right] \tag{11.2.39}$$

where $\lambda_K = P(Y) = \pi P_{K1} + (1 - \pi)P_{K2}$.

Equation (11.2.39) yields

$$\mathcal{L}_{\rm K} = \begin{cases} \frac{\pi(1-p_{\rm K1})}{1-\lambda_{\rm K}} & \text{if } p_{\rm K1} < p_{\rm K2} \\ \frac{\pi p_{\rm K1}}{\lambda_{\rm K}} & \text{if } p_{\rm K1} > p_{\rm K2} \end{cases} \quad (11.2.40)$$

Guerriero and Sandri (2007) provided comparisons between K-model and U-model as follows:

(i) For a given level of $\mathcal{PP}$, Kuk's model with $p_{\rm K1} > p_{\rm K2}$ is equivalent to the U-model with $p_{\rm G1} = p_{\rm K1} - p_{\rm K2}$ and $p_{\rm G2} = p_{\rm K2}/(1 - p_{\rm K1} - p_{\rm K2})$.
(ii) For a given U-model and $\mathcal{L}$-level of $\mathcal{PP}$, there exists at least one K-model with $p_{\rm K1} < p_{\rm K2}$ that can be made more efficient than the U-model.

$\mathcal{PP}$ **for** $c > 1$

In this case, privacy protection with respect to each of the possible responses $x(= 0, 1, \ldots, c)$ is considered. Let

$$\begin{aligned} P(x = j|A) &= f_A(j) = \binom{c}{j} p_{\rm K1}^j (1 - P_{K1})^{c-j} \\ \text{and} & \qquad\qquad \text{for } j = 0, 1, \ldots, c \\ P(x = j|\overline{A}) &= f_{\overline{A}}(j) = \binom{c}{j} p_{\rm K2}^j (1 - P_{K2})^{c-j} \end{aligned} \quad (11.2.41)$$

Then the LW and $\mathcal{L}$-measures of privacy protections for Kuk's model are

$$g(j|A) = g_{\rm K}(j|A) = \frac{f_A(j)}{f_{\overline{A}}(j)} \text{ for } j = 0, 1, \ldots, c \quad (11.2.42)$$

and

$$\mathcal{L}_{\rm K} = \underset{j\in(0,\ldots,c)}{\text{Max}} \left\{ \frac{\pi f_A(j)}{\pi f_A(j) + (1-\pi) f_{\overline{A}}(j)}, \frac{(1-\pi) f_{\overline{A}}(j)}{\pi f_A(j) + (1-\pi) f_{\overline{A}}(j)} \right\} \quad (11.2.43)$$

11.2.4.7 Christofides Model

In Christofides's (2003) model (in short, the CH-model) described in Sect. 5.6.1, each respondent reports an integer $j(= 1, 2, \ldots, k)$ with a probability $p_j (p_j \geq 0, \sum p_j = 1)$ if he or she belongs to the non-sensitive group $\overline{A}$; otherwise, if he or she belongs to A, should report $k + 1 - j$ as his or her RR. In this case, a respondent can produce k different responses $1, \ldots, k$. Here also, the level of protection against each of the responses as in the K-model is proposed as follows:

LW-Measure

$$g(t|A) = g_{\rm CH}(t|A) = \frac{P(t|A)}{P(t|\overline{A})} = \frac{P_{k+1-t}}{P_t} \mathit{for}\ t = 1, \ldots, ks \quad (11.2.44)$$

$\mathcal{L}$-measure

$$\mathcal{L} = \mathcal{L}_{CH} = \underset{t\in(1,\ldots,k)}{Max}\left\{\frac{\pi P_{k+1-t}}{\pi P_{k+1-t} + (1-\pi)P_t}, \frac{(1-\pi)P_t}{\pi P_{k+1-t} + (1-\pi)P_t}\right\} \quad (11.2.45)$$

Guerriero and Sandri (2007) have shown that (i) For $k = 2$, CH-model is $\mathcal{PP}$ equivalent to the W-model with $P_W = p_1$ and (ii) For $k = 3$, CH-model can be made more efficient and better L-protective by choosing parameters $\theta_{CH} = (p_1, p_2)$ appropriately.

11.2.5 Other Measures of $\mathcal{PP}$

Flinger et al. (1977) provided with a measure like Lanke (1976) as

$$\mathcal{F} = \frac{1 - Max[P(A|Y), P(A|N)]}{1-\pi} = \frac{1-\mathcal{L}}{1-\pi} \quad (11.2.46)$$

Greenberg et al. (1977) defined the hazard of an individual A as a measure of $\mathcal{PP}$. They proposed the following two measures of $\mathcal{PP}$:

$$P(Y|A)P(A|Y) + P(N|A)P(A|N) \text{ and } P(Y|\overline{A})P(A|Y) + P(N|\overline{A})P(A|N) \quad (11.2.47)$$

Greenberg et al. (1977) proposed alternative measures of hazard as follows:

$$P(Y|A)P(A|Y) \text{ and } P(Y|\overline{A})P(A|Y) \quad (11.4.48)$$

Loynes (1976) extended the Leysieffer and Warner (1976) $\mathcal{PP}$ for polychotomous populations.

Ljungqvist (1993) proposed a utilitarian approach for $\mathcal{PP}$ for dichotomous populations under the assumptions that (i) respondents feel discomfort from being thought of as belonging to the sensitive group and (ii) respondents prefer to answer truthfully than to lie, as long as the cost of discomfort is not too high.

Nayak (1994) and Nayak and Adeshiyan (2009) considered a dichotomous population where each of the respondents belongs either to a sensitive category A or to the non-sensitive category $\overline{A}$ and each of the respondents selected in an SRSWR sample either responded "yes = Y" or "no = N" using Warner's (1965) RR device. Since respondents' privacy increases as the posterior probabilities $P(A|Y)$ and $P(A|N)$ decrease, Nayak pointed out that the model R_1 should be better than the model R_2 for estimating π if

$$P_{R_1}(A|Y) \le P_{R_2}(A|Y), P_{R_1}(A|N) \le P_{R_2}(A|N) \text{ and } V_{R_1}(\hat{\pi}_1) \le V_{R_2}(\hat{\pi}_2) \quad (11.2.49)$$

where $\hat{\pi}_1$ and $\hat{\pi}_2$ are unbiased estimators of π for the models R_1 and R_2 respectively. However, they didn't point out which estimator of π should be selected among several possible unbiased estimators.

Nayak and Adeshiyan (2009) also considered a dichotomous population where each respondent is allowed to provide any one of the $k(\geq 1)$ responses $c_1, \ldots, c_k$; $c_i \neq c_j$. Let z denotes the response variable with $\alpha_i = P(z = c_i|A)$ and $\beta_i = P(z = c_i|\overline{A})$. They proposed a measure of privacy protection as

$$\text{Max}\left\{\frac{\alpha_1}{\beta_1}, \ldots, \frac{\alpha_k}{\beta_k}\right\} \tag{11.2.50}$$

Quatember (2009) considered a dichotomous population where the response from the ith unit z_i is either 0 (no) or 1 (= yes). In this case $P(z_i = 1) = \alpha + \beta y_i$ where $y_i = 1(0)$ if the ith unit possesses the attribute $A(\overline{A})$ where α and β are constants. Quatember's measure of $\mathcal{PP}$ is

$$\lambda_j = \frac{\text{Max}\{(z_i = j|A), (z_i = j|\overline{A})\}}{\text{Min}\{(z_i = j|A), (z_i = j|\overline{A})\}}; j = 0, 1 \tag{11.2.51}$$

where $1 \leq \lambda_j \leq \infty$.

The larger λ_j indicates a larger loss of privacy protection with respect to the answer $z_i = j$. For the direct questioning method, $\lambda_j = \infty$. Privacy is 100% protected (i.e., $\lambda_0 = \lambda_1 = 1$) if $\alpha = 0$. But in this case, the RR procedure becomes non-informative. The expressions of λ_j in terms of α and β are

$$\lambda_0 = \frac{\text{Max}(1 - \alpha - \beta, 1 - \beta)}{\text{Min}(1 - \alpha - \beta, 1 - \beta)} \text{ and } \lambda_1 = \frac{\text{Max}(\alpha + \beta, \beta)}{\text{Min}(\alpha + \beta, \beta)} \tag{11.2.52}$$

For given values of λ_0 and λ_1, one finds $\alpha = \frac{(\lambda_0 - 1)(\lambda_1 - 1)}{\lambda_0\lambda_1 - 1}$ and $\beta = \frac{\lambda_0 - 1}{\lambda_0\lambda_1 - 1}$ that yield a RR technique that satisfies the RR model (11.2.52). Since different randomization devices can yield the same model (11.2.52) with the given values of α and β, there exists more than one randomization technique that yields the same level of protection while maintaining the same degree of efficiency.

Chaudhuri and Saha (2004) used the Bayesian method in measuring $\mathcal{PP}$ for a dichotomous sensitive variable y where $y_i = 1$ if the ith unit possesses the sensitive attribute A and $y_i = 0$ if the ith unit belongs to $\overline{A}$, complement of A. Since the variable y is sensitive, a randomized response $z_i = Q_j$, $(j = 1, \ldots, M)$ is obtained from each of the selected respondents independently where M is the total number of possible responses. Let $p_j(\geq 0, \sum_{j=1}^{M} p_j = 1)$ be the probability of obtaining a response $z_i = Q_j$ from the ith respondent given that the respondent belongs to the sensitive group A, i.e., $p_j = P(z_i = Q_j|y_i = 1) = P(z_i = Q_j|i \in A)$. Let the prior probability of $P(y_i = 1) = P(i \in A) = L_i$ and $P(y_i = 0) = P(i \in \overline{A}) = 1 - L_i$ where $L_i(\in [0, 1])$ is unknown. The posterior probability of classifying the ith person to the sensitive group A given the ith respondent supplied a response $z_i = Q_j$ is

$$\begin{aligned} J_i(Q_j) &= P(y_i = 1|z_i = Q_j) = P(i \in A|z_i = Q_j) \\ &= \frac{P(z_i = Q_j|i \in A)P(i \in A)}{P(z_i = Q_j|i \in A)P(i \in A) + P(z_i = Q_j|i \in \overline{A})P(i \in \overline{A})} \end{aligned} \tag{11.2.53}$$

Chaudhuri and Saha (2004) defined "response-specific" jeopardy measure for the ith unit for the response Q_j as

$$J_i(Q_j) = \frac{L_i(Q_j)/L_i}{[1 - L_i(Q_j)]/(1 - L_i)} \text{ for } j = 1, \ldots, M;\, i \in U \tag{11.2.54}$$

and the average measure of jeopardy is

$$\overline{J}_i = \frac{1}{M} \sum_{j=1}^{M} \frac{L_i(Q_j)/L_i}{[1 - L_i(Q_j)]/(1 - L_i)} \tag{11.2.55}$$

For Warner's model, $y_i = 1(0)$ if the ith unit $\in A(\overline{A})$; $z_i = Q_1 = 1$ if the ith unit answers "yes" and $z_i = Q_2 = 0$ if the answer is "no"; $P(z_i = 1|i \in A) = p$, $P(z_i = 0|i \in A) = 1 - p$, $P(z_i = 1|i \in \overline{A}) = 1 - p$ and $P(z_i = 0|i \in \overline{A}) = p$. In this case

$$\begin{aligned} L_i(0) &= \frac{p(i \in A)p(z_i = 0|i \in A)}{[p(i \in A)p(z_i = 0|i \in A) + p(i \in \overline{A})p(z_i = 0|i \in \overline{A}]} \\ &= \frac{L_i(1 - p)}{L_i(1 - p) + (1 - L_i)p}, \\ L_i(1) &= \frac{p(i \in A)p(z_i = 1|i \in A)}{[p(i \in A)p(z_i = 1|i \in A) + p(i \in \overline{A})p(z_i = 0|i \in \overline{A}]} \\ &= \frac{L_i p}{L_i p + (1 - L_i)(1 - p)}, \end{aligned}$$

$$\begin{aligned} J_i(1) &= \frac{L_i(1)/L_i}{[1 - L_i(1)]/(1 - L_i)} = \frac{p}{1 - p}, \\ J_i(0) &= \frac{L_i(0)/L_i}{[1 - L_i(0)]/(1 - L_i)} = \frac{1 - p}{p} \end{aligned}$$

and

$$\overline{J}_i = \frac{1}{2}\left[\frac{p}{1 - p} + \frac{1 - p}{p}\right]. \tag{11.2.56}$$

Chaudhuri and Saha (2004) also provided expressions of $\mathcal{PP}$ for the Kuk (1990), Mangat and Singh (1990), and Christofides's (2003) RR devices.

11.3 SRSWR Sampling for Discrete Variables

Bose (2013) proposed a measure of $\mathcal{PP}$ for a discrete sensitive variable X that can take a finite number of known values, $x_1, \ldots, x_m$, e.g., number of criminal convictions, number of induced abortions, number of months spent in correctional services, etc., where the Prob$(X = x_j) = \theta_j$ is unknown. A sample of size n is selected from the population U by the SRSWR method, and each of the selected respondents is asked to disclose either his or her true value x_j with probability p or to select one of the m possible values $x_1, \ldots, x_m$ with probability $(1-p)/m$ as his or her RR by using some randomization device.

Bose assumed that all the values $x_1, \ldots, x_m$ are sensitive and the $i(\in U)$th respondent would feel comfortable if the reception of his or her true value x_j did not deviate much from the randomized response z_i provided, i.e.,

$$\alpha_{jk} = \left|\Pr ob(x = x_j | z_i = x_k) - \Pr ob(x = x_j)\right|; j, k = 1, \ldots m$$

should be as small as possible. The proposed measure of $\mathcal{PP}$ is

$$\alpha = \max_{1 \le j \le m} \alpha_{jk} \tag{11.3.1}$$

Thus, the privacy protection $\alpha = \alpha_0$ would guarantee that the discrepancies between the true and revealing probabilities would be at most α_0 for all respondents. Thus, a randomization device provides a lower value of α gives a higher level of privacy protection. Bose also considered the $\mathcal{PP}$ when all values of X were not stigmatized.

11.4 SRSWR Sampling for Qualitative and Quantitative Characteristics

Anderson (1977) proposed measures of $\mathcal{PP}$ that are applicable to both qualitative and quantitative characteristics. He considered a general RR model where the embarrassing characteristic $y(\in \Omega_y)$ follows some unknown distribution function $F(y)$. Since the character y is sensitive, the sampled individuals provided a response $R = r(\in \Omega_r)$ with a known probability density $h(r|y)$ which depends on the sensitive y value of the respondent. The density $h(r|y)$ is called the response density. The marginal distribution of r is given by

$$g(r) = \int h(r|y)dF(y) \tag{11.4.1}$$

The posterior distribution of y given r is

$$k(y|r) = \frac{h(r|y)f(y)}{g(r)}, r \in \Omega_r \tag{11.4.2}$$

is called the revealing density, which depends on response distribution $h(r|y)$ and unknown population distribution $F(y)$ with density $f(y)$.

Let a respondent with the true value y provide a randomized response $R = r$. Then, the revealing density $k(y|r)$ provides information about y and the discrepancy between $f(y)$ and $k(y|r)$ is the amount of invasion of privacy caused by the randomized response $R = r$. The protection of privacy caused by the response r is small if the discrepancy is small. Alternatively, if the discrepancy is large, privacy is well protected. Hence, Anderson provided a general measure of $\mathcal{PP}$ is

$$\mathrm{A} = E[V(y|r)] \tag{11.4.3}$$

However, one can use the following relative measures for measures of $\mathcal{PP}$

$$\mathrm{A}_1 = V(y|r)/V(y) \text{ and } \mathrm{A}_2 = E\{V(y|r)\}/V(y) \tag{11.4.4}$$

In practice, all y-values are not sensitive, e.g., illegal abortion. Let the set A_y denote the sensitive values of y. In this situation $P(A_y|R = r)$, the conditional probability of A_y given $R = r$ is treated as a measure of the $\mathcal{PP}$. Hence, one should demand that an RR procedure should provide the minimum level of protection while maintaining a certain degree of sampling efficiency of the estimator used, i.e.,

$$\underset{r}{\text{Min}}\, V(y|R = r) \geq c_1 \text{ and } \underset{r}{\text{Max}}\, P(A_y|R = r) \leq c_2 \,\forall\, y \in \Omega_y \tag{11.4.5}$$

where c_1 and c_2 are suitably chosen positive constants.

Example 11.4.1 (Anderson, 1977) Suppose we want to estimate π, proportion of tax evaders in a population by using Warner's RR model, where a respondent is asked to choose a card from a pack of two types of cards bearing the statements "I am a tax evader" and "I am not a tax evader" with proportions P and $1 - P$. The respondent should answer $r = 1$ if the statement matches his status; otherwise, he or she should answer $r = 0$. Let $y = 1$ if the respondent is a tax evader and $y = 0$ otherwise. In this case

$$h(1|y = 1) = 1 - h(0|y = 1) = P,$$
$$h(1|y = 0) = 1 - h(0|y = 0) = 1 - P \text{ and } A_y = \{y = 1\}.$$

Then,

$$\begin{aligned} P(A_y|r) &= P(y=1|r) \\ &= \frac{\pi P(r|y=1)}{\pi P(r|y=1)+(1-\pi)P(r|y=0)} \end{aligned} \tag{11.4.6}$$

For protection of privacy, one should keep $\underset{r}{\text{Max}}\ P(A_y|r) \le c_2$.

11.5 Measures of $\mathcal{PP}$ Under a Finite Population Setup

Consider a finite population $U = (U_1, \ldots, U_i, \ldots, U_N)$ of N identifiable units from which a sample s of n units is selected with probability $p(s)$ by a suitable sampling design p. Let the inclusion probabilities of the ith, and $i, j(\neq i)$th units be π_i and π_{ij} respectively. Let y_i denotes the value of the study variable y for the ith unit. In this case, the parameter of the population $\mathbf{y} = (y_1, \ldots, y_i, \ldots, y_N)$ is a fixed point in the N dimensional Euclidean space $\Omega = \{-\infty < y_i < \infty | i \in U\}$ (Cassel et al., 1977).

Consider the general RR model (3.9.1) where the revised RR r_i satisfies the model

$$E_R(r_i) = y_i,\ V_R(r_i) = \phi_i(y_i) \text{ and } C_R(r_i, r_j) = 0 \text{ for } i \neq j \in U \tag{11.5.1}$$

For this model (11.5.1), the Horvitz–Thompson (1952, HT in brief) estimator for the population mean $\overline{Y}$ is

$$\hat{\overline{Y}}_{\text{HT}} = \frac{1}{N}\sum_{i\in s}\frac{y_i}{\pi_i} \tag{11.5.2}$$

The variance of $\hat{\overline{Y}}_{\text{HT}}$ is

$$\begin{aligned} \text{Var}(\hat{\overline{Y}}_{\text{HT}}) &= V_p[E_R(\hat{\overline{Y}}_{\text{HT}})] + E_p[V_R(\hat{\overline{Y}}_{\text{HT}})] \\ &= \frac{1}{N^2}\left[\sum\sum_{i<\ j\in U}(\pi_i\pi_j - \pi_{ij})\left(\frac{y_i}{\pi_i} - \frac{y_j}{\pi_j}\right)^2 + \sum_{i\in U}\frac{\phi_i(y_i)}{\pi_i}\right] \end{aligned} \tag{11.5.3}$$

For the consideration of $\mathcal{PP}$, one should feel that their confidentiality is well protected if $\phi_i(y_i)$ is large enough for every $i \in U$ (vide Anderson, 1977). On the other hand, a statistician's interest is to minimize $\phi_i(y_i)$ to get an efficient estimator that controls the sampling variance. As a result, there is a conflict of interest between respondents and statisticians. So, we need to balance their interests. Hence, one can compromise by setting

$$\underset{i\in U}{\text{Min}}\ \phi_i(y_i) \ge \alpha \tag{11.5.4}$$

to provide a guarantee to the respondents that their confidentiality should be maintained to a certain level α (a suitably chosen constant), and at the same time, statisticians should be allowed to minimize $\sum\limits_{i \in U} \frac{\phi_i(y_i)}{\pi_i}$ subject to (11.5.4) to maintain the efficiency of the estimators. For a quantitative variable y, (11.5.4) is unit dependent. Hence, one may consider

$$\underset{i \in U}{\text{Min}} \frac{\phi_i(y_i)}{\overline{Y}} \geq \alpha_0 \tag{11.5.5}$$

as a measure of $\mathcal{PP}$. A higher value of α_0 indicates a higher level of $\mathcal{PP}$ is provided.

Comparison of Two RR Models

Consider the RR models R_1 and R_2 where the revised RR r_{1i} and r_{2i} satisfy.

$$E_R(r_{ki}) = y_i,\ V_R(r_{ki}) = \phi_{ki}(y_i) \text{ and } C_R(r_{ki}, r_{kj}) = 0 \text{ for } i \neq j, k = 1, 2$$

For a given sampling strategy (sampling design and estimator), the RR model R_1 will be better than R_2 if the model R_1 maintains the same level of $\mathcal{PP}$ as R_2, and at the same time, the sampling variance of the estimator under the model R_1 is less than that of R_2

$$\text{i.e., (i) } \frac{\phi_{1i}(y_i)}{\overline{Y}}, \frac{\phi_{2i}(y_i)}{\overline{Y}} \geq \alpha_0 \ \forall i \in U \text{ and (ii) } \sum_{i \in U} \frac{\phi_{1i}(y_i)}{\pi_i} \leq \sum_{i \in U} \frac{\phi_{2i}(y_i)}{\pi_i} \tag{11.5.6}$$

11.6 Arnab's Method of Comparison

To compare the performance of various RR techniques, Leysieffer and Warner (1976), Lanke (1976), and others recommended that the variances of the estimators of the parameter of interest for the different models should be compared, keeping the level of $\mathcal{PP}$ fixed at a certain level. Using the proposed methods, Bhargava and Singh (1999, 2000, 2002), and Guerriero and Sandri (2007), among others, concluded that Warner's (1965) model and Mangat and Singh's (1990) are equally efficient for a given level of $\mathcal{PP}$. In a similar vein, Hussain et al. (2016) demonstrated that their model outperforms the Singh and Grewal (2013) model unconditionally and the Kuk (1990) model in some situations.

Arnab (2021, 2023) pointed out that such comparisons are unfair because they are not made on an equal footing. For example, Warner's model requires drawing only one card , while Kuk's model requires drawing $c(\geq 1)$ cards. Thus, Warner's model is comparable with Kuk's model only if $c = 1$, otherwise, it is not. The number of trials in Mangat and Singh (1994), Singh and Grewal (2013) and Hussain et al.'s (2016) RR techniques is not fixed; they are random variables. Hence, for the sake of comparisons of various RR models, Arnab (2021, 2023) pointed out that one should compare the variances of the estimators for the different models while keeping (i) the level of $\mathcal{PP}$ and (ii) the expected number of trials fixed to a certain level. For simplicity, let us ignore the level of $\mathcal{PP}$ and consider the SRSWR method of sampling procedure only.

Comparison of Warner's, Kuk's, Singh, and Grewal, and Mangat and Singh's Method

For comparison of the above RR techniques, let us consider the situation where a sample s of size n is selected by the SRSWR method.

Warner' (1965) Model

In this model, each of the selected respondents is directed to perform Warner's trial γ times independently with parameter P_{w}. Let $z_i(j)$ be the RR obtained from the ith ($\in s$) respondent in the $j(= 1, \ldots, \gamma)$th trial. In this case, an unbiased estimator of the population proportion π is

$$\overline{r}_{\mathrm{w}} = \frac{1}{n}\sum\nolimits_{i\in s}\overline{r}_{\mathrm{w}i} \tag{11.6.1}$$

where $\overline{r}_{\mathrm{w}i} = \frac{1}{\gamma}\sum\limits_{i=1}^{\gamma}\frac{z_i(j)-(1-P_{\mathrm{w}})}{2P_{\mathrm{w}}-1}$, $P_{\mathrm{w}} \neq \frac{1}{2}$ and $\sum_{i\in s}$ denotes the sum of the units in s with repetition.

The variance of $\overline{r}_{\mathrm{w}}$ is

$$\mathrm{Var}(\overline{r}_{\mathrm{w}}) = \frac{1}{n}\left[\pi(1-\pi) + \frac{\phi_{\mathrm{w}}}{\gamma}\right] \tag{11.6.2}$$

where $\phi_{\mathrm{w}} = P_{\mathrm{w}}(1-P_{\mathrm{w}})/(2P_{\mathrm{w}}-1)^2$.

Kuk's (1990) Model

In this model, each of the selected individuals draws γ cards independently, with the success probability $\theta_1\,(\theta_2)$ if he or she belongs to the sensitive (non-sensitive) group. The proposed unbiased estimator of π is

$$\bar{r}_{\mathrm{k}} = \frac{1}{n}\sum\nolimits_{i \in s} r_{\mathrm{k}i} \tag{11.6.3}$$

where $r_{\mathrm{k}i} = \frac{z_i - \theta_2}{\theta_1 - \theta_2}$, $\theta_1 \neq \theta_2$ and z_i be the proportion of success obtained by the ith respondent.

The variance of $\bar{r}_{\mathrm{k}}$ is

$$\mathrm{Var}(\bar{r}_{\mathrm{k}}) = \frac{1}{n}\left[\pi(1-\pi) + \frac{\phi_{\mathrm{k}}}{\gamma}\right] \tag{11.6.4}$$

where $\phi_{\mathrm{k}} = \frac{\pi\theta_1(1-\theta_1) + (1-\pi)\theta_2(1-\theta_2)}{(\theta_1 - \theta_2)^2}$.

Singh and Grewal's (2013) *Model*

In this method, each of the selected respondents is directed to perform Bernoulli trials with the probability of success $\theta_g(\theta_g^*)$ if he or she possesses the sensitive (non-sensitive) attribute until he or she gets a success for the first time. The respondent should report the total number of trails made by his or her response. In this case, an unbiased estimator of π is

$$\bar{r}_{\mathrm{sg}} = \frac{1}{n}\sum\nolimits_{i \in s} r_{sgi}$$

where $r_{gi} = \frac{\theta_g^* z_{gi} - 1}{\theta_g^* - \theta_g}\theta_g$, $\theta_g^* \neq \theta_g$ and z_{gi} be the RR provided by the ith respondent.

The variance of $\bar{r}_{\mathrm{sg}}$ is

$$\mathrm{Var}(\bar{r}_{sg}) = \frac{\pi(1-\pi)}{n} + \frac{\phi_{sg}}{n}$$

where $\phi_g = \frac{\pi(1-\theta_g)\theta_g^{*2} + (1-\pi)(1-\theta_g^*)\theta_g^2}{(\theta_g^* - \theta_g)^2}$.

In the Singh and Grewal model, the number of trials is not fixed. It is a random number. The expected number of trials is $\gamma_{\mathrm{sg}} = \pi\frac{1}{\theta_g} + (1-\pi)\frac{1}{\theta_g^*}$. Suppose that each respondent is asked to repeat Singh and Grewal's RR procedure $k_{\mathrm{sg}} = \gamma/\gamma_{\mathrm{sg}}$ (assuming an integer) times, so that the expected number of trials performed by each respondent is γ, a fixed number. Let $r_{sgi}(j)$ be RR provided by the ith respondent at the $j(= 1, \ldots, k_{\mathrm{sg}})$th repetition. In this case, an unbiased estimator of π is

$$\bar{\bar{r}}_{\mathrm{sg}} = \frac{1}{n}\sum\nolimits_{i \in s} \bar{r}_{sgi}, \bar{r}_{sgi} = \frac{1}{k_{\mathrm{sg}}}\sum_{j=1}^{k_{\mathrm{sg}}} r_{sgi}(j) \tag{11.6.5}$$

The variance of $\bar{\bar{r}}_{sg}$ is

$$\mathrm{Var}(\bar{\bar{r}}_{sg}) = \frac{\pi(1-\pi)}{n} + \gamma_{sg}\frac{\phi_{sg}}{n\gamma} \tag{11.6.6}$$

Mangat and Singh's (*1990*) Two Stage Model

In this model, at the first stage, each of the respondents is asked to choose one of the two statements (i) "I belong to the sensitive group A" and (ii) "Do the RR device" with probabilities T and $1 - T$ by using a suitable randomization device, e.g., a random number. The respondent should answer "yes" or "no" if he or she chooses statement (i). Otherwise, if the respondent chooses statement (ii), then he or she should answer "yes" or "no" by using Warner's method with the parameter P. In this two-stage model, those respondents who belong to the sensitive group A and choose the statement "(i)" do only one trial (select a random number once). Respondents who belong to the group A and choose the statement "(ii)" or those who belong to the non-sensitive group $\bar{A}$ have to do two trials while choosing the statement "(i)" and while executing the Warner Method. In this case, the number of trails is not fixed. The expected number of trials is

$$\gamma_{ms} = \pi T \times 1 + \{\pi(1-T) + (1-\pi)\} \times 2 = 1 + (1-\pi)$$

Let each of the selected respondents perform Mangat and Singh's trials $k_{ms} = \gamma/\gamma_{ms}$ (assuming an integer) times and let $\bar{z}_{msi}$ be the average number of "yes" answers obtained by the ith respondent. In this case, an estimator of π becomes

$$\bar{r}_{ms} = \frac{1}{n}\sum\nolimits_{i\in s} \bar{r}_{msi} \tag{11.6.7}$$

where $\bar{r}_{msi} = \frac{\bar{z}_{msi}-(1-P_{ms})}{(2P_{ms}-1)}$ and $P_{ms} = T + (1-T)P$.

The variance of $\bar{r}_{ms}$ is

$$\mathrm{Var}(\bar{r}_{ms}) = \frac{\pi(1-\pi)}{n} + k_{ms}\frac{\phi_{ms}}{n\gamma};\ \phi_{ms} = \frac{P_{ms}(1-P_{ms})}{(2P_{ms}-1)^2} \tag{11.6.8}$$

Equations (11.6.2) and (11.6.3) yield Warner's model will be more equal or less efficient than Kuk's model for a fixed number of trails γ if

$$\mathrm{Var}(\bar{r}_w) - \mathrm{Var}(\bar{r}_k) <, = \text{ or } > 0$$

$$\text{i.e., if } \phi_w = \frac{P_w(1-P_w)}{(2P_w-1)^2} <, = \text{ or } > \phi_k = \frac{\pi\theta_1(1-\theta_1) + (1-\pi)\theta_2(1-\theta_2)}{(\theta_1-\theta_2)^2} \tag{11.6.9}$$

In particular, if $\theta_1 = P_{\mathrm{w}}$ and $\theta_2 = 1 - P_{\mathrm{w}}$, Kuk's method is equally efficient to Warner's model.

Equations (11.6.2) and (11.6.3) indicate that Warner's model would be more, equal, or less efficient than Singh and Grewal's (2013) model if

$$\mathrm{Var}(\overline{r}_{\mathrm{w}}) - \mathrm{Var}(\overline{r}_{\mathrm{sg}}) <, = \text{or} > 0$$

$$\text{i.e.,} \quad \frac{P_{\mathrm{w}}(1-P_{\mathrm{w}})}{(2P_{\mathrm{w}}-1)^2} <= \text{or} > \gamma_{\mathrm{sg}} \frac{\pi(1-\theta_g)\theta_g^{*2} + (1-\pi)(1-\theta_g^*)\theta_g^2}{(\theta_g^* - \theta_g)^2} \tag{11.6.10}$$

Similarly, Warner's model will be more, less, or equally efficient than Mangat and Singh's (1990) model if

$$\mathrm{Var}(\overline{r}_{\mathrm{w}}) - \mathrm{Var}(\overline{r}_{\mathrm{ms}}) <, = \text{or} > 0$$

$$\text{i.e.,} \quad \text{if } \phi_{\mathrm{w}} = \frac{P_{\mathrm{w}}(1-P_{\mathrm{w}})}{(2P_{\mathrm{w}}-1)^2} <, = \text{or} > k_{\mathrm{ms}}\phi_{\mathrm{ms}} = k_{\mathrm{ms}} \frac{P_{\mathrm{ms}}(1-P_{\mathrm{ms}})}{(2P_{\mathrm{ms}}-1)^2} \tag{11.6.11}$$

It is mentioned in Theorem 11.2.1 that Mangat and Singh's (1990) model is equally efficient as Warner's (1965) model under the same level of $\mathcal{PP}$ if $P_{\mathrm{w}} = P_{\mathrm{ms}}$. Substituting $P_{\mathrm{w}} = P_{\mathrm{ms}}$ in Eq. (11.6.11), we note that

$$\frac{\mathrm{Var}(\overline{r}_{\mathrm{w}})}{\mathrm{Var}(\overline{r}_{\mathrm{ms}})} = \frac{1}{k_{\mathrm{ms}}} < 1 \text{ as } k_{\mathrm{ms}} > 1 \tag{11.6.12}$$

Equation (11.6.12) yields the following theorem.

Theorem 11.6.1 Under a fixed level of $\mathcal{PP}$, Warner's model is more efficient than Mangat and Singh's (1990) model whenever $P_{\mathrm{w}} = P_{\mathrm{ms}}$ and the total expected number of trials per respondent is fixed.

The expressions (11.6.9), (11.6.10), and (11.6.11) do not provide any meaningful theoretical comparisons among the performances of Warner's, Kuk's, and Sing and Grewal's models in general, as they involve several unknown parameters. Numerical investigations by Arnab (2021, 1923) reveal that none of the proposed models is best in all situations. However, the modified Kuk's model performed well in most situations. The majority preferred this model because of its easy execution and the fact that the respondent did not need to answer embarrassing questions.

11.7 Discussions

The main purpose of the RR technique is to collect sensitive or personal data indirectly so that the respondents' true responses remain undisclosed to the interviewers and third parties. At the same time, efficient estimates of the parameters of interest can be obtained from the collected data. The two objectives, maintaining $\mathcal{PP}$ and obtaining efficient estimates, conflict with each other. Hence, as a compromise, one seeks to find an RR technique that maximizes the efficiency of the estimators while keeping the $\mathcal{PP}$ of the data at a certain level. Several measures of $\mathcal{PP}$ are available in the literature. The measures of $\mathcal{PP}$ that are available in the literature are not adequate as they exclude several important factors like the degree of complexity of the RR devices, total cost, level of embarrassment, and cooperation from the respondents (Christofieds, 2010; Guerriero & Sandri, 2007). For example, in some methods, like Warner's method, Mangat's method, and Singh's method, respondents are required to answer the unpleasant question "Do you belong to sensitive group A?" in certain stages. In some of the two-stage methods, some respondents are required to draw one card, and others are required to draw more than one card. In some RR techniques, they are forced to give "yes" or "no" answers to specific questions. Such RR techniques create suspicion in the respondents and get little cooperation from them.

Most measures of $\mathcal{PP}$ are limited to the SRSWR sampling design. The Bayesian method of $\mathcal{PP}$ was provided by Chaudhuri and Saha (2004), Chaudhuri (2011), Chaudhuri and Christofides (2013), and Quatember (2012). Bhargava and Singh (1999, 2000, 2002), and Guerriero and Sandri (2007), among others, found that Warner's model and Mangat and Singh's model are equally efficient for a given level of $\mathcal{PP}$, which is not true as both methods are not based on the same number of trials. It is shown that Warner's model performs better than Mangat and Singh's model for a given level of $\mathcal{PP}$ when the total number of trials per respondent is fixed. Theoretical comparisons among different methods rarely yield any meaningful results. Hence, researchers provide comparisons by using simulations and empirical studies.

References

Anderson, H. (1975a). *Efficiency versus protection in RR designs for estimating proportions*. Technical Report 9, University of Lund, Sweden.

Anderson, H. (1975b). *Efficiency versus protection in a general RR model*. Technical Report, 10, University of Lund, Sweden.

Anderson, H. (1977). Efficiency versus protection in a general RR model. *Scandinavian Journal of Statistics, 4*, 11–19.

Arnab, R. (2017). *Survey sampling theory and applications*. Academic Press.

Arnab, R. (2021). Randomized response trial using geometric distribution. *Journal of the Indian Society of Agricultural Statistics, 75*, 195–204.

Arnab, R. (2023). Randomized response trial using negative binomial distribution. To appear in *Communications in Statistics-Computation and Simulation*. https://doi.org/10.1080/03610918.2021.2006710

Bhargava, M., & Singh, R. (1999). On relative efficiency of certain randomized response strategies. *Journal of the Indian Society of the Agricultural Statistics, 52*, 245–253.
Bhargava, M., & Singh, R. (2000). A modified randomization device for Warner's model. *Statistica Anno, LX*(2), 315–321.
Bhargava, M., & Singh, R. (2002). On the efficiency comparison of certain randomized response strategies. *Metrika, 55*, 191–197.
Bose, M. (2013). Respondent privacy and estimation efficiency in randomized response surveys for discrete-valued sensitive variables. arXiv:1303.5172v1
Cassel, C. M., Särndal, C. E., & Wretman, J. H. (1977). *Foundations of inference in survey sampling*. Wiley.
Chang, H.-J., & Liang, D.-H. (1996). A two-stage unrelated randomized response procedure. *Australian Journal of Statistics, 38*, 43–51.
Chang, H.-J., Wang, C.-L., & Huang, K.-C. (2004). On estimating the proportion of a qualitative sensitive character using randomized response sampling. *Quality and Quantity, 38*, 675–680.
Chaudhuri, A. (2011). *Randomized response and Indirect questioning techniques in surveys*. Chapman & Hall/CRC.
Chaudhuri, A., & Christofied, T. C. (2013). *Indirect questioning in sample surveys*. Springer.
Chaudhuri, A., & Mukerjee, R. (1988). *Randomized response: Theory and techniques*. Marcel Dekker.
Chaudhuri, A., & Saha, A. (2004). Utilizing covariates by logistic regression modeling in improved estimation of population proportions bearing stigmatizing features through randomized response in complex surveys. *Journal of the Indian Society of the Agricultural Statistics, 58*, 190–211.
Christofides, T. C. (2003). A generalized randomized response technique. *Metrika, 57*, 195–200.
Christofides, T. C. (2010). Comments on a method of comparisons of randomized response techniques. *Journal of Statistical Planning and Inference, 140*, 574–575.
Flinger, M. A., Policello, G. E., & Singh, J. (1977). A comparison of two RR survey methods with consideration for the level of respondent protection. *Communications in Statistics-Theory and Methods., 6*, 1511–1524.
Greenberg, B. D., Abul-Ela, A. L. A., Simmons, W. R., & Horvitz, D. G. (1969). The unrelated question randomized response model: Theoretical framework. *Journal of the American Statistical Association, 64*, 520–539.
Greenberg, B. G., Kuebler, R. R., Abernathy, J. R., & Horvitz, D. G. (1977). Respondent hazards in unrelated question randomized response model. *Journal of Statistical Planning and Inference, 1*, 53–60.
Guerriero, M., & Sandri, M. F. (2007). A note on comparison of some randomized response procedures. *Journal of Statistical Planning and Inference, 64*, 520–539.
Horvitz, D. G., & Thompson, D. J. (1952). A generalization of sampling without replacement from a finite population. *Journal of the American Statistical Association, 47*, 663–685.
Hussain, Z., Shabbir, J., Prevez, Z., Shah, S. F., & Khan, M. (2016). Generalized geometric distribution of order k: A flexible choice to randomize response. *Communications in Statistics-Theory and Methods., 46*, 4708–4721.
Kuk, A. Y. (1990). Asking sensitive question indirectly. *Biometrika, 77*, 436–438.
Lanke, J. (1975). On the choice of unrelated questions in Simmons' version of RR. *Journal of the American Statistical Association, 68*, 525–530.
Lanke, J. (1976). On the degree of protection in randomized interviews. *International Statistical Review, 44*, 197–203.
Leysieffer, R. W., & Warne, S. L. (1976). Respondent jeopardy and optimal designs in RR models. *Journal of the American Statistical Association, 71*, 649–656.
Ljungqvist, L. (1993). A unified approach to measure of privacy protection in randomized response models: A utilitarian perspective. *Journal of the American Statistical Association, 88*, 97–103.
Loynes, R. M. (1976). Asymptotically optimum randomized response procedures. *Journal of the American 33 Statistical Association, 71*, 924–928.

Mangat, N. S. (1994). An improved randomized response strategy. *Journal of the Royal Statistical Society B, 56*, 93–95.

Mangat, N. S., & Singh, R. (1990). An alternative randomized response procedure. *Biometrika, 77*, 439–442.

Mangat, N. S., Singh, R., & Singh, S. (1992). An improved U-question randomized response strategies. *Calcutta Statistical Association Bulletin, 42*, 277–281.

Nayak, T. K. (1994). On randomized response surveys for estimating a proportion. *Communications in Statistics-Theory and Methods, 23*, 3303–3321.

Nayak, T. K., & Adeshiyan, S. A. (2009). A unified framework for analysis and comparison of randomized response surveys of binary characteristics. *Journal of Statistical Planning and Inference, 139*, 2757–2766.

Quatember, A. (2009). A standardization of randomized response strategies. *Survey Methodology, 35*, 143–152.

Quatember, A. (2012). An extension of the standardized randomization technique to a multi-stage setup. *Statistical Methods and Applications, 21*, 475–484.

Singh, S., & Grewal, I. (2013). Geometric distribution as randomization device: Implemented to the Kuk's model. *International Journal of Contemporary Mathematical Sciences, 8*, 243–248.

Warner, S. L. (1965). Randomize response: A survey technique for eliminating evasive answer bias. *Journal of the American Statistical Association, 60*, 63–69.

Chapter 12
Bayesian Approaches for Randomized Response Models

Abstract In this chapter, Bayesian methods for analyzing RR data have been introduced. The works of Winkler and Franklin (1979), Unikrishnan and Kunte (1999), Barabesisi and Marcheselli (2006), O'Hagan (1987), and Bar-Lev et al. (2003) have been reviewed. Bayes estimators for Warner's (1965), Horvitz et al.'s (1967) unrelated question model, Devore's (1977), and the Mangat and Singh (1990) models with appropriate priors have been obtained.

Keywords Bayes estimator · Bayesian approach · Classical approach · Complex sampling · Frequentist approach · Interval estimates · Maximum likelihood · Moment estimates · Point estimates · Posterior distribution · Prior distribution · Uniform prior

12.1 Introduction

The frequentist approach often uses the moment method or, more rarely, the maximum likelihood (MLE) method for estimating population parameters such as population proportions or means from randomized response (RR) survey data. In the moment method, point or interval estimates of the proportion π may take inadmissible values outside of the parameter space [0, 1] especially when π is small or large. In addition, MLE estimates could be zero or one, which may not provide useful results. Furthermore, the likelihood method cannot be employed for complex sampling designs. All these disadvantages are serious since small values of π generally occur in real surveys (Barabesi and Marchaeselli, 2010). The Bayesian approach seems quite appropriate in this situation since the Bayes and credible intervals assume values within the parameter space. In addition, prior knowledge about sensitive attributes, e.g., illegal substance use, alcohol abuse, and incidents of domestic violence, is generally available in many official publications or past surveys.

The Bayesian approach in the RR survey was introduced by Winkler and Franklin (1979) for estimating π from Warner's (1965) model. Under the Bayesian approach,

R. Arnab, *Indirect Methods of Data Collection and Analysis from Surveys*, Indian Statistical Institute Series, https://doi.org/10.1007/978-981-97-6005-3_12

the Bayes estimator of the unknown parameter of interest is obtained as a combination of the sample information and the prior information. The approach was extended in several directions by various researchers. Important contributions include O'Hagan's (1987) work on Bayesian linear estimators, and Migon and Tachibana's (1997) MAPLE routine to reduce the burden of computation of Bayes estimates. Pitz (1980) provided a Bayes estimator using a uniform prior for the unknown population proportion π. Unikrishnan and Kunte (1999) proposed Bayesian analysis and its extension to polychotomous populations for estimating from Warner's (1965) and Symmons' (1965) models. Oh (1994), and Unikrishnan and Kunte (1999) proposed computer insensitive techniques such as the Gibbs sampler for computing Bayes estimates. Bar-Lev et al. (2003) extend Winkler and Franklin's procedure to several RR models, including Horvitz et al.'s (1967), Devore's (1977), Flinger et al.'s (1977), and Mangat and Singh's (1990) model. Kim et al. (2006) applied Bayesian estimation to Mangat's (1994) procedure. Barabesisi and Marcheselli (2006) introduced a Bayesian approach to Franklin's (1989) procedure, while Tian et al. (2009) used a Bayesian method for non-randomized response models. Barabesi and Marcheselli (2010) proposed a Bayesian approach for the joint estimation of π and the sensitivity level T of the stigmatizing attribute in a two-stage RR procedure based on Huang's (2004) model. Song and Kim (2012) proposed a Bayesian approach for modeling the RR sum of score variables. Narjis and Shabbir (2021) presented a Bayesian analysis for estimating population proportion and the sensitivity level of the questions for the optional RR technique.

Applications of the Bayesian method for estimating population proportion π based on real data were considered by several researchers. For example, Winkler and Franklin (1989) considered data relating to the incidence of induced abortions in Taichung, Taiwan, reported by Liu and Chow (1976); Song and Kim (2012) considered RR data from a Dutch survey on non-compliance with social security regulations in 2004; Nandaram and Yu (2017) studied college cheating in a university; Narjis and Shabbir (2021) used victims of sexual abuse of women by a friend or family member collected by Chhabra et al. (2016).

Some of the important contributions are listed below in detail.

12.2 Warner and Franklin's Approach

In Warner's (1965) and Franklin's (1979) approaches, a sample of n respondents was selected from a population by the SRSWR procedure. Each of the selected respondents is asked to provide an answer of "yes" or "no" by using Warner's RR model. In Warner's model, a respondent chooses one of the two questions Q1: "Do you belong to the sensitive group A?" or Q2: "Do you belong to the group $\overline{A}$ (complement of A)?" with probabilities $P(\neq 1/2)$ and $1 - P$ respectively. The probability of obtaining a "yes" answer for Warner's model is

$$\lambda_{\mathrm{w}} = \pi P + (1 - \pi)(1 - P) \tag{12.2.1}$$

Since the sampling process is Bernoulli with parameter (n, P), the likelihood function is

$$l(\lambda_W) = \binom{n}{n_W} \lambda_W^{n_W} (1 - \lambda_W)^{n-n_W} \text{ for } 1 - P \leq \lambda_W \leq P \tag{12.2.2}$$

where n_W is the total number of "yes" answers obtained in the sample.

Equation (12.2.2) yields the moment estimate of π as

$$\hat{\pi}_W = \frac{1}{2P - 1}\left[\hat{\lambda}_W - (1 - P)\right]; \hat{\lambda}_W = \frac{n_W}{n} \tag{12.2.3}$$

Obviously, the moment estimate $\hat{\pi}_W$ can take values outside of the parameter space [0, 1]. Singh (1978) provided the MLE of π as

$$\hat{\pi}_{mle} = \begin{cases} 0 & \text{for } \hat{\pi}_W < 0 \\ \hat{\pi}_W & \text{for } 0 < \hat{\pi}_W < 1 \\ 1 & \text{for } \hat{\pi}_W > 1 \end{cases} \tag{12.2.4}$$

Although the proposed MLE $\hat{\pi}_{mle}$ takes only admissible values of π, it is a biased estimate of π and its mean-square error is not in the closed form.

For the Bayesian analysis, we multiply the likelihood $l(\lambda_W)$ by an appropriate prior distribution $\xi(\pi)$ of π to obtain the posterior distribution

$$\xi(\pi|\lambda) \propto l(\lambda_W)\xi(\pi) \tag{12.2.5}$$

The posterior distribution (12.2.5) generally does not have a manageable closed form. Hence, the standard options are (Unnikrishan and Kunte, 1999): (a) using a standard distribution to approximate the posterior distribution (Winkler and Franklin, 1979); (b) using numerical integration (Berger, 1985) and (c) using computer-intensive techniques (e.g., Gibbs sampler).

Winkler and Franklin (1979) considered $\xi(\pi)$ as beta distribution

$$\xi(\pi) = \frac{1}{B(\alpha, \beta)}\pi^{\alpha-1}(1 - \pi)^{\beta-1}; \ \alpha, \beta > 0 \tag{12.2.6}$$

and obtained posterior distribution as

$$\begin{aligned} \xi(\pi|\lambda_W) \propto \pi^{\alpha-1}(1 - \pi)^{\beta-1}[(2P - 1)\pi + (1 - P)]^{n_W} \\ [P - (2P - 1)\pi]^{n-n_W}; \ 0 \leq \pi \leq 1 \end{aligned} \tag{12.2.7}$$

After simplification, the expression (12.2.7) becomes

$$\xi(\pi | n_{\mathrm{w}}, n) = \sum_{t=0}^{n} w_t \xi(\pi | \alpha + t, \beta + n - t);\ 0 \leq \pi \leq 1 \tag{12.2.8}$$

where

$$w_t = w_t^* / \sum_{r=0}^{n} w_r^* \tag{12.2.9}$$

with

$$w_t^* = \binom{n}{t} \frac{B(\alpha + t, \beta + n - t)}{B(\alpha, \beta)} \sum_{j=0}^{\min(r,t)} \binom{t}{j} \binom{n-t}{r-j} P^{n-t-r+2j} (1-P)^{t+r-2j} \tag{12.2.10}$$

The posterior distribution of π is a mixture of beta distributions. Winkler and Franklin (1979) provided a simple interpretation of weights w_js. In the case t be the actual number of respondents in the sensitive group A, the predictive distribution of t is a beta-binomial distribution:

$$g(t|n, \alpha, \beta) = \binom{n}{t} \frac{B(\alpha + t, \beta + n - t)}{B(\alpha, \beta)};\ t = 0, 1, \ldots, n \tag{12.2.11}$$

Winkler and Franklin (1979) noted that each sum (12.2.8) is the product of the posterior probability that exactly t of the n respondents actually belong to the group A and the posterior distribution of π given that exactly t of the n respondents actually belong to the group A. Under the squared error loss function, the Bayes estimator, the posterior mean of π comes out as

$$\hat{\pi}_B = \frac{1}{\alpha + \beta + n} \sum_{t=0}^{n} (\alpha + r) w_t \tag{12.2.12}$$

The posterior distribution $\xi(\pi | n_{\mathrm{w}}, n)$ of π and hence the Bayes estimator $\hat{\pi}_B$ are not easy to compute in practice because of the involvement of the weights w_t in (12.2.9). Even if the weights are determined, finding the posterior distribution may be tedious. Difficulties increase with the sample size n. Winkler and Warner suggested two approaches to approximations. The first one is to approximate the prior distribution and use an exact likelihood function, while the second one is to use an exact prior distribution and approximate the likelihood function.

12.3 O'Hagan's Approach

O'Hagan (1987) proposed an alternative Bayesian analysis method for randomized response data using the Bayes linear estimator. Under the superpopulation models, the best linear unbiased estimator (BLUE) proposed by Royall (1976) is mathematically equivalent to the Bayes linear estimator. The proposed Bayes linear estimator is distribution-free. For the proposed estimator, neither prior distributions nor likelihoods need to be fully specified. Only the first two order moments are required. The basic result is straightforward.

Warner's Model

Consider a finite population $U = (1, \ldots, i, \ldots, N)$ of N identifiable units. Here we define $y_i = 1$ if the ith unit possesses a certain sensitive attribute A, otherwise, $y_i = 0$. The parameter of interest is $\pi = \frac{1}{N} \sum\limits_{i \in U} y_i$. From the population U, a sample of n units is selected by the SRSWR method. Each of the selected respondents is asked to choose one of the following two questions and answer truthfully "yes" or "no".

Q1: "Do you possess the attribute A?" with probability P.

and Q2: "Do you not possess the attribute A?" with probability $1 - P$.

Let

$$z_i = \begin{cases} 1 \text{ if the } i\text{th respondent answers "yes"} \\ 0 \text{ if the } i\text{th respondent answers "no"} \end{cases}$$

In order to construct a Bayes linear estimator, O'Hagan assumed that the vector $\mathbf{y} = (y_1, \ldots, y_i, \ldots, y_N)$ has an exchangeable prior distribution ξ. For the exchangeable prior distribution ξ, the mean, variance of y_i, and co-variance of y_i and $y_j (i \neq j)$ are respectively.

$$\begin{aligned} E_\xi(y_i) &= \mu, V_\xi(y_i) = \sigma^2 = \mu(1-\mu) \\ \text{and } Cov_\xi(y_i, y_j) &= \rho\mu(1-\mu); -1/(N-1) \leq \rho \leq 1 \end{aligned} \tag{12.3.1}$$

Then, the mean, variance of z_i, and the co-variance of z_i and $z_j (j \neq i)$ are given by

$$\begin{aligned} E(z_i) &= E_\xi[E_R(z_i)] \\ &= E_\xi\big[(2P-1)y_i + (1-P)\big], \\ &= (2P-1)\mu + (1-P) \end{aligned}$$

$$V(z_i) = E_\xi[V_R(z_i)] + V_\xi[E_R(z_i)]$$

$$= E_\xi\{P(1-P)\} + V_\xi\{(2P-1)y_i + (1-P)\}$$
$$= P(1-P) + (2P-1)^2\mu(1-\mu)$$

and

$$\begin{aligned} Cov(z_i, z_j) &= E_\xi\left[Cov_R(z_i, z_j)\right] + Cov_\xi\left[E_R(z_i), E_R(z_j)\right] \\ &= Cov_\xi\left[(2P-1)y_i + (1-P), (2P-1)y_j + (1-P)\right] \\ &= (2P-1)^2 Cov_\xi(y_i, y_j) \\ &= (2P-1)^2\rho\mu(1-\mu);\ i \neq j \end{aligned}$$

O'Hagan proposed Bayes linear estimator of π as

$$\begin{aligned} \hat{\pi}_{\mathrm{LB}} &= \frac{1}{n}\sum_{i=1}^{n} z_i \\ &= \alpha * \hat{\pi}_{\mathrm{w}} + (1-\alpha*)\mu \end{aligned} \tag{12.3.2}$$

where $\alpha* = (2P-1)\frac{\delta}{\beta}$, $\delta = Cov(\overline{z}, \pi) = \frac{1}{N}[1 + (N-1)\rho](2P-1)^2\mu(1-\mu)$ and

$$\beta = \frac{1}{n}[1 + (n-1)\rho]\left[P(1-P) + (2P-1)^2\mu(1-\mu)\right].$$

The estimator (12.3.2) is the weighted average of the prior mean μ and the classical moment estimator (Warner's estimator) $\hat{\pi}_{\mathrm{w}}$. The weight $\alpha^* \in [0, 1]$ increases with n. Even if the whole population is sampled $n = N$, α^* remains strictly less than unity. This indicates that randomized response surveys do not provide complete information.

12.4 Bar-Lev et al.'s Approach

Bar-Lev et al. (2003) presented a common Bayesian approach for Warner's (1965), Horvitz et al.'s (1967) unrelated question, Devore's (1977), and Mangat and Singh's (1990) models. They proposed Bayes estimates for the population proportion π using a common conjugate prior. For each of the proposed models, respondents were asked to answer "yes" or "no" following the instructions of the respective RR model. The probability of obtaining a "yes" answer for each of the proposed models can be expressed as

$$\theta = \pi\alpha^* + \beta^* \tag{12.4.1}$$

where the constants α^* and β^* are subject to $0 \leq \beta^* < \alpha^* + \beta^* \leq 1$.

Equation (12.4.1) yields

$$\pi = \frac{\theta - \beta^*}{\alpha^*}, 0 \leq \pi < 1 \tag{12.4.2}$$

Let n_k be the total number of "yes" answers obtained from a sample of n respondents. Since, n_k follows $B(n, \theta)$ distribution, the likelihood function of θ is.

$$l(\theta|n_k) = \binom{n}{n_k} \theta^{n_k}(1-\theta)^{n-n_k};\ \theta \in [\beta^*, \alpha^* + \beta^*] \tag{12.4.3}$$

Expression (12.4.3) suggests that the appropriate prior for θ is a truncated Beta distribution of the form

$$\xi(\theta) = \frac{1}{B_{\beta^*,\alpha^*+\beta^*}(\alpha, \beta)} \theta^{\alpha-1}(1-\theta)^{\beta-1};\ \alpha, \beta > 0,\ \theta \in [\beta^*, \alpha^* + \beta^*] \tag{12.4.4}$$

where

$$B_{\beta^*,\alpha^*+\beta^*}(\alpha, \beta) = \int\limits_{\beta^*}^{\beta^*+\alpha^*} x^{\alpha-1}(1-x)^{\beta-1}dx \tag{12.4.5}$$

Equation (12.4.3) yields the distribution of π as

$$h(\pi) = \frac{\alpha^*}{B_{\beta^*,\alpha^*+\beta^*}(\alpha, \beta)} (\pi\alpha^* + \beta^*)^{\alpha-1}(1 - \pi\alpha^* - \beta^*)^{\beta-1};\ \alpha, \beta > 0 \tag{12.4.6}$$

The density $h(\pi)$ is a truncated beta distribution.

The posterior densities of θ and π for a given value of n_k are respectively.

$$\xi(\theta|n_W) = \frac{1}{B_{\beta^*,\alpha^*+\beta^*}(\alpha, \beta)} \theta^{n_W+\alpha-1}(1-\theta)^{n-n_W+\beta-1};$$
$$n_W = 0, 1, \ldots, n;\ \theta \in [\beta^*, \alpha^* + \beta^*]. \tag{12.4.7}$$

and

$$h(\pi|n_W) = \frac{1}{B_{\beta^*,\alpha^*+\beta^*}(\alpha, \beta)} (\pi\alpha^* + \beta^*)^{n_W+\alpha-1}(1 - \pi\alpha^* - \beta^*)^{n-n_W+\beta-1};$$
$$n_W = 0, 1, \ldots, n\ \ 0 \leq \pi \leq 1 \tag{12.4.8}$$

The Bayes estimators of θ and π with respect to the squared error loss functions were provided by Bar-Lev et al. (2003) are

$$\hat{\theta}_{\text{BB}} = E(\hat{\theta}|n_{\text{w}}) = \frac{B_{\beta^*,\alpha^*+\beta^*}(n_{\text{w}} + \alpha + 1, n - n_{\text{W}} + \beta)}{B_{\beta^*,\alpha^*+\beta^*}(n_{\text{w}} + \alpha, n - n_{\text{W}} + \beta)} \tag{12.4.9}$$

and

$$\hat{\pi}_{\text{BB}} = E(\hat{\pi}|n_{\text{w}}) = \frac{\hat{\theta}_{\text{BB}} - \beta^*}{\alpha^*} \tag{12.4.10}$$

The posterior variances of $\hat{\theta}_{\text{BB}}$ and $\hat{\pi}_{\text{BB}}$ are

$$\begin{aligned}\text{Var}(\hat{\theta}_{\text{BB}}|n_{\text{w}}) &= \alpha^{*2}\text{Var}(\hat{\pi}_{\text{BB}}|n_{\text{w}}) \\ &= \frac{B_{\beta^*,\alpha^*+\beta^*}(n_{\text{w}} + \alpha + 2, n - n_{\text{W}} + \beta)}{B_{\beta^*,\alpha^*+\beta^*}(n_{\text{w}} + \alpha, n - n_{\text{W}} + \beta)} \\ &\quad - \left[\frac{B_{\beta^*,\alpha^*+\beta^*}(n_{\text{w}} + \alpha + 1, n - n_{\text{W}} + \beta)}{B_{\beta^*,\alpha^*+\beta^*}(n_{\text{w}} + \alpha, n - n_{\text{W}} + \beta)}\right]^2\end{aligned}$$

12.5 Comparison of Classical and Bayes Estimators

Consider the simple situation where each of the selected respondents provides an answer of "yes" or "no". Let t be the total number of "yes" answers obtained in a sample of size n selected by the SRSWR method. Let $\hat{\theta}_{\text{M}}$, $\hat{\theta}_{\text{ML}}$ and $\hat{\theta}_{\text{B}}$ be the moment, maximum likelihood, and Bayes estimates of the parameter θ. The performances of the estimators $\hat{\theta}_{\text{k}}$, k = M, ML, B can be judged by the magnitudes of the bias and mean-square errors (MSE) of $\hat{\theta}_{\text{k}}$ using the approach proposed by Chaubey and Li (1969). The expressions of bias and MSE of the estimators are as follows:

$$\left.\begin{aligned}\text{Bias}(\hat{\theta}_{\text{k}}) &= \sum_{t=0}^{n}(\hat{\theta}_{\text{k}} - \theta)\binom{n}{t}\lambda^t(1-\lambda)^{n-t} \\ \text{MSE}(\hat{\theta}_{\text{k}}) &= \sum_{t=0}^{n}(\hat{\theta}_{\text{k}} - \theta)^2\binom{n}{t}\lambda^t(1-\lambda)^{n-t}\end{aligned}\right\}; \text{k} = \text{M, ML, B}$$

where λ is the probability of obtaining a "yes" answer from a respondent. Theoretical comparisons of Bias($\hat{\theta}_{\text{k}}$) and MSE($\hat{\theta}_{\text{k}}$) are difficult in most situations; hence, they go for numerical comparisons.

References

Barabesi, L., & Marcheselli, M. (2006). A practical implementation and Bayesian estimation in Franklin's randomized response procedure. *Communications in Statistics-Simulation and Computation, 35*, 563–573.

Barabesi, L., & Marcheselli, M. (2010). Bayesian estimation of proportion and sensitivity level in randomized response procedures. *Metrika, 72*, 75–88.

Bar-Lev, S. K., Bobovitch, E., & Bukai, B. (2003). A common conjugate prior structure for several randomized response models. *TEST, 12*, 101–112.

Berger, J. O. (1985). *Statistical decision theory and Bayesian analysis*. Springer

Chaubey, Y. P., & Li, W. (1969). Comparison between maximum likelihood and Bayes methods of estimation for binomial probability with sample composting. *Journal of the Official Statistics, 11*, 379–390.

Chhabra, A., Dass, B. K., & Gupta, S. (2016). Estimating prevalence of sexual abuse by an acquaintance with an optional unrelated question RRT model. *Carolina Journal of Mathematics and Statistics, 2*, 1–9.

Devore, J. L. (1977). A note of the randomized response technique. *Communications in Statistics-Theory and Methods, 6*, 1525–1529.

Flinger, M. A., Policello, G. E., & Singh, J. (1977). A comparison of two survey methods with consideration for the level of respondent protection. *Communications in Statistics-Theory and Methods, 6*, 1511–1524.

Franklin, L. A. (1989). A comparison of estimators for randomized response sampling with continuous distribution from a dichotomous population. *Communications in Statistics-Theory and Methods, 18*, 489–505.

Horvitz, D. G., Shah, B. V., Simmons, W. R. (1967). The unrelated question RR model. In *Proceedings of the Social Statistics Section of the American Statistical Association* (pp. 65–72). ASA.

Huang, K. C. (2004). A survey technique for estimating the proportion and sensitivity in a dichotomous finite population. *Statistica Neeralandica, 58*, 75–82.

Kim, J.-M., Tebbs, J. M., & An, S.-W. (2006). Extensions of Mangat's randomized response model. *Journal of Statistical Planning and Inference, 136*, 1554–1567.

Liu, P. T., & Chow, L. P. (1976). The efficiency of multiple trial randomized response technique. *Biometrics, 32*, 607–618.

Mangat, N. S. (1994). An improved randomized response strategy. *Journal of the Royal Statistical Society: Series B, 56*, 93–95.

Mangat, N. S., & Singh, R. (1990). An alternative randomized response procedure. *Biometrika, 77*, 439–432.

Migon, H. S., & Tachibana, V. M. (1997). Bayesian approximation in randomized response model. *Computational Statistics and Data Analysis, 24*, 401–409.

Narjis, G., & Shabbir, J. (2021). Bayesian analysis of optional unrelated question randomized response model. *Communications in Statistics-Theory and Methods, 50*, 4203–4215.

Nandram, B., & Yu, Y. (2017). Bayesian analysis of sparse counts under the unrelated question design. *Joint Statistical Meeting, Survey research Methods section, American Statistical Association* (pp. 1162–1175).

Oh, M. (1994). Bayesian analysis of randomized response models: A Gibbs sampling approach. *Journal of the Korean Statistical Society, 23*, 463–482.

O'Hagan, A. (1987). Bayes linear estimators for randomized response model. *Journal of the American Statistical Association, 30*, 109–122.

Pitz, G. F. (1980). Bayes analysis of random response models. *Psychological Bulletin, 87*, 209–212.

Royall, R. M. (1976). The linear least square prediction approach in two-stage sampling. *Journal of the American Statistical Association, 71*, 657–664.

Singh, J. (1978). A note on maximum likelihood estimation from randomized response models. In *Proceedings of the Survey Research Methods Section of the American Statistical Association* (pp. 282–283).
Song, J. J., & Kim, J.-M. (2012). Bayesian analysis of randomized response sum score variables. *Communications in Statistics-Theory and Methods., 41*, 1875–1884.
Tian, G.-L., Yuen, K. C., & Tang, M.-L. (2009). Sample surveys with sensitive question: A randomized response approach. *The American Statistician, 63*, 9–16.
Unnikrishnan, N., & Kunte, S. (1999). Bayesian analysis for randomized response models. *Sankhyā, b., 61*, 422–432.
Warner, S. L. (1965). Randomized response: A survey technique for estimating evasive answer bias. *Journal of the American Statistical Association, 60*, 63–69.
Winkler, S. L., & Franklin, L. (1979). Warner's randomized response model: A Bayesian approach. *Journal of the American Statistical Association, 74*, 207–213.

Chapter 13
Miscellaneous Applications of RR Techniques

Abstract In this section, the nominative technique, negative question surveys, and RR techniques for the estimation of nonlinear functions of population parameters such as the ratio of two characters, regression coefficient, correlation coefficient, and coefficient of variation are proposed. Applications of RR techniques in multistage sampling and estimation of rare sensitive attributes by using a Poisson distribution are presented. Optimal sampling strategies for the various superpopulation models are also presented.

Keywords Coefficient of variation · Correlation coefficient · First stage units · Mean-square errors · Multistage sampling · Nominative technique · Nonlinear functions · Poisson distribution · Ratio estimator

13.1 Nominative Techniques

The nominative technique was proposed by Miller (1985). It may be considered a variant of the multiplicity method proposed by Sirken (1975) for collecting data such as rare illnesses, behavior, or experience where respondents are asked to serve as informants reporting about other persons such as their brothers, sisters, siblings, akin, and close friends as well as their own. For gathering information relating to socially unacceptable illnesses or behaviors, respondents are not asked to provide their own or a close friend's status to protect their privacy. In this situation, they are rather asked to provide information about distant but known people. Details are given by Sudman et al. (1977) and Arnab (2017).

The "nominative technique" is an indirect method of collecting data relating to the possession of sensitive or stigmatizing attributes A, such as deviant behavior, gambling, and cheating in examinations, among others. In this method, the selected respondents are not required to provide any information relating to their status of possession of the attribute A. Instead, they are asked to provide answers to the following two questions of their close friends or relatives (nominees):

R. Arnab, *Indirect Methods of Data Collection and Analysis from Surveys*, Indian Statistical Institute Series, https://doi.org/10.1007/978-981-97-6005-3_13

Q1. How many of your close friends possess the attribute A? Just say the number.
Q2. If the answer to Q1 $\geq$ 1, give only the number of other people, including yourself, who know about each of your friends (nominees) membership in the attribute A separately.

The selected respondent is called the "informant" and close friends whose information is provided are called "nominees".

Let a sample s of size n be selected from a population $U = (1, .., i, .., N)$ of N units by using a probability sampling design p. The inclusion probabilities for the ith, and the pair i and $j(\neq i)$th units are π_i and $\pi_{ij}(> 0)$ respectively. Define,

$$y_i = \begin{cases} 0, & \text{if the unit } i \notin A \\ 1, & \text{if the unit } i \in A \end{cases},$$

$$x_{ij} = \begin{cases} 1, & \text{if the unit } i \text{ serves as a nominee of the unit } j \in U \\ & 0, \text{ otherwise} \end{cases} \text{ and } x_{ii} = 0 \quad (13.1.1)$$

Let G_i be the set of nominees associated with the informant $i(\in U)$ and H_j be the set of informants associated with the nominee $j(\in U)$. Then, the ith respondent (informant) provides a response $z_i = \sum_{j \in U} x_{ij} y_j = \sum_{j \in G_i} y_j$ to question Q1. If $z_i \geq 1$, then the ith informant provides an answer to Q2 as $f_{j|i} = \sum_{i \in U} x_{ij} = \sum_{i \in H_j} x_{ij}$ for $j \in H_i$. The answers to Q2 $f_{j|i}$ and $f_{j|k}$ about the nominee j given by the informants i and $k(\neq i)$ may not be the same in reality. Here we assume (i) there is at least one informant for a given nominee and (ii)$f_{j|i} = f_j$ is independent of i. Let us define,

$$w_i = \sum_{j \in U} \frac{x_{ij}}{f_j} y_j \text{ and } w_i = 0 \text{ if } f_j = 0. \quad (13.1.2)$$

The Horvitz-Thompson estimator of the population proportion $\pi = \frac{1}{N} \sum_{i \in U} y_i$ is

$$\hat{\pi}_{\text{M}} = \frac{1}{N} \sum_{i \in s} \frac{w_i}{\pi_i} \quad (13.1.3)$$

The estimator $\hat{\pi}_{\text{M}}$ is unbiased for π as

$$\begin{aligned} E(\hat{\pi}_{\text{M}}) &= \frac{1}{N} E\left[\sum_{i \in s} \frac{1}{\pi_i} \sum_{j \in U} \frac{x_{ij}}{f_j} y_j\right] \\ &= \frac{1}{N} \sum_{j \in U} y_j \frac{1}{f_j} \sum_{i \in U} x_{ij} \\ &= \pi \end{aligned}$$

Hence, we obtain the following theorem:

Theorem 13.1.1

(i) $\hat{\pi}_{\mathrm{M}} = \frac{1}{N}\sum\limits_{i\in s}\frac{w_i}{\pi_i}$ is an unbiased estimator of π

(ii) The variance of $\hat{\pi}_{\mathrm{M}}$ is

$$\mathrm{Var}(\hat{\pi}_{\mathrm{M}}) = \frac{1}{N^2}\sum\sum_{i< j\in U}(\pi_i\pi_j - \pi_{ij})\left(\frac{w_i}{\pi_i} - \frac{w_j}{\pi_j}\right)^2$$

(iii) An unbiased estimator of $\mathrm{Var}(\hat{\pi}_{\mathrm{M}})$ is

$$\hat{\mathrm{Var}}(\hat{\pi}_{\mathrm{M}}) = \frac{1}{N^2}\sum\sum_{i< j\in s}\frac{\pi_i\pi_j - \pi_{ij}}{\pi_{ij}}\left(\frac{w_i}{\pi_i} - \frac{w_j}{\pi_j}\right)^2$$

The nominative technique is simple to implement and decreases nonresponse. Based on real data (heroin use), Miller (1985) pointed out that the nominative method boosts cooperation and respondents' confidence in disclosing confidential status.

13.2 Negative Question Survey Methodology

The "negative question surveys" were introduced by Espanda (2006) for collecting data relating to sensitive and/or stigmatizing issues. Suppose that a population comprises $k(\geq 2)$ mutually exclusive and exhaustive categories. Let a sample s of n units (respondents) be selected from the population by the SRSWR method. Each of the selected respondents is asked to answer a sensitive question, e.g., "Q : To which of the $k(\geq 2)$ categories do you belong?" Since the categories are mutually exclusive and exhaustive, only one category is true or applicable to a respondent. The remaining $k-1$ categories do not apply to him or her. Esponda and Guerrero (2009) termed such a direct questioning method a "positive question survey". In such a positive survey, respondents do not feel comfortable disclosing their categories to the interviewer because of the sensitivity of the question.

Espanda (2006), Esponda and Guerrero (2009), and Esponda et al. (2016) proposed an alternative method of interviewing termed "the negative question survey" where the same question Q : is framed in a negative way, e.g., "$\overline{Q}$: To which of the k categories do you not belong?". In this case, each respondent has only one category that is false or non-applicable, and the remaining $k-1$ categories are applicable to him or her. In the negative survey, a respondent should answer only one out of the $k-1$ categories that might be applicable to him or her with a certain preassigned probability. Thus, the negative survey protects the privacy of the respondents because they need not disclose their true category to the interviewer. From such a negative question survey, it is possible to estimate the population proportion of individuals belonging to each of the categories.

Positive survey Question:		**Negative survey Question:**	
Which of the following residency statuses do you have? (State the category only.)		Which of the following residency statuses do you not have? (State one category only.)	
Category	Residency Status	Category	Residency Status
1	Citizen	1	Citizen
2	Permanent Resident	2	Permanent Resident
3	Temporary Resident	3	Temporary Resident
4	No residency (illegal resident)	4	No residency (illegal resident)

Fig. 13.1 Positive and negative survey questions

Estimation of Proportion

Let,

π_j = Proportion of individuals in the population belonging to the category j; $\sum_{j=1}^{k} \pi_j = 1$.

$p_{i|j}$ = Probability that an individual provides his or her response as category "j" given that he or she.

really belongs to the category "i".

$p_{i|i} = 0$ and $\sum\limits_{j=1}^{k} p_{i|j} = 1$.

Then,

λ_i = Probability of obtaining a response category "i" from a respondent selected at random

$$= \sum_{j=1}^{k} p_{i|j}\pi_j$$

$$= \sum_{j=1}^{k-1} p_{i|j}\pi_j + (1 - \pi_1 - .. - \pi_{k-1})p_{i|k} \tag{13.2.1}$$

$$\text{i.e. } \lambda_i - p_{i|k} = \sum_{j=1}^{k-1} (p_{i|j} - p_{i|k})\pi_j$$

$$\text{i.e. } \xi_i = \sum_{j=1}^{k-1} q_{i|j}\pi_j; \tag{13.2.2}$$

where $\xi_i = \lambda_i - p_{i|k}$ and $q_{i|j} = p_{i|j} - p_{k|j}$

Equation (13.2.2) can be expressed as

$$\boldsymbol{\xi} = \boldsymbol{Q\pi} \tag{13.2.3}$$

where $\boldsymbol{\xi} = (\xi_1, .., \xi_i, ..., \xi_{k-1})'$, $\boldsymbol{\pi} = (\pi_1, .., \pi_i, ..., \pi_{k-1})'$ and $\boldsymbol{Q} = ((q_{i|j}))$ is a $k-1 \times k-1$ matrix

In this case, $\boldsymbol{Q}$ is a full rank $k-1$ matrix, a unique solution of $\boldsymbol{\pi}$ is obtained as

$$\boldsymbol{\pi} = \boldsymbol{Q}^{-1}\boldsymbol{\xi} \tag{13.2.4}$$

Let n_i = the total number of respondents who reported that they belonged to the category "i" in the sample s Then, the vector $\boldsymbol{n} = (n_1, .., n_i, ..., n_k)$ follows a multinomial distribution with parameters $n(=\sum_{i=1}^{k} n_i)$ and $\boldsymbol{\lambda} = (\lambda_1, .., \lambda_i, ..., \lambda_k)'$.

Denoting $\hat{\lambda}_i = \frac{n_i}{n}$ = proportion of respondents in the sample s who provided their category as "i", one obtains by applying the properties of the multinomial distributions

$$E(\hat{\lambda}_i) = \lambda_i,\ V(\hat{\lambda}_i) = \frac{\lambda_i(1-\lambda_i)}{n} \text{ and } Cov(\hat{\lambda}_i, \hat{\lambda}_j) = -\frac{\lambda_i\lambda_j}{n}$$

Substituting $\lambda_i = \hat{\lambda}_i$, an estimate of ξ_i is obtained as $\hat{\xi}_i = \hat{\lambda}_i - p_{k|j}$ which intern provides an unbiased estimator of $\boldsymbol{\pi}$ as

$$\hat{\boldsymbol{\pi}}_{\text{SG}} = \boldsymbol{Q}^{-1}\hat{\boldsymbol{\xi}} \tag{13.2.5}$$

and $\hat{\pi}_{\text{SG}k} = 1 - \hat{\pi}_{\text{SG}1} - \ldots - \hat{\pi}_{\text{SG}k-1}$

the variance of $\hat{\boldsymbol{\pi}}_{\text{SG}}$ is

$$\begin{aligned}\text{Var}(\hat{\boldsymbol{\pi}}_{\text{SG}}) &= \boldsymbol{Q}^{-1}\text{Var}(\hat{\boldsymbol{\xi}}')(\boldsymbol{Q}^{-1})^{\text{T}} \\ &= \boldsymbol{Q}^{-1}\boldsymbol{D}_{\lambda}(\boldsymbol{Q}^{-1})^{\text{T}}\end{aligned} \tag{13.2.6}$$

where $\boldsymbol{D}_{\lambda} = \frac{1}{n}[Diag(\boldsymbol{\lambda}) - \boldsymbol{\lambda\lambda}']$ and $Diag(\boldsymbol{\lambda})$ is a diagonal matrix with $\lambda_1, .., \lambda_{k-1}$ as its diagonal elements.

Equations (13.2.5) and (13.2.6) yield the following theorem analogous to Theorem 2.2.1.

Theorem 13.2.1 For a nonsingular $\boldsymbol{Q}$,

(i) $\hat{\boldsymbol{\pi}}_{\text{SG}} = \boldsymbol{Q}^{-1}\hat{\boldsymbol{\xi}}$ is an unbiased estimator of $\hat{\boldsymbol{\pi}}$ and $\hat{\pi}_{\text{SG}k} = 1 - \sum_{j=1}^{k-1}\hat{\pi}_{\text{SG}j}$ is an unbiased estimator of π_k.

(ii) The variance of $\hat{\pi}_{SG}$ is $Var(\hat{\pi}_{SG}) = \boldsymbol{Q}^{-1}\boldsymbol{D}_{\lambda}(\boldsymbol{Q}^{-1})^T$

(iii) An unbiased estimator of $Var(\hat{\pi}_{SG})$ is $\hat{Var}(\hat{\pi}_{SG}) = \boldsymbol{Q}^{-1}\hat{\boldsymbol{D}}_{\lambda}(\boldsymbol{Q}^{-1})^T$

where $\hat{\boldsymbol{D}}_{\lambda} = \frac{1}{n}\left[Diag(\hat{\boldsymbol{\lambda}}) - \hat{\boldsymbol{\lambda}}\hat{\boldsymbol{\lambda}}'\right]$.

Remark 13.2.1 The expression of the variance of $\hat{\pi}_{SG}$ provided by Espanda (2006), Esponda and Guerrero (2009), and later by Chaudhuri and Christofied (2013), Theorem 6.6, page 146 is incorrect because they derive the results assuming the $k \times k$ matrix $\boldsymbol{P} = ((p_{i|j}))$ is nonsingular, which is incorrect as the sum of the elements of each row of $\boldsymbol{P}$ is one.

The choice of $p_{j|i}$ is not easy. Espanda (2006), and Esponda and Guerrero (2009) did not provide any guidelines of choosing $p_{j|i}$ values. If all possible alternative answers are chosen with equal probabilities $1/(k-1)$, an unbiased estimator of π_i is obtained as

$$Var(\hat{\pi}_{SGi}) = \frac{\pi_i(1-\pi_i)}{n}\left(1 + \frac{k-2}{\pi_i}\right) \tag{13.2.7}$$

Since $Var(\hat{\pi}_{SGi}) \geq \frac{\pi(1-\pi)}{n}$, we conclude that the negative survey estimate $\hat{\pi}_{SGi}$ is less efficient than the direct survey estimate. The variance $Var(\hat{\pi}_{SGi})$ increases with the number of alternatives k. Negative survey methodology is relatively new. Applications of negative surveys in real life are few, and the validity of such surveys has not yet been studied. Further research is expected to be undertaken in the future.

13.3 Estimation of Nonlinear Parametric Functions

From survey data, one often estimates complex or nonlinear functions of the population parameters, such as population variance, population ratio of two characters, population regression coefficient, population coefficient of variation, and correlation coefficient, amongst others. Several authors, e.g., Arnab (2017), Särndal et al. (1992), and Wolter (1985), have extensively discussed methods for estimating complex functions, variances, and unbiased estimators of their variances for the DR surveys. But for the RR surveys, where the study variable is not directly available from the respondents, the general theory of estimation of nonlinear or complex functions is almost non-existent. In this section, we have proposed a few methods for estimating complex functions. The nonnegativity properties of the proposed estimators are also discussed.

Consider a finite population $U = (1, .., i, ..., N)$ of N identifiable units. The values of the study and auxiliary variables y and x of the $i(\in U)$th unit of the population will be denoted by y_i and x_i respectively. Let $\theta_k = \theta_k(x, y)$, $k = 1, 2, ...$ be a parametric function of y and x. For example, $\theta_1 = N = \sum_{i \in U} 1$, $\theta_2 = X = \sum_{i \in U} x_i$, $\theta_3 = Y = \sum_{i \in U} y_i$, $\theta_4 = \sum_{i \in U} x_i y_i$ etc.

In this section, we will consider methods of estimating the parametric function $f(\boldsymbol{\theta}) = f(\theta_1, .., \theta_k)$ by selecting a sample s.of size n using a sampling design p with probability $p(s)$. Let π_i and $\pi_{ij} (j \neq i)$ be the inclusion probabilities of the ith, and ith and jth units of the population. Let $\hat{\theta}_k$ be an unbiased or consistent estimator of θ_k so that an unbiased or consistent estimator of $f(\boldsymbol{\theta})$ is

$$T(\hat{\boldsymbol{\theta}}) = f(\hat{\theta}_1, .., \hat{\theta}_k) \tag{13.3.1}$$

Assuming $T = T(\hat{\boldsymbol{\theta}})$ is a differentiable of $\hat{\theta}_j$ s, for $j = 1, .., k$, one may express $T(\hat{\boldsymbol{\theta}})$ by using Taylor series expansion as

$$T = f(\hat{\theta}_1, .., \hat{\theta}_k)\big|_{\hat{\theta}_j=\theta_j} + \sum_{j=1}^{k} (\hat{\theta}_j - \theta_j) \left.\frac{\partial f(\hat{\theta}_1, .., \hat{\theta}_k)}{\partial \hat{\theta}_j}\right|_{\hat{\theta}_j=\theta_j} + R_e \tag{13.3.2}$$

where R_e is the error or remainder term.

Assuming R_e is negligible, one may approximate

$$T \cong f(\boldsymbol{\theta}) - \sum_{j=1}^{k} w_j(\hat{\theta}_j - \theta_j) \tag{13.3.3}$$

where $w_j = \frac{\partial f(\hat{\theta}_1,..,\hat{\theta}_k)}{\partial \hat{\theta}_j}\big|_{\hat{\theta}_j=\theta_j}$.

The properties of T were provided by Arnab (2017) and Wolter (1985) as follows:

Theorem 13.3.1

(i) $E(T) \cong f(\theta_1, ..., \theta_k)$
(ii) The variance of T is

$$\begin{aligned}\text{Var}(T) &\cong \sum_{j=1}^{k} w_j^2 V(\hat{\theta}_j) + \sum_{i\neq}^{k}\sum_{j=1}^{k} w_j w_k Cov(\hat{\theta}_j, \hat{\theta}_k)\\ &= \mathbf{W}'\Sigma\mathbf{W}\end{aligned}$$

(iii) An approximate unbiased estimator of Var(T) is

$$\begin{aligned}\hat{\text{Var}}(T) &\cong \sum_{j=1}^{k} \hat{w}_j^2 \hat{V}(\hat{\theta}_j) + \sum_{i\neq}^{k}\sum_{j=1}^{k} \hat{w}_j \hat{w}_k \hat{Cov}(\hat{\theta}_j, \hat{\theta}_k)\\ &= \hat{\mathbf{W}}'\hat{\Sigma}\hat{\mathbf{W}}\end{aligned}$$

where $\boldsymbol{W}' = (w_1, .., w_k)$,$\boldsymbol{\Sigma} = ((\sigma_{ij}))$ = variance covariance matrix of $\hat{\theta}_j$'s, $V(\hat{\theta}_j) = \sigma_{jj}$, $Cov(\hat{\theta}_j, \hat{\theta}_k) = \sigma_{jk}$, $\hat{\mathbf{W}}' = (\hat{w}_1, .., \hat{w}_k)$ = an unbiased or approximately unbiased estimator of $\boldsymbol{W}'$,$\hat{V}(\hat{\theta}_j) = \hat{\sigma}_{jj}$ and $C\hat{o}v(\hat{\theta}_j, \hat{\theta}_k) = \hat{\sigma}_{jk}$ are unbiased estimators of σ_{jj} and σ_{jk}, and $\hat{\boldsymbol{\Sigma}} = ((\hat{\sigma}_{jk}))$ is an unbiased estimator of $\boldsymbol{\Sigma}$.

In case θ_j is of the form $\theta_j = \sum_{i \in U} f_i(j)$, one can choose a linear homogeneous unbiased estimator of θ_j as

$$\hat{\theta}_j = \sum_{i \in s} b_{si} f_i(j) \tag{13.3.4}$$

Equations (13.3.3) and (13.3.4) yield

$$\begin{aligned} T &\cong f(\boldsymbol{\theta}) + \sum_{j=1}^{k} w_j \left(\sum_{i \in s} b_{si} f_i(j) - \theta_j \right) \\ &= f(\boldsymbol{\theta}) + \sum_{i \in s} b_{si} h_i - \sum_{j=1}^{k} w_j \theta_j \end{aligned} \tag{13.3.5}$$

where

$$h_i = \sum_{j=1}^{k} w_j f_i(j) \tag{13.3.6}$$

The variance of T can be expressed as

$$\begin{aligned} \mathrm{Var}(T) &\cong \mathrm{Var}\left(\sum_{i \in s} b_{si} h_i \right) \\ &= \sum_{i \in U} (\alpha_{ii} - 1) h_i^2 + \sum_{i \neq} \sum_{j \in U} (\alpha_{ij} - 1) h_i h_j \end{aligned} \tag{13.3.7}$$

where

$$\alpha_{ii} = \sum_{s \supset i} b_{si}^2 p(s) \text{ and } \alpha_{ij} = \sum_{s \supset i,j} b_{si} b_{sj} p(s) \tag{13.3.8}$$

Equation (13.3.7) yields an unbiased estimator of Var(T) as

$$\hat{\mathrm{Var}}(T) = \sum_{i \in s} \frac{\alpha_{ii} - 1}{\pi_i} \hat{h}_i^2 + \sum_{i \neq} \sum_{j \in s} \frac{\alpha_{ij} - 1}{\pi_{ij}} \hat{h}_i \hat{h}_j \tag{13.3.9}$$

where $\hat{h}_j$ is an unbiased or consistent estimator of $h_j, j \in s$.

13.3.1 Estimation of Nonlinear Functions in RR Surveys

Let us consider the situation where the study variable is sensitive and no direct response is obtainable from the respondent. Hence, a RR is obtained from each of the respondents by using a certain randomization device. Let the revised RR be obtained from the i th respondent that satisfies the following general model proposed by Arnab (1990) as follows:

$$E_R(r_i) = y_i,\ V_R(r_i) = \phi_i \text{ and } C_R(r_i, r_j) = 0 \quad \text{for } i \neq j \tag{13.3.10}$$

As usual, an unbiased estimator of ϕ_i will be denoted by $\hat{\phi}_i$.

By replacing y_i with r_i in the expression of $\hat{\theta}_j$, an unbiased or consistent estimator of θ_j is obtained for the RR survey as $\hat{\theta}_{jr} = \hat{\theta}_j|_{y_i=r_i}$. In this case, a consistent or unbiased estimator of $f(\boldsymbol{\theta})$ is obtained as

$$T(\hat{\boldsymbol{\theta}}_r) = f(\hat{\theta}_{1r}, .., \hat{\theta}_{kr}) \tag{13.3.11}$$

In case θ_j is of the form $\theta_j = \sum_{i\in U} f_i(j)$, one can estimate θ_j by

$$\hat{\theta}_{jr} = \sum_{i\in s} b_{si} f_i(j)|_{y_i=r_i} = \sum_{i\in s} b_{si} f_{ir}(j) \tag{13.3.12}$$

where $f_{ir}(j) = f_i(j)|_{y_i=r_i}$.

Using Taylor series expansion, one can approximate $T(\hat{\boldsymbol{\theta}}_r)$ as

$$T(\hat{\boldsymbol{\theta}}_r) \cong f(\boldsymbol{\theta}) + \sum_{i\in s} b_{si} h_{ir} - \sum_{j=1}^{k} w_j \theta_j \tag{13.3.13}$$

where $h_{ir} = \sum_{j=1}^{k} w_j f_{ir}(j)$.

Writing,

$$\text{(i)} E_R(h_{ir}) = h_i,\ V_R(h_{ir}) = \psi_i \text{ and } C_R(h_{ir}, h_{jr}) = 0 \quad \text{for } i \neq j \tag{13.3.14}$$

and (ii) an unbiased estimator of ψ_i as $\hat{\psi}_i$, the following theorem is obtained.

Theorem: 13.3.2

(i) $E\left[T(\hat{\boldsymbol{\theta}}_r)\right] = f(\boldsymbol{\theta})$

(ii) The variance of $T(\hat{\boldsymbol{\theta}}_r)$ is

$$\text{Var}\left[T(\hat{\boldsymbol{\theta}}_r)\right] = \left[\sum_{i\in U} (\alpha_{ii} - 1) h_i^2 + \sum_{i\neq} \sum_{j\in U} (\alpha_{ij} - 1) h_i h_j\right] + \sum_{i\in U} \alpha_{ii} \psi_i$$

(iii) Two unbiased estimators of $\text{Var}\left[T(\hat{\boldsymbol{\theta}}_r)\right]$ are

$$(\text{a})\text{V}\hat{\text{a}}\text{r}_1\left[T(\hat{\boldsymbol{\theta}}_r)\right] = \left[\sum_{i\in s}\frac{\alpha_{ii}-1}{\pi_i}h_{ir}^2 + \sum_{i\neq}\sum_{j\in s}\frac{\alpha_{ij}-1}{\pi_{ij}}h_{ir}h_{jr}\right] + \sum_{i\in s} b_{si}\hat{\psi}_i$$

and

$$(\text{b})\text{V}\hat{\text{a}}\text{r}_2\left[T(\hat{\boldsymbol{\theta}}_r)\right] = \left[\sum_{i\in s}\frac{\alpha_{ii}-1}{\pi_i}(h_{ir}^2 - \hat{\psi}_{ir}) + \sum_{i\neq}\sum_{j\in s}\frac{\alpha_{ij}-1}{\pi_{ij}}h_{ir}h_{jr}\right] + \sum_{i\in s}\frac{\alpha_{ii}}{\pi_i}\hat{\psi}_i$$

Proof

$$\begin{aligned}
\text{(i) } E\left[T(\hat{\boldsymbol{\theta}}_r)\right] &= E_pE_R\left\{T(\hat{\boldsymbol{\theta}}_r)\right\} \\
&= E_p\left[T(\hat{\boldsymbol{\theta}})\right] \\
&= f(\boldsymbol{\theta})
\end{aligned}$$

$$\begin{aligned}
\text{(ii) Var}\left[T(\hat{\boldsymbol{\theta}}_r)\right] &= V_p\left[E_R\left\{T(\hat{\boldsymbol{\theta}}_r)\right\}\right] + E_p\left[V_R\left\{T(\hat{\boldsymbol{\theta}}_r)\right\}\right] \\
&= V_p\left\{T(\hat{\boldsymbol{\theta}})\right\} + E_p\left[\textstyle\sum_{i\in s} b_{si}^2\psi_i\right] \\
&= \textstyle\sum_{i\in U}(\alpha_{ii}-1)h_i^2 + \sum_{i\neq}\sum_{j\in U}(\alpha_{ij}-1)h_ih_j + \sum_{i\in U}\alpha_{ii}\psi_i
\end{aligned}$$

$$\text{(iii) (a)}E\{\text{V}\hat{\text{a}}\text{r}_1[T(\theta_r)]\} = E_p\left[\begin{array}{c}\displaystyle\sum_{i\in s}\frac{\alpha_{ii}-1}{\pi_i}(h_i^2+\psi_i)+ \\ \displaystyle\sum_{i\neq}\sum_{j\in s}\frac{\alpha_{ij}-1}{\pi_{ij}}h_{ir}h_{jr}\end{array}\right] + E_p\left(\sum_{i\in s}b_{si}\psi_i\right)$$

$$= \text{Var}[T(\theta)]$$

$$\text{(b)}E\{\text{V}\hat{\text{a}}\text{r}_2[T(\theta_r)]\} = E_p\left[\sum_{i\in s}\frac{\alpha_{ii}-1}{\pi_i}h_i^2 + \sum_{i\neq}\sum_{j\in s}\frac{\alpha_{ij}-1}{\pi_{ij}}h_ih_j\right] + \sum_{i\in s}\frac{\alpha_{ii}}{\pi_i}\psi_i$$

$$= \text{Var}[T(\theta)]$$

13.3.2 Nonnegative Properties of Variance Estimators

The estimator $\text{V}\hat{\text{a}}\text{r}_1\left[T(\hat{\boldsymbol{\theta}}_r)\right]$ is nonnegative if both the variance $\text{V}\hat{\text{a}}\text{r}(T)$ based on DR and $\hat{\psi}_i$ are nonnegative. On the other hand, $\text{V}\hat{\text{a}}\text{r}_2\left[T(\hat{\boldsymbol{\theta}}_r)\right]$ can be negative even if both $\text{V}\hat{\text{a}}\text{r}(T)$ and $\hat{\psi}_i$ are nonnegative. Chaudhuri (2011), Chaudhuri and Chistofied (2013, page 48), and Adhikary (2015) used $\text{V}\hat{\text{a}}\text{r}_2\left[T(\hat{\boldsymbol{\theta}}_r)\right]$ type estimators which often take negative values. Arnab (1994) proposed $\text{V}\hat{\text{a}}\text{r}_1\left[T(\hat{\boldsymbol{\theta}}_r)\right]$ type estimators.

13.3.3 An Example

Consider the population ratio $\theta = R = Y/X$ of the totals of two characters y and x. Here,

$$\theta_1 = Y = \sum_{i\in U} f_i(1), \theta_2 = X = \sum_{i\in U} f_i(2), f_i(1) = y_i, f_i(2) = x_i, f(\boldsymbol{\theta}) = \frac{\theta_1}{\theta_2} = R.$$

$$T(\hat{\boldsymbol{\theta}}) = f(\hat{\boldsymbol{\theta}}) = \frac{\hat{\theta}_1}{\hat{\theta}_2} = \hat{R};\ \hat{\theta}_1 = \sum_{i\in s} b_{si} y_i, \hat{\theta}_2 = \sum_{i\in s} b_{si} x_i;$$

$$w_1 = \frac{\partial f(\boldsymbol{\theta})}{\partial \hat{\theta}_1}|_{\hat{\theta}_j=\theta_j} = \frac{1}{\theta_2}, w_2 = \frac{\partial f(\boldsymbol{\theta})}{\partial \hat{\theta}_2}|_{\hat{\theta}_j=\theta_j} = -\frac{\theta_1}{\theta_2^2};$$

$$h_i = \sum_{j=1}^{2} w_j f_i(j) = \frac{y_i}{\theta_2} - \frac{\theta_1}{\theta_2^2} x_i = \frac{y_i - Rx_i}{X};$$

$$\hat{\boldsymbol{\theta}}_r = \frac{\sum_{i\in s} b_{si} r_i}{\sum_{i\in s} b_{si} x_i}, h_{ir} = \frac{r_i - \hat{R}x_i}{X}, V_R(h_{ir}) = \frac{1}{X^2}\phi_i = \psi_i$$

$$\hat{\boldsymbol{\theta}}_r = T(\hat{\boldsymbol{\theta}}_r) \cong f(\boldsymbol{\theta}) + \sum_{i\in s} b_{si} h_{ir} - \sum_{j=1}^{k} w_j \theta_j$$

13.3.3.1 Arbitrary Sampling Scheme

For an arbitrary sampling scheme, we may choose $b_{si} = 1/\pi_i$. In this case $\hat{\boldsymbol{\theta}}_r$ reduces to $\hat{\boldsymbol{\theta}}_r = \sum_{i\in s} \frac{r_i}{\pi_i} / \sum_{i\in s} \frac{x_i}{\pi_i}$ and $\alpha_{ii} = 1/\pi_i$. Properties of $\hat{\boldsymbol{\theta}}_r$ are obtained from the Theorem 13.3.1 as follows:

(i) $E(\theta_r) = R$

(ii) $\text{Var}(\theta_r) = \frac{1}{X^2}\left[\begin{array}{l} \sum_{i\in U}\left(\frac{1}{\pi_i} - 1\right)(y_i - Rx_i)^2 \\ +\sum_{i\neq}\sum_{j\in U}\left(\frac{\pi_{ij}}{\pi_i\pi_j} - 1\right)(y_i - Rx_i)(y_j - Rx_j) + \sum_{\in U}\frac{\phi_i}{\pi_i} \end{array}\right]$

(iii) $(a)\hat{\text{Var}}_1(\theta_r) \cong \frac{1}{X^2}\left[\begin{array}{l} \sum_{i\in s}\frac{1-\pi_i}{\pi_i^2}(r_i - \hat{R}x_i)^2 \\ +\sum_{i\neq}\sum_{j\in s}\frac{\pi_{ij} - \pi_i\pi_j}{\pi_i\pi_j\pi_{ij}}(r_i - \hat{R}x_i)(r_j - \hat{R}x_j) - \sum_{i\in s}\frac{1}{\pi_i}\hat{\phi}_i \end{array}\right]$

$$\text{(b)}\hat{\text{Var}}_2(\theta_r) \cong \frac{1}{X^2}\left[\begin{array}{l}\sum_{i\in s}\left(\frac{1}{\pi_i}-1\right)\frac{\left\{(r_i-\hat{R}x_i)^2-\hat{\phi}_i\right\}}{\pi_i} \\ +\sum_{i\neq}\sum_{j\in s}\left(\frac{\pi_{ij}}{\pi_i\pi_j}-1\right)\frac{(r_i-\hat{R}x_i)(r_j-\hat{R}x_j)}{\pi_{ij}}+\sum_{i\in s}\frac{\hat{\phi}_i}{\pi_i^2}\end{array}\right]$$

13.3.3.2 Simple Random Sampling Without Replacement

For the SRSWOR sampling, $\pi_i = \frac{n}{N}$, $\pi_{ij} = \frac{n(n-1)}{N(N-1)}$ and $\hat{R} = \hat{\theta}_r = \sum_{i\in s} r_i / \sum_{i\in s} x_i$.

Properties of $\hat{\theta}_r$ are as follows:

(i) $E(\hat{\theta}_r) \cong R$,

$$\text{(ii)}\quad \text{Var}(\hat{\theta}_r) \cong \frac{1}{X^2}\left[\frac{N-n}{Nn}\frac{1}{N-1}\sum_{i\in U}(y_i-Rx_i)^2+\frac{N}{n}\sum_{i\in U}\phi_i\right]$$
$$= \frac{1}{X^2}\left[\frac{N-n}{Nn}(S_y^2+R^2S_x^2-2RS_{xy})+\frac{N}{n}\sum_{i\in U}\phi_i\right]$$

where $S_x^2 = \frac{1}{N-1}\sum_{i\in U}(x_i-\overline{X})^2$, $S_y^2 = \frac{1}{N-1}\sum_{i\in U}(y_i-\overline{Y})^2$, $S_{xy} = \frac{1}{N-1}\sum_{i\in U}(y_i-\overline{Y})(x_i-\overline{X})$, $\overline{X} = \sum_{i\in U} x_i/N$ and $\overline{Y} = \sum_{i\in U} y_i/N$.

$$\text{(iii)}\quad \hat{V}\text{ar}_1(\hat{R}) \cong \frac{1}{X^2}\left[\frac{N-n}{Nn}\frac{1}{n-1}\sum_{i\in s}(u_i-\overline{u}_s)^2+\frac{N}{n}\sum_{i\in s}\hat{\phi}_i\right]$$
$$= \hat{V}\text{ar}_2(\hat{R})$$

where $u_i = r_i - \hat{R}x_i$ and $\overline{u}_s = \sum_{i\in s} r_i/n$.

13.4 Multistage Sampling

In multistage sampling, at first the entire population U is divided into a number of mutually exclusive and exhaustive groups $U_1, .., U_i, ..., U_N$. Each group is called a cluster, or a first-stage unit (fsu), or a primary stage unit (psu). The ith fsu U_i is further subdivided into M_i disjoint subgroups $U_{i_1}, ..., U_{i_j}, ..., U_{i_{M_i}}$. The subgroup U_{i_j} is called a second stage unit (ssu). The ssus are subdivided further into third stage units, and so on. Division is repeated until a stage is reached from which statistical information is collected. Such a unit is called the ultimate stage unit.

Multistage sampling is commonly used in large-scale surveys since the sampling frame of the ultimate units is not readily available in most situations. In multistage sampling, only the sampling frames of the selected first-stage units are required to select second-stage units, while sampling frames for the selected second-stage units are necessary for selecting third-stage units, and so on. Thus, multistage sampling is much less expensive than uni-stage sampling, in which the ultimate stage units are chosen in a single stage. However, as per efficiency considerations, multistage sampling is less efficient than uni-stage sampling.

Consider two-stage sampling, where at the first stage, a sample s of n units is selected with probability $p(s)$. The inclusion probabilities for the ith , and ith and jth $(j \neq i)$ units are denoted by π_i and π_{ij} respectively. If the ith unit U_i is selected in the sample s, a sub-sample s_i of m_i ssus is selected from U_i with probability $p(s_i)$ The inclusion probabilities of the kth, kth and lth $(k \neq l)$ units of U_i are denoted by $\pi_{k|i}$ and $\pi_{kl|i}$ respectively. Let y_{ij} be the value of the jth ssu of the ith fsu of the study variable y. Then the population total of y is $Y = \sum_{i=1}^{N} Y_i$ where $Y_i = \sum_{j=1}^{M_i} y_{ij}$ is the total of the ith fsu of $U_i(= 1, .., N)$.

13.4.1 Use of RR Techniques for Multistage Sampling

Here we suppose that the study variable y is sensitive, and hence no direct response (DR) y is obtained from the respondents. Hence, a suitable randomization device is used for the collection of data from the selected sample. Let z_{ij} be the RR obtained from the jth ssu (respondent) of the ith fsu by using some suitable randomization device. Let the revised response r_{ij} obtained from z_{ij} satisfies the model:

$$E_R(r_{ij}) = y_{ij},\ V_R(r_{ij}) = \phi_{ij} \text{ and } C_R(r_{ij}, r_{kl}) = 0 \text{ for } (i, j) \neq (k, l), \qquad (13.4.1)$$

Let an unbiased estimator of ϕ_{ij} be $\hat{\phi}_{ij}$.

13.4.1.1 Estimation of Total

An unbiased estimator of the total Y is given by

$$\hat{Y}_r = \sum_{i \in s} b_{si} \hat{Y}_{ri} \qquad (13.4.2)$$

where b_{si} s are constants subjected to $\sum_{s \supset i} b_{si} p(s) = 1\, \forall\, i \in U$,$\hat{Y}_{ri} = \sum_{j \in s_i} b_{sj|i} r_{ij}$ is an unbiased estimator of Y_i and constants $b_{sj|i}$ satisfy $\sum_{s_i \supset j} b_{sj|i} p(s) = 1\, \forall\, j \in U_i$.

Define,

$$\alpha_i = \sum\nolimits_{s \supset i} b_{si}^2 p(s),\ \alpha_{ij} = \sum\nolimits_{s \supset i,j} b_{si} b_{sj} p(s),\ \beta_{j|i} = \sum\nolimits_{s_i \supset j} b_{sj|i}^2 p(s_i),\ \beta_{jk|i} = \sum\nolimits_{s_i \supset j,k} b_{sj|i} b_{sk|i} p(s_i)$$
$$\sigma_i^2 = V_2(\hat{Y}_i) = \sum\nolimits_{j \in U_i} (\beta_{j|i} - 1) y_{ij}^2 + \sum\nolimits_{j \neq} \sum\nolimits_{k \in U_i} (\beta_{jk|i} - 1) y_{ij} y_{ik} \text{ and } \phi_{i\cdot} = \sum\nolimits_{j \in U_i} \phi_{ij} \beta_{ji}\,.$$

Theorem 13.4.1

(i) $\hat{Y}_r$ is an unbiased estimator of Y
(ii) The variance of $\hat{Y}_r$ is

$$\text{Var}(\hat{Y}_r) = \left[\sum_{i=1}^{N} (\alpha_i - 1) Y_i^2 + \sum_{i \neq}^{N} \sum_{j=1}^{N} (\alpha_{ij} - 1) Y_i Y_j\right] + \sum_{i=1}^{N} \alpha_i \left(\sigma_i^2 + \phi_{i\cdot}\right)$$

(iii) An unbiased estimator of $\text{Var}(\hat{Y}_r)$ is

$$\hat{\text{Var}}(\hat{Y}_r) = \hat{V}_1(r) + \hat{V}_2(r)$$

where $\hat{V}_1(r) = \sum\limits_{i \in s} \frac{\alpha_i - 1}{\pi_i} \hat{Y}_{ri}^2 + \sum\limits_{i \neq} \sum\limits_{j \in s} \frac{\alpha_{ij} - 1}{\pi_{ij}} \hat{Y}_{ri} \hat{Y}_{rj}$ and

$$\hat{V}_2(r) = \sum_{i \in s} \frac{1}{\pi_i} \left[\hat{Y}_{ri}^2 - \left\{\sum_{j \in s_i} \frac{1}{\pi_{j|i}} \left(r_{ij}^2 - \hat{\phi}_{ij}\right) + \sum_{j \neq} \sum_{k \in s_i} \frac{1}{\pi_{jk|i}} r_{ij} r_{ik}\right\}\right]$$

Proof Let E = overall expectation, $E_p(V_p)$ = expectation with respect to the sampling design p, $E_1(V_1)$ = unconditional expectation for the selection of fsu s and $E_2(V_2)$ = conditional expectation over ssus for a given fsu. Then,

$$\begin{aligned}
\text{(i)}\quad E(\hat{Y}_r) &= E_p\left[\sum_{i \in s} b_{si} E_R(\hat{Y}_{ri})\right] \\
&= E_p\left[\sum_{i \in s} b_{si} \sum_{j \in s_i} b_{sij} y_{ij}\right] \\
&= E_1\left[\sum_{i \in s} b_{si} E_2\left(\sum_{j \in s_i} b_{sij} y_{ij} | s\right)\right] \\
&= E_1\left[\sum_{i \in s} b_{si} Y_i\right] \\
&= Y
\end{aligned}$$

(ii) $\mathrm{Var}(\hat{Y}_r) = V_p\left[E_R(\hat{Y}_r)\right] + E_p\left[V_R(\hat{Y}_r)\right]$

$$= V_p\left(\sum_{i\in s} b_{si}\hat{Y}_i\right) + E_p\left(\sum_{i\in s} b_{si}^2 \sum_{j\in s_i} b_{sij}^2\phi_{ij}\right)$$

$$= \left[V_1\left\{\sum_{i\in s} b_{si}E_2(\hat{Y}_i|s)\right\} + E_1\left\{\sum_{i\in s} b_{si}^2 V_2(\hat{Y}_i|s)\right\}\right]$$

$$+E_1\left[\sum_{i\in s} b_{si}^2\left\{E_2\left(\sum_{j\in s_i} b_{sij}^2\phi_{ij}|s\right)\right\}\right]$$

$$= \left[V_1\left(\sum_{i\in s} b_{si}Y_i\right) + E_1\left(\sum_{i\in s} b_{si}^2\sigma_i^2\right)\right]$$

$$+E_1\left[\sum_{i\in s} b_{si}^2\left(\sum_{j\in U_i}\phi_{ij}\beta_{ij}\right)\right]$$

$$= \left[\sum_{i\in U}(\alpha_i - 1)Y_i^2 + \sum_{i\neq j}\sum_{\in U}(\alpha_{ij} - 1)Y_iY_j\right]$$

$$+\sum_{i\in U}\alpha_i(\sigma_i^2 + \phi_{i\cdot})$$

(iii) Noting

$$E\left[\hat{V}_1(r)\right] = E_P\left[\sum_{i\in s}\frac{\alpha_i - 1}{\pi_i}(\hat{Y}_i^2 + \sum_{j\in s_i} b_{si}^2\phi_{ij}) + \sum_{i\neq}\sum_{j\in s}\frac{\alpha_{ij} - 1}{\pi_{ij}}\hat{Y}_i\hat{Y}_j\right]$$

$$= E_1\left[\sum_{i\in s}\frac{\alpha_i - 1}{\pi_i}(Y_i^2 + \sigma_i^2 + \phi_{i\cdot}) + \sum_{i\neq}\sum_{j\in s}\frac{\alpha_{ij} - 1}{\pi_{ij}}Y_iY_j\right]$$

$$= \left[\sum_{i\in U}(\alpha_i - 1)(Y_i^2 + \sigma_i^2 + \phi_{i\cdot}) + \sum_{i\neq}\sum_{j\in U}(\alpha_{ij} - 1)Y_iY_j\right]$$

$$= \mathrm{Var}(\hat{Y}_r) - \sum_{i\in U}(\sigma_i^2 + \phi_{i\cdot})$$

and

$$E\left[\hat{V}_2(r)\right] = E_p\sum_{i\in s}\frac{1}{\pi_i}\left[\hat{Y}_i^2 + \sum_{j\in s_i} b_{sij}^2\phi_{ij} - \left\{\sum_{j\in s_i}\frac{1}{\pi_{j|i}}y_{ij}^2 + \sum_{j\neq}\sum_{k\in s_i}\frac{1}{\pi_{jk|i}}y_{ij}y_{ik}\right\}\right]$$

$$= E_1 \sum_{i \in s} \frac{1}{\pi_i} \left[(Y_i^2 + \sigma_i^2) + \phi_{i\cdot} - \left\{ \sum_{j \in U_i} y_{ij}^2 + \sum_{j \neq} \sum_{k \in U_i} y_{ij} y_{ik} \right\} \right]$$

$$= \sum_{i=1}^{N} (\sigma_i^2 + \phi_{i\cdot}),$$

one can verify $E\left[\hat{V}\mathrm{ar}(\hat{Y}_r)\right] = \mathrm{Var}(\hat{Y}_r)$.

13.4.2 Special Cases

13.4.2.1 Arbitrary Sampling Design

For the choice of $b_{si} = 1/\pi_i$ and $b_{sij} = 1/\pi_{j|i}$, one obtains $\alpha_i = 1/\pi_i$, $\alpha_{ij} = \pi_{ij}/(\pi_i \pi_j)$, $\beta_{j|i} = 1/\pi_{j|i}$, $\hat{Y}_{ri} = \sum_{j \in s_i} \frac{r_{ij}}{\pi_{j|i}}$, $\beta_{jk|i} = \frac{\pi_{jk|i}}{\pi_{j|i}\pi_{k|i}}$, $\sigma_i^2 = \sum_{j \in U_i} (\frac{1}{\pi_{j|i}} - 1) y_{ij}^2 + \sum_{j \neq} \sum_{k \in U_i} (\frac{\pi_{jk|i}}{\pi_{j|i}\pi_{k|i}} - 1) y_{ij} y_{ik}$ and $\phi_{i\cdot} = \sum_{j \in U_i} \frac{\phi_{ij}}{\pi_{j|i}}$. The Theorem 13.4.1 reduces to.

Theorem 13.4.2

(i) $\hat{Y}_r = \sum_{i \in s} \frac{1}{\pi_i} \sum_{j \in s_i} \frac{r_{ij}}{\pi_{j|i}}$ is an unbiased estimator for the total Y

(ii) $\mathrm{Var}(\hat{Y}_r) = \left[\sum_{i \in U} \left(\frac{1}{\pi_i} - 1 \right) Y_i^2 + \sum_{i \neq} \sum_{j \in U} \left(\frac{\pi_{ij}}{\pi_i \pi_j} - 1 \right) Y_i Y_j \right] + \sum_{i \in U} \frac{1}{\pi_i} (\sigma_i^2 + \phi_{i\cdot})$

(iii) $\hat{V}\mathrm{ar}(\hat{Y}_r) = \hat{V}_1(r) + \hat{V}_2(r)$

$$\hat{V}_1(r) = \sum_{i \in s} \frac{1 - \pi_i}{\pi_i^2} \hat{Y}_{ri}^2 + \sum_{i \neq} \sum_{j \in s} \frac{\pi_{ij} - \pi_i \pi_j}{\pi_i \pi_j \pi_{ij}} \hat{Y}_{ri} \hat{Y}_{rj};$$

$$\hat{V}_2(r) = \sum_{i \in s} \frac{1}{\pi_i} \left[\hat{Y}_{ri}^2 - \left\{ \sum_{j \in s_i} \frac{1}{\pi_{j|i}} (r_{ij}^2 - \hat{\phi}_{ij}) + \sum_{j \neq} \sum_{k \in s_i} \frac{1}{\pi_{jk|i}} r_{ij} r_{ik} \right\} \right].$$

13.4.2.2 SRSWOR Sampling

For the SRSWOR sampling of both the stages, one finds $\pi_i = \frac{n}{N}$, $\pi_{ij} = \frac{n(n-1)}{N(N-1)}$; $\pi_{j|i} = \frac{m_i}{M_i}$, $\pi_{jk|i} = \frac{m_i(m_i - 1)}{M_i(M_i - 1)}$, $\hat{Y}_{ri} = M_i \overline{r}_i$, $\overline{r}_i = \frac{1}{m_i} \sum_{j \in s_i} r_{ij}$, $\sigma_i^2 = M_i^2 \frac{(1-f_i)}{m_i} S_{yi}^2$, $\overline{Y}_i = \frac{1}{M_i} \sum_{\in U_i} y_{ij}$ and $\phi_{i\cdot} = \frac{M_i}{m_i} \sum_{\in U_i} \phi_{ij}$.

In this case Theorem 13.4.1 becomes.

Theorem 13.4.3

(i) $\hat{Y}_r = \frac{N}{n}\sum_{i\in s} M_i \bar{r}_i$ is an unbiased estimator for the total Y

(ii) $\mathrm{Var}(\hat{Y}_r) = N^2\left[(1-f)\frac{S_Y^2}{n} + \frac{1}{nN}\sum_{i\in U}(\sigma_i^2 + \phi_{i\cdot})\right]$

with $S_Y^2 = \frac{1}{N-1}\sum_{i\in U}(Y_i - \overline{Y})^2$, $\overline{Y} = Y/N$ and $f = n/N$.

(iii) $\hat{\mathrm{Var}}(\hat{Y}_r) = \hat{V}_1(r) + \hat{V}_2(r)$

where
$\hat{V}_1(r) = \frac{N^2(1-f)}{n}\frac{1}{n-1}\sum_{i\in s}(\hat{Y}_{ri} - \overline{Y}_r)^2, \overline{Y}_r = \frac{1}{n}\sum_{i\in s}\hat{Y}_{ri}$ and

$$\hat{V}_2(r) = \frac{N}{n}\sum_{i\in s}\left[\hat{Y}_{ri}^2 - \left\{\frac{M_i}{m_i}\sum_{j\in s_i}(r_{ij}^2 - \hat{\phi}_{ij}) + \frac{M_i(M_i-1)}{m_i(m_i-1)}\sum_{j\neq}\sum_{k\in s_i} r_{ij}r_{ik}\right\}\right].$$

13.4.2.3 PPSWR Sampling

Consider a two-stage sampling where, in the first stage, a sample s of n fsus is selected by the PPSWR (probability proportion to size with replacement) method of sampling with a normed size measure P_i attached to the ith ($i \in U$) unit. Let $U_{(j)}$ be the fsu that selected at the $j(= 1, ..., n)$ th draw with probability $P_{(j)}$. From the selected fsu $U_{(j)}$, a sub-sample $s_{(j)}$ of $m_{(j)}$ ssus is selected by the PPSWR method using a normed size measure $P_{(j)g}$ attached to the g th ssu of $U_{(j)}$. Let $r_{(j)g}$ be the revised RR obtained from the gth ssu of the selected fsu $U_{(j)}$ that satisfies the model

$$E_R(r_{g(j)}) = y_{g(j)},\ V_R(r_{(j)g}) = \phi_{(j)g} \text{ and } C_R(r_{g(j)}, r_{h(j)}) = 0 \text{ for } g \neq h$$

If the jth draw produces kth fsu, then $U_{(j)} = U_k, P_{(j)} = P_k$, $Y_{(j)} = Y_k$ and $M_{(j)} = M_k$. In this case, an unbiased estimator of $Y_{(j)} = \sum_{g=1}^{M_{(j)}} y_{(j)g}$ is

$$\hat{Y}_{r(j)} = \frac{1}{m_{(j)}}\sum_{g=1}^{m_{(j)}}\frac{r_{(j)g}}{P_{(j)g}} \tag{13.4.3}$$

Theorem 13.4.4 For the PPSWR sampling,

(i) $\hat{Y}_r = \frac{1}{n}\sum_{j=1}^{n}\frac{\hat{Y}_{r(j)}}{P_{(j)}}$ is an unbiased estimator of Y

(ii) $\mathrm{Var}(\hat{Y}_r) = \frac{1}{n}\left[\sum_{i=1}^{N} P_i\left(\frac{Y_i}{P_i} - Y\right)^2 + \sum_{i=1}^{N}\frac{\sigma_{Pi}^2 + \phi_{Pi\cdot}}{P_i}\right]$

where $\phi_{Pi\cdot} = \frac{1}{m_i}\sum_{j=1}^{M_i}\frac{\phi_{ij}}{P_{ij}}$ and $\sigma_{Pi}^2 = \frac{1}{m_i}\left(\sum_{j=1}^{M_i}\frac{y_{ij}^2}{p_{ij}} - Y_i^2\right)$.

(iii) $\hat{\mathrm{Var}}(\hat{Y}_r) = \frac{1}{n(n-1)}\sum_{j=1}^{n}\left(\frac{\hat{Y}_{r(j)}}{P_{(j)}} - \hat{Y}_r\right)^2$

Proof

(i) $$E(\hat{Y}_r) = E_p\left[\frac{1}{n}\sum\nolimits_{j=1}^{n}\frac{E_R(\hat{Y}_{r(j)})}{P_{(j)}}\right]$$
$$=E_1\left[\frac{1}{n}\sum\nolimits_{j=1}^{n}\frac{Y_{(j)}}{P_{(j)}}\right]$$
$$=Y$$

(ii) $$\mathrm{Var}(\hat{Y}_r) = V_P\left(\frac{1}{n}\sum_{j=1}^{n}\frac{E_R(\hat{Y}_{r(j)})}{P_{(j)}}\right) + E_P\left(\frac{1}{n^2}\sum_{j=1}^{n}\frac{V_R(\hat{Y}_{r(j)})}{P^2_{(j)}}\right)$$
$$=V_p\left(\frac{1}{n}\sum_{j=1}^{n}\frac{\hat{Y}_{(j)}}{P_{(j)}}\right)$$
$$+E_p\left(\frac{1}{n^2}\sum_{j=1}^{n}\frac{1}{P^2_{(j)}}\sum_{j\in s_i}\frac{1}{m^2_{(j)}}\sum_{g=1}^{m_{(j)}}\frac{\phi_{(j)g}}{P^2_{(j)g}}\right)$$
$$=\left[V_1\left(\frac{1}{n}\sum_{j=1}^{n}\frac{Y_{(j)}}{P_{(j)}}\right) + E_1\left(\frac{1}{n^2}\sum_{j=1}^{n}\frac{1}{P^2_{(j)}}\sigma^2_{P(j)}\right)\right]$$
$$+E_1\left(\frac{1}{n^2}\sum_{j=1}^{n}\frac{1}{P^2_{(j)}}\frac{1}{m_{(j)}}\sum_{g=1}^{M_{(j)}}\frac{\phi_{(j)g}}{P_{(j)g}}\right)$$

where $$\sigma^2_{P(j)} = \frac{1}{m_{(j)}}\sum_{j=1}^{M_{(j)}}\frac{y^2_{(j)g}}{P_{(j)g}} - Y^2_{(j)}$$

$$=V_p\left(\frac{1}{n}\sum_{j=1}^{n}\frac{\hat{Y}_{(j)}}{P_{(j)}}\right)$$
$$+E_p\left(\frac{1}{n^2}\sum_{j=1}^{n}\frac{1}{P^2_{(j)}}\sum_{j\in s_i}\frac{1}{m^2_{(j)}}\sum_{g=1}^{m_{(j)}}\frac{\phi_{(j)g}}{P^2_{(j)g}}\right)$$
$$=\left[V_1\left(\frac{1}{n}\sum_{j=1}^{n}\frac{Y_{(j)}}{P_{(j)}}\right) + E_1\left(\frac{1}{n^2}\sum_{j=1}^{n}\frac{1}{P^2_{(j)}}\sigma^2_{P(j)}\right)\right]$$
$$+E_1\left(\frac{1}{n^2}\sum_{j=1}^{n}\frac{1}{P^2_{(j)}}\frac{1}{m_{(j)}}\sum_{g=1}^{M_{(j)}}\frac{\phi_{(j)g}}{P_{(j)g}}\right)$$

where $$\sigma^2_{P(j)} = \frac{1}{m_{(j)}}\sum_{j=1}^{M_{(j)}}\frac{y^2_{(j)g}}{P_{(j)g}} - Y^2_{(j)}$$

(iii) Noting that $\frac{\hat{Y}_{r(j)}}{P_{(j)}}$ are independently identically distributed random variable, one finds

13.4.2.4 SRSWR Sampling

Let a sample s of n fsus be selected from a population U by the SRSWR method. If the $j(=1, .., n)$th draw selects fsu $U_{(j)}$, then a sub-sample $s_{(j)}$ of $m_{(j)}$ ssus are selected from it by the SRSWR method again. Let $r_{g(j)}$ be the RR obtained from g th ssu of the subsample $s_{(j)}$. Then an unbiased estimator of the population total is

$$\hat{Y}_r = \frac{N}{n}\sum_{j=1}^{n}\hat{Y}_{r(j)} \tag{13.4.4}$$

where $\hat{Y}_{r(j)} = \frac{M_{(j)}}{m_{(j)}}\sum_{g=1}^{m_{(j)}} r_{g(j)}$.

Substituting $P_i = 1/N$ and $P_{ij} = 1/M_i$ in Theorem 13.4.3, we have.

Theorem 13.4.5 For the SRSWR sampling,

(i) $\hat{Y}_r = \frac{N}{n}\sum_{j=1}^{n}\hat{Y}_{r(j)}$ is an unbiased estimator of Y

(ii) $\text{Var}(\hat{Y}_r) = \frac{N^2}{n}\left[\sigma_Y^2 + \frac{1}{N}\sum_{i=1}^{N}(\sigma_{wi}^2 + \phi_{wi\cdot})\right]$

where $\sigma_Y^2 = \frac{1}{N}\sum_{i=1}^{N}(Y_i - \overline{Y})^2$, $\sigma_{wi}^2 = \frac{M_i^2}{m_i}\sigma_{yi}^2$, $\phi_{wi\cdot} = \frac{M}{m_i}\sum_{j=1}^{M_i}\phi_{ij}$ and $\sigma_{yi}^2 = \frac{1}{M_i}\sum_{j=1}^{M_i}(y_{ij} - \overline{Y}_i)^2$.

(iii) $\hat{\text{Var}}(\hat{Y}_r) = \frac{1}{n(n-1)}\sum_{j=1}^{n}(\hat{Y}_{r(j)} - \hat{Y}_r)^2$.

13.5 Poisson Distribution in RR Surveys

Land et al. (2012) proposed a method of estimating λ_A, the mean of the number of persons in a family possessing a rare sensitive attribute A such as AIDS, having witnessed a murder, girls being raped by their own fathers, etc. in a certain population U. In order to estimate λ_A, Land et al. considered situations where λ_B, the average number of people in the same population U possess another rare, unrelated, but non-sensitive attribute, B such as the number of twins, doctors in a family, and so on. In order to estimate λ_A, a large sample of size n was selected from the population U by the simple random sampling with replacement (SRSWR) method. Land et al. (2012) used the randomized response (RR) unrelated question method for the collection of data, which was proposed by Greenberg et al. (1969). In this method, each of the selected respondents was requested to answer truthfully "yes" or "no" to one of the following two questions:

Q_A: Do you possess the attribute A?
Q_B: Do you possess the attribute B?

Selected respondents should answer the question Q_A if he or she was born in the first quarter of a year, i.e., between 01 January and 30 April; otherwise, answer Q_B. It is assumed that P, the proportion of people born between January and April is known. Let,

$$y_i = \begin{cases} 0, & \text{if the ith respondent answers "no"} \\ 1, & \text{if the ith respondent answers "yes"} \end{cases} \tag{13.5.1}$$

Then the probability of obtaining a "yes" answer from a respondent is

$$\text{Prob}(y_i = 1) = \theta_0 = P\pi_A + (1 - P)\pi_B \tag{13.5.2}$$

Land et al. (2012) assumed that the sample size n is so large, and attributes A and B are rare enough to satisfy

$$n\pi_A \to \lambda_A \text{ and } n\pi_B \to \lambda_B \text{ as } n \to \infty \tag{13.5.3}$$

Treating $y_1, y_2, .., y_n$ a random sample from a Poisson distribution with parameter $\lambda = \lambda_A + \lambda_B$, Land et al. (2012) proposed the Likelihood function of λ as

$$L(\lambda|y_1, y_2, .., y_n) = \frac{e^{-\lambda}\lambda^{\sum_{i=1}^{n} y_i}}{\prod_{i=1}^{n} y_i} \tag{13.5.4}$$

Finally, from Eq. (13.5.4), Land et al. (2012) derived the following results:

Theorem 13.5.1

(i) $\hat{\lambda}_A = \frac{1}{P}\left[\overline{y} - (1 - P)\lambda_B\right]$ is an unbiased estimator of λ_A
(ii) $\text{Var}(\hat{\lambda}_A) = \frac{\lambda_A}{nP} + \frac{(1-P)\lambda_B}{nP^2}$
(iii) $\hat{\text{Var}}(\hat{\lambda}_A) = \frac{\overline{y}}{nP^2}$ is an unbiased estimator of $\text{Var}(\hat{\lambda}_A)$.

Remark 13.5.1 Here, the random variable $y_i (i = 1, .., n)$ can take only two values 0 and 1, with respective probabilities P and $1 - P$. Hence, the random variable y_i clearly follows the Bernoulli distribution with a parameter P. So, treating y_i s as Poisson variables is clearly incorrect. Hence, the likelihood function presented by Land et al. (2012) is incorrect. Subsequently, the results derived by Land et al. (2012), and Chaudhuri and Christofied (2013) are also incorrect.

13.6 Optimality Results for RR Surveys Under the Superpopulation Model

As usual, we assume that the study variable y is sensitive and that no direct information from the respondent is available. Hence, a randomized response z is obtained from each of the selected respondents in the sample s independently by using a suitable randomization device. Let z_i be the RR obtained from the $i(\in s)$ th respondent and r_i be the revised RR, an estimate of y_i obtained as a function of z_i. We assume that r_i follows the following general model proposed by Arnab (1990):

$$E_R(r_i) = y_i, V_R(r_i) = \phi_i(y_i) = \phi_i \text{ and } C_R(r_i, r_j) = 0 \text{ for } i \neq j \tag{13.6.1}$$

Under the superpopulation model ξ, the parameter vector $\mathbf{y} = (y_1, ..., y_i, ..., y_N)$ is a random variable whose mean, variance, and covariance operators will be denoted by E_ξ, V_ξ and C_ξ respectively. The overall expectation, variance, and covariance with respect to the RR model (13.6.1) and the superpopulation model ξ will be represented by

$$\begin{aligned} E_{\xi R}(r_i) &= E_\xi[E_R(r_i)] = E_\xi(y_i) \\ &= \mu_i, \\ V_{\xi R}(r_i) &= V_\xi[E_R(r_i)] + E_\xi[V_R(r_i)] \\ &= V_\xi(y_i) + E_\xi(\phi_i) \\ &= \sigma_i^2 + \phi_{\xi i} \text{ where } \phi_{\xi i} = E_\xi(\phi_i) \end{aligned} \tag{13.6.2}$$

and

$$\begin{aligned} C_{\xi R}(r_i, r_j) &= C_\xi[E_R(r_i), E_R(r_j)] + E_\xi[C_R(r_i, r_j)] \\ &= C_\xi(y_i, y_j) \\ &= \sigma_{ij} \end{aligned}$$

Let $e(s, y)$ be an unbiased estimator of the population total $Y = \sum_{i \in U} y_i$ for the DR. Then, replacing y_i by its estimate r_i in $e(s, y)$, an unbiased estimator of Y for the RR survey is obtained as

$$e_r = e(s, r) = e(s, y)|_{y_i = r_i} \tag{13.6.3}$$

The estimator e_r satisfies $E_p\{E_R(e_r)\} = Y$ and the class of all unbiased estimators for the RR surveys will be denoted by C_{rp}.

The variance of e_r is

$$\text{Var}(e_r) = E_\xi E_p E_R(e_r - Y)^2$$

Assuming the operators E_ξ, E_p and E_R are commutative, we obtain

$$\begin{aligned}\mathrm{Var}[e_r] &= E_\xi E_R\left[V_p(e_r - Y) + \{E_p(e_r - Y)\}^2\right] \\ &= E_\xi E_R\left[V_p(e_r) + \left(\sum_{i\in U} r_i - Y\right)^2\right] \\ &= E_{\xi R}\left[V_p(e_r)\right] + \sum_{i\in U} E_\xi(\phi_i) \qquad (13.6.4)\end{aligned}$$

Using the results from the DR surveys given in Sect. 4.6, we now derive the following optimality results for the RR surveys.

13.6.1 Product Measure Model

For the RR surveys, the product measure model M1 given in Sect. 4.6.1. reduces to

$$\text{Model MR1}: E_{\xi R}(r_i) = \mu_i,\ V_{\xi R}(r_i) = \sigma_i^2 + \phi_{\xi i}, \text{ and } C_{\xi R}(r_i, r_j) = 0 \text{ for } i \neq j \qquad (13.6.5)$$

Using Eq. (4.6.3), one obtains

$$E_\xi E_p E_R[e_r - Y]^2 \geq \sum_{i\in U} \frac{\sigma_i^2 + \phi_{\xi i}}{\pi_i} - \sum_{i\in U} \sigma_i^2\, \forall e_r \in C_{rp} \qquad (13.6.6)$$

Equations (13.6.5), (13.6.6), and Theorem 4.6.1 lead to.

Theorem 13.6.1 For the model MR1, and a sampling design p with inclusion probability $\pi_i > 0\, \forall\, i \in U$.

$$E_\xi E_p E_R(e_r - Y)^2 \geq \sum_{i\in U} \frac{\sigma_i^2 + \phi_{\xi i}}{\pi_i} - \sum_{i\in U} \sigma_i^2 = E_\xi E_p(e_{r0} - Y)^2\, \forall\, e_r \in C_{rp} \qquad (13.6.7)$$

where

$$e_{r0} = \sum_{i\in s} \frac{r_i - \mu_i}{\pi_i} + \sum_{i\in U} \mu_i \qquad (13.6.8)$$

The estimator e_{r0} may be termed as the generalized regression predictor for the RR model analogous to the generalized regression predictor for the DR survey. The estimator e_{r0} is applicable if μ_i s are known.

13.6.1.1 Special Cases

Minimizing the RHS of (4.2.7) with respect to π_i subject to $\sum_{i\in U}\pi_i = n$, the optimum values of π_i's are obtained as $\pi_{\xi\sigma_i} = n\sqrt{\phi_{\xi i}+\sigma_i^2}/\sum_{i\in U}\sqrt{\phi_{\xi i}+\sigma_i^2}$. We define a sampling design p for which (i) $\pi_i = \pi_{\xi\sigma_i}$ and (ii) $\sum_{i\in s}\frac{\mu_i}{\pi_{\xi\sigma_i}} = \sum_{i\in U}\mu_i\ \forall s$ with $p(s) > 0$ as $p_{\xi n\sigma\mu}$. In this situation, we obtain the following results analogous to Theorem 4.6.2 as follows.

Theorem 13.6.2 Under model MR1, the strategy $h_{r1} = (p_{\xi n\sigma\mu}, e_{r\mu\sigma HT})$ is optimal in the class of strategies $H_1 = (p, e_r)$; $p \in P_n$ and $e_r \in C_{rp}$ i.e.,

$$E_\xi E_p E_R(e_r - Y)^2 \ge \frac{1}{n}\left(\sum_{i\in U}\sqrt{\phi_{\xi i}+\sigma_i^2}\right)^2 - \sum_{i\in U}\sigma_i^2 = E_\xi E_{p_{\xi n\sigma\mu}}(e_{r\mu\sigma HT} - Y)^2\ \forall p \in P_n,\ e_r \in C_{rp}$$

where $e_{r\mu\sigma HT} = \sum_{i\in s}\frac{r_i}{\pi_{\xi\sigma_i}}$.

Under the RR surveys, the model M11 defined in Sect. 4.6.1 reduces to

$$\text{MR11}: E_{\xi R}(r_i) = \mu_i = \beta x_i,\ V_{\xi R}(r_i) = \sigma_i^2 + \phi_{\xi i}, \text{ and } C_{\xi R}(r_i, r_j) = 0 \ \text{ for } i \ne j \tag{13.6.9}$$

where x_i is the value of the auxiliary variable x for the ith unit.

Let p_{nx} be the class of IPPS sampling design based on fixed effective sample size n with inclusion probability $\pi_i = nx_i/X$ for the ith$(i \in U)$ unit.

Theorem 13.6.3 Under the model MR11, the strategy $h_2 = (p_{nx}, e_{rHTx})$ is the optimal in the class of strategies $H_{R1} = (p, e_r)$; $p \in P_n$ and $e_r \in C_{rp}$ i.e.,

$$E_\xi E_{p_{nx}} E_R(e_r - Y)^2 \ge \frac{X}{n}\sum_{i\in U}\frac{\sigma_i^2+\phi_{\xi i}}{x_i} - \sum_{i\in U}\sigma_i^2 = E_\xi E_{p_{nx}}(e_{rHTx} - Y)^2\ \forall e_r \in C_{rp}$$

where $e_{rHTx} = \frac{X}{n}\sum_{i\in s}\frac{r_i}{x_i}$.

Consider the RR models with $\phi_{\xi i} = \lambda x_i^2$ where λ is a positive constant. In this case, Theorem 13.6.3 reduces to.

Theorem 13.6.4 Under the model MR11 with $\phi_{\xi i} = \lambda x_i^2$.

$$E_\xi E_{p_{nx}} E_R(e_r - Y)^2 \ge \frac{(\sigma^2+\lambda)X^2}{n} - \sigma^2\sum_{i\in U}x_i^2 = E_\xi E_{p_{nx}}(e_{rHTx} - Y)^2\ \forall e_r \in C_{rp}$$

Finally, consider the model MR11 with $\phi_{\xi i} = \phi_0 =$ constant. In this case, p_{nx} reduces to p_0 where $\pi_i = \pi_0 = n/N$. In this case, the above Theorem 13.6.4 reduces to.

Theorem 13.6.5 Under the model MR11 with $\phi_{\xi i} = \phi_0$, one finds.

$$E_\xi E_{p_0} E_R(e_r - Y)^2 \geq N\left[\frac{N}{n}(\sigma^2 + \phi_0) - \sigma^2\right] = E_\xi E_{p_0}(\overline{r}_s - Y)^2 \,\forall e_r \in C_{rp}$$

where $\overline{r}_s = \frac{1}{n}\sum_{i\in s} r_i$.

13.6.2 Equi-Correlated Model

Consider the equi-correlation model defined in (4.6.7):

$$\text{Model M2}: E_\xi(y_i) = \beta x_i,\ V_\xi(y_i) = \sigma^2 x_i^2 \text{and } C_\xi(y_i, y_j) = \rho\sigma^2 x_i x_j \qquad (13.6.10)$$

For the RR survey, the model M2 reduces to

$$\text{MR2}: E_{\xi R}(r_i) = \beta x_i,\ V_{\xi R}(r_i) = V_\xi(y_i) + E_\xi(\phi_i) = \sigma^2 x_i^2 + \phi_{\xi i}$$

$$\text{and } C_{\xi R}(r_i, r_j) = C_\xi\left[E_R(r_i), E_R(r_j)\right] + E_\xi\left[C_\xi(r_i, r_j)\right] = \rho\sigma^2 x_i x_j$$

The class of linear p-unbiased estimator C_{lu} of the total Y for the DR survey consists of the estimators of the form

$$e_l(s, y) = a_{ls} + \sum_{i\in s} b_{lsi} y_i \qquad (13.6.11)$$

where the constants a_{ls} and b_{lsi} s satisfy the conditions $\sum_s a_{ls} p(s) = 0$ and $\sum_{s\supset i} b_{lsi} p(s) = 1\,\forall\, i \in U$. Replacing y_i by r_i in (4.2.10), the linear unbiased estimator C_{lu} of Y for the RR survey is obtained as

$$e_l(s, r) = e_l(s, y)|_{y_i = r_i} = a_{ls} + \sum_{i\in s} b_{lsi} r_i \qquad (13.6.12)$$

In this situation,

$$\begin{aligned} E_\xi E_p E_R[e_l(s, r) - Y]^2 &= E_\xi\left[V_p\{E_R(e_l(s, r))\} + E_p\{V_R(e_l(s, r))\}\right] \\ &= E_\xi\left[V_p\{e_l(s, y)\} + E_p\left(\sum_{i\in s} b_{si}^2 \phi_i\right)\right] \end{aligned}$$

$$= E_\xi\left[V_p\{e_l(s, y)\}\right] + \sum_{i \in U} \phi_{\xi i} \sum_{s \supset i} b_{si}^2 p(s)$$

Now noting that $\sum_{s \supset i} b_{si}^2 p(s) \geq 1/\pi_i$, one finds.

Theorem 13.6.6 Under model MR2 with $\phi_{\xi i} = \lambda x_i^2$, $\hat{Y}_{rxHT} = \frac{1}{n}\sum_{i \in s} \frac{r_i}{p_i}$ based on a sampling design $p_{\pi x}$ is the optimal in the sense

$$E_\xi E_p E_R[e_l(s, r) - Y]^2 \geq (1 - \rho)\sigma^2 \left[\frac{X^2}{n} - \sum_{i \in U} x_i^2\right]$$

$$+ \lambda \frac{X^2}{n} = E_\xi E_p E_R\left[\hat{Y}_{rxHT} - Y\right]^2 \forall p \in P_n$$

Special Case

For the model MR2 with $x_i = 1 \forall i \in U$ the estimator $\hat{Y}_{rxHT} = \frac{N}{n}\sum_{i \in s} r_i = N\bar{r}_s$ becomes optimal for the sampling designs for which $\pi_i = n/N$.

13.6.3 Construction of Optimal RR Techniques

Arnab (1998, 2017) defined the optimal RR technique as one that provides the minimum variance under a given superpopulation model. He proposed the method of construction of the optimal RR techniques for Eriksson's (1973a, 1973b) and Chaudhuri's (1987) RR models. The techniques are given as follows.

13.6.3.1 Modifications of Eriksson's (1973a, 1973b) Technique

For the modified Eriksson's (1973a, 1973b) model described in Sect. 3.3, the $i(\in U)$ th respondent should report the true response y_i with probability q_i and a scrambled response $x_i Q_j$ with probability q_j where Q_j is a random sample selected from a set of numbers $\boldsymbol{Q} = (Q_1, ..., Q_k)$ with probability q_j and x_i is the value of the auxiliary variable x associated with the i th unit.

$$\text{i.e. } z_i = \begin{cases} y_i, \text{ with probability } q_i \\ Q_j x_i, \text{ with probability } q_j \end{cases} ; i = 1, .., k;\ j = \sum_{j=1}^{k} q_j = 1 - q_i \qquad (13.6.13)$$

The model (4.2.13) yields

$$E_R(z_i) = q_i y_i + x_i \sum_{j=1}^{k} q_j Q_j \text{ and } E_R(z_i^2) = q_i y_i^2 + x_i^2 \sum_{j=}^{k} q_j Q_j$$

Writing,

$$r_i = \frac{z_i - x_i \sum_{j=1}^{k} q_j Q_j}{q_i}$$

one finds,

$$E_R(r_i) = y_i \text{and } V_R(r_i) = \phi_i = \frac{1}{q_i^2}\left[q_i y_i^2 + x_i^2 \sum_{j=1}^{k} q_j Q_j - \left\{q_i y_i + x_i \sum_{j=1}^{k} q_j Q_j\right\}^2\right]$$

Here we can easily verify that $\phi_{\xi i} = E_\xi(\phi_i) \propto x_i^2$.

Similarly, one can modify Chaudhuri's (1987) technique so that $\phi_{\xi i}$ becomes proportional to x_i^2. For details, readers are referred to Arnab (1998, 2017).

13.6.4 Random Permutation (RP) Model

For the RP model, the probability of realizing the population vector

$$\boldsymbol{y} = (y_1 = h_{j_1}, \ldots, y_i = h_{j_i}, \ldots, y_N = h_{j_N}) \tag{13.6.14}$$

is $1/N!$ for every permutation of $(j_1, .., j_i, ..., j_N) \in (1, .., i, .., N)$ where $h_1, .., h_i, .., h_N$ are fixed numbers. For this RP model $E_\xi(y_i) = \overline{Y} = \sum_{i \in U} h_i/N = \overline{h}$, $V_\xi(y_i) = \frac{1}{N}\sum_{i \in U}(h_i - \overline{h})^2 = \sigma_h^2$ and $C_\xi(y_i, y_j) = -\sigma_h^2/(N-1)$ for $i \neq j$. Hence, under the RP model, $\overline{r}_s$ based on the SRSWOR is the optimal strategy for estimating the population mean $\overline{Y}$ for those RR models for which $E_\xi(\phi_i)$ is constant. Details are given by Arnab (2017).

13.7 Case Studies

Various indirect interviewing (IQ) techniques, e.g., randomized response techniques (RRT), nonrandomized response techniques (NRR), item count techniques (ICT), and item sum techniques (IST), have been proposed to guarantee anonymity and minimize the respondent's privacy. Some of their applications are presented here.

13.7.1 Incidents of Child Abuse

Srivastava et al. (2015) provided case studies relating to the extent of sexual abuse of children aged 11–17 years in Uttar Pradesh, India, in 2011. The data were obtained from the project "*State-level study on status of assessment of child sexual abuse in Uttar Pradesh*". The children belonging to the age group 11–17 were classified into four categories: (i) Children in a family environment; (ii) School-going children; (iii) Working children; and (iv) Children in institutional care. Random samples selected from the four groups were 162, 167, 512, and 630, respectively. From each of the selected samples, information about "severe sexual abuse" (experienced in the past sexual assault, or was made to fondle or exhibit private parts, or he/she exhibited his/her private parts, was photographed in the nude), "mild sexual abuse" (kissed forcibly, or somebody tried to kiss/molest one anytime, or was forced to view pornographic video/photographs), and "no sexual abuse" were collected.

The direct questioning method (DQ or DR) of data collection was used for the children belonging to the family environment and working children, while Eriksson's model B (1973) (see Sect. 1.5) was used for the school-going children and children in institutional care groups. The selected children were divided into two groups: Group 1 and Group 2. Children belonging to Group 1 (Group 2) were asked to pick one ball at random from an urn containing 4 red, 3 green, and 3 yellow (3 red, 4 green, and 3 yellow). Children were asked to answer "yes" ("no") if the statement associated with the color matches (do not match) with his or her own status.

Colour of the ball	Statements
Red	*I was a victim of* ***severe sexual assault***
Green	*I was a victim of* ***minor sexual assault***
Yellow	*I never experienced* ***any sextual assault***

The survey revealed the following summary table relating to the overall prevalences (in percentage) of sexual abuse experienced by the different groups of children. For details, readers are referred to Srivastava et al. (2015), Table 13.1.

In the report "Study on Child Abuse India 2007" by the Ministry of Women and Child Development, the Government of India relating to children of the age group 5–18 years revealed that the incidents of overall sexual abuse for the working and

Table 13.1 Child sexual abuse

Child sexual abuse in Uttar Pradesh, India				
Type of sexual abuse (%)	Family environment	Working	School going*	In institutional care *
Severe sexual abuse	24.1	16.6	27.6	18.4
Mild sexual abuse	53.7	55.1	80.2	88.7

*Estimates based on RR method

institutional care were 49.92 and 47.08. So, we can roughly say that the RR technique performed well and provided higher estimates in some situations.

13.7.2 Illicit Drug Use

Mielecka-Kubień and Wójcik (2022) applied forced RR (F-model, Bourch, 1971, see Sect. 1.5) and Liu and Chow's model (L-C model, 1976, see Sect. 1.5) for estimating the prevalence of illicit drug use among high school students in the Silesian *voivodship* (Poland). For the study, 200 high school students, aged 17–18, from Dąbrowa Górnicza, Poland, were selected in two interviews 10 days apart. In the first interview, they were served an anonymous questionnaire (AQ) to obtain a direct response (DR) from the students. In the second interview, the F-model and L-C model were used.

13.7.2.1 F-Model

For the forced randomized response (F-model), respondents were asked to draw a ball at random from an urn containing 15 blue, 6 red, and 9 green balls. If the selected ball was blue, then the respondent had to answer "yes" or "no" to a sensitive question relating to the use of certain drugs. If the drawn ball was red, then the respondent had to answer "no," and if the drawn ball was green, then the respondent had to answer "yes." Since there were four sensitive questions, the respondent had to perform forced RR independently for each of the questions. The four questions set by Mielecka-Kubień and Wójcik (2022) are as follows:

Q1. Have you used marijuana or hashish (cannabis) during the last 12 months?
Q2. Have you used new psychoactive substances during the last 12 months?
Q3. In your lifetime, have you used amphetamines, methamphetamines, tranquillisers, or sedatives, ecstasy, cocaine, crack, LSD, "magic mushrooms"?
Q4. In your lifetime, have you used opioids such as heroin or morphine?

From the data, the estimated prevalences of the four types of substance use based on the F-model and AQ methods are as follows in Table 13.2.

From the above table, we note that the estimates of prevalences for all types of drugs based on the F-models are always higher than those of the AQ method. Hence, the F-model produces more reliable estimates of the sensitive attribute. But the F-model estimates are less precise than the AQ estimates.

13.7.2.2 Liu-Chow's Method

The Liu-Chow RR (L-CRR) method was applied to obtain the Illicit drug use, to obtain the following questions:

Table 13.2 Estimated prevalences of drug users

Drugs	Prevalence (SE)	
	AQ method	F-model
Use of marijuana or hashish (cannabis)	0.220 (0.029)	0.357(0.071)
Use of new psychoactive substances	0.015 (0.009)	0.160 (0.069
Use of amphetamines, methamphetamines, and the like	0.025 (0.011)	0.150 (0.068)
Use of heroin or morphine	0.015 (0.009)	0.047 (0.066)

SE = standard error

Q5. On how many experiences (if any) during the last 12 months have you used marijuana or hashish (cannabis)?

Q6. On how many experiences (if any) during the last 12 months have you used new psychoactive substances?

For getting answers to the questions, respondents were classified into eight disjoint categories depending on the total number of experiences of use of drugs as follows Table 13.3.

For the L-CRR method, an urn with 40 blue balls and 10 white balls bearing numbers 1 to 7 was taken. The numbers on the white ball represent possible categories of respondents. Respondents were asked to draw a ball at random from the urn and disclose his or her true category if the drawn ball was blue. If the drawn ball was white, the respondents had to report the number written on the ball. Respondents performed L-CRR method two times independently for each question (Table 13.4).

Prevalences for the non-sensitive category "not at all" for both the drugs cannabis and psychoactive substances for the AQ method are higher than the L-CRR method.

Table 13.3 Total number of experiences in seven categories

Total number of experiences	Not at all	1–2	3–5	6–9	10–19	20–39	40 or more
Associate number	1	2	3	4	5	6	7

Table 13.4 Prevalence of substance uses

Total number of experiences	Not at all	1–2	3–5	6–9	10–19	20–39	40 or more
Cannabis use							
AQ method	0.078	0.125	0.040	0.010	0.025	0.005	0.015
L-CRR method	0.630	0.175	0.077	0.036	0.036	0.017	0.030
Psychoactive substances							
AQ method	0.985	0.010	0.000	0.005	0.000	0.000	0.000
L-CRR method	0.906	0.004	0.028	0.023	0.023	−0.013	0.029

But, for all other sensitive categories, the L-CRR method provides much higher prevalences than the AQ method. Thus, the L-CRR method performs better than the anonymous questionnaire method. It should be noted that the estimated prevalence of psychoactive substances in the age group 20–39 is negative.

13.7.3 Induced Abortions

Lara et al. (2004) applied the unrelated question randomized response technique (U-model, Sect. 1.3) to estimate the proportion of induced abortion attempts among women in Mexico. The method was compared with three alternative methods, e.g., face-to-face interview (FTF), audio computer-assisted self-interview (ACASI), and self-administered questionnaire (SAQ). Three independent samples of women aged 15–55 years of sizes 1480, 612, and 1000 were selected from hospital patients in Mexico, in Chiapas, a rural area of Mexico, and a house-to-house study in Mexico in 2001. Convenient sampling was used for the selection of samples from hospital patients and Chiapas, while multistage sampling was used for the house-to-house survey. 370 women from each of the four methods of interview (FTF, ACASI, SAQ, and RRT) were interviewed. For the Chiapas region, the sizes were 313, 105, 91, and 103, while for the house-to-house method, sample sizes were 250 each.

Lara et al. (2004) applied the U-model with the known probability method to get an answer to the sensitive question "Did you ever interrupt a pregnancy?". For this U-model, the interviewer showed two pieces of paper to the respondents, one of which bore the question "Did you ever interrupt a pregnancy?" and the other, "Were you born in April?". Then he folded the papers and placed them in an opaque bag along with other similar papers. Then the interviewer shook the bag and asked the respondent (woman) to pick one piece of paper and answer "yes" or "no."

U-model performed the best compared to other methods, as it provided the highest estimates of abortion attempts. The attempted rates of abortion in the hospital, Chiapas, and household samples were highest for the U-model (22, 36, and 18%), followed by the SAQ (19, 10, and 11%). The ACASI and FTF interviews yielded much lower abortion attempts (13, 10, 9%), and (12, 11, 7%), respectively.

13.7.4 Use of Cannabis

Cobo et al. (2016) conducted a survey relating to the extent of the consumption of cannabis cigarettes and the number of days consumed in a year among Spanish university students using direct questioning (DQ) and randomized response technique (RRT) methods. The target population for this survey was students at the University of Granada and the University of Murcia. A stratified random sample of 1146 students was selected, taking the university as a stratum. The sample was divided into two half-samples at random. Participants in sub-sample 1 were asked to provide sensitive questions by using RRT, while participants in sub-sample 2 were asked to provide DQ methods.

The same questionnaire was served to both sub-samples. The questionnaire includes academic questions, basic demographic questions, and two sensitive questions. The sensitive questions were as follows:

Q1: How many cannabis cigarettes did you consume last year?
Q2: Over the past 90 days, how many days did you consume cannabis?

Students belonging to sub-sample 1 were asked to provide RR using Bar-Lev et al.'s method (2004, see Sect. 3.7, BRR), and the students of sub-sample 2 were asked to provide answers using the DQ method.

Each student in sub-sample 1 was uploaded with the app "Baraja Española" to his or her mobile phone. "Baraja Española" is a pack of 40 cards divided into four suits bearing numbers 1 to 7 and three figures. Students were asked to touch the screen of the phone and see the card on the screen. If it is a face card, the students should provide the true answers to Q1 and Q2; otherwise, he or she should multiply the number given on the screen with the true value to them as their randomized response (RR).

The authors analyzed the data using a software package they developed. Their main findings were as follows:

(i) The DQ method reveals that the estimated mean number of cannabis cigarettes consumed in the previous year was 3, while for the BRR, the number was 7.
(ii) The estimated consumption of cannabis was one of the previous 90 days under the DQ method, while it was seven under the BRR. Hence, we conclude that the BRR outperforms the DQ method, as the former method always provides higher prevalence estimates.

13.7.5 Personal Hygiene Behavior During the COVID-19 Pandemic

During the COVID-19 pandemic, billions of people around the world were advised by the WHO to change their behaviors to slow down the spread of the virus. Various protective measures were implemented. One of them was regular and thorough hand washing. Mieth et al. (2021) conducted a survey to estimate the proportion of people who complied with the regulation in Germany.

A sample of 1434 individuals aged between 18 and 88 years was collected online through social media such as Facebook and WhatsApp. Participants were divided into three groups: the direct response (DR) group (G1) and the indirect response group (G2), consisting of 491 and 943 individuals, respectively. The indirect response group G2 was further divided into two groups. Crosswise Model (CWM, see Sect. 10.6.2) Group and Extended Crosswise Model (ECWM) Group.

The DR question group is to provide an answer of "yes" or "no" to the sensitive question:

Q1: Do you wash your hands regularly and sufficiently long (at least 20 seconds) with soap and water?

Participants in the CWM group received the sensitive question Q1 along with a non-sensitive question Q2 as follows:

Q1: Do you wash your hands regularly and sufficiently long (at least 20 seconds) with soap and water?
Q2: Is your mother's birthday in May, June, or July?

Participants in the ECWM group received the sensitive question Q1 along with a different non-sensitive question Q3 as follows:

Q1: Do you wash your hands regularly and sufficiently long (at least 20 seconds) with soap and water?
Q3: Is your mother's birthday in August, September, October, November, December, January, February, March, or April?

Respondents of CWM should report a "yes" answer as their RR if the answers of (Q1, Q2) were (yes, yes) or (no, no). The respondents should provide a "no" answer otherwise, i.e., if the answers to (Q1, Q2) were (yes, no) or (no, yes). Similarly, respondents in ECWM should report a "yes" answer as their RR if the answers of (Q1, Q3) were (yes, yes) or (no, no) and "no" if the answers of (Q1, Q32) were (yes, no) or (no, yes).

Estimates of the proportion of people who followed hygiene π through hand washing based on CWM and ECWM were very close. A combined estimate of the non-compliance $1 - \pi$ based on CWM and ECWM was 74.1% (SE 3.0). The DR estimate of π was obtained at 94.5% (SE = 1.0). Hence, the prevalence of non-compliance and sensitive information based on the crosswise model is much higher than the DR method. Hence, the crosswise model outperformed the DR method for the estimation of non-compliance.

13.7.6 Racism in Universities

A survey was conducted to investigate the extent of racism among students at the University of Calabria, Italy, in 2016. A snowball sample of 624 Italian and foreign students was collected by using personal contacts and social networks. All the selected students were invited to fill out an online questionnaire. The questionnaire started with a few non-sensitive questions related to the everyday lives of the students, their studies, and their contacts. Besides these, two direct sensitive questions (DQ) and their two nonrandomized response versions, the crosswise model (CWM) and the triangular model (TWM, see Sect. 10.6.1), were placed in the middle and at the end of the questionnaire.

Direct sensitive questions:
Please answer "yes" or "no" to the following questions:

Q1: Usually, do you have any negative attitudes towards other races or ethnic groups?
Q2: Do you feel racist toward people other than your race or ethnicity?

For the CWM and TRM models, they presented the following statements to the respondents and requested that they read questions A to D carefully and answer as per direction.

Statements:
A: *I was born between October and December, and I am a racist*
B: I was born between January and September, and I am a racist
C: *I was born between October and December, and I am not a racist*
D: *I was born between January and September, and I am not a racist*

Q3: CWM version: The respondent should answer "yes" if the statements A or D match his or her status. Otherwise, if statements B or C match, the respondent should answer "no."

Q4: TRM version: The respondents should answer "yes" if at least one of the statements A, B, or C matches his or her status. Otherwise, if the statement D matches, he or she should answer "no."

At the end of the period scheduled for the data collection fieldwork, 624 questionnaires were completed and 55 were found incomplete. Based on the 569 questionnaires filled out, the following results were obtained:

The proportions of "yes" answers obtained from the direct questions Q1 and Q2 were 0.178 and 0.125, respectively. This confirms that Q2 is more stigmatized than Q1. The prevalences of racism estimated from the CWM and TRM models were, respectively, 0.217 and 0.269. Both estimates were higher than the direct estimates. Perri et al. (2023) recommended the TRM over the CWM, as the former produces higher prevalence estimates.

13.7.7 Keeping Too Much Change, Freeriding, Shoplifting, Marijuana Use, Driving Under Influence and Infidelity

Coutts and Jann (2011) compared the effectiveness of Bourch's (1971) forced RRT (F-model) in a computer-assisted setting and that of the UCT (see Sect. 10.2). The effectiveness of the methods was measured by the five indicators: (1) the efficiency of the estimates they provided; (2) understanding the procedures (3) maintaining the anonymity of their answers; (4) the time required to read the instructions and answer the sensitive questions; and (5) the amount of nonresponse induced by the techniques.

A random sample of 2,075 respondents was selected from 10,000 invited panel members recruited from the German "Sozioland" access panel by Respondi AG (see www.sozioland.de) between August 1 and September 30, 2007. Respondents were randomly assigned to eight groups once they had activated the questionnaire: a group for direct questioning (DQ) with a probability of 30%; five groups associated with each of the five variants of F-model with a probability of 10%; and two groups of UCT with a probability of 10% each.

Coutts and Jann (2011) considered the following six sensitive questions with varying sensitivity:

(i) *Have you ever received too much change and knowingly kept it? (Keeping too much change)*
(ii) *Have you ever knowingly used public transportation without buying a ticket? (Freeriding)*
(iii) *Have you ever deliberately taken an article from a store without paying for it? (Shoplifting)*
(iv) *Have you used marijuana in the past month? (Marijuana use)*
(v) *Have you ever driven a car although your blood alcohol was almost certainly over the legal limit? (Driving under influence, DUI)*
(vi) *Have you ever cheated on your partner?(Infidelity).*

For the F-model, respondents were asked to directly answer "yes" or "no" to the sensitive question with a probability $p(=1/2$, in this design) and to answer a forced "yes" with a probability $1 - p$ (probability of a forced "no" was zero). Cutts and Jann (2011) proposed the five variants of F-model with respect to the five methods of randomization: (i) manual coin toss, (ii) electronic coin toss, (iii) banknote serial numbers, (iv) telephone numbers, and (v) banknote serial numbers (with an option to use a telephone number).

The 2075 participants were allocated to the eight groups as follows: direct questioning (643), F-model with manual coin toss (185), F-model with electronic coin toss (201), F-model with bank notes (194), F-model with telephone (218), F-model with bank notes (or phone number 236), UCT-G1(210), and UCT-G2 (188).

Respondents in UCT-G1 are asked to perform three UCTs with sensitive questions (iii), (v), and (vi) and four neutral or innocuous questions, while respondents in UCT-G2 are asked to perform three UCTs with sensitive questions (i), (ii), and (iv) and the same four neutral questions as in UCT-G1.

Respondents had been invited by e-mail to participate in the survey. The questionnaire started with a set of basic demographic questions and some questions about the respondent's living conditions. Respondents belonging to the direct response group (DR) were asked to answer all six sensitive questions.

Respondents belonging to the (v) F-model were asked to answer by using the randomization devices six times for each of the six sensitive questions. Respondents belonging to the UCT-G1 and UCT-G2 groups were asked to perform UCT three times with the associated sensitive questions. Respondents using either an F-model or the UCT were also asked to provide ratings about their understanding of the instructions for using the F-model or UCT and the anonymity of their answers.

From the data analysis, it seems that the highest prevalence of the sensitive attributes "keeping too much change" is provided by the F-model (combined 56.1%), "freeriding" by the UCT (76%), "shoplifting" by the direct questioning (23.4%), "marijuana use" by the UCT (32.5%), "DUI" by the direct questioning (29%), and "infidelity" by the UCT (35.9%). F-model produces negative estimates in two situations.

With regards to the precision of prevalences, direct questioning is the best (the smallest SE), perhaps possibly having a larger sample size. The second place is occupied by the F-model (combined). The UCT is the worst-providing highest SEs all the situations.

As per "quality of measures" for the different techniques, the electronic coin toss is the best (92.9%), followed by the UCT (91.8%), "trust" by the UCT (28.6%), and the minimum "time taken" by the DR method (28 s), followed by the electronic coin toss (97 s).

From the above findings, we conclude that none of the proposed three methods is the best in all situations. We find that the UCT is better than the F-model with respect to understanding of the applications, guaranteed anonymity, and higher estimates (less bias) of prevalences, while the F-model is much better than the UCT with respect to the efficiency of the estimates.

13.7.8 Cannabis Consumptions and Sexual Behavior

Perri et al. (2018) conducted a survey to investigate certain aspects of cannabis consumption and sexual behaviors among students at the universities of Granada and Murcia, Spain, during the academic year 2015/2016.

A stratified random sample of 2398 students enrolled in different faculties was selected by proportional allocation using degree programs and year of degree as strata. Three methods of data collection, e.g., direct questioning or response (DR), Barlev's RR technique (BRRT), and item sum technique (IST), were used for the collection of sensitive data relating to consumption of cannabis and certain aspects of sexual habits. The allocated sample sizes for the respondents to the groups DR, BRRT, and IST were 492, 613, and 1293, respectively.

All students received the same questionnaire, covering academic items and personal characteristics. A total of four sensitive questions, two of which related to the use of cannabis and the remaining two to sexual behavior, were set for the survey. The sensitive questions were as follows:

Q1: How many cannabis cigarettes did you consume last year?
Q2: Over the past 90 days, how many days did you consume cannabis?
Q3: Over the past 90 days, how many times have you had trouble stopping your sexual behavior when you knew it was inappropriate?
Q4: Over the past 90 days, how many times has sex been an escape from your problems?

Respondents were allocated to the DR method and asked to answer all the sensitive questions. The students assigned to the BRR were requested to install the application "Baraja Espan˜ola". After running the application, students would receive a figure with a probability of 3/10 and a number with a probability of 7/10 on the screen.

The student should supply the true answer if a figure is obtained. Otherwise, if a number was obtained, the student should provide an answer after multiplying it by the number. The student should repeat the BRR four times independently for each of the four sensitive questions. Respondents allocated to the IST group were further divided into two groups, SL-group and LL-group, comprising 520 and 773 students, respectively. Respondents in the SL group were given a list of the following four innocuous questions and were asked to provide a total of the scores of the following four questions:

Q5: What was your general mark in the Selectivity exam, without counting specific subjects? (Value between 0 and 10)
Q6: What was your Selectivity mark counting specific subjects? (Value between 0 and14)
Q7: What is the number of subjects in which you have enrolled in the academic year?
Q8: What is the final digit of your mobile phone number?"

Respondents belonging to the LL-group were given four sets of questions comprising one sensitive and one innocuous question, such as (Q1, Q5), (Q2, Q6), (Q3, Q7), and (Q4, Q8). Students were asked to provide total scores against each of the four sets without revealing individual scores.

As far as non-response rates are concerned, the IST yielded the lowest non-response rates for each of the four sensitive questions. Except Q1, the non-response rates in the BRR were much lower than the DR method in all situations. As expected, the DR method yields much lower estimates for all four sensitive characteristics than those of the IST and BRR methods. Estimates of Q2 and Q3 for BRR are higher than those of IST, while for the other two sensitive questions, Q1 and Q4, IST is higher.

With regards to the accuracy of the estimates (standard errors, SEs), except Q4, the DR method provides more precise estimates (SEs) than both the BRR and IST methods. Except for Q4, IST produces lower SEs than those of the BRR. However, comparisons provided by authors are not totally valid as all the methods are not based on the same sample size.

13.7.9 Statistical Software: Packages and Modules for Randomized Response Data

The researchers have proposed several statistical software programs to ease statistical computations originating from indirect data collection methods. Hox and Lensvelt-Mulders (2004) presented "Mplus", Sehra (2008) provided SAS code to perform analysis of RR survey data, Jann (2011) proposed Stata module rrlogit to fit logistic regression, R programs were proposed by Blair et al. (2015), Heck and Moshagen (2018), Rueda et al. (2016) and Tian and Tang (2013) developed the R package RRTCS. For further details, readers are referred to Le et al. (2023).

13.7.10 Conclusions

Various indirect interviewing (IQ) techniques, e.g., RRT, NRR, UCT and IST, have been proposed to guarantee anonymity and minimize the respondent's privacy. The performances of the proposed techniques are evaluated according to the prevalence estimates they provided, the ease of their use, respondent trust, and rate of nonresponse (Coutts & Jann 2011).

Goodstadt and Gruson (1975), Lara et al. (2004), Tracy and Fox (1981), and van der Heijden et al. (2000), among other researchers, reported that RRT can be effectively used to provide improved prevalence estimates and increase responses for some sensitive surveys such as child abuse, drug use, abortion, employee theft, welfare fraud, and premature sign-offs on audits in comparison to the reporting of the same deeds in response to direct questioning surveys. The effectiveness of the UCT/IST in providing higher estimates than the DQ for estimating sensitive items by Anderson et al. (2007), Dalton et al. (1994), Tsuchiya et al. (2007), Wimbush and Dalton (1997).

Finding an RRT that is handy, trustworthy, and can be used correctly in the absence of interviewer supervision is not easy. Respondents may often provide wrong responses, even if the randomization procedure is correct. NRR, UCT, and IST do not require a randomization procedure. Hence, they are much cheaper and easier to apply. The performance of NRR should be the same as that of RRT as the former is a variant of RRT that does not require randomization. Some studies show that UCT produces much higher estimates than the RRT. As per efficiency, estimates based on DQ are more efficient than those of RRT and UCT, while RRT is more precise than UCT. Thus, it is difficult to find a practice that is acceptable to all.

References

Adhikary, A. (2015). Mean square estimation in randomized response surveys. *Model Assisted Statistics and Applications, 10*, 397–409.

Anderson, D., Angela, M., Simmons, S., Uzanne, M., & Mitchell, E. (2007). Effect of response format on endorsement of eating disordered attitudes and behaviors. *International Journal of Eating Disorders., 40*, 903.

Arnab, R. (1990). On commutativity of design and model expectations in randomized response surveys. *Communications in Statistics-Theory and Methods, 19*, 3751–3757.

Arnab, R. (1994). Non-negative variance estimator in randomized response surveys. *Communications in Statistics-Theory and Methods, 23*, 1743–1752.

Arnab, R. (1998). Randomized response surveys: Optimum estimation of finite population total. *Statistical Papers, 39*, 4045–4408.

Arnab, R. (2017). *Survey sampling theory and applications*. Academic Press

Bar-Lev, S., Bobovitch, E., & Boukai, B. (2004). A note on randomized response models for quantitative data. *Metrika, 60*, 255–260.

Blair, G., Imai, K., & Zhou, Y. (2015). Design and analysis of the randomized response technique. *Journal of the American Statistical Association, 110*, 1304–1319.

Bourch, R. (1971). Assuring confidentiality of responses in social research: A note on strategies. *The American Sociologist, 6*, 308–411.

Chaudhuri, A. (1987). Randomized response surveys of finite populations: A unified approach with quantitative data. *Journal of Statistical Planning and Inference, 15*, 157–165.

Chaudhuri, A. (2011). *Randomized response and indirect questioning techniques in surveys*. Chapman & Hall, CRC Press, Taylor & Francis Group.

Chaudhuri, A., & Christofied, T. (2013). *Indirect questioning in sample surveys*. Springer.

Cobo, B., Rueda, M., & Torrecillas, F. (2016). Application of randomized response techniques for investigating cannabis use by Spanish university students. *International Journal of Methods in Psychiatric Research*. https://doi.org/10.1002/mpr.1517

Coutts, J., & Jann, B. (2011). Sensitive questions in online surveys: Experimental results for the randomized response technique (RRT) and the Unmatched Count technique (UCT). *Sociological Methods & Research, 40*(1), 169–193.

Dalton, D., Wimbush, J., & Catherine, M. (1994). Using the Unmatched Count Technique (UCT) to estimate base rates for sensitive behavior. *Personnel Psychology, 47*, 817–828.

Eriksson, S. A. (1973a). A new model for randomized response. *International Statistical Review., 48*, 29–32.

Eriksson, S. A. (1973b). A new model for randomized response. *International Statistical Review, 41*, 101–113.

Espanda, F. (2006). Negative surveys. arXiv:ST/0608176v1

Esponda, F., & Guerrero, V. M. (2009). Surveys with negative questions for sensitive items. *Statistics and Probability Letters, 79*, 2456–2461.

Esponda, F., Huerta, K., & Guerrero, V. M. (2016). A statistical approach to provide individualized privacy for *Survey, POIS one, 11*, e0147314.

Goodstadt, M., & Gruson, V. (1975). The randomized response technique: A test of drug use. *Journal of the American Statistical Association, 70*, 814–818.

Greenberg, B. G., Abul-Ela, A.-L., Simmons, W. R., & Horvitz, D. G. (1969). The unrelated question RR model: Theoretical framework. *Journal of American Statistical Association, 64*, 520–539.

Heck, D., & Moshagen, M. (2018). Rreg: An R package for correlation and regression analyses of randomized response data. *Journal of Statistical Software, 85*, 1–29.

Hox, J., & Lensvelt-Mulders, G. (2004). Randomized response analysis in Mplus. *Structural Equation Modeling, 11*, 615–620.

Jann, B. (2011). Stata module to estimate logistic regression for randomized response data. https://ideas.repec.org/c/boc/bocode/s456203.html

Land, M., Singh, S., & Sedory, S. A. (2012). Estimation of rare sensitive attribute using Poisson distribution. *Statistics, 46*, 351–360.

Lara, D., Strickler, J., Olavarrieta, C. D., & Ellertson, C. (2004). Measuring induced abortion in Mexico a comparison of four methodologies. *Sociological Methods & Research, 32*(4), 529–558. https://doi.org/10.1177/0049124103262685

Le, T.-N., Lee, S.-M., Tran, P.-L., & Li, C.-S. (2023). Randomized response techniques: A systematic review from the pioneering work of warner (1965) to the present. *Mathematics, 11*, 1–26.

Liu, P., & Chow, L. (1976). A new discrete quantitative randomized response model. *Journal of the American Statistical Association., 71*, 72–73.

Mielecka-Kubień, Z., & Toniszewski, M. (2022). Estimation of illicit drug use among high school students in the Silesian *voivodship* (Poland) with the use of the randomized response technique. *Mathematical Population Studies, 29*(2), 47–57.

Mielecka-Kubień, Z., & Toniszewski, M. Estimation of illicit drug use among high school students in the Silesian *voivodship* (Poland) with the use of the randomized response technique. *Mathematical Population Studies, 29*(2), 47–57. https://doi.org/10.1080/08898480.2021.1893009

Mieth, L., Mayer, M., Hoffmann, A., Buchner, A., & Bell, R. (2021). Do they really wash their hands? Prevalence estimates for personal hygiene behaviour during the COVID-19 pandemic based on indirect questions. *Mieth et al. BMC Public Health, 21*(12). https://doi.org/10.1186/s12889-020-10109-5

Miller, J. D. (1985). The nominative technique: A new method of estimating heroin prevalence. *NIDA Research Monograph, 57*, 104–124.

Perri, P., Manoli, E., & Christofides, T. (2023). Assessing the effectiveness of indirect questioning techniques by detecting liars. *Statistical Papers, 64*, 1483–1506. https://doi.org/10.1007/s00362-022-01352-6

Rueda, M., Cobo, B., & Arcos, A. (2016). RRTCS: An R package for randomized response techniques in complex surveys. *Applied Psychological Measurement, 40*, 78–80.

Särndal, C. E., Swensson, B., & Wretman, J. (1992). *Model assisted survey sampling*. Springer.

Sehra, S. (2008). Two-stage optional randomized response models. Master's thesis, The University of North Carolina, Greensboro, NC, USA

Sirken, M. G. (1975). Network surveys of rare and sensitive conditions. In *Advances in Health Survey Research Methods. National Center on Health Statistics Research Proceedings Series* (pp. 31–32).

Srivastava, R., Nigam, A. K., & Singh, N. (2015). Application of randomized response techniques in estimation of prevalence of child sexual abuse. *Statistics and Applications*. ISSN 2454-7395(online)

Study on Child Abuse India 2007, 2007 Study on Child Abuse India 2007 (2007). *Ministry of Women and Child Development, Govt. India.*

Sudman, S., Blair, E., Bradburn, N. M., & Stocking, C. (1977). Estimates of threatening behavior based on reports of friends. *The Public Opinion Quarterly, 41*, 61–264.

Tracy, P., & Fox, J. (1981). The validity of randomized response for sensitive measurements. *American Sociological Review, 46*, 187–200.

Tsuchiya, T., Hirai, Y., & Ono, S. (2007). A study of the properties of the item count technique. *Public Opinion Quarterly, 71*, 253–272; van den Hout, A., & Kooiman, P. (2006). Estimating the linear regression model with categorical covariates subject to randomized response. *Computational Statistics & Data Analysis, 50*, 3311–3323.

van der Heijden, P., Ger van Gils, G., Bouts, J., & Hox, J. (2000). A comparison of randomized response, computer-assisted self-interview, and face-to-face direct questioning: Eliciting sensitive information in the context of welfare and unemployment benefit. *Sociological Methods & Research, 28*(4), 505–537.

Wimbush, J., & Dalton, D. (1997). Base rate for employee theft: Convergence of multiple methods. *Journal of Applied Psychology, 82*, 756–763.

Wolter, K. M. (1985). *Introduction to variance estimation*. Springer.

Author Index

R. Arnab, *Indirect Methods of Data Collection and Analysis from Surveys*, Indian Statistical Institute Series, https://doi.org/10.1007/978-981-97-6005-3

Subject Index

R. Arnab, *Indirect Methods of Data Collection and Analysis from Surveys*, Indian Statistical Institute Series, https://doi.org/10.1007/978-981-97-6005-3

T

U

V

W

The manufacturer's authorised representative in the EU is Springer Nature Customer Service Centre GmbH, Europaplatz 3, 69115 Heidelberg, Germany. If you have any concerns regarding our products, please contact ProductSafety@springernature.com

Printed and bound by CPI Group (UK) Ltd, Croydon, CR0 4YY
01/10/2026
02225950-0017